P9-CAD-951

WILDLIFE-HABITAT RELATIONSHIPS

WILDLIFE–HABITAT RELATIONSHIPS

Concepts & Applications

Second Edition

Michael L. Morrison,

Bruce G. Marcot, &

R. William Mannan

The University of Wisconsin Press

The University of Wisconsin Press
2537 Daniels Street
Madison, Wisconsin 53718

3 Henrietta Street
London WC2E 8LU, England

5 4 3 2 1

Printed in the United States of America

Library of Congress Cataloging-in-Publication Data
Morrison, Michael L.
Wildlife–habitat relationships : concepts & applications /
Michael L. Morrison. Bruce G. Marcot, and R. William Mannan.—
2nd ed.
458p. cm.
Includes bibliographical references and index.
ISBN 0-299-15640-0 (cloth: alk. paper)
1. Habitat (Ecology) 2. Animal ecology. I. Marcot, Bruce G.
II. Mannan, R. William. III. Title.
QH541.M585 1998
591.7—dc21 97-9445

We gratefully dedicate this volume
to the community of field biologists throughout the world
who daily tend to the inheritance of succeeding generations

Contents

Figures, Tables, and Boxes

Figures

Tables

Boxes

Preface

The purpose of our first edition was to advance from the point where the many fine, but introductory, texts in wildlife biology left off. This purpose has not changed. We have further developed this new edition to incorporate the many new ideas that have come our way since the first edition appeared. These ideas came from several sources. First, we took to heart the independent reviews that appeared in scientific journals. Second, our friends and colleagues showed us their hidden talents as book critics; we also attended to these comments. Finally, we each have tapped into new experiences and studies to present the most current findings, concepts, and visions for future development in research and management. Thus, this new edition.

This book is intended for advanced undergraduates, graduate students, and practicing professionals with a background in general biology, zoology, wildlife biology, conservation biology, and related fields. An understanding of statistics through analysis of variance and regression is helpful but not essential. Land managers will especially benefit from this book because of its emphasis on the identification of sound research and the interpretation and application of results.

Our approach combines basic field zoology and natural history, evolutionary biology, ecological theory, and quantitative tools. We think that a synthesis of these topics is necessary for a good understanding of ecological processes and, hence, good wildlife management. We attempt to draw on the best and recent examples of the topics we discuss, regardless of the species involved or its geographic location. We do concentrate on terrestrial vertebrates from temperate latitudes, because this is where the majority of literature has been developed. However, because it is the concepts that are important, the specific examples are really of secondary importance; therefore, our writing can be used by anyone from any location. We did try, however, to bring in examples from amphibians, reptiles, birds, and mammals (both large and small) to help individuals from different backgrounds better understand the application of concepts to their particular interests.

We emphasize the need for critical evaluation of methodologies and their applications in wildlife research. Management decisions all too often are based on data of unknown reliability, that is, on research conducted with biased methods, low sample sizes, and inappropriate analyses. We understand also that, all too often, managers are faced with making decisions using unreliable data. The lack of subsequent properly designed studies, as with monitoring, validation, and adaptive management research, forces a vicious cycle. This neither needs to persist, nor should it.

To aid both the student and the professional, we have tried to explain fundamental concepts of ecological theory and assessment so that the use of more advanced technical tools is more acceptable, more often sought, and more appropriately applied. Ultimately, the success of conservation efforts depends on gathering, analyzing, and interpreting reliable information on species composition, communities, and habitat. We hope that this book encourages such rigor. We have strived not to sound pedantic when we reemphasize seemingly basic concepts; but unfortunately, the current literature often contains much poorly designed research.

Format for the Second Edition

The first edition forms the core of this new work. We have, however, completely revised the presentation into what we think is a more logical format. And we have added much new material to supplement that previously offered.

The first edition treated its seven chapters as primarily separate entities. To better organize fundamental principles into similar sections, we have divided this new edition into three major parts. "Concepts of Wildlife-Habitat Relationships" lays the foundation upon which the remainder of the book is constructed. Here we discuss the historical background and philosophical attitudes that have shaped the wildlife profession and led to the way research should be approached. Next we review the evolutionary background against which the current distribution, abundance, and habits of animals have developed. The final chapter in this first section is new to this edition and discusses habitat relationships from the perspective of vegetation ecology and population biology.

The second major part, "The Measurement of Wildlife-Habitat Relationships," forms the heart of the book. Here we first discuss fundamental approaches to study design and experimental methodologies, reviewing the philosophy of various ways of gaining reliable knowledge. This initial chapter is new to this edition. The absence of this material was noted by several reviewers of our first edition; we hope that we have now covered this important topic more fully. Next we spend considerable space—two chapters—reviewing the many methods that have been used to develop wildlife-habitat relationships, including field methods, data analysis, sampling biases, and data interpretation. We also incorporate discussion of multivariate statistics here (chapter 6), which was treated in more detail as a separate chapter in the first edition. We reduced the emphasis on methods of multivariate analyses because we cannot really do justice to the mathematics of the analyses in this book, and because we think that we can be more effective by emphasizing the concept of multivariate analyses, proper sampling methods, and interpretation of results. Next we cover behavioral sampling and analysis in wildlife research. This chapter is a generalized version of the first edition's emphasis on foraging studies. We realized after publication that in emphasizing foraging behavior we were inadvertently ignoring many important studies in behavioral ecology. The final three chapters in this part represent major revisions of chapters presented in the first edition. Habitat fragmentation, study of metapopulations, and landscape ecology are all gaining more emphasis from researchers and managers alike. We present an updated discussion of these topics, now divided into two chapters because of the major recent advances made in this field of study. Likewise, the final chapter in this part reviews and updates the multifaceted topic of model development.

The final part of this edition, "The Management of Wildlife Habitat," contains two chapters that were conspicuously absent from the earlier publication. We first discuss the new, emerging area of ecosystem management. Further, we discuss or illustrate the need to consider our management goals in an evolutionary context; and we briefly cover the topic of adaptive resource management. In the final chapter we lay a framework for advancing our understanding of wildlife through modified approaches to habitat relationships, a call for greater emphasis on the synthetic field of restoration ecology, and a plea for changes in our educational system. We present this material partly as a prescription and partly as a "null model" upon which we can debate the best means of advancing our profession. Oh yes, we have also added an author index as requested by several reviewers.

Thus, this new edition contains 12 chapters compared with 7 in the first edition. This fulfills the prediction made by Allen Fitchen, director of the University of Wisconsin Press, who noted that he had never seen a smaller second edition when we confidently told him we did not see the need to add pages.

Acknowledgments

First and foremost, we thank Allen Fitchen of the University of Wisconsin Press for encouraging us to put our thoughts onto paper and for shepherding this project through to completion. We were especially pleased with the detailed review of the first edition conducted by students of Dr. Robert J. Cooper, Department of Biology, University of Memphis. We also thank the many independent reviewers of our first edition for their insights; we tried to follow your advice. One or more chapters of this new edition were reviewed by Richard Lancia, Larry Irwin, Guy McPherson, Linnea Hall, Tim Max, David Cleaves, Danny Lee, and William Matter; we thank you all for your insights. Linnea Hall was instrumental in helping pull the book together—assembling the tables and figures and securing the copyright releases—we are grateful for her assistance (and for tolerating the senior author).

There are numerous individuals who have helped shape our views of wildlife biology and science in general; we cannot list them all. We especially appreciate dialogues with our graduate students over the years. In addition, discussions with many ecologists and managers, domestically and internationally, helped us identify recent scientific advances and critical management issues. We also thank the numerous authors whom we cited in this book for their research efforts and insightful analyses.

The first edition of our book was dedicated to Drs. E. Charles Meslow and Jack Ward Thomas, "who taught us that wildlife conservation truly succeeds when practiced with honor, rapport, and rigor." Since the first edition appeared, both men have gone on to even greater accomplishments: Dr. Meslow sharing his knowledge with a wider audience through his work with the Wildlife Management Institute, and Dr. Thomas as chief of the U.S. Forest Service. We assume that our recognition of them led to such advancements! Seriously, although we chose other biologists who are very important to us for our current dedication, we want to reemphasize the enormous positive impact that these two individuals have made on the three of us in shaping our careers and lives. Thanks again.

PART 1

Concepts of Wildlife-Habitat Relationships

1 The Study of Habitat: A Historical and Philosophical Perspective

Some birds live on mountains or in forests, as the hoopoe and the brenthus. . . . Some birds live on the sea-shore, as the wagtail. . . . Web-footed birds without exception live near the sea or rivers or pools, as they naturally resort to places adapted to their structure.

—Aristotle, *Historia animalium*

Introduction

An animal's habitat is, in the most general sense, the place where it lives. All animals, except humans, can live in an area only if basic resources such as food, water, and cover are present and if the animals have adapted in ways that allow them to cope with the climatic extremes and the competitors and predators they encounter. Humans can live in areas even if these requirements are not met, because we can modify environments to suit our needs or desires and because we potentially have access to resources such as food or building materials from all over the world. For these reasons, humans occupy nearly all terrestrial surfaces of the earth, but other species of animals are restricted to particular kinds of places.

The distribution of animal species among environments and the forces that cause these distributions frequently have been the subjects of human interest, but for different reasons at different times. The primary purposes of this introductory chapter are to review some of the reasons why people study the habitats of animals and to outline how these reasons have changed over time. We also introduce the major concepts that will be addressed in this book.

Curiosity about Natural History

Throughout recorded history humans, motivated by their curiosity, have observed and written about the habits of animals. The writings of naturalists were, for centuries, the only recorded sources of information about animal-habitat relationships. Aristotle was among the first and best of the early naturalists. He observed animals and wrote about a wide variety of subjects, including breeding behavior, diets, migration, and hibernation. Aristotle also noted where animals lived and occasionally speculated about the reasons why.

> A number of fish also are found in sea-estuaries; such as the saupe, the gilthead, the red mullet, and, in point of fact, the greater part of the gregarious fishes. . . . Fish penetrate into the Euxine [estuary] for two reasons, and firstly for food. For the feeding is more abundant and better in quality owing to the amount of fresh river-water that discharges into the sea. . . . Furthermore, fish penetrate into this sea for the purpose of breeding; for there are recesses there favorable for spawning, and the fresh and exceptionally sweet water has an invigorating effect on the spawn. (Aristotle [344 B.C.] 1862)

Interest in natural history waned after Aristotle's death. Politics and world conquest were the focus of attention during the growth of the Roman Empire, and interest in religion and metaphysics suppressed creative observation of the natural world during the rise of Christendom (Beebe 1988). As a result, little new information was documented about animals and their habitats for nearly 1700 years after the death of Aristotle. Yet, as Klopfer and Ganzhorn (1985) noted, painters in the medieval and pre-Renaissance periods still showed an appreciation for the association of specific animals with particular features of the environment. "Fanciful renderings aside, peacocks do not appear in drawings of moors nor moorhens in wheatfields" (Klopfer and Ganzhorn 1985:436). Similar appreciation is seen in the artwork from India, China, Japan (e.g., Sumi paintings) and elsewhere during this period. Thus, keen observers noticed relationships between animals and their habitats during the Dark Ages, but few of their observations were recorded.

The study of natural history was renewed in the seventeenth and eighteenth centuries. Most naturalists during this period, such as John Ray (seventeenth century) and Carl Linnaeus (eighteenth century), were interested primarily in naming and classifying organisms in the natural world (Eiseley 1961). Explorers made numerous expeditions into unexplored or unmapped lands during this period, often with the intent of locating new trade routes or identifying new resources. Naturalists usually accompanied these expeditions or proceeded on their own, collecting and/or recording information about the plants and animals they observed. Many Europeans during this period also collected feathers, eggs, pelts, horns, and other parts of animals for "collection cabinets." Some cabinets were serious scientific efforts, but most were not. Nevertheless, new facts about the existence and distribution of animals worldwide were gathered during this time, and the advance in knowledge generated considerable curiosity about the natural world.

During the nineteenth century naturalists continued to describe the distribution of newly discovered plants and animals, but they also began to formulate ideas about how the natural world functions. Charles Darwin was among the most prominent of these naturalists. His observations of the distributions of similar species were one set of facts among many that he marshaled to support his theory of evolution by natural selection (Darwin 1859). The work of Darwin is highlighted here, not only because he recorded many new facts about animals, but also (and more important) because the theory of evolution by natural selection forms the framework and foundation for the field of ecology.

Curiosity about Ecological Relationships

In the early 1900s, curiosity about how animals interact with their environment provided the impetus for numerous investigations into what are now called ecological relationships. Interest in these relationships initially led to detailed descriptions of the distribution of animals along environmental gradients or among plant communities. Merriam (1890), for example, identified the changes that occur in plant and animal species on an elevational gradient, and Adams (1908) studied changes in bird species that accompany plant succession. Biologists living in this period postulated that climatic conditions and availability of food and sites to breed were the primary factors determining the distributions of animals they observed (see Grinnell 1917).

Biologists in the early to mid-1900s, however, recognized that the distribution of some animals could not be explained solely on the basis of climate and essential resources. David Lack (1933) was apparently the first to propose that some animals (in this case, birds) recognize conspicuous features of appropriate environments and that these features are the triggers that induce animals to select a place to live. Areas without these features, according to Lack, generally will not be inhabited, even though they might contain all the necessary resources for survival. Lack's ideas gave birth to the concept of habitat selection and stimulated considerable research on animal-habitat relationships during the next 60 years.

Svardson (1949) developed a general conceptual model of habitat selection, and Hilden (1965) later expressed similar ideas. Their models characterized habitat selection as a two-stage process in which organisms first use general features of the landscape to select broadly from among different environments, and then respond to subtler habitat characteristics to choose a specific place to live. Svardson (1949) also suggested that factors other than those associated with the structure of the environment, such as the presence of conspecifics and interspecific competitors, can influence whether an animal stays in a particular place. Habitat selection therefore, has come to be recognized as a complicated process involving several levels of discrimination and a number of potentially interacting factors.

G. Evelyn Hutchinson formally articulated the multivariate nature of the causes of animal distribution in his presentation of the concept of the *n*-dimensional niche (Hutchinson 1957). From the 1960s to the present, development of the concept of the niche, particularly the assessment of the effects of interspecific competition on the distribution of animals, has stimulated many detailed studies of how animals use their habitats. These studies have shown that, for some species, habitat selection is influenced by conspecifics (Butler (1980), interspecific competitors (Werner and Hall 1979), and predators (Werner et al. 1983), as well as by features of the environment that are directly or indirectly related to resources needed for survival and reproduction.

Hunting Animals for Food or Sport

The earliest humans relied, in part, on killing animals for survival, and they undoubtedly recognized and exploited the patterns of association between the animals they hunted and the kinds of places where these animals were most abundant. Similarly, people who later made their living by trapping and hunting or could afford the luxury of hunting for sport knew where to find animals and probably speculated accurately about the habitat features that influenced the abundance of game species. Marco Polo reported, for example, that in the Mongol Empire in Asia, Kublai Khan (A.D.

1215–1294) increased the number of quail and partridge available to him for falconry by planting patches of food, distributing grain during the winter, and controlling cover (Leopold 1933). This advanced system of habitat management suggests a general understanding of the habitat requirements of the target game species, but it is unlikely that the information was obtained through organized studies of habitat use. Also, the men who hunted and trapped for subsistence or sport rarely recorded their knowledge about habitats for posterity.

Not until people began to attempt to apply biology systematically to the management of game as a "crop" in the early 1900s did they realize that "science had accumulated more knowledge of how to distinguish one species from another than of the habits, requirements, and inter-relationships of living populations" (Leopold 1933:20). The absence of information about habitat requirements of most animals and the desire to increase game populations by manipulating the environment stimulated detailed investigations of the habitats and life histories of game species. H. L. Stoddard's work on bobwhite quail (*Colinus virginianus*), published in 1931, and P. L. Errington's work on pheasants, published in 1937, exemplify early efforts of this kind.

Studies similar to Stoddard's have been conducted on most game animals in North America between 1930 and today (e.g. Bellrose 1976; Wallmo 1981; Thomas and Toweill 1982), but many of these studies only summarize general habitat associations and do not identify critical habitat components. During the last two decades the number of hunters has increased while undeveloped land available for managing wild animal populations has decreased. Therefore, the need to manage populations more intensively is great, and detailed knowledge of habitat requirements is essential for this task. Studies of the habitat requirements of game animals continue to be conducted, as one can easily see by reviewing recent scientific journals on wildlife management.

Public Interest and Environmental Laws

Human activities have dramatically disturbed natural environments in North America, and throughout the world, during the 1900s. These disturbances have been associated primarily with the rapid increase in the size of the human population and the exploitation of natural resources, including wild animals, for human use. Interest in wild animals by the general public also increased during this period, and concern about the negative effects of human activities on animal populations and other aspects of the natural environment eventually led to the passage of laws in the United States that were designed to aid management of wild animals or reduce environmental degradation. The following summary pertains to U.S. history; it is beyond the scope of the text to review public interest and environmental law in other nations.

Public interest early in the century focused on "game" animals, and some laws passed in the 1930s reflected this interest. The Migratory Bird Hunting Stamp Act and the Pittman-Robertson Federal Aid in Wildlife Restoration Act, for example, primarily taxed sportsmen and provided funds for management of waterfowl and other hunted species (table 1.1). As noted in the previous section, information needed for management of these species stimulated efforts to describe and quantify their habitats.

An increase in environmental awareness during the 1960s and 1970s broadened the scope of the kinds of animals about which the general public was concerned. Animal species not

Table 1.1. Important legislation in the United States that stimulated study, preservation, or management of animal habitat

Title	Year	Action
Migratory Bird Treaty Act	1929	Provided for the establishment of wildlife refuges
Migratory Bird Hunting Stamp Act	1934	Required a federal migratory bird hunting license; funds used to purchase lands for refuges
Fish and Wildlife Coordination Act	1934	Authorized conservation measures in federal water projects and required consultation with the U.S. Fish and Wildlife Service and states concerning any water project
Pittman-Robertson Federal Aid in Wildlife Restoration Act	1937	Provided an excise tax on sporting arms and ammunition to finance research on a federal and state basis
Multiple Use–Sustained Yield Act	1960	Stipulated that national forests would be managed for outdoor recreation, range, timber, watershed, and wildlife and fish
Wilderness Act	1964	Gave congress authority to identify and set aside wilderness areas
Classification and Multiple Use Act	1969	Mandated multiple use management on lands administered by the Bureau of Land Management
National Environmental Policy Act (NEPA)	1969	Stipulated that environmental impact statements would be prepared for any federal project that affected the quality of the human environment; wildlife habitat considered part of that environment
Endangered Species Conservation Act	1973	Provided protection for species and the habitat of species threatened with extinction
Sikes Act Extension	1974	Directed secretaries of Agriculture and Interior to cooperate with state game and fish agencies to develop plans for conservation of wildlife, fish, and game
Forest and Rangelands Renewable Resources Planning Act (FRPA)	1974	Called for units of the National Forest System to prepare land management plans for the protection and development of national forests
Federal Land Policy and Management Act	1976	Mandated land use plans on lands administered by the Bureau of Land Management
National Forest Management Act	1976	Stipulated that the plans called for in the FRPA would comply with NEPA and that management would maintain viable populations of existing native vertebrates on national forests

Source: Based in part on Gilbert and Dodds 1987:17

hunted for sport and without any other apparent economic utility also were perceived as having value. (The ethical rationales underlying these values are discussed in the next section.) Among the laws passed during this period were the National Environmental Policy Act, the Endangered Species Conservation Act, the Federal Land Policy and Management Act, and the National Forest Management Act (Bean 1977; see also table 1.1). Legislators designed these laws, in part, to ensure that all wildlife species and other natural resources were considered in the planning and execution of human activities on public lands. Knowledge of the nature of the habitats of animal species obviously is required before the effects of an environmental disturbance can be fully evaluated, before a refuge for an endangered species can be designed, or before animal habitats can be maintained on lands managed under a multiple-use philosophy. Biologists responded to the need for information about habitat requirements by studying, often for the first time, numerous species of "nongame" animals and by developing models to help predict the effects of environmental changes on animal populations (Verner et al. 1986).

Public interest in the nonconsumptive use of animals has not waned in recent years. In the United States in 1991, 76.1 million people (over 16 years of age) spent $18.1 billion observing, feeding, or photographing wildlife (U.S. Department of Interior and U.S. Department of Commerce 1993; see also box 1.1). The funding mechanisms for managing animals in the United States, however, have not kept pace with the broadening umbrella of public interest. Many state fish and game agencies have developed nongame management programs that emphasize identifying and managing habitats, but these programs often are limited by inadequate funding, and the sources of funds are, with rare exception, not broad-based or user-related. In Arizona, for example, the nongame program is funded by a fixed percentage of the funds generated by the state lottery.

Efforts to increase the funding base for managing nongame animals and their habitats were initiated in 1995 when support was sought from conservation organizations for a bill called the Fish and Wildlife Diversity Funding Initiative. This bill is analogous to the Pittman-Robertson Federal Aid in Wildlife Restoration (PR) Act in that it calls for a federal tax on

Box 1.1 Feeding wild animals

An examination of the activities in which people participate is one way to assess the kinds of things they value. Feeding wild animals is one human activity in which there is a direct link between people's expenditure of time, energy, and money and their desire to interact with or aid wild animals. It is, therefore, one measure of the value they place on wild animals and, indirectly, on the habitats that the animals occupy. This particular activity is not without problems, because it can cause some animals to become a nuisance or even a potential danger to humans if contact between animals and humans is encouraged. There are also potential problems for the animals being fed, such as promotion of an inappropriate diet, spread of disease, and the reduction of fear of humans. Nevertheless, feeding animals clearly is an activity that demonstrates the concern, interest, and affection of humans for wild animals.

Tucson, Arizona, is a relatively large city with over 600,000 inhabitants and is situated in the Sonoran Desert of the southwestern United States. Over the last 20 years, several studies have been undertaken in Tucson to evaluate the effects of urbanization on the native fauna, particularly birds (e.g., Emlen 1974; Stenberg 1988; Mills et al. 1991; Dawson and Mannan 1994) and to assess the attitudes of people in urban environments toward wildlife (e.g., Ruther and Shaw 1990; Shaw et al. 1992). A survey of the residents of Tucson (Ruther 1987) indicated that 57 percent of households fed wild birds, and 21 percent fed or provided water for animals other than birds. These values were even higher (60 percent and 35 percent, respectively) for households within a mile of what is now Saguaro National Park (Shaw et al. 1992), probably because these people chose to live near the park in part for the potential to see or interact with wildlife. The relatively high level of human interest in wild animals demonstrated by these surveys is part of the foundation of support for environmental laws in the United States (see table 1.1).

outdoor equipment, such as binoculars and tents, used in activities associated with the nonconsumptive enjoyment of wildlife. Funds generated by this act, like those from the PR Act, would be distributed to the states on a matching basis. If this bill is made into federal law, it will allow state agencies to manage more thoroughly the habitats of a wide variety of species, and will certainly stimulate the acquisition of information about those habitats.

Ethical Concerns

Another impetus for studying habitat partly underlies the public interest and environmental laws outlined in the previous section and relates to an ethical concern for the future of wildlife and natural communities. This concern is, in part, a humanistic one, insofar as the health of natural systems affects our use and enjoyment of natural resources in the broadest sense. From a utilitarian viewpoint, the world also is our habitat, and its health directly relates to our own. The ethical concern, however, transcends humanism in that wildlife and natural communities are intrinsic to the world in which we have evolved and now live. Writers of legal as well as ethical literature have argued that nonhuman species have, in some sense, their own natural right to exist and grow (e.g., Stone 1974, 1987). The study of wildlife species and their habitats in this context may deepen our appreciation for and ethical responsibility to other species and natural systems.

Why should we be concerned about species and habitats that offer no immediate economic or recreational benefits? Several rather standard philosophical arguments offer complementary and even conflicting rationales. One viewpoint argues for conserving species and their environments because we may someday learn how to exploit them for medical or other benefits (future option values). Another viewpoint argues for preserving species for the unknown (and unknowable) interests of future generations; we cannot speak for the desires of our not-yet-born progeny who will inherit the results of our management decisions.

Generally, a traditional conflict has pitted ethical humanism against humane moralism. Ethical humanism, as championed by Guthrie, Kant, Locke, More, and Aquinas, argues that animals are not "worthy" of equal consideration; animals are not "up to" human levels in that they do not share self-consciousness and personal interests. In effect, this argument allows us to subjugate wildlife and their habitats. Kant argued as much. He advanced his idea on a so-called deontological theme (from the Latin *deos,* or duty). That is, rights—specifically human rights—allow us to view animals as having less value because they are less rational (or are arational); we humans have the duty to manage species and the freedom to subjugate them.

On the other hand, humane moralism, as championed in part by Stone, Betham (of the animal liberation movement), and Singer, argues that animals deserve the focus of ethical consideration. According to this argument, humans are moral agents. Animals and, by extension, their habitats require consideration equal to that given humans, even if they do not ultimately receive equal treatment.

There also is a third ethical stance, one which may form the central impetus for studying and conserving wildlife and their habitats: an ecological ethic. The ecological ethic, as proposed by Callicott, has been most eloquently advanced by Leopold in his *A Sand County Almanac,* although elements of his philosophy (and much fuller philosophical expositions) can be traced to Henri Berson, Teilhard de Chardin, and John Dewey. The focus of ethical consideration in this view is on both the individual organism and the community in which it resides. Concern for the community essentially constitutes Leopold's ecological ethic, a holistic ethic which is concerned with the relationships of animals with each other and with their environment.

Leopold wrote of soil, water, plants, animals, oceans, and mountains, calling each a natural entity. In his view, animals' functional roles in the community, not solely their utility for humans, provide a measure of their value. By extension, then, in order to act morally, we must maintain our individual human integrity, our social integrity, and the integrity of the biotic community.

Following such an ecological ethic, a concern for the present and future conditions of wildlife and their habitats motivates the writing of this book. The sad history of massive resource depletion, including extinctions of plant and animal species and the large-scale alteration of terrestrial and aquatic environments, must, in our view, strengthen a commitment to further understanding wildlife and their habitats. Understanding is the necessary overture to living truly by an ecological ethic.

Concepts Addressed

This book covers both the theoretical and applied aspects of wildlife-habitat relationships with an emphasis on the theoretical framework under which researchers should study such relationships. An appropriate way to begin a preview of the concepts covered in subsequent chapters is to define the term *habitat.* A review of even a few papers concerned with the subject will show that the term is used in a variety of ways. Frequently *habitat* is used to describe an area supporting a particular type of vegetation or, less commonly, aquatic or lithic (rock) substrates. This use probably grew from the term *habitat type,* coined by Daubenmire to refer to "land units having approximately the same capacity to produce vegetation" (1976:125).

We, however, view habitat as a concept that is related to a particular species, and sometimes even to a particular population, of plant or animal. Habitat, then, is an area with a combination of resources (like food, cover, water) and environmental conditions (temperature, precipitation, presence or absence of predators and competitors) that promotes occupancy by individuals of a given species (or population) and allows those individuals to survive and reproduce. Habitat of high quality can be defined as those areas that afford conditions necessary for relatively successful survival and reproduction over relatively long periods when compared with other similar environments. (We recognize, though, that the habitats of some animals are ephemeral by nature, such as early seral stages or pools of water in the desert after heavy rains.) Conversely, marginal habitat supports individuals, but their rates of survival and reproduction are relatively low, and/or the area is usually suitable for occupancy for relatively short or intermittent periods. Thus, quality of habitat ultimately is related to the rates of survival and reproduction of the individuals that live there (Van Horne 1983), to the vitality of their offspring, and to the length of time the site remains suitable for occupancy.

Understanding why a particular population or species occupies only a specific area in a region or why it occupies only a specific continent often requires knowledge of the organism's ecological relationships, its evolutionary history, the climatic history of the area, and even the history of the movements of landmasses. We provide in chapter 2 an overview of the forces, factors, and processes that determine why animals are found where they are and how they came to be there. The information presented emphasizes that both past and present conditions can play significant roles in defining the habitat of an animal. In short, we provide in chapter 2 the conceptual framework we feel is necessary before the study of habitat can proceed successfully.

Important elements of the habitat of an animal often are provided by the vegetation in an area. Changes in vegetation can, therefore, alter habitat conditions. Understanding how the structure and composition of vegetation influence habitat quality is central to understanding the distribution and abundance of animals. We review in chapter 3 the patterns and processes associated with plant succession and the relationships between animals and vegetative change. We also review factors that affect the dynamics of animal populations. These factors include not only the availability, distribution, and quality of habitat, but also the genetic make-up of populations, the movements of animals, and the presence and influences of other organisms, especially humans.

Studying wildlife-habitat relationships requires knowledge of the scientific method. We review in chapter 4 activities associated with the scientific method and some of the controversial issues about its application in ecology and wildlife science. We also discuss the use of experiments in wildlife science and the strengths and weaknesses of the experimental approach in field and laboratory situations. We end chapter 4 with some general strategies for deciding how to proceed with investigations of wildlife-habitat relationships.

Identifying what constitutes the habitat of a population or species is the impetus underlying many activities in wildlife science and management. Designing studies that identify habitat conditions requires considerable thoughtfulness about the needs and perceptions of the species under investigation, the spatial scale at which the study is to be conducted, and the methods of measuring habitat that will be used. We provide in chapter 5 a review and analysis of what elements of the environment might be measured in studies of habitat and a discussion of the methods commonly used to measure these elements.

The assessment of what to measure in wildlife habitat and how to measure it, in chapter 5, is followed in chapter 6 with a consideration of when to take the measurements. We focus in chapter 6 on the importance of timing in determining what constitutes habitat. Use of resources by animals can vary on several temporal scales, including time of day, stage of breeding cycle, season of the year, and between years. Deciding which scale or scales to address in a study obviously will influence its design. Evaluation of the numerous factors that can influence whether an animal occupies a given area lends itself to the use of multivariate statistical techniques. We end chapter 6 with a review of the use of these techniques in analyzing, understanding, and conceptualizing wildlife-habitat relationships.

Patterns of resource use detected in animal populations are products of the behaviors of individual animals. We present in chapter 7 the theoretical framework that forms the basis for investigations of animal behavior as it relates to habitat. We also review the principal methods used to measure animal behavior and the resources upon which the animals depend. One focus of this chapter is an assessment of diet and foraging behavior, important because an animal's survival and productivity depends so heavily on acquiring food.

Some animals may select habitat through a hierarchic process that begins on broad spatial scales. Furthermore, the distribution of patches of environmental resources (e.g., vegetation types) across the landscape can influence the dynamics of many populations and elements of community structure. We review in chapter 8 some of the basic tenets of landscape ecology. We discuss the heterogeneity of resources in patchy landscapes and the general responses of wildlife to this heterogeneity. We also discuss the management challenges presented by patchy or fragmented environments; this discussion includes a review of the utility and value of retaining remnant patches of natural environments and some level of connectivity between them. In chapter 9 we continue the discussion of wildlife and landscapes, but focus on the specific responses of organisms, species, populations, and communities to landscape dynamics.

Understanding why a species or population occupies a given area may allow some level of prediction about the distribution and abundance of that species or population in other areas or at other times. In chapter 10 we review models used to predict wildlife-habitat relationships and examine how scientific uncertainty affects the use of these models. We also discuss knowledge-based and decision-aiding models, model validation, and the role of models in development of research hypotheses.

The earth and the natural resources on it are changing rapidly, primarily as a result of human use and exploitation. Management of natural resources, including wildlife, in the future likely will require approaches that conceptually force us to think on broader spatial, temporal, and ecological scales. We discuss in chapter 11 the possibility of managing wildlife in an ecosystem context. We suggest in this chapter that the traditional concept of habitat may need to be broadened beyond the basics of food, cover, and water to include ideas such as the ecological roles of other species, abiotic conditions, and natural disturbance regimes. We also suggest that the traditional notion of wildlife may need to be broadened to encompass the full array of biota present in an ecosystem. We propose in this chapter an enhanced approach to depicting, modeling, and predicting the status and condition of wildlife in ecosystems, and advocate the rigorous use of adaptive management as a foundation for land use decisions.

Changes in environmental conditions on earth, advances in technological devices and analytic methods, and the potential need for new philosophical and conceptual approaches in research and management require that wildlife biologists do their best to keep abreast of new ideas. We discuss in chapter 12 several problems that likely will confront wildlife biologists in the coming years and the kinds of information that will be needed to address those problems. We also identify subjects that, in our opinion, need further investigation and suggest approaches for their study.

We noted at the end of the first chapter in the first edition of this book that the ideas we presented would probably evolve. They have, and their evolution is part of the motivation for this second edition. However, the ideas we expressed then about dealing with changes in the world still hold. We hope that, no matter what changes occur, our readers—current or future conservationists in the broadest sense—remain tied to an ecological land ethic and continue the pursuit of providing vital, productive habitats for wildlife and humans alike.

Literature Cited

Adams, C. C. 1908. The ecological succession of birds. *Auk* 25:109–53.

Aristotle. [344 B.C.] 1862. *Historia animalium.* London: H. G. Bohn.

Bean, M. J. 1977. *The evolution of national wildlife law.* Report to the Council on Environmental Quality. U.S. Government Document, stock no. 041-011-00033-5.
Beebe, W., ed. 1988. *The book of naturalists.* Princeton: Princeton University Press.
Bellrose, F. C. 1976. *Ducks, geese and swans of North America.* Harrisburg, Pa.: Stackpole Books.
Butler, R. G. 1980. Population size, social behavior, and dispersal in house mice: A quantitative investigation. *Animal Behavior* 28:78–85.
Darwin, C. 1859. *The origin of species.* New York: Penguin Books.
Daubenmire, R. 1976. The use of vegetation in assessing the productivity of forest lands. *Botanical Review* 42:115–43.
Dawson, J. W., and R. W. Mannan. 1994. *The ecology of Harris hawks in urban environments.* Final report, Arizona Game and Fish Department, Urban Heritage Project (G20058-A). 59 pp.
Eiseley, L. 1961. *Darwin's century.* Garden City, N.Y.: Doubleday, Anchor Books.
Emlen, J. T. 1974. An urban bird community in Tucson, Arizona: Derivation, structure, regulation. *Condor* 76:184–97.
Errington, P. L., and F. N. Hamerstrom. 1937. The evaluation of nesting losses and juvenile mortality of the ring-necked pheasant. *Journal of Wildlife Management* 1:3–20.
Gilbert, F. F., and D. G. Dodds. 1987. *The philosophy and practice of wildlife management.* Malabar, Fla.: Robert E. Krieger.
Green, R. H. 1971. A multivariate statistical approach to the Hutchinsonian niche: Bivalve molluscs in central Canada. *Ecology* 52:543–56.
Grinnell, J. 1917. Field tests of theories concerning distributional control. *American Naturalist* 51:115–28.
Hilden, O. 1965. Habitat selection in birds. *Annales Zoologici Fennici* 2:53–75.
Hutchinson, G. E. 1957. Concluding remarks. *Cold Spring Harbor Symposium on Quantitative Biology* 22:415–27.
Klopfer, P. H., and J. U. Ganzhorn. 1985. Habitat selection: Behavioral aspects. In *Habitat selection in birds,* ed. M. L. Cody, 435–53. New York: Academic Press.
Lack, D. 1933. Habitat selection in birds with special reference to the effects of afforestation on the Breckland avifauna. *Journal of Animal Ecology* 2:239–62.
Leopold, A. 1933. *Game management.* New York: Charles Scribner's Sons.
Leopold, A. 1949. *A Sand County almanac.* Oxford, England: Oxford University Press.
Merriam, C. H. 1890. Results of a biological survey of the San Francisco Mountains regions and desert of the Little Colorado River in Arizona. USDA Bureau of Biology, *Survey of American Fauna* 3:1–132.
Mills, G. S., J. B. Dunning, Jr., and J. M. Bates. 1991. The relationship between breeding bird density and vegetation volume. *Wilson Bulletin* 103:468–79.
Musgrae, R. S., and M. A. Stein. 1993. *State wildlife laws handbook.* Rockville, Md.: Governments Institutes, Inc.
Ruther, S. A. 1987. Urban wildlife conservation in Arizona: Public opinion and agency involvement. M.S. thesis. University of Arizona, Tucson.
Ruther, S. A., and W. W. Shaw. 1990. Public opinion and urban wildlife conservation program development. In *Managing wildlife in the Southwest,* ed. P. R. Krausman and N. S. Smith, 226–29. Phoenix: Arizona Chapter of the Wildlife Society.
Shaw, W. W., A. Goldsmith, and J. Schelhas. 1992. Studies of urbanization and the wildlife resources of Saguaro National Monument. In *Proceedings of the symposium on research in Saguaro National Monument,* ed. C. P. Stone and E. S. Bellantoni, 173–77. Tucson, Ariz.: National Park Service.
Stenberg, K. 1988. Urban macrostructure and wildlife distributions: Regional planning implications. Ph.D. dissertation, University of Arizona, Tucson.
Stoddard, H. L. 1931. *The bobwhite quail: Its habits, preservation, and increase.* New York: Charles Scribner's Sons.
Stone, C. D. 1974. *Should trees have standing? Toward legal rights for natural objects.* Los Altos, Calif.: William Kaufmann, Inc.
Stone, C. D. 1987. *Earth and other ethics.* New York: Harper and Row, Publishers.
Svardson, G. 1949. Competition and habitat selection in birds. *Oikos* 1:157–74.

Thomas, J. W., and D. E. Toweill, eds. 1982. *Elk of North America.* Harrisburg, Pa.: Stackpole Books.

U.S. Department of the Interior, Fish and Wildlife Service, and U.S. Department of Commerce, Bureau of the Census. 1993. 1991 national survey of fishing, hunting, and wildlife-associated recreation. Washington, D.C.: U.S. Government Printing Office.

Van Horne, B. 1983. Density as a misleading indicator of habitat quality. *Journal of Wildlife Management* 7:893–901.

Verner, J., M. L. Morrison, and C. J. Ralph, eds. 1986. *Wildlife 2000: Modeling habitat relationships of terrestrial vertebrates.* Madison: University of Wisconsin Press.

Wallmo, O. C., ed. 1981. *Mule and black-tailed deer of North America.* Lincoln: University of Nebraska Press.

Werner, E. E., and D. J. Hall. 1979. Foraging efficiency and habitat switching in competing sunfishes. *Ecology* 60:256–64.

Werner, E. E., J. F. Gilliam, D. J. Hall, and G. G. Mittelbach. 1983. An experimental test of the effects of predation risk on habitat use in fish. *Ecology* 64:1540–48.

2 The Evolutionary Perspective

Although we may not fully understand the causes and consequences of many populational phenomena, we can be confident that all have an evolutionary explanation.

—E. R. Pianka, *Evolutionary Ecology*

Introduction: Conceptual Framework

Pianka's quotation simply and eloquently states why we should be interested in the evolutionary processes that have formed present-day expressions of an animal's ecology. We want the reader to approach each chapter with a clear understanding of the ecological foundation upon which this book was written. Our writing centers on the notion that the present distribution and abundance of animals are a result of adaptation to a host of biotic and abiotic factors and, in some cases, of serendipity of species and conditions that happen to come into alignment (such as preadaptation). Many of the factors that shaped the form and function of an animal may now be absent or at least of different intensity. The ability of an animal to adjust to future conditions, be they induced by "natural" environmental changes or through human impacts, depends on its preadaptations and its ability through genetic diversity to develop new adaptations to these changes. Thus, current conditions and an animal's response to them may not provide the basis for a complete understanding of the reasons for the pattern observed, and could result in inappropriate recommendations for managing long-term evolutionary potential.

The fields of evolutionary ecology and biogeography should thus play a central role in the field of wildlife-habitat analysis. The development of an understanding of why a population is found in a specific area requires knowledge of the organism's ecological relationships: Why is it associated with specific combinations of edaphic, aquatic, temperature, and biotic regimes? To understand why it is found only in certain areas requires knowledge of climatic history, which may have resulted in the development of isolated, relict populations. Analysis of the evolutionary history of a population itself and of the geological history of the landscape is a necessary prerequisite to understanding why the population is distributed as we observe today (see, e.g., Cox and Moore 1993).

Animals evolve by responding to both biotic and abiotic factors. Gene pools can adapt to forces being imposed by the abiotic environment, to forces being applied by other animals, and to changing conditions in the nonanimal biotic environment (e.g., plants). Such adaptations are usually referred to as coevolution and are certainly of keen interest in how we interpret the ecologies of the animals we study today. Much interest over the past several decades has centered on studying links or webs within groups of animals in an attempt to identify "communities" of organisms. Recently, interest has grown in identifying individual species that play key roles in structuring such communities or at least subsets of species in an area (*sensu* keystone species).

Competitive interactions may play important roles in shaping morphology and behavior, and thus may be manifested in distribution and abundance. Under this assumption, many of the interactions between species likely took place in the past, and much of what we observe today is the result of these interactions. Competition often becomes unfalsifiable, the "ghost of competition past" (Connell 1980). Although we cannot solve this dilemma, we intend to use such controversies to show that determination of factors responsible for animals' presence in an area and their habitat use are complicated, interrelated, and not completely apparent.

Thus, this chapter examines the conceptual framework for viewing wildlife-habitat relationships throughout this book. We argue that, unless theories of biogeography, habitat selection, and community structure are considered when studying and managing for wildlife-habitat relationships, further progress in understanding will be minimal.

Evolutionary Perspective

The distribution and abundance of animals over the landscape and the use of habitat must be considered in light of the geological and geomorphic events that shaped the area. Everyone has been introduced to the concept of the geological timetable: the chart that relates geological developments to the evolution of plants and animals. Here we are most concerned with events shaping our recent past, the Pleistocene to the present (the Recent epoch).

The Pleistocene

The Pleistocene epoch, which began 2 million to 3 million years ago, is thought to have ended about 10,000 years before the present. The Pleistocene was characterized by a series of advances and retreats of the continental ice sheet and glaciers. There were apparently four major advance-retreat cycles in North America: the Nebraskan, Kansan, Illinoian, and Wisconsin. Between these advances of ice were three interglacial periods: the Aftonian, Yarmouth, and Sangamon. Our current Recent epoch is, in fact, probably best viewed as another interglacial period of the Pleistocene (Cox and Moore 1993). These ice sheets obviously had a dramatic impact on the ground cover and the associated environmental conditions; the possible maximum extent of glaciation is shown in figure 2.1. Although vegetation was destroyed, many animals had time to seek and occupy new areas. Of course, such enormous ice sheets do not just suddenly appear. Further, areas close to but not covered by the ice also underwent dramatic physical and environmental changes

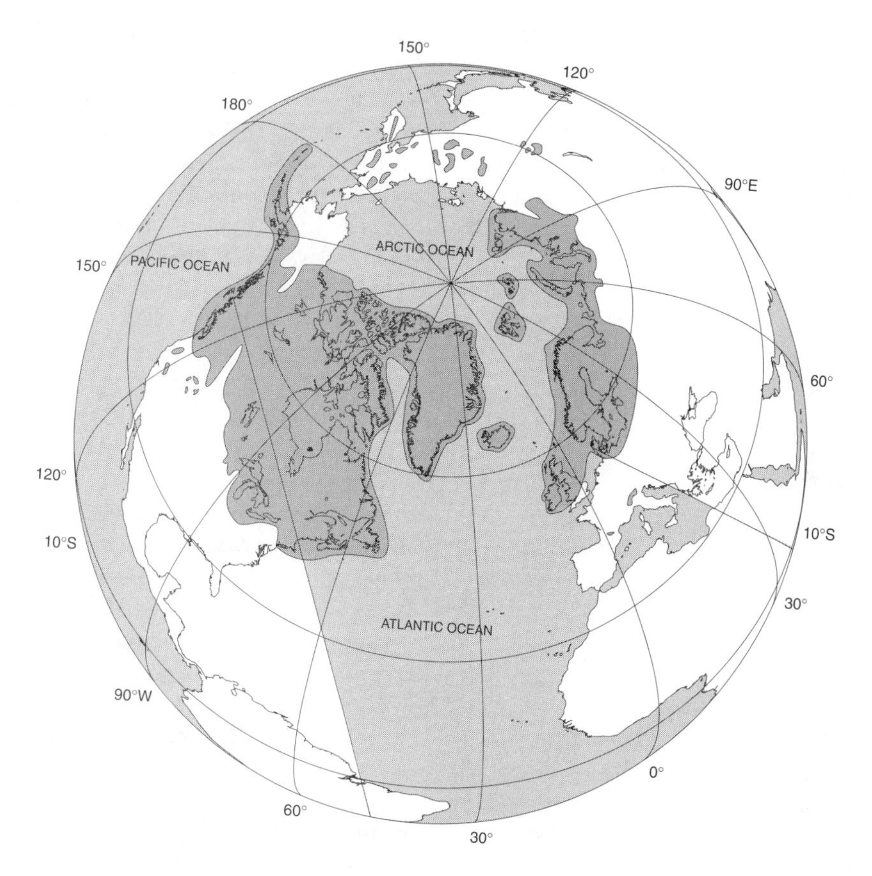

Figure 2.1. The possible maximum extent of glaciation in the Pleistocene in the Northern Hemisphere. C = Cordilleran ice; L = Laurentide ice; S = Scandinavian ice; and A = Alpine ice. (Reproduced from A. Goudie, *Environmental Change,* 3d ed. [Oxford: Clarendon Press ©1992], fig. 2.9, by permission of Oxford University Press)

because of their proximity to the ice fields. Most of the species currently occupying the earth are survivors of the abiotic and biotic influences of the Pleistocene, especially the last several advance-retreat cycles. In New Mexico, for example, the mid- to late Wisconsin fossil evidence suggests a progression of vegetation types, changing initially from semiarid, moderately warm grasslands or grassy woodlands to cooler, more mesic, grassy woodlands; followed by cool, relatively dense sagebrush-grassland-woodland with elements from mixed-coniferous forests (Harris 1993).

Charcoal fragments from tropical rain forests in north Queensland, Australia, indicated that *Eucalyptus* woodlands occupied substantial areas of all the present humid rain forest between about 27,000 B.P. (before present) and 3500 B.P.

Changes in paleoclimates associated with the most recent glaciation provide a plausible explanation for both the *Eucalyptus* expansion and subsequent rain forest reinvasion (Hopkins et al. 1993). Postglacial succession of vegetation has been reconstructed for many areas (e.g., Ritchie 1976; Amundson and Wright 1979; Elias 1992; Nordt et al. 1994).

Although the areas we now call Alaska and Canada were mostly covered with ice, numerous ice-free pockets existed. These ice-free "refugia" were located primarily along the coasts and near-shore islands, with the ocean acting as a mediating influence on air temperatures (Heusser 1977). The result was a chain of forested refugia that extended south below the ice-covered regions. Palynological evidence—deposits of pollen in sediment layers—has allowed scientists to reconstruct the types of plants inhabiting these refugia. In the Pacific Northwest, for example, vegetation on the Olympic Peninsula through much of the middle and late Wisconsin glacial period resembled the edge of the Pacific coast forest in Alaska as it appears today (Heusser 1965, 1977). In New Zealand, warm tropical waters of the southwest equatorial Pacific had a strong ameliorating effect on the mean temperature in the Tasman Sea during the last glaciation. During the last glaciation, northern New Zealand had a climate similar to that of the present (Wright et al. 1995).

Refugia were also present below the southern extent of the ice sheets. Snow levels in the mountains of the continental United States were much lower than those we see today. During times of glaciation in the Cascade Mountains of the Pacific Northwest, for example, summer temperatures were apparently close to freezing, and precipitation and cloudiness were pervasive (Heusser 1977). In the southwestern United States, Wisconsin glaciation lowered the biotic zones by as much as 1200 m, so woodlands and forests occurred over much of the ice-free region (Hubbard 1973; Elias 1992). As the glaciers receded northward and upslope, bare or nearly bare soil was available for colonization by plants. Plants that had survived the glacial period in refugia were the most available for colonization. In turn, animals inhabiting refugia would likely have been the first colonists.

Distribution of Animals during the Pleistocene

The distribution and abundance of animal species existing currently can be linked to the geological events of the Pleistocene outlined above. Birds have received the most research effort because of their ability to travel long distances rapidly, thus potentially moving between refugia by flying over unsuitable areas. The retreat of glaciers opened the way for occupation of vast areas either by species preadapted to the newly developing vegetation or by those able to adjust to the new environmental conditions. The pool of potential colonists is unknown but was probably related to the types of species and their abundances in refugia and in the more southern, ice-free areas. A classic example explaining the occupation of North America by wood warblers (Emberizidae: Parulidae) is presented in box 2.1.

During glacial advances, boreal mammals were apparently widespread in lowlands well south of their present ranges. Concurrent with the movements of these mammals northward during interglacial periods were movements of boreal mammals into montane regions. Here, because of the effect of elevation on climate, cool refugia were present. Many of these montane populations have persisted in boreal refugia far south of the northern range of their closest relatives, and the zonation of mammalian distributions on some mountain ranges in the southwestern United States apparently resulted from Pleistocene faunal movements and subsequent retractions from warm, dry lowlands (e.g., Brown 1978; Vaughan 1986; Elias 1992). Harris (1990) found that many mammals now occurring in the southern Rocky Mountains ranged south into the Chihuahuan Desert in the late Wisconsin. The present ranges of many taxa were probably reached in the early Holocene epoch, but some survived in refugia until the late Holocene (Elias 1992).

Although not critical to our thesis here, there has been debate over whether evolution has been typified by a gradual process (phyletic gradualism) or periodic, dramatic events (e.g., mass extinctions *sensu* punctuated equilibria, which lead to rapid radiation of many taxonomic levels). We will leave this debate to others, but note that Barnosky (1987) concluded that both gradualism and punctuated equilibrium are occurring, depending upon the taxa in question (see also Eldredge and Gould 1972).

Ecological Compromises

Patterns of migration and seasonal changes in habitat use have ramifications for our attempts to determine why animals occupy the areas they do, and they complicate our attempts to predict the response of animals to environmental changes. The majority of ecological studies have focused on animals during breeding periods. This is understandable, given the obvious importance that the production of young plays in survival of the species. However, equally important are nonbreeding areas and movement to and from geographic areas that serve as seasonal centers of activity. Failure to maintain wintering areas and migratory routes is likely to lead to extinction or at least to substantial population declines. Exploring the causes of seasonal movements and migration is a necessary part of determining means of conserving animals (Brower and Malcolm 1991). Additionally, seasonal shifts in diet and habitat use in the same geographic area also are of likely adaptive and evolutionary significance.

Migration of birds is certainly tied to seasonal changes in climate and food resources,

Box 2.1 A classic example to explain the occupation of North America by wood warblers

Mengel (1964) presented what has become a classic scenario to account for the occupation of North America by wood warblers (Emberizidae: Parulidae). By his reasoning, the retreat of glacial ice opened the way for occupation of vast areas by generalist species able to adjust to varying environmental conditions. The pool of potential colonists is unknown but is probably related to the types of species and their abundance in refugia and more southern habitats. Using a precursor of the black-throated green warbler (*Dendrocia virens*) as an example, Mengel postulated that this "pro-*virens*" was able to occupy vast reaches of western North America as the environment warmed (box fig. 2.1). Most warblers apparently evolved in more tropical regions of North, Central, and South America, colonizing western regions of North America by passing through the Caribbean Islands and the southeastern United States. With the subsequent advance of the ice sheets, pro-*virens* was isolated to the north in refugia and to the south in warmer regions of the Southwest. During this glacial period, these isolated populations either adapted to their changing surroundings or became extinct. With the warming and concomitant retreat of the glaciers, the populations were able to expand their range as the forest expanded; many were now distinct species (see Mengel 1970). Although the general premise of Mengel is well accepted, many of the specifics are not (Hubbard 1973; Flack 1976). Similar scenarios have been developed for other species of birds (e.g., Hubbard 1973) and mammals (Smith 1981).

(box continued on following page)

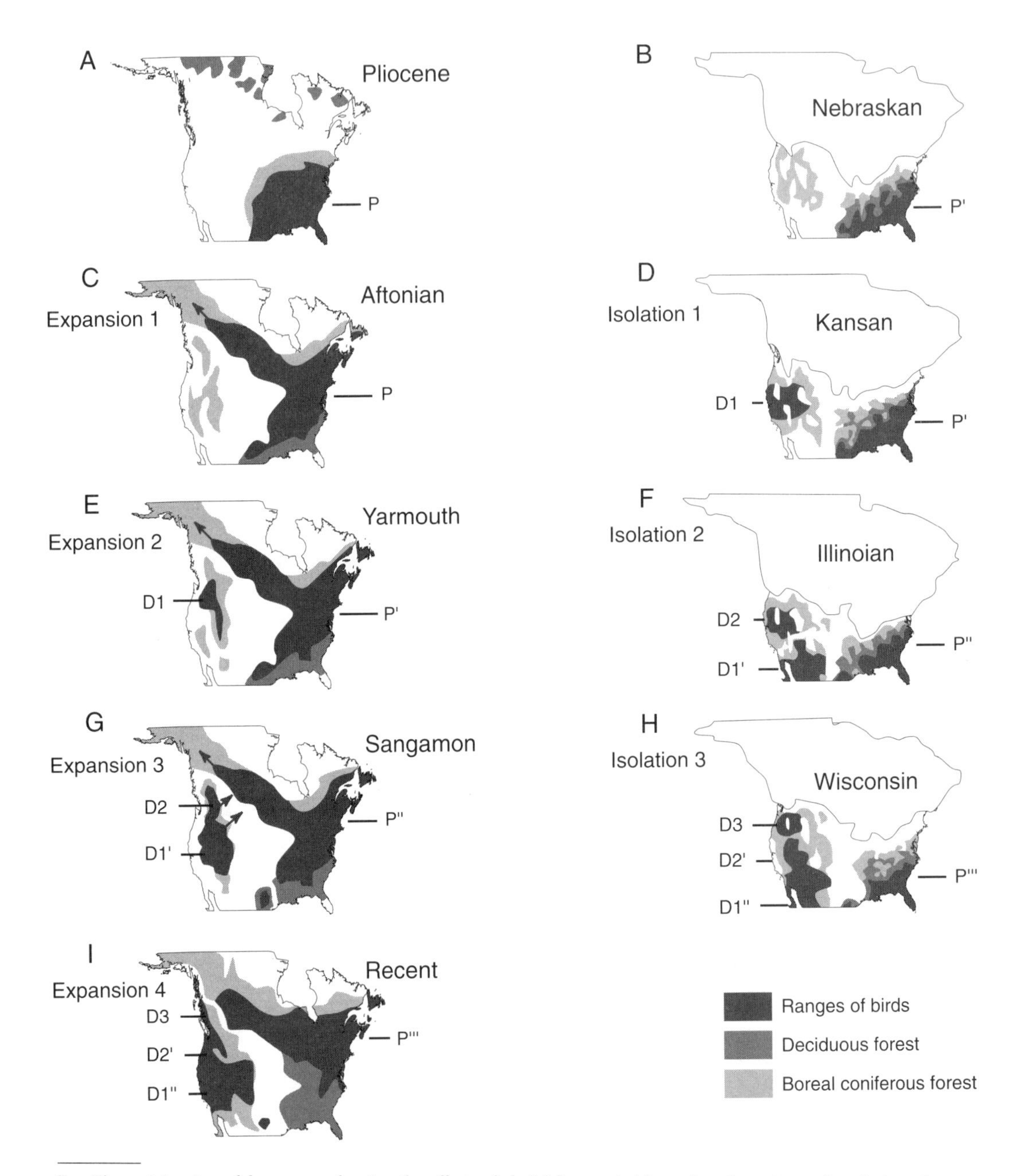

Box Figure 2.1. A model sequence showing the effects of glacial flow and ebb on the adaptation and evolution of a hypothetical ancestral wood warbler and its descendants. The details of glacial boundaries are approximate. P = parental species; D = derivative species; prime marks = number of glacial cycles removed from origin (Reproduced from Mengel 1964 [fig. 4], by permission of the Laboratory of Ornithology, Cornell University)

and many migrations are likely to have resulted from Pleistocene glaciation. Migration can be considered either as an adaptation to exploit resources in newly unglaciated areas or as an attempt to escape the harsh northern winters (Stiles 1980). Migratory birds breeding in North America spend five to seven months on wintering areas that usually bear little resemblance to their breeding grounds; several months are often spent in migration. Because of a long series of increasing range extensions, the global pattern of seasonal migrations must have been established or reestablished in the 20,000 years or so since the maximum extent of the last glacial advance. Most migrations are apparently legacies of the Pleistocene that evolved as adaptations for animals to escape predation, to reduce competition, to exploit periods and areas of resource abundance, to avoid winter cold, or to leave excessively dry- and wet-season areas (see summary in Brower and Malcolm 1991).

The use of different summer and winter areas, plus environmental conditions within areas along the migration path, set morphological, behavioral, and physiological constraints on the abilities of an animal to use either the summer or the winter area efficiently. For example, Morrison (1983) found that, in the western United States, two populations of Townsend's warblers (*Dendroica townsendi*) apparently exist: one a shorter-winged group that winters in the United States; the other, a longer-winged group that migrates to Mexico and Central America (see also Grinnell 1905). Both groups have separate breeding areas. Cody (1985) thought that such morphological adaptations might compromise the habitat- and foraging-site selection of animals during breeding. Finch bills, for example, seem more strongly influenced by their granivorous winter diet than by their mostly insectivorous diet during breeding (Cody 1985). In the Sierra Nevada in California, many small gleaning and hover-gleaning, foliage insectivorous songbirds become flakers and probers of bark during winter in response to changes in food abundance and location (Morrison et al. 1985). Some herbivorous mammals make seasonal migrations in concert with seasonal changes in the local availability of high-quality food; and some smaller mammals (e.g., voles, lemmings) change habitat use on a local scale with seasonal changes in food (Batzli 1994).

The fossil record can sometimes be used to reconstruct the former range of species. Changes in distribution can sometimes be related to changes in paleoclimatic factors. For example, Harris (1993) reconstructed the succession of microtine rodents from the mid- to the late Wisconsin period in New Mexico. For instance, the disappearance of the prairie vole (*Microtus ochrogaster*) and appearance of the sagebrush vole (*Lemmiscus curtatus*) were interpreted as a result of cooling summer temperatures and a shift in seasonality from warm season–dominant to cool season–dominant precipitation (Harris 1993; see box 2.2 for details). Goodwin (1995) reconstructed the Pleistocene distribution of prairie dogs (*Cynomys* spp.) and concluded that fossil evidence supports the historical association of white-tailed prairie dogs (*C. gunnisoni*) with shrub-steppe, whereas the black-tailed prairie dog (*C. ludovicianus*) has a long history in or near its present range as indicated by its use of relatively more complex habitats and its more primitive biology.

Sullivan (1988, in Hafner 1993) described a "biogeographic indicator species" as diagnostic of specific environmental conditions through time. Reconstruction of the biogeographic history of such a species provides a reconstruction of the changing distribution of the habitat to

Box 2.2 Progression of four species of microtine rodents

Harris (1993) developed the progression, derived from the fossil record of New Mexico, in the relationship between four species of microtine rodents (box fig. 2.2). Mid-Wisconsin, semiarid, moderately warm grasslands were replaced by cooler, more mesic, grassy woodlands; these were followed by cool, relatively dense sagebrush (*Artemisia*) -grassland-woodland with elements from mixed-coniferous forest. Concomitant with these environmental changes were changes in the microtine rodents occupying the region. Of the four species discovered, only the prairie vole was present in the lower levels (*levels* refers to the stratigraphic sequence going from older at the lower levels to more recent at the upper levels). In levels 7 to 5, that species coexisted with the Mexican vole; there then followed a brief interval in which only the latter species was present (level 4). Level 3 documents the invasion of the sagebrush vole, followed in level 2 by the only occurrence of the long-tailed vole. Level 2 also shows the peak frequency of the Mexican vole. The record is closed with the sagebrush and Mexican voles being rejoined by the prairie vole. The disappearance of the prairie vole after level 5 and the appearance of the sagebrush vole in level 3 were interpreted not only as a result of cooling summer temperatures, but also as indicating a shift in seasonality from warm season–dominant to cool season–dominant precipitation.

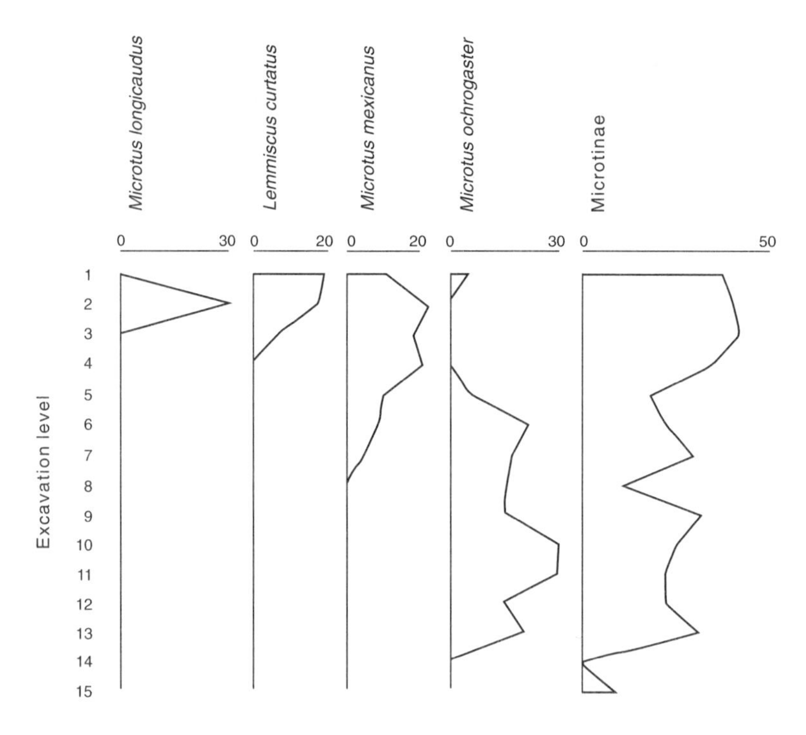

Box Figure 2.2. Abundances (percentages of four species of microtine rodents and of total microtines (including those unidentifiable to species) recovered from 15 excavation levels (Reproduced from Harris 1993 [fig. 4], by permission of the Quaternary Research Center, University of Washington)

which the species is restricted. Thus, this reconstruction provides an idea of the historical biogeography of other recent species that are restricted to the same general conditions. Hafner (1993) offered the Neartic pikas (*Ochotona princeps* and *O. collaris*) as biogeographic indicators of cool, mesic, rocky areas. Fossil pika have been found far from extant populations, which is especially evident in Nevada (see fig. 2.2). Hafner provided the following speculation regarding interpretation of locations having fossil pika but lacking extant populations: If the overall site reconstruction is xeric and at low elevations, then a suitable rocky microsite must be assumed to have existed nearby. If the overall site reconstruction is one of a cool, mesic area but fossil remains are rare, they may represent dispersing individuals; however, suitable mesic habitat must be assumed to have existed nearby.

Reconstructions such as these have obvious ramifications for modern-day research and management: they enable us to understand partly why species change in distribution as

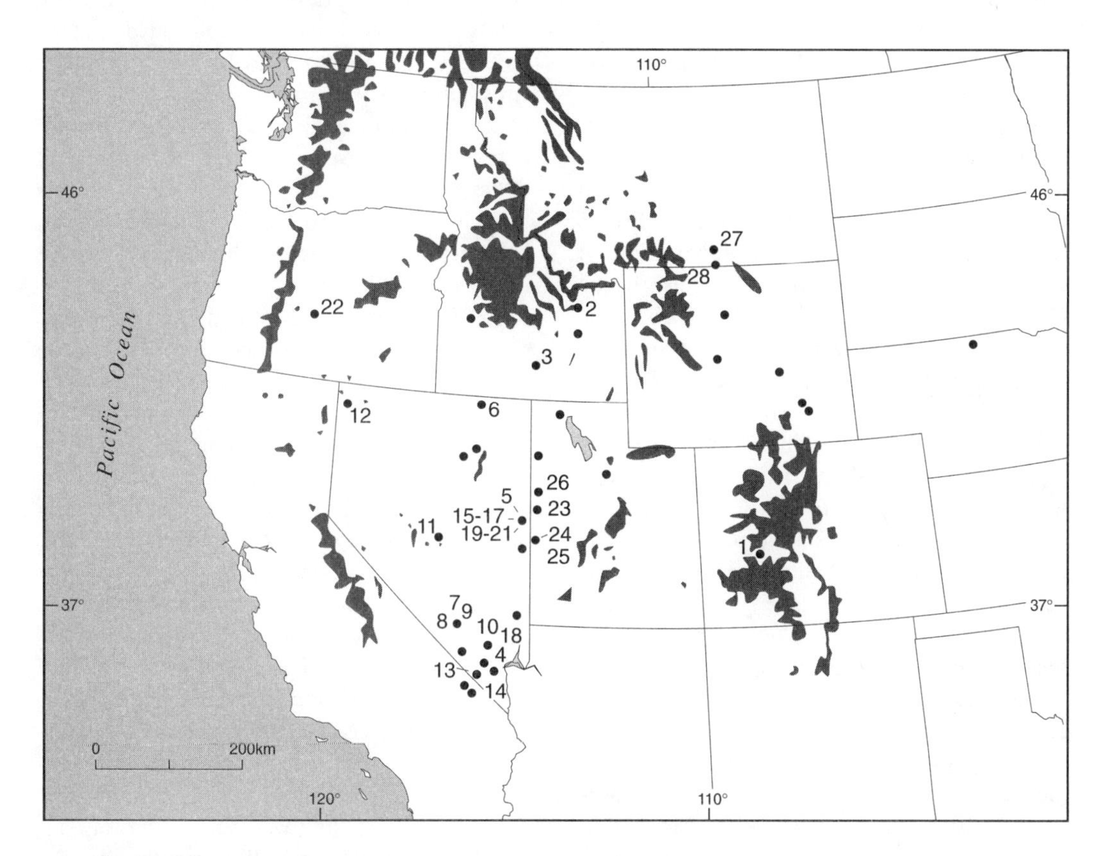

Figure 2.2. Distribution (shaded area) of extant pikas (*Ochotona princeps*) and late Pleistocene-Holocene fossil records (circles) in western North America (Reproduced from Hafner 1993 [fig. 1], by permission of the Quaternary Research Center, University of Washington)

climate and perhaps even interspecific interactions change, and further, they allow us to predict changes that might occur with future natural and human-induced changes in the environment. Of course, these ideas assume that habitat specificity is unchanging, an assumption that should be fully explored in any such analysis. For example, fossil evidence of an unchanging morphology does not mean that physiology has remained static through time.

Post-Pleistocene Events

Biogeographic patterns of animal distribution have been explained by both vicariance (defined as the distribution that results from the replacement of one member of a species pair by the other, geographically) and dispersal. As an example of vicariant distribution, Brown (1978) explained the current distribution of small, nonflying, montane mammals in the Great Basin as resulting from post-Pleistocene climatic and vegetational changes that left these species stranded in woodland and forest habitats on mountaintops and isolated by intervening desert scrub. He concluded that the desert vegetation prevented intermountain colonization. In other regions of the Southwest, however, evidence indicates that post-Pleistocene colonization has dramatically influenced the current distribution of some species (Davis et al. 1988). Davis and his co-workers concluded that both post-Pleistocene dispersal and subsequent colonization as well as vicariant events and subsequent extinction have influenced the assemblages of mammals. They stressed that the degree to which each process influences animal distribution should be considered in explaining current community composition.

As discussed above, the record of vertebrate biogeographic migrations during the past 125,000 years indicates movements of considerable magnitude in response to changing temperature and moisture regimes during glacial advances and retreats. Because climatological changes of different periodicities are occurring continuously, biogeographic migrations continue at present, and records of vertebrate faunal shifts during the last two centuries have been documented for many regions of the world.

Johnson (1994) studied range expansions of 24 species of birds in the contiguous western United States that have occurred since the late 1950s and early 1960s. He concluded that all have enlarged their ranges for reasons apparently unrelated to direct human modifications of the environment. He proposed that, instead, pervasive climatic changes over the past several decades are the most likely explanation. Although climatic warming is probably involved, he concluded that increased summer moisture is the overriding factor. Climatic information from the region has offered support for wetter and warmer summers in recent decades in the contiguous western United States.

As noted by Johnson, the establishment of substantial numbers of breeding birds well outside their normal breeding ranges must alter the biotic relationships in the new areas; this might be especially evident when a large predator is involved (e.g., the barred owl [*Strix varia*] has expanded westward to, and southward down, the Pacific Coast).

Distributions of vertebrates during the Quaternary indicate distinct north-south changes in their distributions in response to glacial conditions. During the Wisconsin advance, for example, many species shifted southward. The presently disjunct southern populations of relatively northern species reflects southward displacement during Wisconsin glaciation. Such range adjustments, in essence, strand populations of species outside what we now consider their normal ranges. Such populations are often referred to as relicts, and they present difficult management problems because they are usually unable to interchange naturally with other populations of their species. Moreover, it is difficult to ascertain if declines in a relict population are due to continuing naturally occurring phenomena or are being hastened by human-induced changes. These situations are further complicated by human-induced changes in land uses that hinder or even prevent the population from undergoing additional biogeographic shifts in response to environmental changes.

Bottlenecks

A population may be influenced in a major way by catastrophic events that occur only infrequently. Termed ecological crunches, these events are thought to drive a population so low in abundance that it is forced through a "bottleneck" that changes the genotype and its phenotypic expression and possibly its ability to adapt to new environments (Wiens 1977). Various studies have identified situations in which genetic difficulties, especially inbreeding resulting from bottlenecks, have been manifested in physiological problems. Probably the most cited study is that on the South African cheetah (*Acinonyx jubatus*) by O'Brien, Goldman et al. (1983) and O'Brien, Wildt, and Bush (1986). They found extreme genetic monomorphism in the two populations they studied and reported low sperm counts, low sperm mobility, a high frequency of abnormally shaped sperm, and concomitant low reproductive output. They thought that these physiological problems and low genetic variability are the result of a severe population bottleneck followed by inbreeding, likely the result of climatic changes in the late Pleistocene. Other examples of species going through apparent population bottlenecks are available (see O'Brien et al. 1983; O'Brien, Wildt, and Bush 1986; Pimm 1991:156–62). However, Pimm (1991:159–62) has questioned the conclusions of O'Brien and colleagues and, in general, the notion that inbreeding will necessarily doom a relatively large population to extinction. It is clear, however, that changes in the distribution and abundance of a species, especially through fragmentation of the species into disjunct populations, will lead to changes in genetic composition. Human activities producing such fragmented populations can lead to increased genetic differentiation among the populations (e.g., Leberg 1991). The impacts that such changes have on the long-term viability of a species are unknown.

Coevolution

Our previous discussion centered on the abiotic factors shaping the types and distributions of plants and animals through geological time. But as an organism evolves to meet the demands of a changing environment, it also has to evolve in light of forces exerted on it by other organisms. This joint evolution of two or more taxa that have close ecological relationships but do not exchange genes has been termed coevolution. Under coevolution, reciprocal selection pressures operate to make the evolution of either taxon partly dependent on the evolution of the other. Coevolution includes various types of population interactions, including some instances of predation, competition, parasitism, commensalism, and mutualism. Although the classic examples of coevolution involve plant-animal interactions, the term rightfully applies to any interdependent phenomena between taxa (Pianka 1983:238; for example, competitive displacement of sibling species).

The interdependence of taxa naturally increases our fascination with the living world; consequently, this interdependence also increases our challenge when trying to understand and then manage these taxa. Nevertheless, it is critical that researchers and managers appreciate the fact that most of our actions will cause some type of reaction by other taxa. Historically, when much of wildlife management centered on featured game or pest species, little consideration was given to the impacts such management had on other organisms (see chapter 1 for a review). Today, however, we understand that professional management—and usually the law—requires some ability to predict more fully the consequences of our management activities. This understanding should begin with careful evaluation of the historical geological and environmental conditions that shaped the plant and animal communities under study (see box 2.1).

As summarized by Franklin (1988), human activities have been responsible for dramatic changes in the composition and structure of vegetated areas throughout the world. For example, most forests in the temperate zone are now secondary (i.e., previously harvested), and tree-species composition has shifted toward commercially valuable species and relatively early successional stages. In addition, standing dead trees (snags), fallen logs, and the litter layers, which are not usually maintained or are heavily impacted in managed forests, are essential to many organisms and biological processes. Efforts to conserve structural and functional diversity within an area are often linked; for example, by maintaining woody debris, we can retain one of the sites of nitrogen fixation along with the microsite conditions often necessary for numerous organisms (see also chapter 10). Changes in patterns of grazing (by domestic and native herbivores) and fire have similar impacts (e.g., Risser 1988). These changes alter the conditions under which species evolve and are especially critical for coevolved relationships (e.g., introduction of exotic species changes the relationship between the habitat and native animals; see chapter 4).

Food Webs

Groups of organisms can be represented by a food web, which is simply a diagram of all the trophic relationships between its component species. A food web is generally composed of many food chains, each of which represents a single pathway of the food web (Pianka 1983). Webs thus depict the flows of matter, nutrients, and energy across and within user-defined boundaries (usually referred to arbitrarily as a community). Since first introduced by Elton (1927), the concepts of food passage and nutrient cycling have now achieved a central place in ecological theory and the structuring of communities. Several excellent books and articles explain and document the development of this theory, which often links the structure of food webs to community stability (Cohen 1978; Lawton 1989; Cohen et al. 1990; Pimm 1982, 1991; Hall and Raffaelli 1993).

Pimm (1982) linked food-web structure and ecosystem function under three principal categories: resilience, which measures how fast species densities return to equilibrium following a perturbation; nutrient cycling, which measures how tightly ecosystems retain the nutrients that make up part of the biomass of their constituent species; and resistance, which measures how productivity or biomass at a particular trophic level changes with variations in feeding rates (e.g., increased or decreased predator pressure). These interrelationships between the food-web structure, ecosystem functions, and successional state envisioned by Pimm are illustrated in figure 2.3.

The apparent link between food webs and ecosystem functioning was thought by Cohen (1990) to relate to our understanding and subsequent management of the environment in several ways. First, environmental toxins accumulate along food chains. Second, knowledge of the food web is necessary for anticipating the consequences of species removals and introductions. Third, an understanding of webs will help in the design of nature reserves and other land use practices. (Additional details on these topics are found below.)

As noted above, the empirical study of food chains and webs has developed into a prolific branch of ecological theory. However, there remains a fundamental disagreement over whether "bottom-up forces" (e.g., nutrient availability, especially plants) or "top-down forces" (e.g., predators, especially carnivores) predominate in populations and serve to structure communities. Hunter and Price (1992) argued that a synthesis of the top-down and bottom-up forces in terrestrial systems requires a model that incorporates the influences of biotic and abiotic heterogeneity. This view thus incorporates the classic debate between supporters of the relative roles that biotic or abiotic factors play in determining changes in animal abundances (e.g., Andrewartha and Birch 1954; Lack 1954). Huner and Price (1992) hypothesized that differences between species within a trophic level (i.e., all species on one level are not indivisible units), differences in species interactions in a changing environment, and changes in population quality (e.g., reproductive success) with population density are as important as determinants of population and community dynamics as the number of levels in a food web and the position of the system along a resource gradient are.

A single bottom-up template has been proposed for viewing the relative contributions of top-down and bottom-up forces (fig. 2.4). In figure 2.4a, variability in climate, soil parameters, decomposers, and plant-soil symbionts determines the initial heterogeneity of form exhibited among primary producers. This pattern of heterogeneity represents the template upon which the complex interactions between species in real populations and communities are superimposed. Figure 2.4b adds back some level of biological reality by allowing species at each trophic level to impact those in trophic

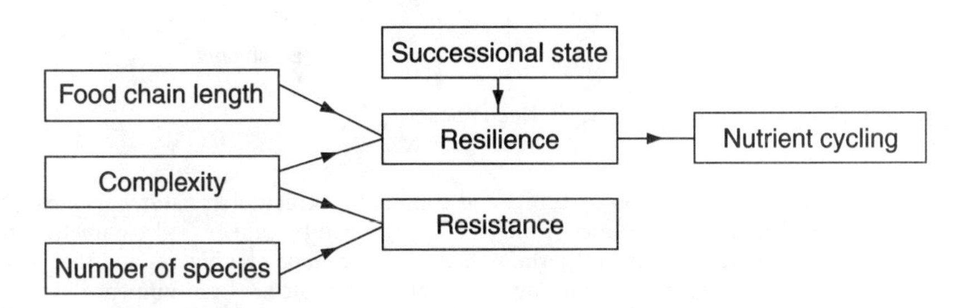

Figure 2.3. A diagram of the causal relationships between food-web structures, successional state, and ecosystem functions (Reproduced from S. L. Pimm, *Food Webs* [New York: Chapman and Hall © 1992], fig. 10.2, by permission of Chapman and Hall)

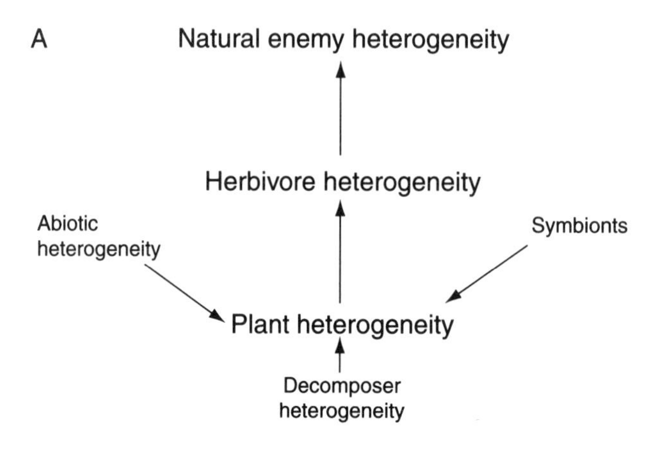

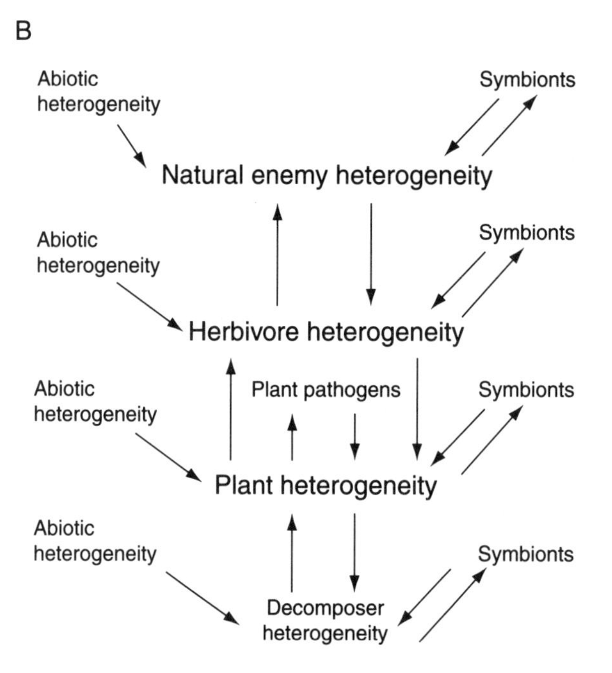

Figure 2.4. Factors influencing population dynamics and community structure in natural systems. (A) A simple model in which variation among primary producers, determined by climate, soil parameters, and symbionts, cascades up the trophic system to determine heterogeneity among herbivores and their natural enemies. (B) With the addition of feedback loops, organisms at any trophic level can influence heterogeneity at any other level by cascading effects both up and down the system. (Reproduced from Hunter and Price 1992 [fig. 1], by permission of the Ecological Society of America)

levels below as well as above them, and by including the effects of abiotic heterogeneity at all trophic levels. Hunter and Price thought that the bottom-up template is compelling because plants form a major component of large-scale patterns over landscapes and geographic regions.

Hunter and Price's model includes the following advantages: It permits the system to be dominated by species or guilds at any trophic level through feedback loops; it can encompass the mechanisms of interactions between species; and it focuses attention on the extensive heterogeneity in natural systems. The model does not, however, consider changes in community structure or interactions between animals and their abiotic environment over evolutionary time (Hunter and Price 1992). Rather than trying to dichotomize systems into either bottom-up or top-down categories, we have concluded that a much more fruitful line of research concerns the extent to which variation at different trophic levels, or in abiotic factors, can influence the relative strengths of bottom-up and top-down forces. The model proposed by Hunter and Price is a useful means of conceptualizing how groups of organisms might be interacting, and provides a starting point for designing studies that test the influence of heterogeneity at different levels in a food web.

Despite ecologists' interest in food webs, there are real difficulties in applying food-web theories to real world management. As summarized by Power (1992) and Hall and Raffaelli (1993), there is a need to resolve methodological issues concerning appropriate spatio-temporal scales, to agree upon operational definitions for concepts like trophic levels, to evaluate assumptions of the variety of available models of top-down and bottom-up forces, and to develop testable hypotheses that address the dynamic feedbacks between adjacent and nonadjacent trophic levels (see also Oksanen et al. 1995). The usefulness of food-web theory as currently practiced also has been questioned by Peters (1991), who concluded that, because food webs represent qualitative, verbal models, they indicate only weakly what we can expect from nature. Because the predictions are not risky (i.e., because they are not specific or quantitative) and because exceptions are explicitly allowed (e.g., see Pimm 1982; also Peters 1991: table 7.3), the models are protected from falsification. Echoing Power's (1992) call for operationalization of definitions, Peters concludes that most tests of food webs are ambiguous and apparent refutations are inconclusive. As discussed in the following section, how we view species interactions can have a substantial impact on how we plan our conservation strategies.

Hall and Raffaelli (1993) presented a sensible discussion of how we might better advance studies of food webs. They stated that, if we are to construct models to help us understand and manage the environment, we need to know which structural properties are important and why. Efforts to document the links in webs ever more completely are probably misplaced in view of the alternative avenues that are available. It is more sensible to focus on specific questions that emerge from food-web models using systems that are amenable to study.

Keystone-Species Concept

Robert T. Paine (1966, 1969) introduced the concept of keystone species to the ecological literature as a result of his pioneering work in intertidal systems. Paine found that removal of a predator high in the food web has substantial impacts on the species composition of the intertidal community. The keystone concept has been greatly expanded over the last three decades and is now applied to a host of species that are thought to have the following two characteristics: Their presence is crucial in maintaining the organization and diversity of their ecological communities; and these species are exceptional, relative to the rest of the community, in their importance. As reviewed by Mills et al. (1993), the keystone-species concept has become entrenched in the ecological literature and is now the center of conservation efforts to maximize biodiversity protection, as species in need of priority protection and as indicators of the management of entire communities and even ecosystems.

The assertion that keystone species are real and play a critical role in structuring ecological communities and driving conservation efforts has become ecological dogma lacking rigorous, quantitative, and empirical description and validation. Such a status for the concept of the keystone requires that a clear operational definition exist for it (Peters 1991). In contrast, in their review, Mills et al. (1993) found that the term is broadly applied, poorly defined, and nonspecific in meaning. More seriously, they found that the concept is largely undemonstrated in nature, even though it has fundamental implications for food-web theory and conservation strategies. They categorized the uses of the keystone concept found in the literature into five main, but not mutually exclusive, categories (table 2.1). Other species have been classified as keystones because they are critical in mutualistic relationships. Gilbert (1980) used the term *mobile links* to identify species that are significant factors in the persistence of several plant species which, in turn, support otherwise separate food webs. Classic examples of mobile links are seed dispersers and pollinators of plants that in turn support numerous other species. In this scenario, organisms that provide critical support to mobile links are labeled keystone mutualists. Mills et al. found that the strict and original definition of keystones was not being used in most studies. In fact, virtually any species that modifies the environment in some way can be labeled a keystone (of some type; see table 2.1). For example, the removal of any herbivore will have an impact on the plants it is no longer eating.

Table 2.1. Categories of presumed keystones and the effects of their effective removal from a system

Keystone category	Effect of removal
Predator	Increase in one or several predators/consumers/competitors, which subsequently extirpates several prey/competitor species
Prey	Other species more sensitive to predation may become extinct; predator populations may crash
Plant	Extirpation of dependent animals, potentially including pollinators and seed dispersers
Link	Failure of reproduction and recruitment in certain plants, with potential subsequent losses
Modifier	Loss of structures/materials that affect habitat type and energy flow; disappearance of species dependent on particular successional habitats and resources

Source: L. S. Mills, M. E. Soule, and D. F. Doak, "The Keystone-Species Concept in Ecology and Conservation," *BioScience* 43(1993): table 1, © 1993 by the American Institute of Biological Sciences

However, the original intent of the keystone concept required the assumption that the frequency of community-importance values is strongly skewed because only a few species have a large impact on the composition or structure of the community (fig. 2.5). In contrast with this assumption, food-web theory holds that

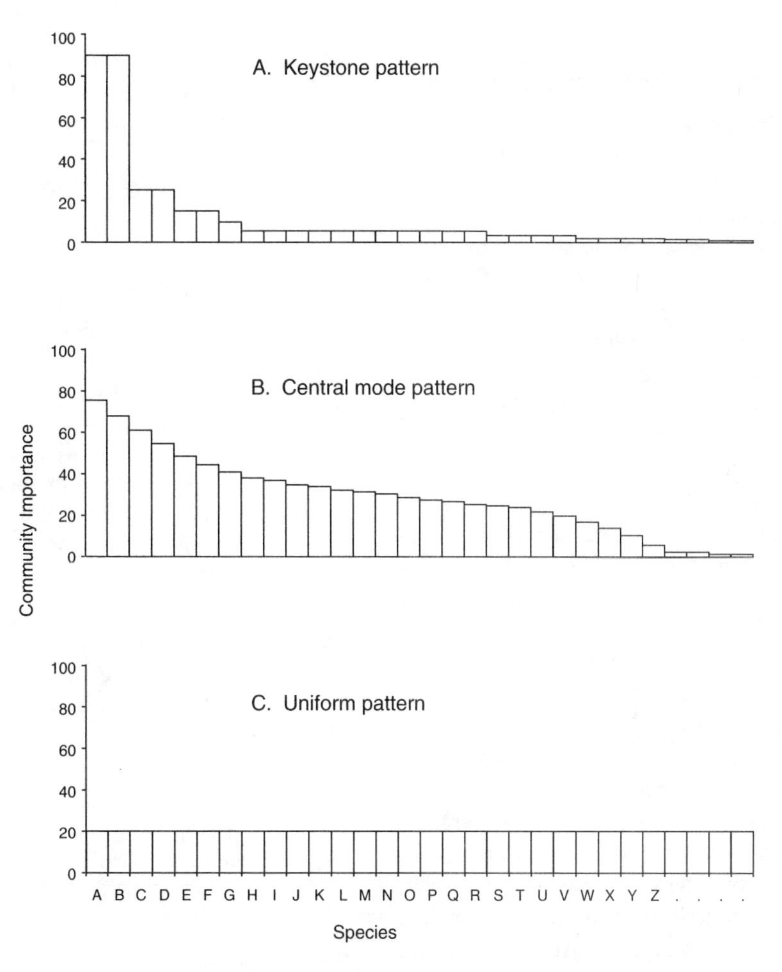

Figure 2.5. Expected distributions of community-importance values (percentage of species lost from a community upon removal of a given species) for a hypothetical community based on the keystone-species model (A) and on food-web theory (B and C). Axes are arbitrarily scaled to demonstrate the general shape of the distributions. (Reproduced from L. S. Mills, M. E. Soule, and D. F. Doak, "The Keystone-Species Concept in Ecology and Conservation," *BioScience* 43 [1993]: fig. 1, © 1993 by the American Institute of Biological Sciences)

the strength of species-by-species interaction is drawn from symmetric or uniform distributions. According to Mills et al., it is difficult to imagine one species having a large effect on species diversity (i.e., having community importance) without its also having strong interactions with other species. Thus, the keystone concept's implicit assumption about interaction strengths appears to be in direct conflict with the more explicit (but not necessarily more realistic) food-web models. This apparent dichotomy implies different patterns of community structure and thus requires different conservation strategies. That is, if many species are of similar importance (as implied from the literature; see fig. 2.5) and each has a unique niche, efforts to protect only a few keystones will fail to protect the rest. Conversely, if only a few species have strong interactions, then detailed knowledge of only a few taxa will be required to protect the many.

Mills et al. concluded that, although the keystone-species concept has been shown to function in a few situations, it should not be promoted as a centerpiece of ecological theory and conservation strategy. Rather, researchers should conduct studies that identify and quantify strengths of interaction (see chapter 10 on key ecological functions of species), rather than assume a keystone-nonkeystone dualism. Countering Mills et al., Menge et al. (1994) concluded that the keystone concept should be retained but placed in a broader context. They thought that communities may be affected by strong or weak predation, and those with strong predation may be under the influence of either keystone or diffuse competition (see also Paine 1995). Given the controversy concerning the keystone concept, we caution against using it in management situations, particularly for identifying conservation priorities, at this time. There is strong motivation to find quick means of explaining the seemingly indescribable complexity of nature, but we have yet to find a shortcut that applies to more than a few situations.

Extinctions and Endangered Phenomena

There is also much to be learned regarding conservation of extant species by analyzing the causes of extinctions. Numerous hypotheses have been proposed to explain the extinction of animals seen in the fossil record (Owen-Smith 1987; see also table 2.2). The *climatic change hypothesis* relates waves of extinction to abrupt changes in climatic conditions. For example, major extinctions of North American mammals took place at the end of the Pleistocene. The *human predation hypothesis* was developed because of the apparent relationship between the pattern of Pleistocene extinctions of large herbivores and the pattern of human occupation. The *keystone herbivore hypothesis* assumes that

Table 2.2. The timing of the major Pleistocene and Holocene mammalian extinctions

Location	Year (B.P.)
North America	11,000
South America	10,000
Northern Eurasia	13,000–11,000
Australia	13,000
West Indies	mid-postglacial
Madagascar	800
New Zealand	900
Africa and Southeast Asia	40,000–50,000

Source: Reproduced from P. S. Martin, "Prehistoric Overkill," in P. S. Martin and H. E. Wright, eds., *Pleistocene Extinctions* (New Haven: Yale University Press © 1967); reproduced by permission of Yale University Press

the extinction of large herbivores (by whatever event) caused a concomitant change in the distribution and abundance of vegetation, which in turn caused the extinction of animals associated with the now extinct or substantially changed vegetation types. For example, the conversion of the open parklike woodlands and mosaic grasslands typical of North America during the Pleistocene to the more uniform forests and prairie grasslands of today could have been a consequence of the elimination of these herbivores (Owen-Smith 1987).

Goudie (1992) provided a good summary of the influence of humans and climate change on animal extinctions, concluding that the combination of the two likely influences the extinction rate of some species. Regardless of the causes of extinction, we need more direct, empirical data on the physical environmental history of the Pleistocene and the biological consequences of environmental changes, with emphasis on species extinctions. If we can increase our knowledge of the Pleistocene record, then we will be better able to evaluate the consequences of contemporary human activities on the biota. Without consideration of the time perspective available from the geological record, a complete evaluation of the current rates of extinction will be at best difficult and at worst misleading (Raup 1988). To a certain degree, the fossil record can provide an understanding of how species appear, disappear, and then reappear in an area (see box 2.2).

Comparing contemporary rates of extinction (and species originations) with those of the past is only one means of placing the current situation in perspective. On local spatial scales and within relatively recent time periods, we can compare known and supposed species distributions with current conditions to develop plans for habitat restoration and species reintroductions. In southern California, for example, Morrison et al. (1994) used museum records (many from early 1900) to help determine what species "should" be in a heavily degraded urban park. They could have taken this process a few steps further by hypothesizing which species disappeared from the park because of natural environmental changes and, in contrast, which species were missing historically but have recently expanded their ranges.

In addition to outright extinction of a population or species, natural and human-induced changes in animal populations and their habitats can cause dramatic changes in the behavior of the surviving individuals. Brower and Malcolm (1991) have coined the term *endangered phenomena* to describe the situation in which a spectacular aspect (e.g., massive migration) of the life history of an organism is endangered (see also chapter 10 on key ecological functions), even though the species itself might not become extinct. They envisioned in the near future an increasing number of species reduced in range and so constrained in numbers that they could no longer exhibit their spectacular life history phenomena. As examples, they include the greatly diminished herds of *Bison* of the North American prairie and endangered migrations of the east African wildebeest (*Connochaete taurinus*), the Atlantic salmon (*Salmo salar*), and numerous North American songbirds. The consequences that such forced changes in life history traits will have on long-term survival of a species are poorly understood, but cannot be assumed to be favorable to maintaining natural ecosystems with native communities and species.

Conclusions

Pimm (1991) made the important distinction between short- and long-term studies and the types of management we do. Most management is concerned about short-term changes in diversity; this is understandable given the human time frame. However, over the longer term, we must realize that changes in a species' density will affect the species' immediate predators and prey and eventually its competitors. Over even longer periods a broader ripple effect often occurs: The predators and prey of this species' predators and prey will be affected. Thus, over the long term we must consider both direct and indirect, including time-lagged, impacts that our management scenarios will have on entire groups of organisms.

The fossil record is generally credited only with allowing the observation of dynamics in species and communities in time spans of 10^4 years or greater and only for selected taxa that can be fossilized. However, the paleoecology record can be used to examine patterns and infer dynamic processes on time scales relevant to many debates in contemporary ecology (DiMichele 1994). As summarized by DiMichele, the paleoecological record reveals that ecosystem structure and taxonomic composition commonly persist or repeatedly recur throughout the intervals of hundreds of thousands to millions of years. DiMichele has asked: Can the Recent era be a time period of dominantly Gleasonian, individualistic dynamics, not characteristic of most of geological time? Can current theory account for paleontological patterns, and further, can it predict future patterns? Answers to these questions will most certainly help place our current ecological views in a perspective that includes both a short- and long-term reality.

We know that species diversity has changed over time and is continuing to change for a variety of reasons. Studying the past helps us set contemporary changes in perspective and assists us in predicting the consequences of future changes in resource use (Pimm 1991). Extant species may have had geographic distributions, coexisted with species, and lived under environmental conditions during the Pleistocene that are all different from those of today. Interpretation of the current distribution and habitat affinities of animals should be evaluated in light of these historical biogeographic and ecological data (Goodwin 1995).

Literature Cited

Amundson, D. C., and H. E. Wright, Jr. 1979. Forest changes in Minnesota at the end of the Pleistocene. *Ecological Monographs* 49:1–16.

Andrewartha, H. G., and L. C. Birch. 1954. *The distribution and abundance of animals.* Chicago: University of Chicago Press.

Barnosky, A. D. 1987. Punctuated equilibrium and phyletic gradualism: Some facts from the Quaternary mammalian record. In *Current mammalogy,* ed. H. H. Genoways, 109–47. New York: Plenum Press.

Batzli, G. O. 1994. Special feature: Mammal-plant interactions. *Journal of Mammalogy* 75:813–15.

Brower, L. P., and S. B. Malcolm. 1991. Animal migrations: Endangered phenomena. *American Zoologist* 31:265–76.

Brown, J. H. 1978. The theory of insular biogeography and the distribution of boreal birds and mammals. *Great Basin Naturalist Memoirs* 2:209–27.

Cody, M. L. 1985. An introduction to habitat selection in birds. In *Habitat selection in birds,* ed. M. L. Cody, 3–56. New York: Academic Press.

Cohen, J. E. 1978. *Food webs and niche space.* Princeton, N.J.: Princeton University Press.

Cohen, J. E. 1990. Food webs and community structure. In *Community food webs: Data and theory,* ed. J. E. Cohen, F. Briand, and C. M. Newman, 1–14. New York: Springer-Verlag.

Cohen, J. E., F. Briand, and C. M. Newman. 1990. *Community food webs: Data and theory.* New York: Springer-Verlag.

Connell, J. H. 1980. Diversity and the coevolution of competitors; or, The ghost of competition past. *Oikos* 35:131–38.

Cox, C. B., and P. D. Moore. 1993. *Biogeography: An ecological and evolutionary approach.* 5th ed. Boston: Blackwell Scientific Publications.

Davis, R., C. Dunford, and M. V. Lomolino. 1988. Montane mammals of the American southwest: The possible influence of post-Pleistocene colonization. *Journal of Biogeography* 15:841–48.

DiMichele, W. A. 1994. Ecological patterns in time and space. *Paleobiology* 20:89–92.

Eldridge, N., and S. J. Gould. 1972. Punctuated equilibria: An alternative to phyletic gradualism. In *Models in paleobiology,* ed. T. J. Schopf, 82–115. San Francisco: Freeman.

Elias, S. A. 1992. Late Quaternary zoogeography of the Chihuahuan Desert insect fauna, based on fossil records from packrat middens. *Journal of Biogeography* 19:285–97.

Elton, C. 1927. *Animal ecology.* New York: Macmillan.

Flack, J. A. D. 1976. *Bird populations of aspen forests in western North America.* Ornithological Monographs no. 19. Washington, D.C.: American Ornithologists' Union.

Franklin, J. F. 1988. Structural and functional diversity in temperate forests. In *Biodiversity,* ed. E. O. Wilson, 166–75. Washington, D.C.: National Academy Press.

Gilbert, L. E. 1980. Food Web organization and the conservation of neotropical diversity. In *Conservation biology: An evolutionary-ecological perspective,* ed. M. E. Soule and B. A. Wilcox, 11–33. Sunderland, Mass.: Sinauer Associates.

Goodwin, H. T. 1995. Pliocene-Pleistocene biogeographic history of prairie dogs, genus *Cynomys* (Sciuridae). *Journal of Mammalogy* 76:100–122.

Goudie, A. 1992. *Environmental change.* 3d ed. Oxford: Clarendon Press.

Grinnell, J. 1905. Status of Townsend's warbler in California. *Condor* 7:52–53.

Hafner, D. J. 1993. North American pika (*Ochotona princeps*) as a late Quaternary biogeographic indicator species. *Quaternary Research* 39: 373–80.

Hall, S. J., and D. G. Raffaelli. 1993. Food webs: Theory and reality. *Advances in Ecological Research* 24: 187–239.

Harris, A. H. 1990. Fossil evidence bearing on southwestern mammalian biogeography. *Journal of Mammalogy* 71:219–29.

Harris, A. H. 1993. Wisconsinan pre-pleniglacial biotic change in southeastern New Mexico. *Quaternary Research* 40:127–33.

Heusser, C. J. 1965. A Pleistocene phytogeological sketch of the Pacific Northwest and Alaska. In *The Quaternary of the United States,* ed. H. E. Wright, Jr., and D. E. Frey, 469–83. Princeton, N.J.: Princeton University Press.

Heusser, C. J. 1977. Quaternary palynology on the Pacific slope of Washington. *Quaternary Research* 8:282–306.

Hopkins, M. S., J. Ash, A. W. Graham, J. Head, and R. K. Hewett. 1993. Charcoal evidence of the spatial extent of the *Eucalyptus* woodland expansions and rainforest contractions in north Queensland during the late Pleistocene. *Journal of Biogeography* 2:0357–72.

Hubbard, J. P. 1973. Avian evolution in the aridlands of North America. *Living Bird* 1:2155–96.

Hunter, M. D., and P. W. Price. 1992. Playing chutes and ladders: Heterogeneity and the relative roles of bottom-up and top-down forces in natural communities. *Ecology* 73:724–32.

Johnson, N. K. 1994. Pioneering and natural expansion of breeding distributions in western North American birds. *Studies in Avian Biology* 15:27–44.

Lack, D. 1954. *The natural regulation of animal numbers.* New York: Oxford University Press.

Lawton, J. H. 1989. Food webs. In *Ecological concepts,* ed. J. M. Cherrett, 43–78. Oxford: Blackwell Scientific.

Leberg, P. L. 1991. Influence of fragmentation and bottlenecks on genetic divergence of wild turkey populations. *Conservation Biology* 5:522–30.

Martin, P. S. 1967. Prehistoric overkill. In *Pleistocene extinctions: The search for a cause,* ed. P. S. Martin and H. E. Wright, 75–120. New Haven, Conn.: Yale University Press.

Menge, B. A., E. L. Berlow, C. A. Blanchette, S. A. Navarrete, and S. B. Yamada. 1994. The keystone species concept: Variation in interaction strength in a rocky intertidal habitat. *Ecological Monographs* 64:249–86.

Mengel, R. M. 1964. The probable history of species formation in some northern wood warblers (Parulidae). *Living Bird* 3:9–43.

Mengel, R. M. 1970. The North American Central Plains as an isolating agent in bird speciation. In *Pleistocene and Recent environments of the central Great Plains,* ed. W. Dort, Jr., and J. K. Jones, Jr., 279–340. Department of Geology Special Publication no. 3, University of Kansas, Lawrence.

Mills, L. S., M. E. Soule, and D. F. Doak, 1993. The keystone-species concept in ecology and conservation. *BioScience* 43:219–24.

Morrison, M. L. 1983. Analysis of geographic variation in the Townsend's warbler. *Condor* 85:385–91.

Morrison, M. L., T. A. Scott, and T. Tennant. 1994. Wildlife-habitat restoration in an urban park in southern California. *Restoration Ecology* 2:17–30.

Morrison, M. L., I. C. Timossi, K. A. With, and P. N. Manley. 1985. Use of tree species by forest birds during winter and summer. *Journal of Wildlife Management* 49:1098–1102.

Nordt, L. C., T. W. Boutton, C. T. Hallmark, and M. R. Waters. 1994. Late Quaternary vegetation and climate changes in central Texas based on the isotopic composition of organic carbon. *Quaternary Research* 41:109–20.

O'Brien, S. J., D. E. Wildt, and M. Bush. 1986. The cheetah in genetic peril. *Scientific American* 254:84–92.

O'Brien, S. J., D. Goldman, C. R. Merril, and M. Bush. 1983. The cheetah is depauperate in genetic variation. *Science* 221:459–62.

Oksanen, T., M. E. Power, and L. Oksanen. 1995. Ideal free habitat selection and consumer-resource dynamics. *American Naturalist* 146:565–85.

Owen-Smith, N. 1987. Pleistocene extinctions: The pivotal role of megaherbivores. *Paleobiology* 13:351–62.

Paine, R. T. 1966. Food web complexity and species diversity. *American Naturalist* 100:65–75.

Paine, R. T. 1969. A note on trophic complexity and community stability. *American Naturalist* 103:91–93.

Paine, R. T. 1995. A conversation on refining the concept of keystone species. *Conservation Biology* 9:962–64.

Peters, R. H. 1991. *A critique for ecology.* Cambridge: Cambridge University Press.

Pianka, E. R. 1983. *Evolutionary ecology.* 3d ed. New York: Harper and Row Publishers.

Pimm, S. L. 1982. *Food webs.* New York: Chapman and Hall.

Pimm, S. L. 1991. *The balance of nature? Ecological issues in the conservation of species and communities.* Chicago: University of Chicago Press.

Power, M. E. 1992. Top-down and bottom-up forces in food webs: Do plants have primacy? *Ecology* 73:733–46.

Raup, D. M. 1988. Diversity crises in the geological past. In *Biodiversity,* ed. E. O. Wilson, 51–57. Washington, D.C.: National Academy Press.

Risser, P. G. 1988. Diversity in and among grasslands. In *Biodiversity,* ed. E. O. Wilson, 176–80. Washington, D.C.: National Academy Press.

Ritchie, J. C. 1976. The late-Quaternary vegetational history of the western interior of Canada. *Canadian Journal of Botany* 54:1793–818.

Smith, C. C. 1981. The indivisible niche of *Tamiasciurus:* An example of nonpartitioning of resources. *Ecological Monographs* 51:343–63.

Stiles, F. G. 1980. Evolutionary implications of habitat relations between permanent and winter resident landbirds in Costa Rica. In *Migrant birds in the neotropics: Ecology, behavior, distribution, and conservation,* ed. A. Keast and E. S. Morton, 421–35. Washington, D.C.: Smithsonian Institution Press.

Sullivan, R. M. 1988. Biogeography of southwestern montane mammals: An assessment of the historical and environmental predictions. Ph.D. dissertation, University of New Mexico, Albuquerque.

Vaughan, T. A. 1986. *Mammalogy.* 3d ed. Philadelphia: Saunders College Publishing.

Wiens, J. A. 1977. On competition and variable environments. *American Scientist* 65:590–97.

Wright, I. C., M. S. McGlone, C. S. Nelson, and B. J. Pillans. 1995. An integrated latest Quaternary (stage 3 to present) paleoclimate and paleoceanographic record from offshore northern New Zealand. *Quaternary Research* 44:283–93.

3 The Vegetation and Population Perspectives

In conservation it is not a question of blueprints for the future. All that is attempted is to provide conditions, *based on our best scientific insight and subject to the present-day social and economic restraints, which will make it possible for an evolutionary succession of organisms to continue, inevitably subject to the social consent of future generations.*

—O. H. Frankel and M. E. Soule, *Conservation and Evolution*

Introduction

In this chapter we explore the perspectives of vegetation and populations. We discuss plants as indicators of the physical environment; vegetation succession; and the correlation between animals and vegetation, including animal-plant interactions and relations of wildlife to floristics and vegetation structure. We then turn attention to the relations of populations and habitats. We focus on spatial and geographic factors that influence habitats and environments, population structure, fitness of organisms, and ultimately viability of populations.

The Vegetation Perspective

Vegetation's central role in the life of an animal is self-evident. Indeed, much of this book is devoted to analyzing the distribution and abundance of plants as part of developing wildlife-habitat relationships. Thus, it is incumbent on researchers and managers to understand the factors determining the health of plants, how plants are arranged into identifiable units, how these units change through time, and so forth. In the following sections we discuss vegetation ecology and relate this ecology to the study of wildlife-habitat relationships.

Plants as Indicators of the Physical Environment

Clements (1920:3–34) traced the development of the indicator concept as applied to plant communities back to the early 1600s. He noted that an understanding of the relationship between soil and plants must have marked the beginnings of agriculture. Many workers in the mid- to late 1800s, including Hilgard, Chamberlin, Shantz, and Weaver, developed the indicator concept in application to plants, particularly in application to the agricultural sciences (Clements 1920). The concept of using plants as indicators of environmental conditions became ingrained in the scientific literature following the pioneering work of these early scientists. Indeed, as noted by E. P. Odum in his classic *Fundamentals of Ecology* (1971:138), "... the ecologist constantly employs organisms as indicators in exploring new situations or evaluating large areas...."

Vegetation Patterns and Processes of Succession

Humans have long been fascinated with understanding and predicting changes in plant distribution and abundance and often depend on such predictions for survival. As noted by Glenn-Lewin et al. (1992), hunter-gatherer societies desired an understanding of the changes in vegetation that were induced by fire, subsequently altering forage for the game animals that were their food source. Modern wildlife conservation depends on our ability to understand the patterns and processes of vegetation change.

Vegetation change as an orderly process is usually referred to as plant succession. Glenn-Lewin et al. (1992) reviewed the historical development of the concept of plant and plant-community succession. The "Clementsian paradigm" developed from the work of Clements (1904, 1916) on the theory of plant succession. Clements' work dominated until the 1940s. His "community-unit theory" held that species form groupings that characterize distinct, clearly bounded types of communities that are often termed associations (Whittaker 1975). Clements viewed succession as an orderly and predictable process in which vegetation change represents the life history of a plant community; the community thus assumes "organismlike" characteristics. These communities were believed to develop to a distinct climax condition, the characteristics of which are controlled by the regional climate (Glenn- Lewin et al. 1992).

The Clementsian views had some early and strong critics (e.g., Gleason, Ramensky, Tansley). Gleason (1939) and Tansley (1935) argued that plant communities result from the overlap in the independent distributions of species with similar environmental needs, and that vegetation change in a region does not necessarily converge toward a similar climax condition. In the 1960s several workers attempted to synthesize these disparate views of succession into a unified theory (e.g., Margalef 1963; Odum 1969). Margalef offered the idea that succession is driven from simple ecosystems to more complex systems with more trophic levels and greater species diversity. Similarly, Odum postulated that ecosystems tend to develop toward greater homeostasis of species composition. As summarized by Glenn-Lewin et al. (1992), this Margalef-Odum synthesis of successional theory was philosophically similar to the Clementsian paradigm.

By the early 1970s ecologists had recognized the inadequacies of the Margalef-Odum synthesis, primarily because of the increased understanding of the need for site-specific data on the proximate causes of vegetation change (Glenn-Lewin et al. 1992; see also Whittaker 1975). Two major trends have occurred in our ideas about succession since the 1970s: first, a

Box 3.1. Whittaker's four hypotheses on distribution of species populations

The eminent plant ecologist Robert H. Whittaker wrote an excellent book on plant ecology, including a thorough development of how plants might come to be associated. Here we present how Whittaker outlined four working hypotheses on ways species populations might be distributed (Whittaker 1975:113–15).

1. Competing species, including dominant plants, exclude one another along sharp boundaries. Other species evolve toward close association with the dominants and toward adaptations for coexisting with one another. Thus, a gradient develops with distinct zones, with each zone having a characteristic assemblage of plants adapted to one another, which gives way at a sharp boundary to another assemblage of species adapted to one another. Such a hypothesis results in relatively discontinuous kinds of plant communities (box fig. 3.1A, panel A).
2. Competing species exclude one another along sharp boundaries, but do not become organized into groups with parallel distributions (box fig. 3.1A, panel B).
3. Competition generally does not result in sharp boundaries between assemblages. Evolution of species toward adaptation to one another will, however, result in the appearance of groups of species with similar distributions. These groups characterize different kinds of plant communities, but the communities intergrade continuously (box fig. 3.1A, panel C).
4. Competition does not usually produce sharp boundaries between species populations, and evolution of species in relation to one another does not produce well-defined groups of species with similar distributions. The centers and boundaries of populations of species are scattered along the environmental gradient (box fig. 3.1A, panel D).

Whittaker concluded that, when we study the manner in which plant populations increase and decrease in abundance along environmental gradients (e.g., interactions between soil moisture, soil texture, ambient temperature, and other factors), the results support the fourth hypothesis presented above and depicted in box figure 3.1A, panel D. Although cases of sharp boundaries between species are known, most populations are distributed as presented in the two panels of box figure 3.1B. Thus, field observations agree with the "individualistic hypothesis" rather than with the various "community-unit theories."

(box continued on following page)

shift away from a search for holistic explanations toward a more reductionist and mechanistic approach that emphasizes proximate causes of vegetation changes; and second, a shift away from equilibrium toward nonequilibrium theories (Glenn-Lewin et al. 1992). What has become known as the individualistic hypothesis has grown in popularity; it contends that plant populations are individually distributed along gradients and that most communities and their component species intergrade along these gradients rather than forming distinct, separate zones (Whittaker 1975). Rather than being a single hypothesis, this individualistic view is actually an umbrella for various views, including hypotheses that consider succession as a gradient in time or resource availability, as the consequence of differential longevity and other population processes, as the result of differences in life history traits, and as a stochastic process (see review by Glenn-Lewin et al. 1992).

In essence, all these ideas view succession as the combined outcome of populations of plants interacting with a fluctuating environment. The specific pattern observed (i.e., community structure and flora) will vary with location and with time in the same location. Whittaker (1975) synthesized various ways in which plants can be distributed in relation to one another and communities; we present his summary in box 3.1. Barbour (1996) provided an interesting historical account of the debate over Gleason's original ideas.

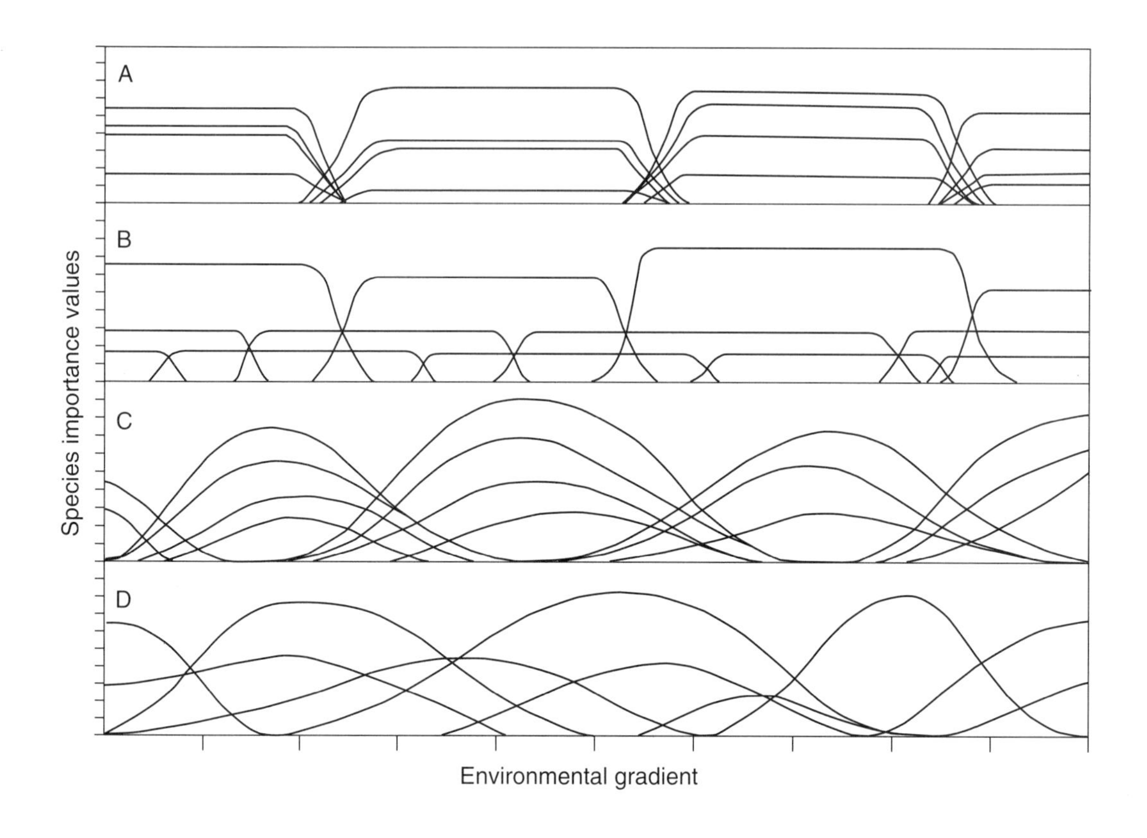

Box Figure 3.1A. Four hypotheses on species distributions along environmental gradients. Each curve represents one species population and the way it might be distributed along the environmental gradient. (From R. H. Whittaker, *Communities and Ecosystems,* 2/E © 1975; reprinted by permission of Prentice-Hall, Inc., Upper Saddle River, N.J.)

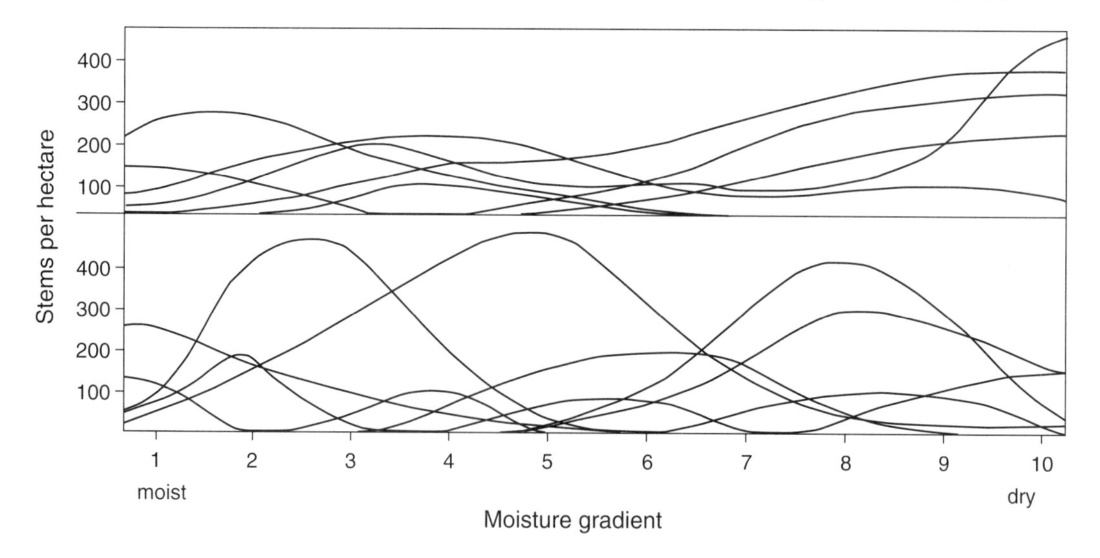

Box Figure 3.1B. Actual distributions of species populations along environmental gradients. Species populations are plotted by densities (number of tree stems per hectare). (From R. H. Whittaker, *Communities and Ecosystems,* 2/E © 1975; reprinted by permission of Prentice-Hall, Inc., Upper Saddle River, N.J.)

Van Hulst (1992) asked, What factors influence future vegetational composition of a site? The answer to this question has obvious ramifications for our study of wildlife-habitat relationships, given the central role that vegetation plays in determining the distribution and abundance of animals. Van Hulst outlined six general categories of factors that influence future vegetation:

- the *present vegetation*
- the present vegetation in the surrounding area, or *immigration of propagules*
- the *past vegetation* (e.g., dormant seeds)
- *present resource levels* (e.g., light, humidity, soil minerals)
- *disturbance levels, including herbivory*
- *stochastic factors* (e.g., climatic variability, supply fluctuations of resources)

Different models are available to help predict the course of succession, depending upon which of the preceding factors are of primary importance in determining the dynamics of the plant community; Van Hulst provided details. Making reliable predictions concerning the future course a wildlife population takes often requires reliable predictions of the future course the vegetative community will take. Thus, it is important that the influence of the above factors on the characteristics of plant communities, including successional patterns, be recognized and understood when developing habitat models.

The Correlation between Animals and Vegetation

Broad-scale relationships between vegetation and animals have long been recognized. The classic "life zone" concept of C. Hart Merriam (1898) used temperature boundaries to describe the limits of major vegetation types (life zones) and their associated animal species. Holdridge (1947) developed his well-known system that used temperature, precipitation, and evapotranspiration to describe major vegetation groupings of the world. In the United States, every state and region has multiple classification systems, many of which are now being revised with the aid of remote sensing data and geographic information systems (GIS). These systems are, in turn, often used to describe the distribution and abundance of wildlife associated with vegetation conditions on broad geographic scales (see chapter 10).

Classically, wildlife habitat was described as containing three basic components, namely, cover, food, and water. Dasmann (1964) further divided cover into two components: habitat requirements and escape cover. These two components are not, of course, mutually exclusive. There was a clear recognition that vegetation (cover) provides essential requirements for animals and that manipulation of this cover could drastically affect wildlife populations.

Of particular interest to wildlife biologists have been the relationships between "edge habitats" and wildlife richness and abundance (see chapter 8). Edge has been described as the joining of two "habitats." We place *habitat* in quotes here because, according to the definitions we use in this book, earlier workers were really referring to the intersection of two vegetation types. This was popularly referred to as the edge effect; Aldo Leopold, in his famous *Game Management* (1933), actually bestowed the effect with law status, namely, his *law of interspersion*. This law stated that the edge between two "habitats" will be more favorable as wildlife habitat than either type considered alone. This effect may be a result of the conjoining of separate environments, each with its associated wildlife communities. By sampling across two vegetation types, including the intersection of the two, one is actually sampling three (or more) vegetation types. It may not be the edge per se that is relevant; what is relevant is that the sample includes more combinations of vegetation and other habitat factors, thus usually resulting in a higher species count than if only one environment were to be sampled. The concept of edge effect is further critiqued in chapters 8 and 9. Recent wildlife texts often repeat the previous view of edge without comment (e.g., Robinson and Bolen 1984:82–83).

Edges, often referred to as ecotones, are scale dependent. As shown in table 3.1, within a single area many edges, or ecotones, can exist depending upon the spatial scale of interest. These edges are influenced by an increasingly complex array of environmental factors as the scale increases in resolution. Such concepts hold extremely important information for researchers hoping to develop wildlife-habitat relationships and for managers hoping to implement management recommendations. A relatively sedentary amphibian will be most severely affected by ecotones at the patch, population, and even plant levels shown in table 3.1, whereas relatively mobile large mammals will be most influenced by landscape patterns. Allowing for both groups of animals, as managers are usually called upon to do today, greatly increases our information needs and complicates management.

Today, the edge concept is no longer viewed as an overriding positive feature of wildlife management. As discussed in chapter 8, increasing edge beyond natural levels leads to fragmented environments, which may cause increased predation by natural and exotic predators and increased rates of avian nest parasitism (see Paton 1994 for review; also Askins 1995; Mills 1995 for mammals; Robinson et al. 1995). This problem is magnified today because fewer and fewer large tracts of undisturbed environments remain. Thus, increasing edge can be seen as a measure of increasing environmental degradation for at least some species. Guthery and Bingham's (1992) critique of Leopold's edge concept noted that few modern researchers and managers recognize that Leopold qualified his principle by stating that it did not generally apply to mobile species and that it was confined to "edge-obligate species." They noted that animals cannot increase abundance above some maximum density; thus, the correlation between animal density and amount of edge has limits. That is, creating edge beyond some

Table 3.1. Ecotone hierarchy for a biome transition area

Ecotone hierarchy	Probable constraints
Biome ecotone	Climate (weather) × topography
Landscape ecotone (mosaic pattern)	Weather × topography × soil characteristics
Patch ecotone	Soil characteristics × biological vectors × species interactions × microtopography × microclimatology
Population ecotone (plant pattern)	Interspecies interactions × intraspecies interactions × physiological controls × population genetics × microtopography × microclimatology
Plant ecotone	Interspecies interactions × intraspecies interactions × physiological controls × plant genetics × microclimatology × soil chemistry × soil fauna × soil microflora, etc.

Source: Gosz 1993: table 1; reproduced by permission of the Ecological Society of America

Note: Each level in the ecotone hierarchy has a range of constraints and interactions between the constraints (× symbolizes these interactions). The primary constraints vary with the scale of the ecotone, with an increase in the number of possible constraints on finer scales.

optimum amount would not result in increased wildlife density.

Beginning in the 1950s, ecologists began to develop formal relationships between various components of vegetation cover and the kinds of animals in an area and their abundances. Most famous is the foliage height diversity–bird species diversity (FHD-BSD) constructs of MacArthur and MacArthur (1961). In figure 3.1 we see that the diversity of birds rises as vegetation becomes increasingly complex vertically. A plethora of studies followed the early work of the MacArthurs, with most researchers finding significant statistical relationships between FHD and BSD, although there were exceptions (Roth 1976). Karr and Roth (1971) suggested that the scatter in a FHD-BSD figure likely results from important, but unmeasured, factors that influence birds. Other workers were able to relate BSD to the heterogeneity, or horizontal "patchiness," of the vegetation (Roth 1976).

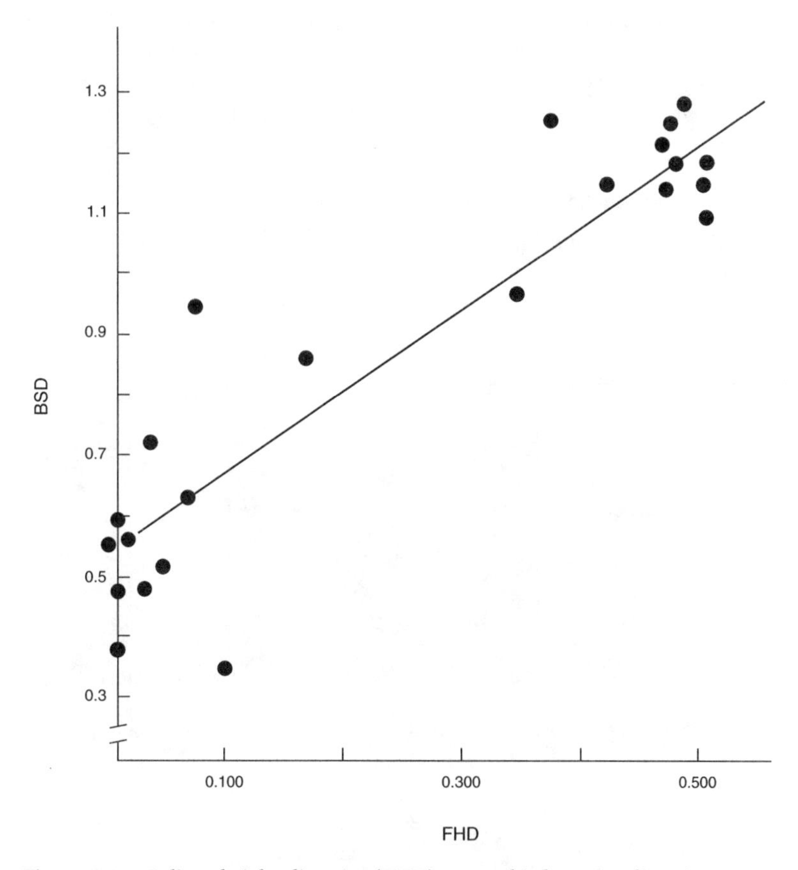

Figure 3.1. Foliage height diversity (FHD) versus bird species diversity (BSD). • represent the study sites. (Reproduced from Willson 1974 [fig. 1], by permission of the Ecological Society of America)

Unfortunately, measures such as FHD collapse detailed information on vegetation into a single number. These relationships depict patterns that occur over broad geographic areas. They do indicate, however, that animals respond to measures of complexity in their environment.

As noted by Dasmann (1964) in his introductory wildlife text, manipulation of plant succession is the principal way of providing the habitat requirements and escape cover required for the support of larger game populations. Thus, what theoretical ecologists had "discovered" had already become a principal tool of wildlife management, albeit for a selected group of preferred species. However, as researchers and management agencies became increasingly concerned about the fate of all animal species, these fundamental principles of ecology and wildlife management were applied to a wide array of species and landscapes. A good indication of the development of researcher and manager interest in this area can be found in the many U.S. Forest Service publications published in the 1970s and thereafter. For example, Thomas et al. (1979) presented many fundamental wildlife-habitat relationships, including that shown in figure 3.2. Here we see that a basic relationship exists whereby the number of animal species increases with advancing seral stages.

Although much is known about succession, the application of this knowledge to resource management has not progressed rapidly (Luken 1990). Luken thought that this is because

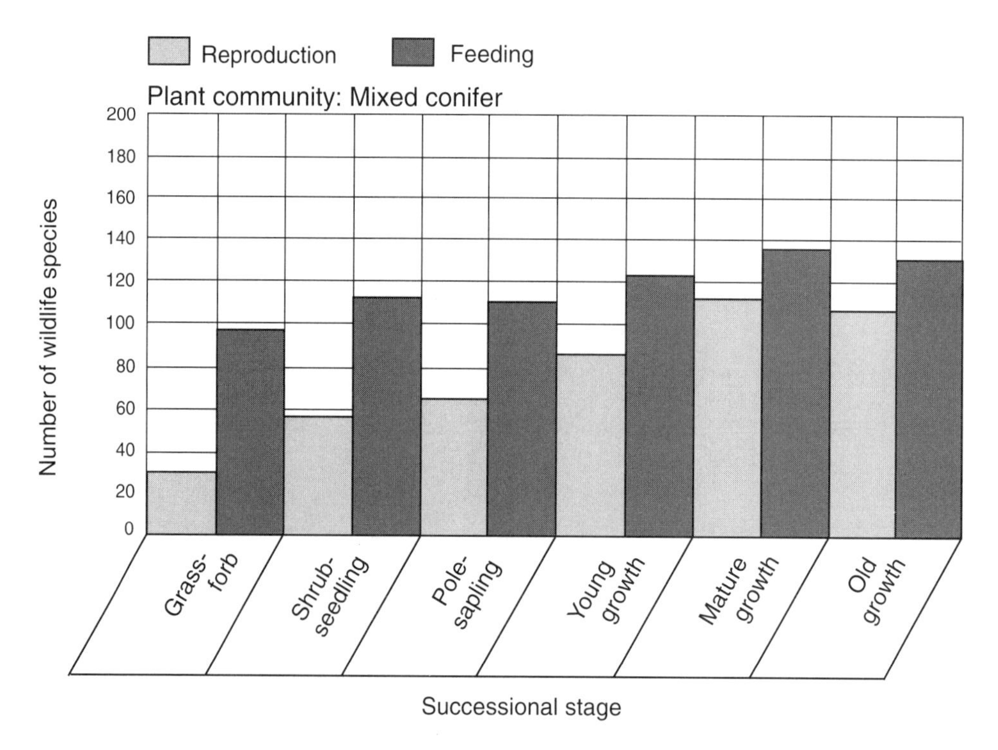

Figure 3.2. Number of wildlife species associated with successional stages in the mixed conifer community. (From Thomas et al. 1979: fig. 14)

ecologists do not want to work in managed systems and that most ecological studies are site-specific and do not lead rapidly to new, robust theories. It is clear, however, that wildlife biologists have long understood the basic relationships between management of seral stages and wildlife populations. This interest traditionally centered on a narrow group of game or pest species. Luken went on to say that the incorporation of succession management into natural resource management decision making is the next step after careful description of succession pathways. The recent (1990s) move by many federal land management agencies to manage at the multiple-species "ecosystem level" is an indication of a move in the direction Luken suggested (see also chapter 11).

Animal-Plant Interactions

As reviewed by Batzli (1994), responses of animals to plant characteristics first involve behavioral and physiological adjustments of individuals to plants. As a result of such adjustments, individual fitness may change, leading to changes in the distribution and abundance of wildlife populations. The resulting activities of these populations then, in turn, influence the plants. Further, animals individually affect soils, nutrient cycles, and other environmental factors (fig. 3.3). The ability of many mammals, for example, to modify the environment is well documented; they have been called ecosystem engineers. Moose, beavers, prairie dogs, Serengeti ungulates, and pocket gophers are classic examples (Batzli 1994).

The influence of animals on vegetation can be much more subtle than that of beavers or prairie dogs. For example, Snyder (1993) found that ponderosa pine (*Pinus ponderosa*) trees that became defoliated as a result of bark feeding by Abert's squirrels (*Sciurus aberti*) showed significant reduction in fitness. The squirrels may actually be agents of natural selection in ponderosa pine populations because they are preferentially attracted to trees with specific genetically determined traits and because such trees show significant reduction in fitness. Bonser and Reader (1995) tested whether the effects of competition and herbivory on plant growth depend on the aboveground biomass of vegetation. They found that competition and herbivory each have a greater effect on plant growth at sites with higher biomass, and that herbivory has less effect than

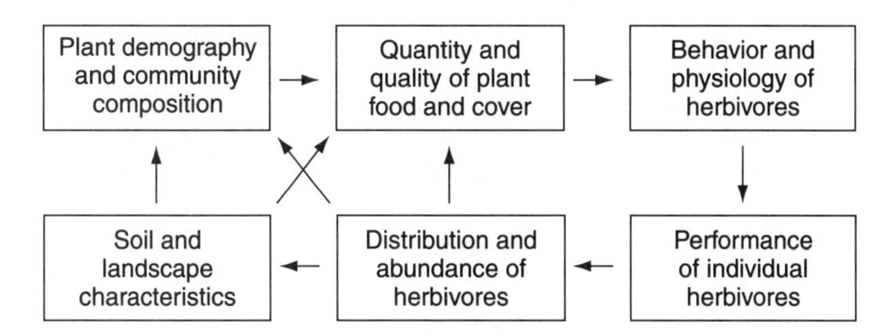

Figure 3.3. Model of interactions between mammalian herbivores and plants. Arrows indicate causal pathways for change, which can be negative or positive. Mammals directly respond to food and cover provided by plants. Mammals influence plants directly (by immediately damaging or facilitating plant success) or indirectly (by affecting substrate or landscape). (From Batzli 1994: fig. 1.)

competition on plant growth at sites with relatively less biomass. The herbivores in their experimental study included voles (*Microtus* spp.), slugs, and snails. Such results show the complex and interactive dynamic between plant biomass, plant competition, and grazing by herbivores (even relatively small ones).

The role of animals in succession management can be viewed in two major ways: first, in the manipulation of seral stages by managers for high and persistent abundances of selected species (the classic function of wildlife biologists); and second, in the control of succession by land managers using grazing animals. The first views animal populations changing as a *result* of succession, while the latter views animal populations as a *catalyst* of succession (Luken 1990:151–52). Luken went on to make a critical point: When managing wildlife habitat, and thus succession, for animals, a phase of succession is not the ultimate goal; rather, the goal is the biotic and abiotic factors in a mix of successional phases that satisfy the resource requirements of the target animal species. Thus, Luken acknowledged the species-specific nature of wildlife habitat, which we are emphasizing throughout this book.

Unfortunately for researchers and managers alike, few absolute generalizations can be made about the influence of animals on succession. As illustrated in figure 3.4, numerous pathways exist by which animals can influence plants and

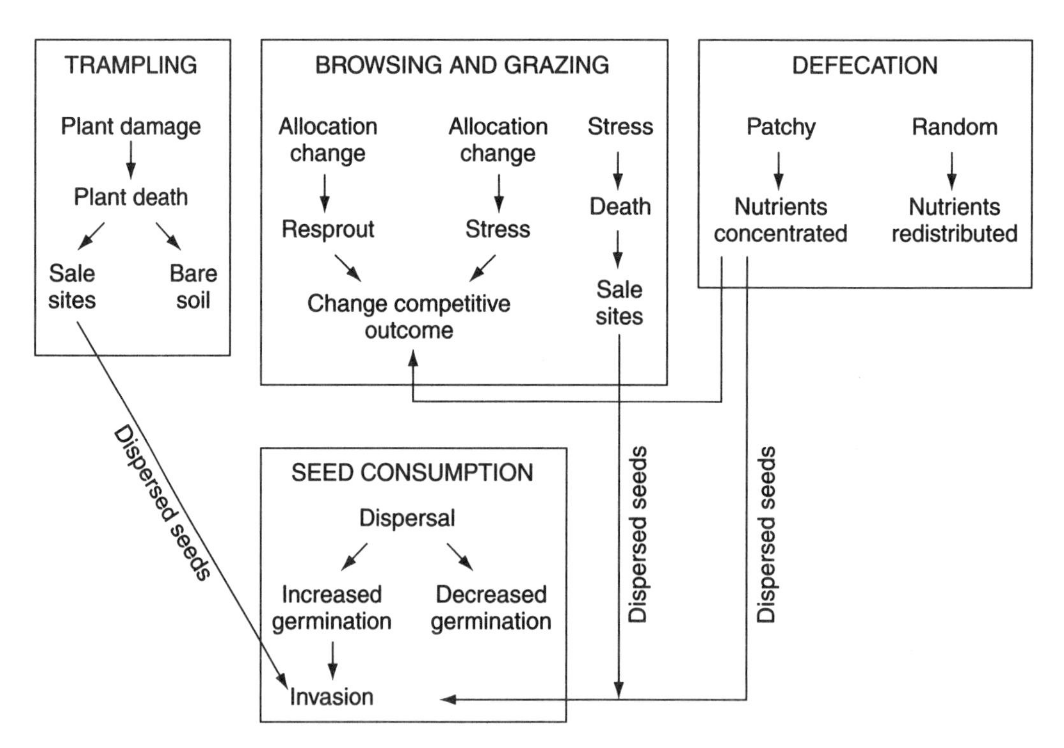

Figure 3.4. Various activities associated with grazing or browsing animals and their possible effects on plants. (Reproduced from J. O. Luken, *Directing Ecological Succession* [New York: Chapman and Hall © 1990], fig. 7.6; by permission of Chapman and Hall)

thus succession, including trampling, eating foliage and seeds, and defecation. Animals also function as dispersers of seeds and other plant disseminules. In addition, there are numerous management activities that can change the path of succession (table 3.2). The course of succession can be generally predicted when a single management impact is applied. However, multiple impacts render prediction highly difficult (e.g., shrub control, followed by burning, followed by grazing by native and domestic animals). It is this complicated interaction between impacts that prevents managers from making accurate predictions concerning the outcome of planned activities on specific sites. And when studying animal-succession relationships or when planning management on the basis of such a study, one must consider the surrounding vegetation types and seral stages and their possible influence on the area of interest (i.e., the broader landscape-level perspective).

Domestic animals have had direct and substantial influences on plant succession. A large

Table 3.2. Anticipated changes in successional stage or condition because of management activities

Management action	Successional stage condition					
Type of control	Grass-forb	Shrub-seedling	Pole-sapling	Young	Mature	Old growth
Shrub control						
Herbicides	>	>	>	<	<	<
Mechanical control	<	<		<	<	<
Controlled burn						
Cold burn	<	<	>	>	<	<
Hot burn	<	<	<	<	<	<
Fertilization	<	>	>	>	>	—
Grazing and browsing (moderate rates)						
Cattle and sheep	<	>	—	—	<	<
Goats	—	<	—	—	<	<
Deer and elk	<	>	—	—	<	<
Planting						
Trees	>	>			>	>
Shrubs	>	<				
Grasses-forbs	>	<		>	>	—
Regeneration cut						
Clearcut				<	<	<
Shelterwood				>	<	<
Seed tree				<	<	<
Salvage			<	<	<	>
Thinning (including single tree selection harvest)		>	>	>	>	—

Source: Thomas et al. 1979: table 3

body of literature has documented the long-term changes caused by the overgrazing of cattle. In southeast Arizona and southwest New Mexico, for example, overgrazing has been implicated in changing the historical semidesert grasslands into mesquite-dominated woodlands with little herbaceous cover; fire suppression has exacerbated these changes (e.g., see review by Bahre 1991).

Domestic herbivores have also been used, however, to help restore natural successional processes and plant communities. In the Netherlands, for example, cattle, ponies, and sheep have been used in nature reserves to restore ecological processes associated with herbivory (see also Savory 1988). Reserve managers in the United States sometimes use livestock to help reduce the cover of exotic plants as part of their attempts to reintroduce native plants. Domestic animals were introduced as "ecological substitutes" for their extinct ancestors. For example, the wild horse (*Equus przewalskii*) disappeared from northern Europe in about 2000 B.C. (Bokdam and de Vries 1992).

Floristics and Vegetation Structure and Relations to Wildlife

Two basic and obvious aspects of vegetation can be distinguished: the structure, or physiognomy; and the taxon of the plant, or floristics. Many authors have initially concluded that vegetation structure and habitat configuration (size, shape, and distribution of vegetation in an area), rather than particular plant taxonomic composition, most determine patterns of habitat occupancy by animals (e.g., see Hilden 1965; Wiens 1969; James 1971; Rotenberry 1985). As Rotenberry observed, however, subsequent studies have shown that plant species composition plays a much greater role in determining occupancy of an area than previously thought. He noted that such comparisons between structure and floristics really involve the spatial scale at which the animal is being examined. The same species that appears to respond to the physical configuration of an area at a broad (landscape) scale might show little correlation with structure at a more localized scale (see our previous discussion in this chapter of the FHD-BSD relationship). Rotenberry thus reached the same conclusion, albeit independently and from a different direction, as that of other workers interested in the relationship between spatial scale and occupancy of an area (e.g., Johnson 1980; Hutto 1985).

Thus, we see that the variables measured and the spatial resolution of those measurements should be based on the scale involved and the level of model refinement required. Simple presence-absence studies of animal abundance at broad scales likely do not require analysis of vegetation on a fine taxonomic level. Broad categorizations by vegetative structural class (e.g., sapling, pole, old growth) are probably adequate. Plant taxonomy becomes increasingly important, however, as our studies become more site specific and call for prediction of wildlife population size or density. Attempts to apply broad, structurally based models of wildlife-habitat relationships to local management situations (e.g., the stand level) usually fail (Block et al. 1994).

The Population Perspective

Why Study Populations?

To conserve wildlife we must ultimately provide for the survival and protection of individual organisms, the populations to which they belong, the communities to which populations appertain, the ecosystems in which communities occur, and the long-term evolutionary potential of species lineages. To this end, management of habitat can provide for conditions in which organisms can maximize their *realized fitness.* Realized fitness is measured as the number of viable offspring produced that in turn find mates and suitable environments and successfully reproduce. Fitness is influenced by the dynamics of interactions of individuals within a population, by interactions between populations and between species, and by interactions between organisms and their habitats and environments. To study and manage habitat successfully thus requires knowledge of population dynamics and behaviors.

A related reason for studying populations is to identify better the factors in the environment that regulate population trends. This extends beyond simply recognizing correlates of population trends; we need to know the true environmental causes. Once such linkages are made, we may be able to specify the most efficient means for conservation or at least predict possible outcomes of our activities.

Populations or Habitats?

If the goal is to provide for the conservation of species and populations, then why study population dynamics and trends if general habitat conditions can be provided? The answer is that, although habitat is essential to the survival of all species, by itself it does not guarantee the long-term fitness and viability of populations.

An example helps to explain. In the intermountain West of the United States, macrohabitat conditions, measured as vegetation cover types and structural stages, for Townsend's big-eared bat (*Corynorhinus townsendii*) are estimated to have increased since the early 1800s by about 3 percent (Marcot 1996). However, populations of this bat likely have *declined* over this period. The reason is that, although the species uses a wide range of macrohabitats, substrates, and roosts, it is particularly vulnerable to human activity. Disturbing females with young adversely affects breeding success, and disturbing winter hibernacula can increase winter mortality (Nagorsen and Brigham 1993). Thus, in this case, the trend in macrohabitats belies the trend in populations, even though providing such habitat is essential to species conservation and restoration. Similarly, the American bison (*Bison bison*) suffered great declines, nearly to extinction, during the nineteenth century because of overhunting, even though for a long time its habitat remained suitable and available. Overharvesting or disturbance can cause populations to follow trajectories quite different from those of their habitat. Thus, it is often desirable to have knowledge of true population distribution, abundance, and trend, as well as that of habitat; both together tell complementary stories.

Population Concepts Tied to Habitat Analysis and Management

Of Populations, Demes, and Distances

The traditional definition of *population* is "a collection of organisms of the same species that freely share genetic material, that is, that interbreed." However, few wild collections of organisms completely interbreed, that is, are completely *panmictic.* A local collection of organisms that have a high likelihood of sharing genetic material through interbreeding is called a *deme.* In wild populations, demes can be isolated or can partly interact, and the simple definition of *population* given here is seldom entirely true. In some literature, the term *subpopulation* is used to refer variously to a deme or to a portion of a population in a specific geographic location or as delineated by nonbiological criteria (e.g., administrative or political boundaries).

Several characteristics of habitats and environments can act to prevent complete panmixia in wild populations. Factors limiting panmixia include (1) dispersal filters and barriers, which limit distances and locations traveled by dispersing individuals, and (2) patchiness of resources through space. Examples of dispersal filters and barriers which tend to isolate demes and populations are the various mountain ranges ringing the Central Valley of California, which probably inhibit dispersal of the kit fox (*Vulpes macrotis*), and, on an even larger geographic scope, the major oceans separating populations of great gray owls (*Strix nebulosa*) between Nearctic (North America) and Palearctic (Eurasia) zoogeographic regions. One example of how patchiness of resources through space can separate demes is the physical separation of aquatic environments that can divide demes of river otters (*Lutra canadensis*) into different river systems and watersheds in North America. Another example is the occurrence of small isolated marshes used as breeding habitats by local colonies and demes of tricolored blackbirds (*Agelaius tricolor*), such as those that occur in portions of the Klamath Basin in northern California and southern Oregon.

Behavioral characteristics of individuals and social structures of breeding groups also can act to select for specific breeding individuals differentially and thereby to dissuade complete panmixia. Examples are mate defense in pronghorns (*Antilocapra americana*) and the territoriality of iguanid lizards such as the western fence lizard (*Sceloporus occidentalis*). Other factors, including resource competition or predation by other species, natural perturbations of environmental conditions, wide geographic distances as compared with vagility of organisms, and genetic structure of the population, can act to limit the geographic area over which organisms interact and can serve to partly or fully isolate populations genetically (see box 3.2).

Partial panmixia and degrees of isolation between populations can result in *metapopulation* structures. A metapopulation occurs when "a species . . . range is composed of more or less geographically isolated patches, interconnected through patterns of gene flow, extinction, and recolonization," and has been termed "a population of populations" (Lande and Barrowclough 1987:106). Often, these component populations have been referred to as *subpopulations.* Metapopulations occur when environmental conditions and species characteristics provide for less than a complete interchange of reproductive individuals and there is greater demographic and reproductive interaction between individuals within a subpopulation than between subpopulations. It is the condition probably most often found in wild animal populations. We discuss these conditions further below and draw some conclusions here and in later chapters on managing habitats for viable metapopulations.

Population Dynamics and Viability

The dynamics of wildlife populations are mediated through their relations with their environments. The older German concept of *Umwelt* nicely describes the set of biotic and abiotic factors that ultimately influence the realized fitness of organisms (e.g., Klopfer 1959) and viability of populations. Today, we would list exogenous as well as endogenous factors that can influence population viability (see table 3.3), including: population demography, population genetics, metapopulation dynamics, environmental stochasticity, species biogeography, evolutionary adaptations and selection mecha nisms, reproductive ecology and behavior, effects of other species, and human activities.

Box 3.2 How distance can act to isolate organisms and populations genetically

Patterns of spatial distribution of organisms can take many forms (box fig. 3.2). First, consider the simplest case of one contiguous population, as in box figure 3.2, part C. Even in contiguously homogeneous habitat, complete panmixia (interbreeding) of individuals within a population can be hindered through isolation-by-distance deme structures of the populations. The area over which a deme is effectively panmictic has been called a neighborhood, which is the space over which genotypes of offspring are not significantly different from those resulting from randomly selected parents in that area. In the case of linearly arrayed habitats, such as shorelines or rivers, a neighborhood, N_L, can be defined as

$$N_L = 2\sqrt{\pi\rho\sigma}$$

where ρ = density of breeding individuals (in the case of linear habitat, number of breeding individuals per unit habitat length), and σ = standard deviation of individual dispersal distances, also called root-mean-square dispersal distance. In the case of a two-dimensional habitat area, a neighborhood, N_A, can be defined as

$$N_A = 4\,\pi\rho\sigma^2$$

In this case, ρ = number of breeding individuals per unit habitat area, and σ is defined as above (Lande and Barrowclough 1987).

In linear habitats, populations can be considered entirely panmictic if $\rho\sigma^2 > L/10$, where L = the total linear range of the population; in areal habitats, populations are panmictic if $\rho\sigma^2 > 1$ (Lande and Barrowclough 1987; Maruyama 1977). Or, panmictic populations must be $< 10\ \rho\sigma^2$ in linear habitats and $\rho\sigma^2$ in two-dimensional habitats.

Thus, a linear neighborhood, N_L, of an entirely panmictic deme of tailed frogs (*Ascaphus truei*) in habitats of montane streams, with $\rho = 1$ adult per m of stream (from Nussbaum et al. 1983:150, in eastern Washington State) and $\sigma = 50$ m (hypothesized value; no data available), is 177.2 m of stream length. Panmictic populations must be $< 10\rho\sigma^2$ or 25 km of stream length; streams greater than this length necessarily contain more than one neighborhood. (The problem of greater differential dispersal downstream than upstream is not considered here and is left to the student as an exercise. Under this model, how would the size of a panmictic population vary with different densities? with different estimates of root-mean-square dispersal distance?)

Compare this with a rare species in a linear habitat. In New England northern hardwood forests, a panmictic population, N_L, of the substantially rarer dusky salamander (*Desmognathus fuscus*), also a stream habitat obligate, with $\rho = 0.02$ adults per m of stream (2 per 100 m stream length) (DeGraaf and Rudis 1990:160) and $\sigma = 50$ m (hypothesized value; no data available), is found in only 3.5 m of stream length. This is substantially less than the mean density of the population. No panmixia could occur unless σ were to exceed 705 m, which is probably unlikely. Thus, dusky salamanders in this location may not be able to achieve panmixia, and populations likely would show intergrades of genomes along stream habitat corridors.

Similar examples can be developed for populations in two-dimensional habitats. This exercise if left for the student.

Other, more complex models of distance separation of breeding organisms deal with how local concentrations of organisms interact in metapopulations. These are discussed later in this chapter.

For discussion of other isolating mechanisms in animals, see Futuyma 1986:112ff.

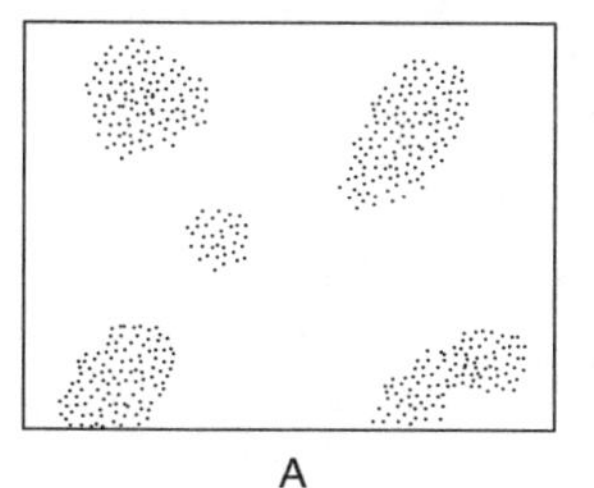

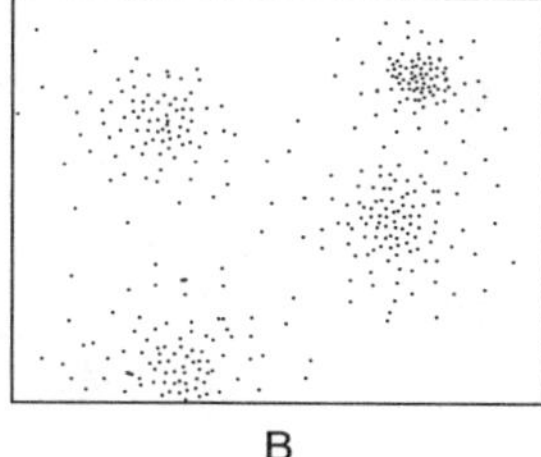

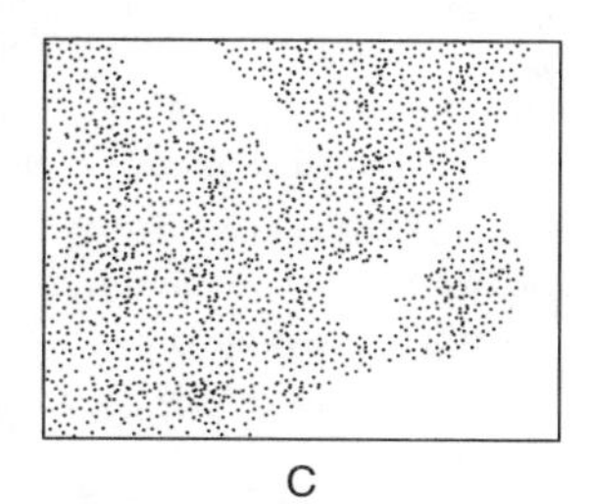

Box Figure 3.2. Some patterns of spatial distribution of organisms; each dot represents an individual. (A) Discrete populations, corresponding to the "island model" if mating is more frequent within than between the populations. (B) Perhaps the most common pattern in nature, ill-defined populations between which density is low. (C) A more or less uniform distribution, corresponding to the simplest "isolation by distance" models, except that regions of unfavorable habitat are acknowledged. (Reproduced from Futuyma 1986:120, by permission of Sinaeur Associates, Inc.)

Table 3.3. Some of the main ecological factors that affect the long-term viability of populations[a]

- *Population demography*
  - Population trend, rate of change
  - Likelihoods of extinction and quasi extinction
  - Vital rates (natality, mortality)
  - Sensitivity of population rate of change (lambda) to age-or stage-class specific vital rates
- *Population genetics*
  - Degree of historical and current loss of heterozygosity in genome
  - Causes for any loss of heterozygosity (inbreeding, genetic drift, etc.)
  - Potential for founder effect to simplify genome in future recolonization and rescue effect dynamics
- *Metapopulation dynamics*
  - Number of linked subpopulations
  - Number of isolated populations
  - Dynamics of future size of populations and numbers of populations
  - Likelihood of loss (extinction) or near loss (quasi extinction) of populations
  - Locations and conditions of existing and historical key links between subpopulations
- *Environmental stochasticity*
  - Types of disturbance events pertinent to persistence of a specific population and species
  - Dynamics of a disturbance event: type, frequency, intensity, location, duration, and effect
  - Effects of the disturbance event on population isolation, population size (direct mortality), ecology (as mediated through effects on other species such as competitors or predators), and population trend and vital rates
  - Descriptions of how stochastic processes have changed, and consequently how their effects have changed throughout the years (e.g., effects on metapopulation dynamics from changes in fire regimes and subsequent changes in forest canopy structure and density; and increases in fire frequency in sagebrush communities in Snake River plains, and the effect on increasing cheatgrass spread and resultant competition with native grasses)
- *Biogeography*
  - Species source pools
  - Zones of spread and dispersal
  - Dispersal rates
- *Evolutionary adaptations and selection mechanisms*
  - Allopatric speciation
  - Isolating mechanisms
  - Species swamping and hybridization
- *Reproductive ecology and behavior*
  - Hybridization
  - Dissemination dynamics
  - Angiosperm pollination vectors and their status
- *Effects of other species*
  - Predation
  - Competition
  - Parasitism, pathogens, and disease
- *Human causes of population declines*
  - Prehistoric (paleoecological), historic, and recent exploitation of populations
  - Invasion of exotic species into human-disturbed environments
    - burning
    - cutting
    - grazing
    - agricultural conversion
    - urban conversion

[a]Such factors should be addressed in a "hard-core" quantitative population viability analysis.

Population viability is the likelihood of the persistence of well-distributed populations for a specified time, typically a century or longer. The term *well-distributed population* refers to the need to ensure that individuals can freely interact where "natural" conditions have permitted. However, all species are limited to some degree in distribution; some species are inherently uncommon, and some are even locally endemic. Thus, "well distributed" needs to be interpreted in the context of the ecological distribution of the species in question.

The time span over which viability should be assessed should be scaled according to the species' life history, body size, longevity, and especially population generation time. We suggest that a rule of thumb may be to use at least 10 generations for gauging lag effects of demographic dynamics and 50 generations for genetic dynamics. Longer time spans can be considered if environmental changes can be predicted beyond that time period. Thus, for a population of parrots with a generation time measured perhaps on the order of a decade, population viability should be projected over a century for demographic factors and five centuries for genetic factors, whereas for a population of voles (with more rapid reproduction, a shorter life span, and a far shorter generation time) it may be projected over only a few years.

Logistic Gaussian population models predict smooth (or predictable stepwise) changes (increases or deceases) to some idealized and long-term asymptote representing carrying capacity. But real-world populations seldom exhibit the simple dynamics of Gaussian population models. Rather, populations are complicated by many exogenous (external) and endogenous (internal) factors. One complicating factor is variation in age-specific survivorship and fecundity. Stochastic (random) variations in these vital rates distributed over reproductive individuals within a given generation, such as expressed by a variance in the number of offspring, have been called *demographic stochasticity.* Stochastic variation in average population values of these vital rates over time, such as expressed by the average number of offspring over successive breeding seasons, has been called *environmental stochasticity.* Demographic stochasticity plays a role in affecting population viability where the effective number of breeding individuals is small and where the range of progeny per reproductive adult is large. Environmental stochasticity plays a role in affecting population viability where external factors influencing mortality or fecundity randomly vary over time, such as food levels or suitability of nest sites as affected by weather conditions. Both demographic and environmental stochasticity can contribute to population trajectories being substantially more complex than what simple models predict, and both can act to lower likelihoods of population persistence over a given time period. In general, smaller populations can be more subject than larger populations to adverse effects of both kinds of stochasticity.

Simple Gaussian population models also assume a fixed and measurable carrying capacity as well as fixed vital rates of fecundity and mortality of organisms in the population. Under such idealized circumstances, even allowing for some degree of demographic and environmental stochasticity, it is easy to identify *threshold effects* in model populations. In population viability modeling, thresholds typically are conditions of the environment that, when changed slightly past particular values, cause populations to crash[1] (Lande 1987; Soule 1980). Such

1. Conversely, some population models refer to explosion thresholds, which are environmental conditions that lead to large increases in population.

threshold conditions have given rise to the concept of *minimum viable populations* (MVPs) (Gilpin and Soule 1986; Lacava and Hughes 1984). In models, an MVP is the smallest population (typically measured in the absolute number of organisms rather than density or distribution of organisms) that can sustain itself over time and below which extinction is inevitable. Through modeling stochastic demography, Shaffer (1983) reported the viability and MVP of grizzly bear (*Ursus arctos horribilis*) populations. Reed et al. (1988) modeled MVPs for populations of red-cockaded woodpeckers (*Picoides borealis*) in the southeastern United States. Many other examples are available in the literature (e.g., Boyce 1992), but these typically constitute modeling results rather than empirical evidence of threshold effects.

So, do actual populations follow such threshold effects? Perhaps they do in rare cases of unchanging environments. But most often, the many factors influencing population viability vary through space and time, rendering identification of specific MVP levels problematic at best and misleading at worst. Such considerations prompted Thomas (1990) to reexamine simple guidelines for MVP size to ensure long-term viability of populations. Thomas concluded that MVP rules of thumb should be adjusted upward from Soule's (1987) suggestion of "low thousands" to at least 5500 for undivided (essentially panmictic) populations, and that future work should focus on better understanding the complicating factors in real-world populations. We concur with Thomas and suggest that the term *minimum viable population* and its concept be abandoned for more realistic models and for empirical studies. We advocate using models of population viability that account for random variations in demographic, environmental, or genetic factors, and that represent location- and case-specific outcomes as likelihoods of population persistence rather than as fixed MVP threshold sizes (e.g., fig. 3.5; also see box 3.3).

Population Demography

Population demography is the proximate expression of a host of factors that influence individual fitness and population viability (table 3.3). The reader is directed to other sources on specific techniques for modeling population demography (e.g., Caughley 1977; Caswell 1989; Gilpin and Hanski 1991).

How is demography influenced by habitat quality and distribution? Many models of population response assume invariant vital rates of fecundity and mortality, or at least fixed probability distributions around mean values of vital rates. However, vital rates may vary substantially in space and time as a function of many factors, including food quality, weather, imbalance in sex ratios, and other factors. In some cases, populations respond to weather conditions, such as harsh winters, with lag effects measured in seasons or years.

In other cases, the causes of variations in vital rates are unknown. One example is the northern spotted owl (*Strix occidentalis caurina*), which for decades has shown quasi-periodic variations in nesting attempts, successful breeding, and production of young. These periods may last 2–4 years, and may be synchronized over broad geographic areas and among owl populations that are likely to be largely isolated demographically. Causes of these fluctuations may be attributed to synchronous variations in prey density, although this is unlikely given the variety of prey taken and the breadth of this subspecies' geographic area. Other possible causes are lag effects of harsh winters and some innate mechanism that shuts off breeding investment for one or a few seasons, although the closely related and sympatric barred owl (*Strix varia*) can be a relatively prodigious breeder in the same habitats

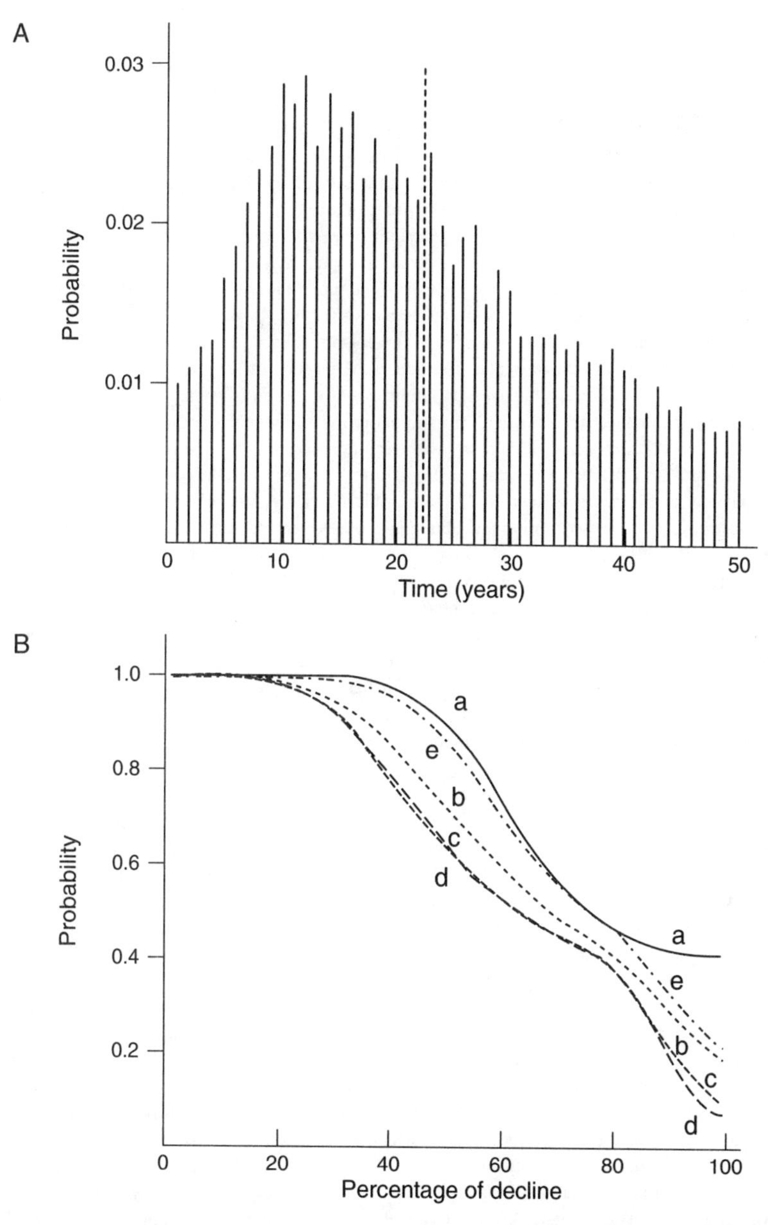

Figure 3.5. Two representations of viability of metapopulations of helmeted honeyeaters (*Lichenostomus melanops cassidix*), an endangered bird endemic to Victoria, Australia. (A) The number of years required for the number of helmeted honeyeater females to fall below 15 (the "quasi-extinction level" set for this particular analysis), given no translocations. The vertical dotted line represents the median of the distribution. The distribution is based on a simulation with 10,000 trials. (B) The probability of a decline in the metapopulation of helmeted honeyeaters within 50 years, as a function of the amount of decline: a, no translocation; b, two adults translocated; c, six adults; d, 10 adults; e, 10 adults with the new population subject to catastrophes. Each curve is derived from simulations with 10,000 trials. (From Akçakaya et al. 1995:174)

Box 3.3 On modeling and analyzing population viability

Population viability has been defined as the likelihood of continued persistence of well-distributed population(s) to a specified future time. A population with "high viability" thus has a high likelihood of continued persistence with a well-distributed population over a long time, say, on the order of a century or longer. This definition, used in numerous wildlife planning analyses, suggests that viability is a probabilistic event and that the likelihood of not achieving specific, desired viability levels is a measure of risk under any given management plan.

Population viability is affected by many factors intrinsic to a population and to its environment (see table 3.3 in text). Those that likely would receive attention in a quantitative analysis include: population demography, population genetics, metapopulation dynamics, environmental stochasticity, biogeography, evolutionary adaptations and selection mechanisms, reproductive ecology and behavior, rates and causes of population declines, and degree of and reasons for endemism and range restriction. Unfortunately, such empirical information is available for very few species.

So where may one turn for information useful in a population viability analysis? Our experience suggests three main sources: (1) contract reports and panel discussions with leading species experts, which will provide information on current population status and suspected reasons for population declines, as well as information on key environmental factors and key ecological functions of species; (2) geographically referenced (GIS) modeling of "key environmental correlates" that most influence the distribution and abundance of selected species; and (3) published empirical studies on species ecology.

Ideally, a population viability analysis should combine empirical studies with simulation modeling of demography, genetics, environmental disturbances, and other factors. Such an approach is often referred to as a quantitative population viability analysis, or PVA, which can be provided for only a few species. In the absence of adequate empirical species studies on which to base a quantitative PVA, the three-pronged approach listed above provides at least a starting point for helping to identify which species might warrant further, more intensive population analyses and for providing base information for such future analyses. Future analyses can be part of an explicit research, development, and application study plan and part of the implementation, inventory, and monitoring phases of any habitat or land management plan.

The GIS modeling approach (item 2 above) can be used to produce maps of past, current, and future potentially suitable environments for the species of interest. Each map can be analyzed for how the abundance, distribution, and trend of potentially suitable environments might influence the likelihood of population persistence (e.g., Boyce et al. 1994; Irwin 1994). This is *not* the same as modeling actual or anticipated population sizes, structures, and distributions, and is not a strict replacement for more intensive analysis. This is an important distinction. By modeling species' environments, results must be interpreted as tentative working hypotheses concerning population response. Only follow-up monitoring of populations, environments, or both, will determine the extent to which potentially suitable environments index true population persistence.

By modeling environments instead of *a priori*–defined habitat for a specific species, one can provide the basis for assessing conditions for a range of species. This can set the stage for further species-specific PVAs and also for broader assessments of biodiversity. As part of the three-pronged approach—particularly focusing on existing population abundance and distribution and on patterns and trends of environments—such modeling can be used to draw inferences about potential population viability response by individual species and about potential persistence of species groups, scarce ecological communities, and other aspects of biodiversity.

Coupling modeling of environmental conditions with knowledge of species demographics has the advantage of helping determine population distribution patterns, such as if a population is potentially large and interbreeding, partly interbreeding (in "metapopulation" patterns), isolated but of relatively large separate populations, or consisting of small and isolated populations. Also useful is the determination of when conditions contributing to a particular viability level are due to natural biophysical conditions or human activities and thus might be ameliorated. But it is important to remember that there is no real substitute for field studies of population and biodiversity conditions.

at the same times. Accurately predicting demographic trends of northern spotted owls through modeling will entail a better understanding of causes of such variations in vital rates and how these rates are affected by habitat and environmental conditions.

How does population size affect likelihood of decline? We might pose three patterns for a size-decline relation, as depicted in figure 3.6. The first pattern (curve A in fig. 3.6), and really the null model, is that population size does not influence rate of change as measured on a per capita basis. That is, large and small populations alike would tend to experience the same annual rate of per capita change, all other factors being equal. This pattern is unlikely, because many factors that influence population viability (table 3.3) have greater influence on small populations than on large populations.

For example, the selective contribution of only some genes to successive generations through genetic drift and through inbreeding depression can become a key factor in small populations. Such sampling increases the de-

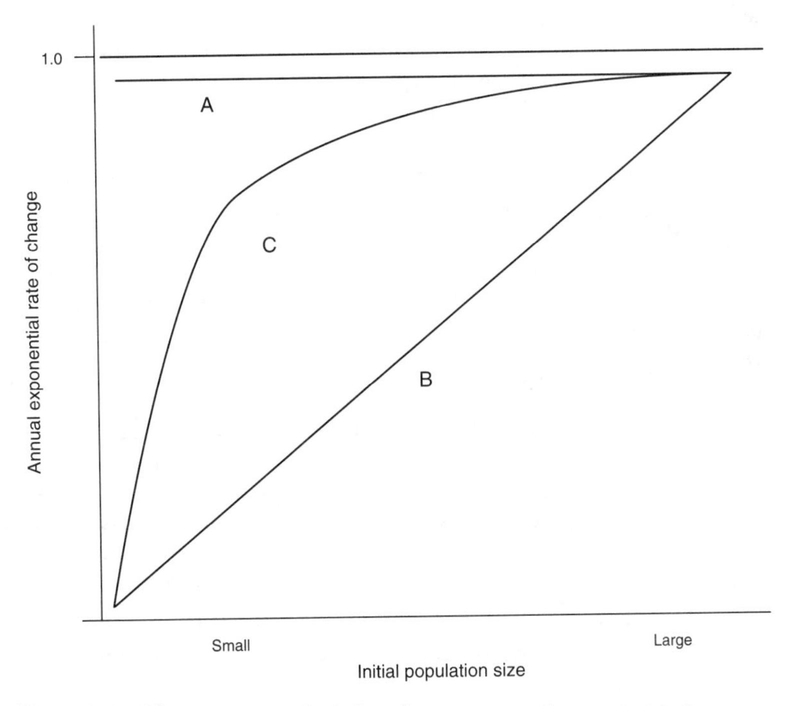

Figure 3.6. Three patterns of relations between annual exponential changes in population and initial size of the population. (A) Populations decline at rates regardless of initial size of the population. (B) Smaller populations decline at a higher rate, and the relation with population size is linear. (C) Smaller populations decline at a higher rate, and the relation with population size is nonlinear (e.g., exponential). In this figure, the ordinate is scaled so that declining populations occur at a rate of <1.0. The heavy horizontal line at 1.0 denotes no population change.

gree of genetic homozygosity of genotypes (through fixation of deleterious and other alleles) and thereby decreases overall diversity of alleles in gene pools. This happens at faster rates in small populations than in large ones. Another reason small populations might decline differently from (potentially faster than) large populations is the Allee effect, in which some degree of social interaction between organisms, afforded by population size or density above some threshold, is necessary to stimulate reproduction. Below the threshold, such facilitation is too infrequent or unlikely to support population numbers.

The second pattern for a size-decline relation (curve B in fig. 3.6) is that there is a linear effect of population size on rate of change. Under this pattern, smaller populations are likely to decline faster than larger populations,[2] but the effect is merely a linear increase in the rate of decline with respect to population size. This pattern is likely to be closer to most real-world cases than the null model of no effect. One argument for a linear relation over no relation has to do more with the simple geometry of increasing difficulty of finding mates in increasingly sparse populations or increasingly fragmented and divided habitats.

However, a linear relation is probably not commonly the case in real-world populations. One major reason is that the effect of a small population size on allelic fixation and on demographic stochasticity is disproportionately greater than that of larger population sizes. Greater demographic stochasticity means greater variation in the number of progeny among breeding individuals of a given generation and among years or breeding intervals. Mathematically, these factors affect smaller populations much more acutely than larger populations, on a per capita basis. It is easier for a small population to dip to zero or below a critical nonzero threshold, often called *quasi-extinction levels* (Ginzburg et al. 1982), than it is for a large population, *ceteris paribus.*

This leads us to the third pattern for a size-decline relation (curve C in fig. 3.6), that is, where smaller populations decline more rapidly than larger ones, and nonlinearly so. Mathematical models of population genetics and demography suggest that this third pattern is often the case, but surprisingly few data on real-world populations are available to test this. In one analysis, we can confirm this pattern empirically for a set of tropical island birds (box 3.4). Surprisingly few other real-world examples from the literature are available, but all seem to fit curve C: examples include birds on the Channel Islands of southern California (Jones and Diamond 1976), breeding birds of Bardsey Isle off Britain (Diamond 1984), and population data of the plant *Astrocaryum mexicanum* (Menges 1991). This pattern (curve C) may also hold for populations of herps, mammals, and other taxonomic groups, as suggested by population modeling (e.g., Diamond 1984), but this needs empirical testing. Under this pattern, small populations are substantially at

2. We summarily reject a linear trend with opposite slope. For reasons stated above, we greatly suspect that small populations are more susceptible to factors adversely affecting viability than large populations are. However, there may also be important cases where very large or dense populations undergo greater decline because of many possible reasons: the carrying capacity of the environment has been exceeded, disease is being transmitted, physiological effects of crowding have shut down the normal breeding cycle, food resources have diminished through overuse, high population size has triggered a devastating predator, pathogen, or parasite, and others. These factors are not to be dismissed in population viability assessments, although our discussion deals with small populations.

greater risk of severe decline and extinction than even moderately small populations are, and the relation is far worse than linear. This has important implications for timing of, for investing funding in, and for developing realistic expectations for species and habitat recovery programs.

Population Genetics

Many references are available on modeling population genetics (e.g., Crow and Kimura 1970; Falconer 1989; Hartl and Clark 1989). However, real-world relations between habitat conditions, population genetics, realized fitness

Box 3.4 A test of the hypotheses of how population size affects the rate of decline for a set of tropical forest birds

Data on census numbers of native Hawaiian forest birds from Scott et al. (1986) provide an opportunity to test empirically the effect of population size on rates of decline. Two censuses were conducted of birds on Kauai, the first in the early 1970s and the second in the early 1980s. Additional censuses were conducted of the Hawaiian crow on the island of Hawaii in 1978 and 1988 and are included in this analysis. Box table 3.4 lists these first and second counts and the years taken (columns a, b, d, e). (Additional data were presented on common 'amakihi [*Hemignathus virens*], but according to Scott et al., the 1981 counts were possibly biased by immigrants from neighboring islands, and thus we do not include this species in this analysis.)

The absolute annual population change, ΔN_t (column f in box table 3.4) is calculated by the formula

$$\Delta N_t = \frac{N_2 - N_1}{t_2 - t_1}$$

where N_t = mean number in study area at year t. This value is weighted by the absolute difference in census population size (counts). A measure of change not so weighted is the annual exponential rate of change, ΔNE_t (column g in box table 3.4), calculated by the formula

$$\Delta NE_t = \left[\frac{N_2}{N_1}\right]^{\frac{1}{t_2 - t_1}}$$

where terms are as defined above. This value expresses the rate of change on the basis of the proportion of the population at year t_1 still remaining at year t_2 and is not influenced by the absolute difference in counts. Values are smallest (<1) for populations undergoing the greatest decline; = 1 for unchanging populations; and > 1 for increasing populations.

Thus, rates of change can be compared between species with different initial population sizes, N_1. This is illustrated for the Hawaiian forest bird species in box figure 3.4A. Lines have been drawn between data points rank-ordered according to initial numbers in the study area (N_1). Note that the ordinate is exaggerated to illustrate the overall shape of the data scatter.

This figure corresponds best with the nonlinear size-decline relation pattern (curve C) in figure 3.6. That is, the annual exponential rate of change is a function of population size, and the most endangered bird species have the greatest rates of per capita annual decline. This may have implications for prioritizing funds and using conservation technologies (e.g., captive breeding) for endangered species recovery by attending to the smallest populations first.

If the nonlinear size-decline relation pattern (curve C of fig. 3.6) approximates an exponential relation, then a log-normal plot of box figure 3.4A should result in a linear function. Box figure 3.4B plots the annual exponential rate of change against $\log_{10}$ of the initial numbers in the study area (shown in column c in box table 3.4).

(box continued on following page)

Results here suggest a more complex relation than simple linearity in this log-normal plot. Note that various nonlinear relations can be posed for curve C in figue 3.6, including logistic, asymptotic, and convergent exponential functions. However, a linear regression of the data in box figure 3.4B is statistically significant ($R^2 = 0.62$, df = 7, $P < 0.02$), and therefore linearity of the log-normal plot should be considered first before advancing to more complex curves. The most striking outlier in the figure is the Hawaiian crow—the only corvid and the only non-Kauai species in this data set. This species occurs in geographic settings and habitats different from those of the rest of these species. It has a particularly steep rate of population decline and may be responding to factors quite different from those of the rest of the species in this group.

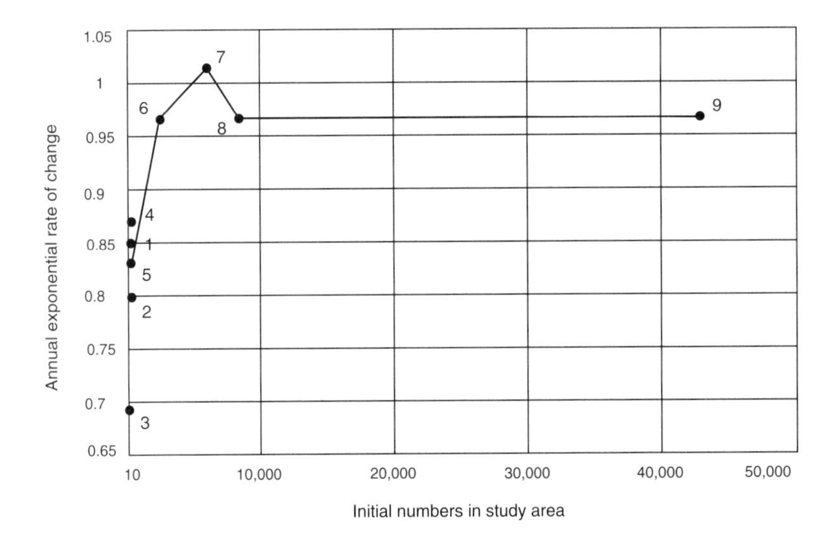

Box Figure 3.4A. Annual exponential rate of population change of native forest birds of Kauai, Hawaiian Islands. Note that the ordinate is truncated at the origin. The numbers at the data points correspond to the species as numbered in box table 3.4. Compare with the hypothetical curves in figure 3.6 in the text. (Calculated from data in Scott et al. 1986)

(*box continued on following page*)

of organisms, and viability of populations are poorly understood at best.

Two aspects of population genetics of major interest to conservation managers are inbreeding depression and genetic drift, often mistakenly assumed to be the same phenomenon. *Inbreeding depression* refers to fixation and adverse phenotypic expression of deleterious alleles in a population, caused by breeding between often increasingly related individuals. Deleterious effects can be manufactured through several genetic mechanisms including increasing the proportion of homozygous recessive genotypes, increasing the proportion of deleterious alleles in overdominance, maintaining the presence of deleterious genetic mutations through successive breeding of the same pedigree line, among others. Senner (1980) described three

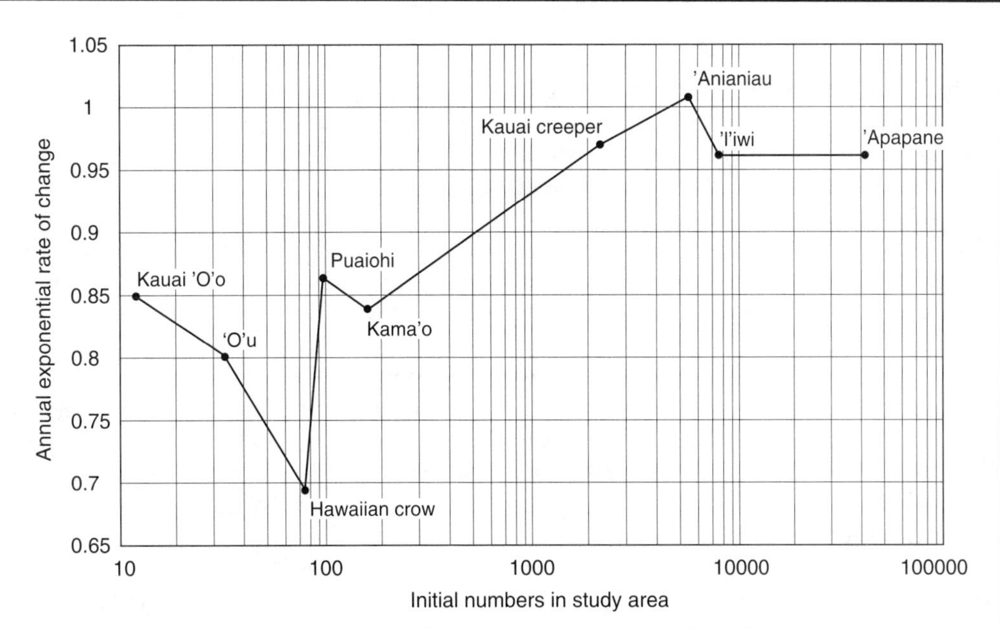

Box Figure 3.4B. Annual exponential rate of population change of native forest birds of Kauai, Hawaiian Islands. Note that the abscissa is log-scaled to emphasize outliers from a linear distribution, such as the Hawaiian crow. (Calculated from data in Scott et al. 1986)

Box Table 3.4. Census count data of native forest birds of Hawaii and an analysis of the exponential rate of population change between census counts

	First count			Second count		Exponential changes	
Species[a]	Year (a)	Mean number in study area(b)	Log of initial numbers in study area (c)	Year (d)	Mean number in study area (e)	Annual population change (e − b)/(d − a) (f)	Annual rate of change (g)
1. Kauai 'o'o, or 'o'o 'a'a	1970	12	1.08	1981	2	−0.91	0.85
2. 'O'u	1970	34	1.53	1981	3	−2.82	0.80
3. Hawaiian crow, or 'alala	1978	76	1.88	1988	2	−7.40	0.70
4. Puaiohi	1970	97	1.99	1981	20	−7.00	0.87
5. Kama'o	1970	173	2.24	1981	24	−13.55	0.84
6. Kauai creeper, or 'akikiki	1971	2,300	3.36	1981	1,649	−65.10	0.97
7. 'Anianiau	1970	5,500	3.74	1981	6,077	52.45	1.01
8. 'I'iwi	1970	7,800	3.89	1981	5,400	−218.18	0.97
9. 'Apapane	1970	43,000	4.63	1981	30,000	−1,181.82	0.97

[a]Scientific names: 1, *Moho braccatus;* 2, *Psittirostra psittacea;* 3, *Corvus hawaiiensis;* 4, *Myadestes palmeri;* 5, *Myadestes myadestinus;* 6, *Oremystis bairdi;* 7, *Hemignathus parvus;* 8, *Vestiaria coccinea;* 9, *Himatione sanguinea*
Source of census count data: Scott et al. 1986

manifestations of inbreeding depression: individual viability depression, which is a failure of young to mature; fecundity depression, which is an increase in sterility; and sex ratio depression, which is an overabundance of males. Models of inbreeding dynamics (e.g., Crow and Kimura 1970; Hartl and Clark 1989) have been used in estimating effects on long-term viability of populations (fig. 3.7).

Inbreeding depression has been implicated as the cause of low fertility in Florida panthers (*Felis concolor coryi*) (Maehr and Caddick 1995), captive populations of the gray wolf (*Canis lupus*) (Laikre and Ryman 1991), and other endangered carnivores. However, there are cases of populations having gone through severe bottlenecks and not suffered any apparent or lasting demographic problems from inbreeding depression. Examples include bird populations as small as 10 pairs persisting for 80 years on the Channel Islands off southern California (Jones and Diamond 1976) and severely reduced populations of northern elephant seals (*Mirounga angustirostris*), which rebounded from about 20 to about 30,000 individuals over 75 years following protection (Bonnell and Selander 1974).

In comparison with inbreeding depression, *genetic drift* is the random change in frequency of alleles through random mating and allelic assortment over generations. Genetic drift may result in the deleterious fixation of alleles in homozygous genotypes, but it can also result in neutral changes and heterozygous conditions as well. Drift may or may not reduce overall genetic diversity of a population, depending on the size of the population, the initial degree of homozygosity, the degree of outbreeding, and other factors. The effects of genetic drift on habitat relations and population vitality—and, conversely, the effects of habitat quality and distribution on the likelihood of deleterious effects from genetic drift—are poorly understood at best but are probably important factors over the long run in small, isolated populations or in demes with little or no outbreeding.

In the 1980s, researchers modeled MVPs by considering only genetic conditions of inbreeding depression and genetic drift. In theory, a minimum viable population supposedly is of a size (number of breeding individuals, assumedly) below which the population is doomed to extinction and at and above which it is secure. One guideline that was proposed was the "50–500 rule," specifying that populations of at least 50 breeding individuals be maintained for ensuring short-term viability and 500 for long-term viability (Gilpin and Soule 1986). However, as reviewed above, such guidelines seldom pertain to real-world situations and were devised mostly on the basis of genetic considerations alone.

Viable population management goals often specify providing for large interconnected wildlife populations and maintaining diverse gene pools. Rather, management goals should begin with understanding the natural conditions of a species in the wild. There are cases where populations are fully or partly isolated under natural conditions, and artificially inducing outbreeding between isolates, such as through captive breeding programs or manipulation of habitats, may contradict natural conditions. In some cases, natural balanced polymorphisms, local ecotypes, and local endemics with unique gene pools have derived from and can be maintained by partial or full reproductive isolation of populations. An example is presented in box 3.5.

Of Systematics and Subspecies

It should be obvious by now that providing habitat and suitable environments for wildlife should attend to more than just the needs of individual species. Although the idea of species

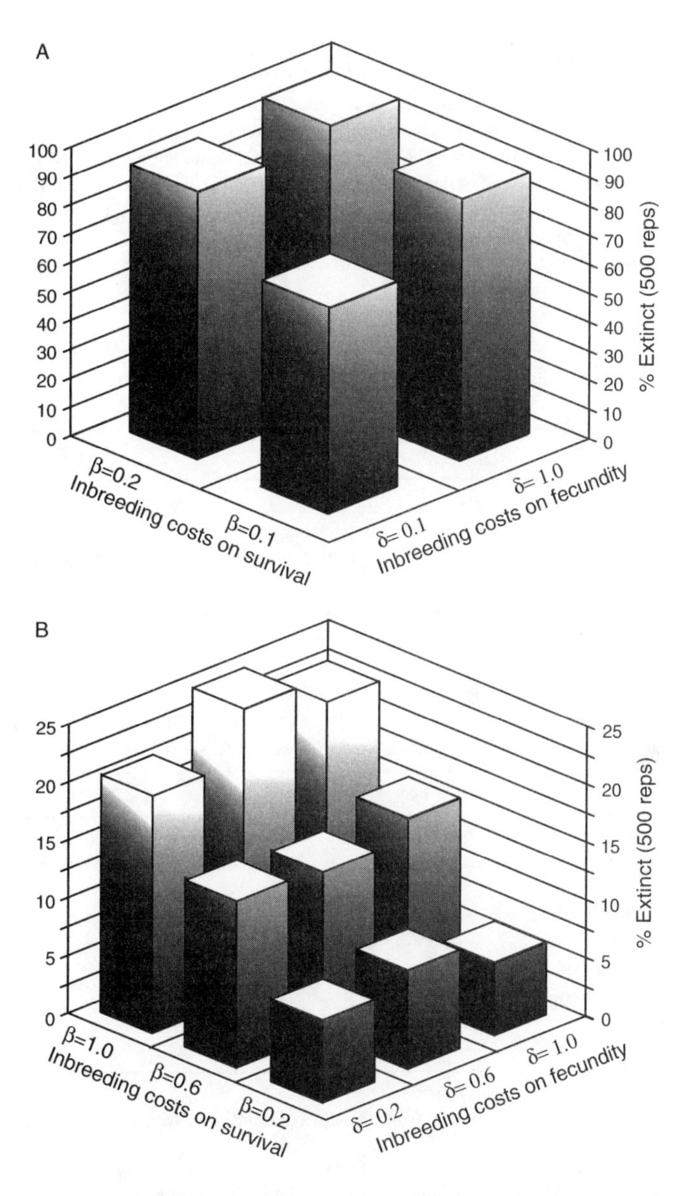

Figure 3.7. An example of an analysis of effects of inbreeding depression and population size on the viability of a hypothetical population with an increasing trend ($\lambda = 1.24$), an equal sex ratio, and an initial population set at (A) $N_{(0)} = 5$ and (B) $N_{(0)} = 80$. In this analysis, inbreeding was modeled as costs on survival and fecundity and expressed as the percentage of populations in replicate model runs that became extinct. In this model, the combined effects of inbreeding costs on both survival and fecundity resulted in a greater extinction risk than that of each cost factor separately, although initial population size greatly mitigated extinction risk. (From Mills and Smouse 1994:421)

Box 3.5 An example of balanced polymorphism in isolated populations of newts

Balanced polymorphism refers to several morphological types or phenotypes recurring in the same, sympatric population. Examples of balanced polymorphisms have often been cited for various species of heliconiid butterflies in the New World tropics. Another, little-known example occurs in isolated populations of northern rough-skinned newts (*Taricha g. granulosa*) in the western United States (Garber and Garber 1978; B. Marcot, unpubl. data). Some subalpine and montane pounds in the Klamath and Siskiyou mountains in northwestern California hold populations of newts that probably are at least partly genetically and demographically isolated from other populations. Individuals of some but not all of these pond populations exhibit a wide diversity of skin melanism, ranging from the normal dark brown or black to very light gray with dark spots (box table 3.5).

Box Table 3.5. Color variations of northern rough-skinned newts observed in two isolated lakes of the Klamath Mountains in northwestern California (Upper Mill Creek Lake, Humboldt County, HBM T9N, R6E, Sec. 34; and Bear Wallow Lake, Del Norte County, HBM T15N, R4E, Sec. 22)

Variant[a]	Dorsum	Ventrum
Upper Mill Creek Lake		
1	black to dark brown	unspotted or unblotched orange
2	black to dark brown	varying from yellow-orange to tangerine orange
3	light blotching consisting of lack or dearth of pigmentation, on matrix of black to dark brown	dark blotches and spots almost obscuring orange on ventrum
4	as with (3) but progressively more of the normally dark matrix is covered by light	as with (3) but dark dorsal color meets around vent
5	as with (3) but even more of the normally dark matrix covered by light than with (4)	as with (4) but orange on ventrum does not extend to ventral tip of tail
Bear Wallow Lake		
1	black to dark brown	unspotted or unblotched orange
2	dark brown matrix heavily covered with black splotches	orange matrix heavily covered with black splotches
3	gray with multiple black splotches	mostly unspotted orange but with black line across ventrum
4	gray with multiple black splotches	mostly unspotted orange
5	gray with one black splotch, or black splotches found anywhere on body	mostly unspotted orange
6	gray	unspotted orange

Source: B. Marcot, unpublished data
[a]Variants are rank-ordered by increasing deviation from the normal condition, with the 1 ranking signifying that condition.

Different ponds (habitat islands; see chapter 9) hold slightly different color morphs (box table 3.5) and proportions of the local population in each morph. For example, in Bear Wallow Lake, approximately half of all newts are of normal coloration, and half belong to one of the morph variants; percentages in other ponds may vary widely.

The causes of the color morphs are unknown; hypothetically, they may be due to selective reproduction, genetic effects of immigrants, length of time isolated, or other factors. These particular populations number approximately 1500–2000 individuals each and are likely to be largely panmictic, so inbreeding and genetic drift probably are not major factors

(box continued on following page)

Box 3.5 Continued

causing the morphs, although the morphs may have a genetic basis. Other causes of the color variants may be some unknown selection mechanism such as a differential vulnerability to predation (although this seems unlikely) or a differential phenotypic response to some environmental condition such as ultraviolet solar radiation. The occurrence of color variants in higher-elevation lakes may suggest a harsher environment and potentially greater influence of ultraviolet radiation. Or it may be that higher-elevation populations are more isolated than lower-elevation populations, and can thus develop color morphs because of that isolation. Only further studies could reveal the true case.

Rough-skinned newts contain toxins in their skin and are aposematic, that is, shunned as prey by most terrestrial predators including garter snakes (*Thamnophis* spp.; B. Marcot, pers. obs.). When disturbed, rough-skinned newts often arch their backs, revealing the bright orange or reddish ventrum as a warning. The color morphs, however, have not been tested for differential vulnerability to sunlight or for differential content of skin toxins; nor have they been tested for how variations in ventrum color affect predation rates and for the success of their aposematic warning behavior.

Despite the unknowns, these are naturally occurring and scientifically interesting polymorphic populations. Their very different individual phenotypes likely would vanish under genetic swamping if artificially forced to interbreed with outside populations, although Garber and Garber (1978) conjectured that the variant forms were already being invaded by normally pigmented individuals from the surrounding lowlands and that current temporal and spatial isolating mechanisms may be insufficient to maintain natural variation (however, this needs validation). Thus, here is a case where naturally occurring, isolated, polymorphic populations are probably best maintained in relative isolation and by not artificially inducing larger effective population sizes by connecting them with outside populations.

is one of the building blocks of ecological science, it is in some ways an artificial construct. Many taxonomic groups, particularly plants, arthropods, and some herps, are replete with biological entities that defy easy classification into the standard Linnean taxonomic system. Reasons are varied. They include unusual or complicated reproductive life histories. One example is that of triploid hybrids of the silvery salamander (*Ambystoma platineum,* cf. *jeffersonianum*) and Tremblay's salamander (*A. tremblayi,* cf. *laterale*) in the northeastern United States (see box 3.6).

Another example occurs with some clonal populations of the Chihuahua whiptail (*Cnemidophorus exsanguis*) and checkered whiptail (*C. tesselatus*) lizards that reproduce by parthenogenesis (females give birth to haploid female offspring without contribution of male genes). In some areas their clones are sympatric (have overlapping ranges) with their parent species and behave as reproductively isolated, separate species, although differences in coloration and scalation are subtle (Stebbins 1966). Checkered whiptail populations tend to be mostly parthenogenic, with males extremely rare.

Why is this important? In the real world of biopolitics it is vital to recognize species, subspecies, and populations clearly and unambiguously in order to focus activities related to conservation, recovery, recreation, hunting and collection, and restoration. But habitat managers sometimes have to struggle with imperfectly classified organisms, and thus encounter problems of correctly identifying habitats and environments deserving conservation activity.

One recent example is that of the red-legged frog (*Rana aurora*) in the western United States. Two western subspecies have been recognized: *R. a. aurora,* which occurs in the northern part of the range of the species, and *R. a. draytonii,* which occurs in wetlands and streams in coastal drainages of central California. *Draytonii* has been extirpated from 70 percent of its former range, whereas *aurora* is

Box 3.6 A case of managing habitats for strange amphibian hybrids

Sometimes the Linnean species classification system fails and leaves us with some strange cases of organisms that may deserve conservation for their unique genetic, ecological, or evolutionary significance. Here is one such story.

In the eastern United States, two closely related "standard" (Linnean) species of *Ambystoma* salamanders occur: the blue-spotted salamander (*A. laterale*) and the Jefferson salamander (*A. jeffersonianum*). Like most ambystomids, these species breed in water and are sexually reproducing and diploid (2N = 28 chromosomes).

Under some conditions, however, they hybridize. But the hybrids are dimorphic (of two forms) and have been given individual names: silvery salamander (*Ambystoma platineum*, cf. *jeffersonianum*) and Tremblay's salamander (*A. tremblayi*, cf. *laterale*). Each hybrid is triploid (3N = 42 chromosomes) and, although genetically unique, is identified only by chromosome count and size of red blood cells taken from living laboratory specimens (Conant 1975). *A. platineum* receives 2N chromosomes from *A. jeffersonianum* and 1N from *A. laterale*, whereas *A. tremblayi* receives 1N from *A. jeffersonianum* and 2N from *A. laterale* (box fig. 3.6). Thus, two recognizably different hybrids result from one parent stock. It is a genetically bizarre quartet indeed, but the story gets even stranger.

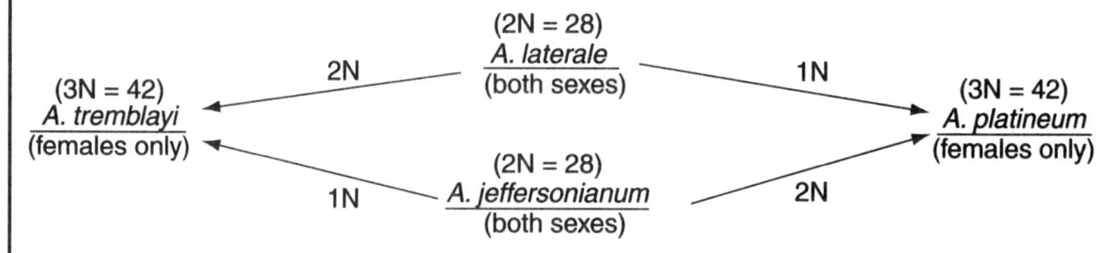

Box Figure 3.6. The complicated breeding arrangement of four species of *Ambystoma*.

Each hybrid consists of only females that reproduce as sexual parasites as they back-cross with males of *A. laterale* or *A. jeffersonianum*. The hybrid females are stimulated to lay eggs by being clasped by the males of the parent stocks and even travel to ponds with the males. There, it appears that the eggs of the hybrids require stimulation of sperm passing through the egg membrane, but the sperm contributes no chromosomes to the "zygote." This is called pseudofertilization.

The hybrids occur widely throughout the northeastern United States and may maintain themselves in nature even where the parent species are absent (such as in central Indiana). That is, in some cases, their eggs may develop even without pseudofertilization (but this needs study). Thus, *A. platineum* and *A. tremblayi* constitute a case of dimorphic triploid hybrids that can maintain themselves by pseudofertilization back-crossing with their parent stock or perhaps even by parthenogenesis in isolation from their parent stock.

Is this a case of sympatric speciation in progress? How well adapted are the hybrids and their progeny to environmental conditions at and beyond the distributional margins of their parent stock? Should the hybrids be given the same conservation interest and status as their parent stock species, particularly for managing the pond and wetland habitats where they occur in isolation from their parent stock? Perhaps this case suggests that the Linnean classification of species and even the genetic definition of species might be allowed some latitude for conservation of genetically unique organisms, in the spirit of maintaining the evolutionary potential of species lineages.

more common where it occurs farther north. Red-legged frogs found in the intervening area of northern California exhibit intergrade characteristics of both subspecies (Hayes and Krempels 1986), and systematic relationships of the two subspecies are not fully known (Hayes and Miyamoto 1984; Green 1985, 1986; Hayes and Krempels 1986). However, significant morphological and behavioral differences between the two subspecies suggest that they may actually be two species in secondary contact (Hayes and Krempels 1986).

The story continues. Recently, the USDI Fish and Wildlife Service has designated *R. a aurora* as a federally threatened taxon in a portion of its range. If *R. a. aurora* is found to be a separate species, this may narrow future conservation options, such as locating appropriate sites for reintroductions, identifying sources for use in augmentation of native or captive populations, and tracking captive breeding pedigrees, should these be useful conservation measures. It would also raise the question of conservation attention to red-legged frogs in the northern California intergrade zone, should they prove to be hybrids between species rather than hybrids between subspecies. Species hybrids—such as the cases of the triploid *Ambystoma* hybrid cited above or potentially the red-legged frog in northern California—may signal undesirable genetic swamping if the species' sympatry is a result of human activities; however, it may also be a natural outcome of sympatric remixing of closely related but previously separated species. There is no clear, single direction that conservation action should follow in such circumstances. Each case should be evaluated on its own.

The case of the red-legged frog may be one of unknown identity awaiting further taxonomic study. But a sufficient number of other cases, particularly among herps, question the utility of species and subspecies as the standard—and often the only—measure of entities deserving special habitat conservation action. In another example, throughout the western United States range several species and a number of described subspecies of garter snakes (*Thamnophis* spp.) which defy simple taxonomic description (see box 3.7). Intergrades abound, varying in morphometries, habitat selection, and coloration (B. Marcot, unpub. data).

One solution to the problem may lie in intensive study of genetic relations between potential species or subspecies using advanced laboratory methods, including comparative nucleic acids, protein sequencing, electrophoresis, immunology, chromosome matching, and studies of relations between individual gene loci, as well as studies of morphometries and breeding relations (Chambers and Bayless 1983). The next step would be to correlate genetic constitution and variation with patterns of *in situ* habitat use and selection. This would help determine taxonomic entities that relate to specific environmental and geographic conditions. However, most wildlife species officially listed as sensitive, threatened, or endangered or as candidates for these lists have not been studied to this degree; nor can we realistically expect funding to study the vast majority of these species. Also, removing individuals from at-risk populations for such research may not be feasible or desirable. Still, molecular methods have proved useful in determining that the red wolf (*Canis rufus*)—an increasingly rare and federally listed endangered carnivore of the southern United States—is probably a hybrid of the coyote (*C. latrans*) and the gray wolf (*C. lupus*) (Roy et al. 1994; Wayne and Jenks 1991). Whether the red wolf will continue to receive federal conservation attention should this finding gain universal acceptance may be more a matter of biopolitics than science (Brownlow 1996). At the least, as with the case of red-legged frogs and garter snakes in northern California, it raises the question of what to do with dubious taxonomic distinctions and well-defined hybrids between species.

Use of such recent laboratory methods in systematics has led Sibley and Monroe (1991) to pose a new and controversial overhaul to taxonomy of birds of the world. Their new taxonomy greatly revises relations of many orders, families, and genera, and in many cases provides molecular evidence of new species previously underscribed. Their approach uses a

Box 3.7 What habitats should be provided for conservation of garter snakes in northwestern California? An example of how a messy taxonomic group can complicate questions of habitat conservation

First and foremost, this question is largely an academic one in northwestern California, because most populations of garter snakes are probably secure there. However, should current taxonomies change from further study, it is unknown if new forms in northwestern California would be identified that also could be viability concerns. In central coastal California, the species of giant garter snake (*Thamnophis gigas*) is listed as threatened, and the subspecies of San Francisco garter snake (*Thamnophis sirtalis tetrataenia*) is listed as endangered by the USDI Fish and Wildlife Service.

Understanding habitat requirements of *Thamnophis* spp. can be important for ensuring that physiological and metabolic requirements and habitat needs for feeding and reproduction are met. For example, Peterson (1987) found that, in eastern Washington, the physical environment usually prevented specimens of *T. elegans vagrans* from attaining their preferred body temperatures and thereby limited when and where they could be active for foraging and reproduction. In northern California, Kephart and Arnold (1982) found that a lowering lake level caused breeding failure of western toads (*Bufo boreas*), which were one of the main prey species taken by terrestrial garter snakes (*T. elegans*) and common garter snakes (*T. sirtalis*); selection of the lake environment for foraging and variations in the lake level were correlated to interannual variation in opportunistic feeding by the snakes. In northeastern California, Huey et al. (1989) found that selection of retreat sites, particularly rocks of intermediate thickness in terrestrial settings, by *T. elegans* was guided largely by behavioral thermoregulation requirements.

Raphael (1988) surveyed *Thamnophis* among terrestrial seral stages of a Douglas-fir (*Pseudotsuga menziesii*) forest in the Klamath Mountains of northwestern California. His data suggested that *T. sirtalis* was most abundant in early brush-sapling and pole stages and, in decreasing abundance, also in old-growth, late brush-sapling, and mature forest stages; and *T. elegans* occurred mostly in the brush-sapling stage and, in decreasing abundance, also in late brush-sapling, pole, sawtimber, old-growth, and mature forest stages. Thus, at least these two species occupied overlapping habitats, but they found optimal forest conditions in different successional stages.

In the western United States, taxonomy of the four garter snake species and their sundry subspecies present there is uncertain and needs much study (Storm and Leonard 1995; Johnson 1995). Taxonomy is particularly uncertain between the western terrestrial garter snake (*T. elegans*) and the western aquatic garter snake (*T. couchi*), and between the currently recognized subspecies of common garter snake (*T. sirtalis*) (Nussbaum et al. 1983). Some species of garter snakes in northwestern California are largely aquatic (*T. couchii*), others are terrestrial or riparian (*T. ordinoides*), and still others broadly occupy all kinds of habitats (*T. elegans*). Habitat selection varies even among subspecies of some species (*T. sirtalis, T. elegans*) (Nussbaum et al. 1983; Stebbins 1966).

There may be rather widespread hybridization or intergrades between some of the western species, and there certainly is considerable variation among many subspecies. Morphometric characteristics of one sample of individuals from the Klamath Mountains suggest the presence of such intergrades, because individuals there vary widely in scale counts, dorsal and ventral coloration, and habitat selection (box table 3.7.A). Existing identification keys (Nussbaum et al. 1983; Stebbins 1966) seem to be inadequate to describe the current taxonomic situation.

To test this situation, one of us (B. Marcot) evaluated characteristics of specimens taken in the field, comparing them with published characteristics, using the following methods: (1) The published field characteristics of garter snake species and subspecies potentially found in northwestern California were listed in a table (box table 3.7B). (2) Then observed meristics of a field sample of 26 garter snakes taken from this area (Klamath Mountains) were listed, using the same set of attributes (see box table 3.7A). (3) Then an identification key for species and subspecies was built (see box table 3.7C) on the basis of the published characteristics. The key was built by using an example-based expert system (1st-CLASS Expert Systems, Inc.®, Wayland, Mass.). The expert system applies an induction-based optimization algorithm (the iterative dichotomizer three, or ID3, induction algorithm) (Quinlan 1986; Shapiro 1987) that creates the most efficient rule to distinguish best between known outcomes of examples. In this case, outcomes were garter snake species and subspecies, and examples were the published characteristics for their identification. The ID3 algorithm, rather than a multivariate statistical approach, was chosen for this analysis because: it can explicitly handle uncertain or unknown

(*box continued on following page*)

Box Table 3.7A. Meristics and descriptions of garter snakes (*Thamnophis* spp.) captured in the Klamath Mountains of northwestern California

		Coloration				Scalation					
Specimen	Habitat	Dorsal line	Lateral line	Matrix and miscellaneous	Ventrum	Upper/lower labials	6th & 7th upper labials	No. of midbody dorsal scales	No. of neck scales	Internasals	Chin shields
1	aquatic (pond)	yellow	yellow	no red	gray-blue	8/10	—	21	—	—	—
2	riparian (pond)	narrow, dull yellow	yellow	—	—	8/10	—	19	—	not pointed	—
3	aquatic (pond)	—	—	red on side	—	7/10	—		—	not pointed	—
4	terrestrial	yellow, 3 scales wide	merges with ventrum	red on side	—	7/10	—	19	—	not pointed	—
5	riparian & aquatic (pond)	pale yellow, 3 scales wide	distinct, yellow	no red	—	8/10	—	19	—	—	—
6	terrestrial	bright yellow, 3 scales wide	none	slight orange-red on sides and ventrum	—	8/10	—	19	21	not pointed	—
7	riparian (pond)	orange-yellow	diffuse orange	—	diffuse red-orange	8/10	—	19	21	not pointed	—
8	aquatic (pond)	pale yellow, 3 scales wide	—	no red on back or on lateral line	light reddish diffused band down center	8/10	—	19	21	not pointed	—
9	terrestrial	pale yellow, 3 scales wide	weak, pale yellow, merges with ventrum	dark red spots on side	no red; cream white (pale and unmarked)	7/10	—	19	21	not pointed	—
10	terrestrial	yellow-orange, 3 scales wide	merges with ventrum	red spots on sides	cream-blue, no red	7/10	—	19	21	not pointed	—
11	aquatic	distinct, yellow	yellow, distinct	—	cream-yellow	8/10	—	19	21	not pointed	
12	riparian by spring	slightly more than 1 scale wide	merges with ventrum	blue-gray flecks on dorsum; no red anywhere; large eyes	creamy blue-gray	8/10	—	19	21	not pointed	about equal

(*box continued of following page*)

Box Table 3.7A Continued

Specimen	Habitat	Coloration: Dorsal line	Lateral line	Matrix and miscellaneous	Ventrum	Scalation: Upper/lower labials	6th & 7th upper labials	No. of midbody dorsal scales	No. of neck scales	Internasals	Chin shields
13	aquatic	distinct, dull yellow	distinct	red spots on side	cream-white, no red	7/10	—	19	21	not pointed	
14	riparian (pond)	distinct, yellow	rather distinct, with red	much red on side; side of head reddish	cream, no red	8/10	higher than wide	19	21	not pointed	rear > front
15	riparian (pond)	distinct, yellow	merges with ventrum	red spots on side	cream-white	7/10	—	19	21	not pointed	about equal
16	riparian (pond)	ochre	ochre, rather distinct	no red	cream-orange	8/10	—	19	21	not pointed	about equal
17	terrestrial	red-orange	distinct, pale yellow, on 2d & 3d scale rows, extends onto upper labials	head top is brown	cream-gray w/ central reddish flecks	8/10	square	17	19	not pointed	equal
18	terrestrial	ochre, 1 scale wide	ochre-yellow, on 2d & 3d scale rows, extends onto upper labials	no red; back black speckled with light blue	cream-orange	8/10	higher than wide	19	23	more or less elongated	rear = front, middle shorter
19	aquatic	dull yellow, 1 scale wide	more or less distinct, on 2d & 3d scale rows, extends onto upper labials	no red	cream-yellow with central indistinct gray band	8/10	—	19	21	more or less elongated	front < rear
20	aquatic	ochre stripe, 1 - 1/2 scales wide	ochre stripe, 2 scales wide, extends onto upper labials	no red; faint blue flecks on black matrix	cream (salmon)	8/10	—	19	—	elongated	—
21	terrestrial	prominent yellow, 1 or more scales wide	merges with ventral line; on 2d & 3d scale rows	no red	cream-gray	8/10	higher than wide	21	21	not pointed	equal

(*box continued on following page*)

Box Table 3.7A Continued

		Coloration				Scalation					
Specimen	Habitat	Dorsal line	Lateral line	Matrix and miscellaneous	Ventrum	Upper/ lower labials	6th & 7th upper labials	No. of midbody dorsal scales	No. of neck scales	Internasals	Chin shields
22	riparian (pond)	pale yellow-orange, 1 scale wide	more or less confluent with ventrals; on 2d & 3d scale rows; extends onto upper labials	no red; black back sprinkled with light blue	cream-orange	8/10	not higher than wide	19	23	not pointed	more or less equal
23	riparian (pond)	yellow, prominent, 3 scales wide	yellow-white on 2d scale row, merges with ventral	red blotches on black matrix	cream (no red)	7/10	higher than wide	19	21	not pointed	equal
24	aquatic (creek)	orange, 1 scale wide	orange, more or less distinct, on 2d & 3d scale rows; extends weakly onto rear of head	no red	cream with cream-orange stripe down center	8/10	higher than wide	19	23	not pointed	front < rear
25	terrestrial	prominent	—	red spots above lateral line on dorsal scales 4 & 5, extending anteriorly to approx. 2 cm behind head	—	7/10	—	19	21	not pointed	split; more or less equal
26	terrestrial	prominent	—	red spots above lateral line on dorsal scales 4 & 5, extending anteriorly to approx. 2 cm behind head	—	7/10	—	17	17	elongated	split; more or less equal

Source: B. Marcot, unpublished data

(*box continued on following page*)

Box Table 3.7B. Descriptions of currently recognized species and subspecies of garter snakes (*Thamnophis* spp.) potentially occurring in the Klamath Mountains of northwest California

		Coloration				Scalation					
Taxon[a]	Habitat	Dorsal line	Lateral line	Matrix and miscellaneous	Ventrum	Upper/ lower labials	6th & 7th upper labials	No. of midbody dorsal scales	No. of neck scales	Inter- nasals	Chin shields
A1	diverse; ponds, marshes, ditches, streams, meadows, woods; near water & aquatic	broad, well-defined borders, bright yellow	well defined, on 2d & 3d scale rows; buff or yellow, on 2d & 3d scale rows	slaty or brownish; top of head black; red blotches on sides between stripes; eyes large	yellow-green or yellow-blue; bluish gray or dusky posteriorly, becoming pale on throat; no black	7/10; uppers buff or yellow; lowers cream colored	—	—	—	—	—
B	diverse; damp near water; terrestrial & aquatic	well-defined	present, on 2d & 3d scale rows	—	—	8/10	enlarged	19–23	—	broader than long, not pointed anteriorly	equal length
B1	seeks shelter on land	typically bright yellow, broad	olive yellow with red flecks	matrix reddish brown or olive brown; bright red or orange flecks on belly and sides, incl. lateral lines	pale green or blue; with red flecks (also on sides)	7 or 8/10	as with B?	usually 19, rarely 21	—	as with B?	as with B?
B2	near water or in wet meadows, but not aquatic	well-defined, yellow or orange- yellow, broad, sharp-edged	well-defined, yellow or buff	dorsum blackish, with light flecks; no red	pale and unmarked or with light spotting of dusky, rarely marked with black	as with B?, uppers pale gray or tan	as with B?	21, sometimes 19	—	as with B?	as with B?
B3	rocky streams	well-defined, yellow to brown, uneven	well-defined	dorsum dark	light gray, often w/ black or slate	as with B?; uppers olive gray	as with B?	23, sometimes 21		as with B?	as with B?

(*box continued on following page*)

Box Table 3.7B Continued

		Coloration				Scalation					
Taxon[a]	Habitat	Dorsal line	Lateral line	Matrix and miscellaneous	Ventrum	Upper/lower labials	6th & 7th upper labials	No. of midbody dorsal scales	No. of neck scales	Inter-nasals	Chin shields
C	rivers, streams, other aquatic	lacking or weak; when present, narrow & dull yellow near neck	when present, on 2d & 3d scale rows, faint, blending into ventrum color	blotched; pale gray matrix	white to gray, marked w/ dark	8/10; uppers pale gray to white w/ black on posterior margins	not enlarged	21		narrower than long & pointed anteriorly	about equal length
C1	permanent streams with rocky beds & swift clear water	narrow, dull	as with C?	obvious dark markings in checks, on pale gray matrix	light, unmarked	8/ usually 10	as with C?	as with C?		as with C?	as with C?
C2	ponds, small lakes, slow streams	yellow to orange	as with C?	dark olive to black matrix	blotches w/ golden or pale salmon	as with C?	as with C?	as with C?		as with C?	as with C?
D	terrestrial; meadows, clearings	well-defined, yellow, orange, red, blue, or white; or faint or lacking	pale yellow & distinct or faint; sides flecked with white	black, brown, greenish, or bluish matrix	yellowish, olive, or slate, often w/ red or black blotches	7/8 or 9; uppers same color as ventrum		17 (rarely 19)	17		

Sources: Stebbins 1966; Nussbaum et al. 1983

[a]Taxa:

A—*Thamnophis sirtalis,* common garter snake

A1—*T. s. fitchi,* valley garter snake

B—*T. elegans,* western terrestrial garter snake

B1—*T. e. terrestris,* coast garter snake

B2—*T. e. elegans,* mountain garter snake

B3—*T. e. biscutatus,* Klamath garter snake (reportedly small range in south-central Oregon; Nussbaum et al. 1983)

C—*T. couchii,* western aquatic garter snake

C1—*T. c. hydrophilus,* Oregon garter snake

C2—*T. c. aquaticus,* aquatic garter snake

D—*T. ordinoides,* northwestern garter snake

(*box continued on following page*)

Box Table 3.7C. An identification key for species and subspecies of garter snakes (*Thamnophis* spp.) occurring in the Klamath Mountains, northwestern California, based on field marks (habitat, coloration, and scalation)

Rule for knowledge base GARTER SNAKE

- Number of lower labial scales?
  - 8 . . . . . . *T. ordinoides*
  - 9 . . . . . . *T. ordinoides*
  - 10: Presence of lateral line?
    - Distinct: number of upper labials?
      - 7: Habitat?
        - Terrestrial . . . . . . *T. elegans terrestris*
        - Riparian . . . . . . *T. sirtalis fitchi*
        - Aquatic . . . . . . *T. sirtalis fitchi*
      - 8: Number of midbody dorsal scales?
        - 17 . . . . . . no identification
        - 19: Ventrum color?
          - Yellow-green . . . . . . *T. elegans*; *T. e. terrestris*
          - Yellow-blue . . . . . . *T. elegans*; *T. e. terrestris*
          - Pale & unmarked . . . . . . *T. elegans*
          - Pale & light spots . . . . . . *T. elegans*
          - Light gray . . . . . . *T. elegans*
          - White . . . . . . *T. elegans*
          - Blotches w/ gold, salmon: . . . . . . *T. elegans*
          - Olive . . . . . . *T. elegans*
          - Slate . . . . . . *T. elegans*
        - 20 . . . . . . *T. elegans*
        - 21: Habitat?
          - Terrestrial . . . . . . *T. elegans*; *T. e. elegans*
          - Riparian . . . . . . *T. elegans*
          - Aquatic: dorsal line color?
            - Bright yellow . . . . . . *T. elegans*
            - Dull yellow . . . . . . *T. elegans*; *T. e. biscutatus*
            - Brown . . . . . . *T. elegans*; *T. e. biscutatus*
            - Orange-yellow . . . . . . *T. elegans*
            - Orange . . . . . . *T. elegans*
            - Red . . . . . . *T. elegans*
            - Blue . . . . . . *T. elegans*
            - White . . . . . . *T. elegans*
        - 22 . . . . . . no identification
        - 23: Dorsal line color?
          - Bright yellow . . . . . . *T. elegans*
          - Dull yellow . . . . . . *T. elegans*; *T. e. biscutatus*

(*box continued on following page*)

Box Table 3.7C Continued

Rule for knowledge base GARTER SNAKE	
Brown	*T. elegans* *T. e. biscutatus*
Orange-yellow	*T. elegans*
Orange	*T. elegans*
Red	*T. elegans*
Blue	*T. elegans*
White	*T. elegans*
Weak: ventrum (belly) color?	
Yellow-green or pale green	no identification
Yellow-blue or pale blue	no identification
Pale and unmarked	no identification
Pale with light spotting or dusky	no identification
Light gray, with black or slate	*T. couchii* *T. c. hydrophilus*
White to gray, marked with dark	*T. couchii*
Blotches with golden or pale salmon	*T. c. hydrophilus*
Olive	no identification
Slate	no identification
Absent	*T. couchii*

Source: Compiled from field guides (see box table 3.7B)
Note: See box table 3.7B for species' common names. This key was built in an expert system by using an optimization algorithm (see text).

characteristics; it can deal with "fuzzy" or overlapping categorical descriptions, such as coloration patterns; it can be used to explain classification results easily and to explore alternative descriptions; and it produces a simple and easily understood classification rule tree. (4) Next, the expert system for each of the 26 field sample observations was run to determine the degree to which the key can identify each field specimen. This helped identify which field specimens represented outliers, potential hybrids, or undescribed forms, and the specimens' characteristics contributing to these findings.[1] Note that the operational expert identification system is more complex than the static identification key presented in box table 3.7C. The expert system allows for answers of uncertainty, from which it then prompts for additional characteristics not covered in the static identification key.

Results of running the expert identification system for each field specimen are presented in box table 3.7D. Overall, there were four main categories of results:

1. Identification to subspecies. Eight samples (31 percent) were fully identified to subspecies (samples 4, 10, 13, 15, 22, 23, 25, 26). The taxa that seemed to be identifiable for field specimens according to published characteristics were *T. elegans terrestris* and *T. sirtalis fitchi.*
2. Identification only to species. Eleven samples (42 percent) were identified to species only, where subspecies were also included in the key (samples 1, 2, 5, 7, 11, 12, 14, 16, 19, 21, 24). This occurred with *T. elegans,* which had three subspecies as potentially occurring in the study area and which were included in the expert system identification key.

1. A similar approach using a neural network model trained to recognize taxa on the basis of published characteristics resulted in similar findings but failed to provide explanations for mismatches of field specimens.

(*box continued on following page*)

Box 3.7 Continued

3. Conflicting identification to two different taxa. Three samples (12 percent) resulted in mixed identification (samples 3, 6, 20). These specimens keyed out, respectively, to the following pairs of taxa: *T. sirtalis fitchi* and *T. couchii; T. elegans terrestris* and *T. elegans;* and *T. couchii hydrophilus* and *T. elegans.* Thus, two of these conflicts represented potentially mixed characteristics between species, and one was an uncertainty of subspecies identification (based on coloration). Sample 3 may have resulted in confusing identification because of incomplete information on coloration of this specimen. Sample 20 is less easily explained and may represent a hybrid or, as with sample 6, a subspecies or morph with characteristics not adequately described in published guides.
4. No identification. Four samples (15 percent) could not be identified to any known taxa (samples 8, 9, 17, 18). For these specimens, the key factors resulting in no identification seem to include degree of prominence of the lateral lines[2] or ventrum color (samples 8, 9) or the full set of observed coloration and scalation characteristics (samples 17, 18). These specimens may represent real cases of interspecific hybrids or undescribed subspecific forms or morphs.

How should habitats be identified for the taxa represented by the field specimens? Result 1 above poses no ambiguities; *T. elegans terrestris* is mostly terrestrial, and *T. sirtalis fitchi* occupies a diversity of riparian, wetland, and aquatic habitats and terrestrial habitats near water. Result 2, confusion of subspecies of *T. elegans,* may pose a problem if these subspecies are found to have different conservation needs, because habitat requirements of these subspecies do vary considerably. Results 3 and 4, confusion of identification to species or no identification, clearly pose difficulties in identifying habitats for management focus.

Obviously, more sampling and quantitative studies are needed. The field sample analyzed here probably does not represent the full set of taxa and individual variation present in this geographic area, but nonetheless helps pose tantilizing questions of confusing taxonomy and potential habitat selection of *Thamnophis* taxa. Additional meristics and electrophoretic examinations of other field specimens need to be collected, as well as observations of breeding behaviors and other life history attributes.

In conclusion, messy taxonomic groups pose potential problems for habitat management. The abundance status of the species and many subspecies of western garter snakes is largely unknown. Are current designations of species and subspecies of western garter snakes appropriate? Are some currently described or unknown subspecies rare in this geographic area? Should they be listed? Are there undescribed morphs not given subspecific status that need to be recognized for conservation of genetic diversity (see box 3.4)? Given uncertainties in taxonomy, what habitats should be provided, and where should they be provided? Might swamping or homogenization of species gene pools caused by interspecific hybridization affect their status, and, if known, should it be controlled? In this case, only further field studies can help resolve these questions.

2. Specimen 8 can key out to *T. elegans* if the lateral line is assumed to be weak, but this was not the case.

(box continued on following page)

method called DNA-DNA hybridization.[3] This estimates genetic similarity between complete genomes by measuring the amount of heat required to melt the hydrogen bonds between the base pairs that form the links between the two strands of the double helix of duplex DNA. The comparison may be between the two DNA strands of an individual or between individuals representing different levels of genetic and taxonomic divergence (Sibley and Monroe 1991). Results suggested by Sibley and Monroe include some unexpected genetic diversity within some bird groups and designation of some major subdivisions of class Aves (into

3. The term *hybridization* here does not refer to breeding between species, although the technique certainly can help unmask such situations.

Box Table 3.7D. Identification of specimens of garter snakes (*Thamnophis* spp.) to species or subspecies, based on applying meristics of observed specimens (box table 3.7A) to a rule-based expert system (box table 3.7C) that was based on identifying characteristics of taxa from leading field guides (box table 3.7B)

Specimen number[a]	Identification by use of expert system rule[b]	Comments
1	*T. elegans*	expert system rule cannot distinguish subspecies of this specimen
2	*T. elegans*	expert system rule cannot distinguish subspecies of this specimen
3	*T. sirtalis fitchi*, or *T. couchii*	incomplete information on coloration of this specimen
4	*T. elegans terrestris*	identifiable
5	*T. elegans*	expert system rule cannot distinguish subspecies of this specimen
6	*T. elegans terrestris* or *T. elegans*	coloration characteristics are a poor fit to described marks
7	*T. elegans*	expert system rule cannot distinguish subspecies of this specimen
8	no identification or *T. elegans*	poor fit to rule; keys out to *T. elegans* if dorsal line information is ignored
9	no identification or *T. elegans terrestris*	poor fit to rule; keys out to *T. elegans terrestris* if dorsal and lateral line information is ignored
10	*T. elegans terrestris*	identifiable
11	*T. elegans*	expert system rule cannot distinguish subspecies of this specimen
12	*T. elegans*	expert system rule cannot distinguish subspecies of this specimen
13	*T. sirtalis fitchi*	identifiable
14	*T. elegans*	expert system rule cannot distinguish subspecies of this specimen
15	*T. sirtalis fitchi*	identifiable
16	*T. elegans*	expert system rule cannot distinguish subspecies of this specimen
17	no identification	cannot identify this specimen to any known taxa
18	no identification	cannot identify this specimen to any known taxa
19	*T. elegans*	expert system rule cannot distinguish subspecies of this specimen
20	*T. sirtalis fitchi*, or *T. elegans*	identification depends on distinctiveness of lateral line; expert system rule cannot distinguish *T. elegans* subspecies of this specimen
21	*T. elegans*	expert system rule cannot distinguish subspecies of this specimen
22	*T. sirtalis fitchi*	identifiable
23	*T. sirtalis fitchi*	identifiable
24	*T. elegans*	expert system rule cannot distinguish subspecies of this specimen
25	*T. elegans terrestris*	identifiable
26	*T. elegans terrestris*	identifiable

[a]Specimen number corresponds to box table 3.7A.
[b]Identification is based on applying the expert system rule in box table 3.7C to the specimen data in box table 3.7A.

newly termed "parvclasses," namely: Ratitae, Galloanserae, Turnicae, Coliae, and Passerae). Most controversial is the change in order Ciconiiformes, which, according to the new taxonomy, now includes several "traditional" orders of Charadriiformes (shorebirds), Falconiformes (hawks, falcons), Podicipediformes (grebes), Pelecaniformes (pelicans, cormorants, etc.), Ciconiiformes (storks, herons, etc.), Sphenisciformes (penguins), Gaviiformes (loons), and Procellariiformes (petrels, albatrosses, etc.). Whether the Sibley and Monroe systematics of birds becomes the new standard and how it might influence identification of at-risk bird taxa and habitat conservation activities remain to be seen.[4]

Still other taxonomic designations have focused habitat conservation attention on taxonomic distinctions finer than subspecies. Plant conservationists, as well as foresters transplanting commercially useful trees, have long focused on *ecotypes*—populations in gene pools adapted to local environmental (usually climatic) conditions. The concept has recently been used in a study of two ecotypes, mountain and northern, of woodland caribou (*Rangifer tarandus caribou*) recently translocated to augment a population of woodland caribou in northern Idaho (Warren et al. 1996). After translocation, the two ecotypes of this one subspecies responded with different patterns of habitat selection, dispersal, and mortality. This prompted Warren et al. (1996) to conclude that a larger number of individuals may need to be translocated to sustain a new or recipient population when donor subpopulations (ecotypes) must be used that do not closely resemble the habitat-selection characteristics of extant or extinct resident subpopulations. Another conclusion we can draw is that donor populations should be taken, if possible, from habitats and environments most closely matching those of the recipient population. This was an unexpected complication of an augmentation strategy for an endangered vertebrate, and one not well studied in most other programs of augmentation, translocation, and introduction.

Metapopulations in Patchy Habitats and Implications for Habitat Management

The distribution, abundance, and dynamics of a population in a landscape are influenced by species attributes, habitat attributes, and other factors. Species attributes include movement and dispersal patterns, habitat specialization, demography including density-dependence relations, and genetics of the populations. Habitat attributes include quality, size, spacing, connectivity, and fragmentation of habitat patches (also see chapter 9) and the resulting availability and distribution of food, water, and cover. Other factors include a host of environmental conditions such as weather, hunting pressure, and influences from other species.

Much work in simulation modeling has focused on the dynamics of habitat occupancy by organisms in metapopulations (e.g., Gilpin and Hanski 1991; Possingham et al. 1992) through the use of geographically referenced or spatially explicit population models (see chapter 10). In

4. Another classification system worthy of historical mention—although it has pertained mostly to plants—is that of the genecological classification system of Turesson (1922). This system focused on genetic similarity between species in a classification hierarchy (listed here from lowest to highest levels) of ecophenes, ecotypes, ecospecies, and coenospecies (also see Daubenmire 1974). The approach was intended to augment the traditional Linnean taxonomic system, although the correspondence between systems may prove complex. Perhaps there are some seeds of utility of the genecological system for taxonomically difficult wildlife species complexes such as the garter snakes.

general, occupancy of habitats by organisms is usually more variable in suboptimal environments, peripheral locations, smaller habitat patches, and patches with greater edge effects adversely affecting required resources, although density-dependent relations may mediate these trends. Also, the effects of competitors, predators, disease, and other nonhabitat factors must not be discounted, although few spatially explicit population models currently include them.

Many models of metapopulations in patchy environments suggest that not all suitable habitats will be occupied at any one time. This in turn suggests that habitats be conserved even if they seem unoccupied in any one year and, thus, that monitoring wildlife use of habitats in conditions listed above should proceed for more than one year or season, perhaps for several. Concluding absence of a species where it is actually present is a Type II error that can be corrected with adequate sample size and monitoring duration to increase the power of the statistical evaluation (see fig. 3.8). At the same time, the staunch preservation of every instance of a potentially suitable patch may also be unwise if it is clear that the patch does not serve a function in maintaining the population or in maintaining an interesting genetic variant of the species of interest. The appropriate approach, discussed further below, depends in part on the size and fragility of the population and its habitats and on conservation objectives.

In the sections that follow, we review concepts and examples of dynamics of populations and habitats. This sets the stage for discussing habitat modeling later in chapter 10.

Dynamics of Populations and Habitats

Discerning Population Requirements from Patterns of Habitat Selection and Specialization

The distribution and density of a population is ultimately determined by the collective responses of its individual organisms. It should be remembered that descriptions of population- or species-habitat relations are really just a statistical—even just a categorical—summary of the behavior of individuals. One can describe a central tendency of a statistical population, such as the typical or average habitat conditions in which a particular species is found. But descriptions of variation around that central tendency provide insights into much more interesting biological phenomena. In fact, it is often the spread, rather than the central tendency, that tells us the most about

		PREDICTION	
		presence:	absence:
OBSERVATION	presence:	(OK)	Type II (power)
	absence:	Type I (confidence)	(OK)

Figure 3.8. Kinds of error in predictions of species presence in habitats

how populations and species occur in and select their environments.

The degree of variation in selection of habitats between individuals of a population describes the degree of specialization or specificity of habitat use. This is often termed habitat or niche breadth (or habitat use breadth), and many statistical measures have been devised to measure it (e.g., Colwell and Futuyma 1971; Feinsinger and Spears 1981; Petraitis 1981). Habitat breadth is really a collective property of a population resulting from either variation in habitat selection between individuals of the population at a particular time or variation over time of individuals' habitat selection patterns. Often in studies lacking individual demarcation, these two components are not differentiated. Thus, the resulting estimates of habitat breadth are unpartitioned combinations of individual and temporal variations in habitat selection. If the studies are used to identify habitats for conservation management, then it is not critical that the two components be differentiated, because conservation activities might as well target the entire range of variation anyhow, including individual and temporal variations. But the contributions to and causes of the observed overall habitat breadth will not be known with certainty.

Habitat specialists—those with narrow habitat breadth, or that are *stenotopic*, as compared with those with wide habitat breadth, or that are *eurytopic*—may be more at risk of local declines or extirpation should the environment change suddenly. Many descriptions of species at risk from regional climate changes use habitat breadth as the measure of risk (e.g., Marcot et al., in press). For example, in the U.S. inland West, the white-tailed ptarmigan (*Lagopus leucurus*) is found principally in alpine tundra communities, and the American pipit (*Anthus rubescens*) occurs regularly in alpine tundra and only one other vegetation community, making these two stenotopic species particularly locally vulnerable to upper-elevation warming and loss of alpine tundra habitats (Marcot et al., in press). Eurytopic species found in alpine tundra may use other communities as well and, thus, are not as subject to local threat.

It is generally not advisable to combine observations of habitat use patterns of individuals from different ecotypes, populations, geographic areas, or ecoregions (Ruggiero et al. 1988). Combining individuals in this way could mask the gamma diversity of habitat selection patterns that vary substantially across geographic areas. Providing habitat for one ecotype or for organisms in one portion of their range may be senseless for another ecotype or range portion if the organisms there select for different environmental cues (see the example of the woodland caribou cited above).

Distribution Patterns of Populations

The overall distribution and local abundance of many wildlife species are related in time and space. Many species have a "bull's-eye" pattern of distribution, with their greatest areas of abundance toward the middle of their overall ranges and peripheral portions of the range in marginal conditions, although for many reasons the areas of highest density may not occur exactly in the center of the range. Such distributions typically reflect several aspects of biophysical conditions and species ecology: (1) the geographic range of suitable biophysical conditions, (2) the range of tolerance of biophysical characteristics by the species, and (3) the occurrence of marginally suitable conditions at the peripheries of the geographic range that often act as sink habitat to hold nonreproductive individuals or individuals that have spread from higher-abundance areas during good reproductive years. (Sinks can also be habitats

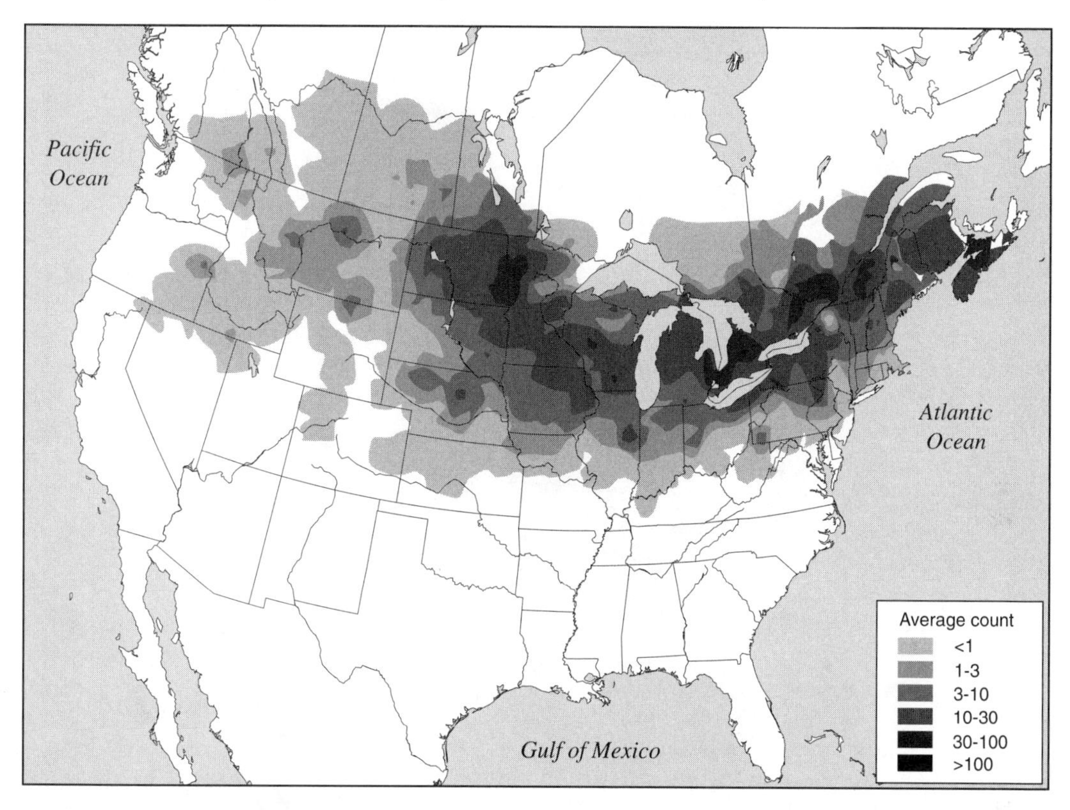

Figure 3.9. Current distribution of bobolinks (*Dolichonyx oryzivorus*) in North America based on bird counts per Breeding Bird Survey route. Note that highest densities occur more or less centrally within the overall range. (From Sauer et al. 1996)

where mortality and emigration exceed natality and immigration.) A few examples of species with a bull's-eye distribution are the bobolink (*Dolichonyx oryzivorus*) in North America (fig. 3.9) and the sable (*Martes zibellina*) and Siberian weasel (*M. sibirica*) in southern areas of the Russian Far East (figs. 3.10, 3.11).

In studying censuses of birds from the North American Breeding Bird Survey, Brown et al. (1995) noted that such highly clumped abundance distributions resemble distributions such as the negative binomial canonic lognormal, used to characterize the abundance distributions of species within local ecological communities. They hypothesized that the spatial variation in abundance largely reflects how well local sites meet resource requirements of a species. They concluded that patterns of spatial and temporal variation in abundance should be factors considered when designing nature reserves and conserving biological diversity.

The sable and Siberian weasel are particularly interesting in that they are closely related species that have evolved very different habitat associations, possibly to avoid resource competition. The sable is larger-bodied than the Siberian weasel and occurs in the mixed conifer-hardwood taiga and montane forests of the

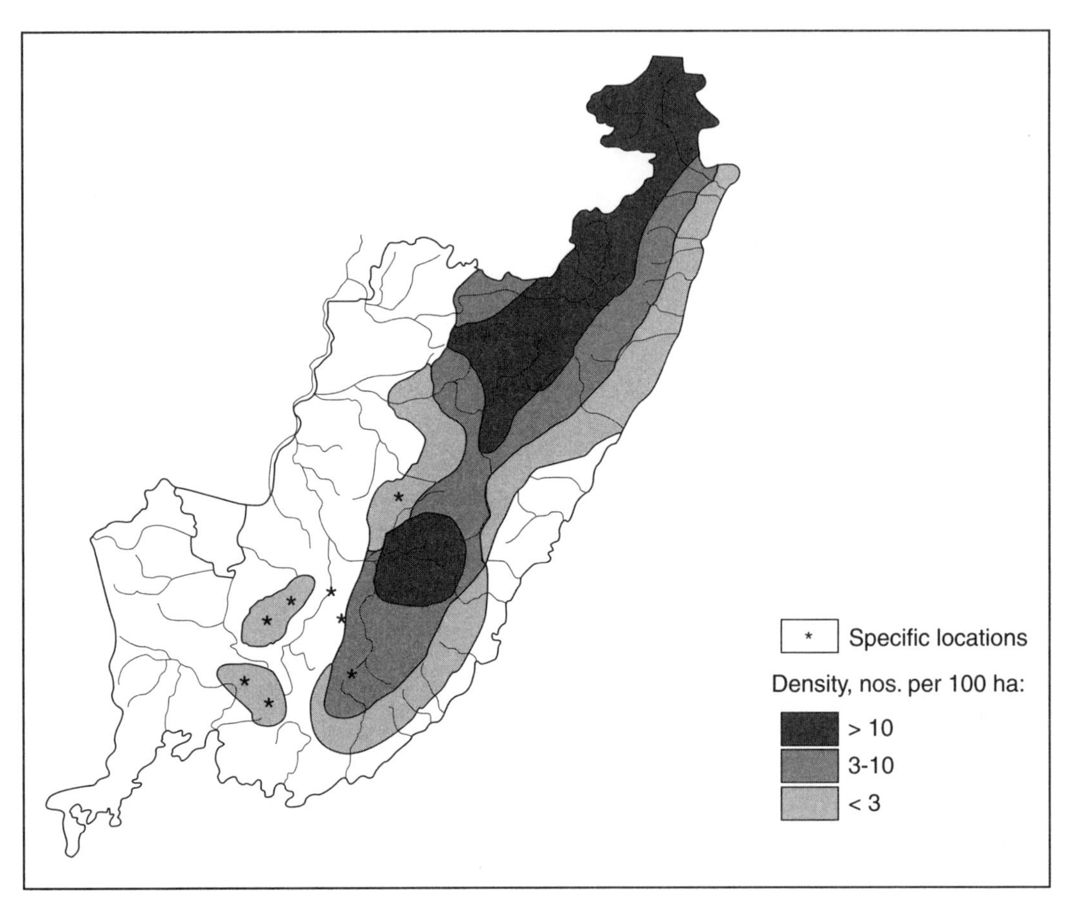

Figure 3.10. Distributional range of the sable (*Mustela zibellina*) in Primorski Krai, Russian Far East. (From the Pacific Institute of Geography, Vladivostok, Russia)

Sichote-Alin' Mountains in the southern area of the Russian Far East. The more numerous Siberian weasel finds optimal conditions on lower-elevation slopes and in bottomlands. Although the overall distributions of the two species overlap, there is little to no overlap of areas of the highest density of each species.

The bull's-eye distribution pattern is not always the rule, however. The abundance distribution of some species is truncated where biophysical conditions come to an abrupt halt, such as along mountain ranges, large rivers, or other major dispersal barriers, or at the edges of continents. An example is the wrentit (*Chamaea fasciata*), a species of chaparral, brush, and thickets, whose distribution truncates at its greatest density along the Pacific coast of the United States (fig. 3.12). Thus, one cannot always assume that peripheral distributions of a species mean marginal

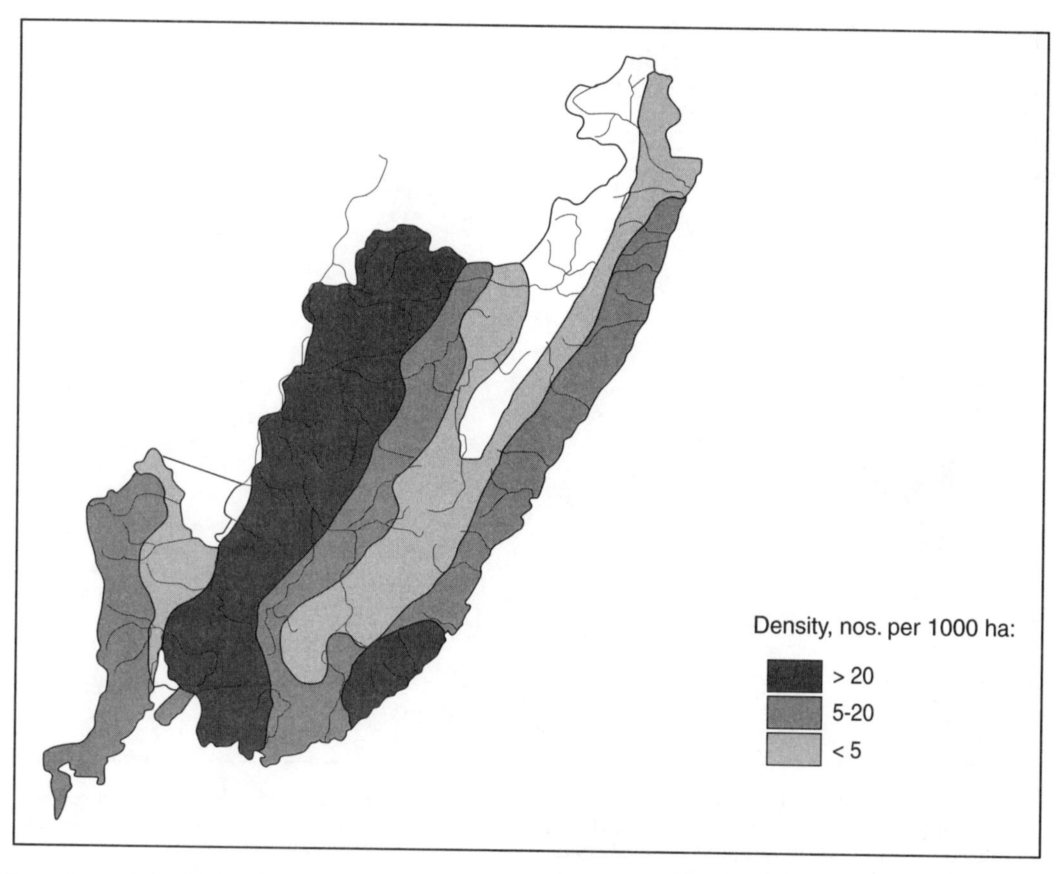

Figure 3.11. Distributional range of the Siberian weasel (*Mustela sibirica*) in Primorski Krai, Russian Far East. Compare with distribution of sable in fig 3.10. (From the Pacific Institute of Geography, Vladivostok, Russia)

environmental conditions and lowest population densities.

Identifying areas of high density of organisms or areas of high environmental suitability (remembering that these are not necessarily always synonymous) can be important for management purposes. Wolf et al. (1996) reviewed the success of 421 avian and mammalian translocation programs in North America, Australia, and New Zealand and found that release of organisms into the core of their historical range and into habitat of high quality were two key factors contributing to success. Other factors included the use of native game species, a greater number of released animals, and an omnivorous diet. Nor should it be forgotten that many factors other than those listed here also contribute to population viability.

Populations of most organisms ebb and flow over time. Changes can be simple trends or

Figure 3.12. Current distribution of wrentits (*Chamaea fasciata*) in North America based on bird counts per Breeding Bird Survey route. Note that highest densities occur at the edge of the overall range, truncated by the continent's margin. (From Sauer et al. 1996)

complex variations. Causes can be density-independent, such as from population-wide effects of weather, or density-dependent, such as from density-induced variations in natality. Density-dependent relations are often difficult to study but lend themselves well to modeling (Boucher 1985; Ginzburg et al. 1990). To detect density dependence, Pollard et al. (1987) devised a method using the correlation coefficient between the observed population changes and population size, with a randomization procedure to define a rejection region for the hypothesis of density independence. Numerous other approaches to detecting and modeling density-dependent population behavior have been proposed.

Dynamics of Populations Affected by Movements of Organisms

Various kinds of movement of wildlife through their habitats and environments impart particular dynamics to their populations. Movements particularly important for habitat management include the following:

- dispersal—one-way movement, typically of young away from natal areas
- migration—seasonal, cyclic movement typically across latitudes or elevations to track resources or to escape harsh conditions changing by season of year
- home range movement—movement throughout a more or less definable and known space over the course of a day to weeks or months, to locate resources
- eruption—irregular movement into geographic areas normally not occupied, as a response to severe weather or sudden availability of high-quality resources

Dispersal. Dispersal, particularly of young away from natal areas, is an innate and genetically programed adaptation to avoid inbreeding or resource competition with confamilials and to locate mates. Dispersal patterns, especially timing and distances, may be density-dependent (but see below). There is often a wide variation in dispersal distances and patterns traveled among individuals even of the same cohort or family year; this may have adaptive significance in separating young and ensuring that at least some individuals reach suitable environments and locate mates. Dispersal patterns also can vary by year, gender (Marzluff and Balda 1989), time (Morton 1992), and species interactions including competition (Rodgers and Klenner 1990). A few examples will demonstrate the variations in dispersal patterns among species.

Allen and Sargeant (1993) reported that, in North Dakota, mean recovery distances of red foxes (*Vulpes vulpes*) tagged as pups increased with age and were greater for males than for females, but dispersal distance was not related to population density. This was interpreted as meaning that populations can be augmented, but also that disease can be transmitted, across long distances regardless of fox population density. Neither males nor females showed any uniformity in dispersal direction, although littermates did. Beier's (1995) study of the dispersal of juvenile cougars (*Felis concolor*) in fragmented habitat in southern California revealed that the animals would use habitat corridors located along natural travel routes with adequate woody cover and low human density, including a roadway underpass. As with the red fox study, littermates dispersed in the same direction.

In a study of fishers (*Martes pennanti*) in Maine, Arthur et al. (1993) reported that a greater proportion of young males than young females dispersed early in the year, but a greater proportion of females than males dispersed by year's end. The mean dispersal distance did not differ between sexes, although was only 11 km. Arthur et al. conjectured that the relatively short dispersal distances may limit fishers in recolonizing areas in which they have been extirpated and may reduce interchange between isolated populations.

Among raptors, dispersal distances of fledgling eastern screech-owls (*Otus asio*) were not significantly correlated with either the dispersal date or the number of days that juveniles remained in natal territories, and dispersal direction was random (Belthoff and Ritchison 1989). Other studies or reviews of dispersal ecology are available (Gliwicz 1988; Greenwood and Harvey 1982; Murray 1988; Taylor 1990).

Buechner (1987) modeled vertebrate dispersal patterns and, on the basis of comparisons of parameter values and deviations from the geometric pattern between groups within taxa, concluded that males and females may follow different patterns of dispersal. She also indicated that mammals and birds show consistent differences in dispersal distributions.

These results are relevant to discussions of natal philopatry, inbreeding avoidance, and proximate mechanisms of dispersal (e.g., Koenig and Hooge 1992). They can also influence spacing selection of habitat reserves.

One aspect of dispersal important for habitat management is recognition of dispersal barriers or strong filters. Not many studies have empirically identified species-specific dispersal barriers or filters, although they could be critical in determining the occupancy of distributed habitat patches and persistence of low-density and small populations in the wild. For example, in one study, Allen and Sargeant (1993) found that a four-lane interstate highway altered dispersal directions of red foxes and apparently caused distant travel from natal areas.

Similarly, few studies have demonstrated use of habitat corridors, which has served as one of the backbones of some habitat conservation strategies (Harris 1988). Haas (1995) found that movements of three migratory bird species among riparian woodlands and shelterbelt woodlots in North Dakota, although rare, occurred more frequently between sites connected by wooded corridors than between unconnected sites. Haas concluded that knowledge of patterns of fledgling, natal, and breeding dispersal of birds in patchy environments would substantially aid decisions about reserve design and protection or construction of habitat corridors. Hill (1995) found that species of ants, butterflies, and dung beetles—bioindicator groups—responded differentially to linear vegetation corridors in the lowland rain forest of northeastern Australia. Hill concluded that habitat corridors can aid dispersal of only some species and that rain forest–interior obligates would not use the linear corridor strips.

Dispersal distances are important parameters for use in some models of genetic diversity and deme size (box 3.2). Commonly, different age or sex groups have different patterns of habitat selection during different parts of their life cycles or seasons, including during dispersal. Understanding age- and sex-specific dispersal patterns and distances helps determine the set of dispersal habitats collectively required by all members of a population. Studies of habitat selection only during breeding tend to miss much of the range of habitats used and required by wildlife. In this spirit, dispersal distances also can be used to help establish spacing patterns of habitat conservation areas (e.g., Lamberson et al. 1994; see also figure 3.13).

Migration. Periodic (typically seasonal and annual) occurrence of organisms among geographic locations and habitats is another type of movement by wildlife. Migration of herding ungulates, large mammalian carnivores, some raptors, many Neotropical migratory songbirds, some amphibians, and other taxa can take place over short or long distances and latitudes. Long-distance migration may have arisen in long-term response to glacial fluxes and associated changes in climate and resource conditions (Chaney 1947), to avoid local competition for scarce resources, or because of other factors, but responses are not well understood for most species. Most references concur that long-distance migration has evolved in response to enormous seasonal differences in availability of food and other ("ultimate") factors, and that competition, predator avoidance, and glaciation are associated but secondary ("proximate") factors.

The migration status of birds bears closer scrutiny. One category of migration is that of latitudinal migration, during which some shorebirds, raptors, and songbirds incur the wrath of inclement weather, habitat alteration, pesticides, direct persecution, and other as-

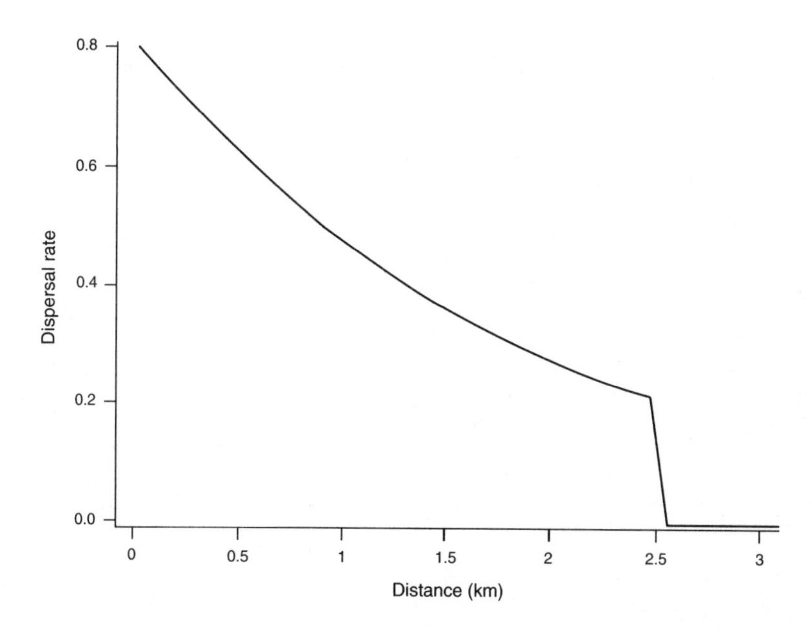

Figure 3.13. Rates (proportion) of successful dispersal by nonbreeding female helmeted honeyeaters as a function of dispersal distance between habitat patches. Such data are used in demographic modeling of metapopulation viability. (From Akçakaya et al. 1995:173)

saults over vast distances to reach wintering or breeding grounds (Terborgh 1989). As early as 1970, Aldrich and Robbins (1970) warned of potential declines in long-distance Neotropical migratory songbirds. Their warning has been borne out in more recent studies (e.g., Askins et al. 1990).

Migration status of terrestrial wildlife, particularly birds, can be defined hierarchically depending on type of movement and distance traveled, as presented in table 3.4. Four general classes of movement are elevational migrants, latitudinal migrants, permanent residents, and nomads (eruptive or irregular movement into new areas). Elevational migrants include species that move downslope during winter so that they move out of or into a particular area.

An example of *elevational migration* is the seasonal movement of the snowcap (*Microchera albocoronata*) and the three-wattled bellbird (*Procnais tricarunculata*) along the Zona Protectora La Selva corridor that links two reserves (the eastern lowland Reserva Biologica La Selva, with the central highland Reserva Forestal Cordillera Volcanica Central and Parque Nacional Braulio Carrillo) along the central cordillera of Costa Rica (Stiles and Clark 1989). The snowcap breeds in the highlands (at 900 m) during December–June and then moves into the lowlands (at 100 m) during July–October (thus, it would be coded as an E1 or E2 species, table 3.4, depending on the location of interest). The bellbird follows similar patterns but breeds higher up (≥2000 m) during March–

Table 3.4. A classification of migration status of wildlife species

Migration class	Code	Seasons present[a] Sp	B	Su	F	W
Elevational migrants						
Downslope movement during winter						
Breed at studied elevation, move lower in winter	E1		X	X	X	X
Breed at higher elevation, move into study area during winter	E2	X			X	X
Upslope movement during postbreeding season						
Breed in bottomlands, move upslope during postbreeding	E3			X	X	
Latitudinal migrants						
Displacement migrants	L1	X	X	X	X	X
Neotropical migrants and low-latitude Nearctic migrants	L2		X	X	X	
High-latitude Nearctic migrants	L3				X	X
Permanent residents	PR	X	X	X	X	X
Nomads	NOM			any		

Source: Marcot 1985

[a]Sp = spring, B = breeding, Su = summer (postbreeding dispersal), F = fall, W = winter. Seasons shown here include a breeding season distinguished from spring (postwinter movement and settling period) and summer (postbreeding dispersal period). Seasons pertinent to each migration class may differ depending on geographic location.

August and moves to the coastal lowlands (<100 m) during August–January. In northern California, some elevational migrants move upslope after breeding in valleys or lowlands (code E3, table 3.4); these include western meadowlarks (*Sturnella neglecta*) and Brewer's blackbirds (*Euphagus cyanocephalus*).

In mountains of western North America, some individual *latitudinal migrants* may displace other individuals (displacement latitudinal migrants) as they move north or south, such as the dark-eyed junco (*Junco hyemalis*) (code L1); others are Neotropical migrants that travel between mid- and low latitudes annually, such as the flammulated owl (*Otus flammeolus*) (code L2); and still others move between high boreal and midlatitudes, such as the yellow-rumped warbler (*Dendroica coronata*) (code L3).

Permanent residents consist of individuals that occupy a location year-round (code PR), such as northern spotted owls, although some spotted owls in northwest Washington and the California spotted owl (*Strix occidentalis occidentalis*) on western slopes of the Sierra Nevada may become elevational migrants, moving to lower slopes during winter. Note that both permanent residents and displacement migrants may occur in all seasons in a location, but the sources of the individuals differ. Finally, some species are *nomads,* who move irregularly and eruptively in any season (code NOM), such as red crossbills (*Loxia curvirostra* species complex).

The use of such a movement classification system can aid habitat management by determining: (1) which species are likely to occur in an area in a given season, and thus the resources and habitats required during that season; (2) the number of species expected in an area over seasons, and thus the collective resources and habitats required; and (3) the need for considering habitat conservation in other regions beyond the area of immediate interest. It may also aid in identifying habitat corridors used during movement, and thus habitats and geographic areas needing potential conservation focus. This was the case with the designation of Costa Rica's Zona Protectora La Selva corridor cited above.

The literature on migration biology and physiology is rich and varied. The practitioner is particularly directed to sources reviewing management implications (Finch 1991; Finch and Stangel 1992; Martin and Finch 1995).

Home range movement. Although home ranges are not a parameter of populations per se, how organisms establish and use them is nonetheless important for managing habitat for populations. In recent years, many new algorithms have been devised for estimating home range area (e.g., Geissler and Fuller 1985; Samuel and Garton 1985; Spenser et al. 1990; Larkin and Halkin 1994; Worton 1995). At the same time, controversy has ensued over appropriate metrics. Different algorithms estimate different aspects of spatial use. Examples are: the overall area circumscribed by movement of an organism in a specified time period (measured or estimated by use of minimum or maximum convex polygon algorithms); the geographic areas of concentrated occurrence of an organism (harmonic mean or Ford-Krumme algorithms); and the general area most used by an organism (95 percent confidence ellipse).

The controversies over which algorithm to use have focused not so much on technical discourses as on implications for identifying the total area of economically valuable and scarce habitats to protect for viable populations of uncommon species. Examples include preservation or restoration of old longleaf pine (*Pinus palaustris*) forests in the southeastern United States for red-cockaded woodpeckers and old-growth conifer forests in the Pacific Northwest for northern spotted owls, pileated woodpeckers (*Dryocopus pileatus*), and mustelids. As it turns out, not surprisingly, different home range algorithms measure different aspects of home range and organisms' use of space and resources, so their results are not identical. It is incumbent upon the researcher to understand better what aspect of space and resource usage would best pertain to a particular management question, and then use the algorithm that best fits the organism's biology and the statistical data; it is incumbent upon the habitat manager to understand the algorithms better and better focus questions of habitat use and allocation.

Further, the researcher should not focus on just measuring home range sizes and amounts of habitat within the home range without equitable treatment of other factors that could influence home range size and habitat selection, such as food supplies, density of conspecifics, effect of body size, competitors, predators, and landforms. In a sense, habitat selection is an optimization process that involves all these (and other) factors; and home range is an expression of the process (L. Irwin, pers. comm.). This has important implications for management. Larger-than-average home ranges may or may not mean that insufficient habitat quality or amount has been provided.

Eruptions. Some species undergo periodic eruptions[5] in distribution and dispersal. In North America during periods of extreme

5. Formally, the term *eruption* refers to populations that suddenly exceed their normal boundaries or densities beyond a given area, whereas *irruption* refers to invasion into a particular area from outside. Think of eruptions being caused by emigration and irruptions caused by immigration. Eruptions can be caused by population growth (numeric response; see text below) or by movement of individuals (functional response).

northern winter conditions, some species typical of higher-latitude boreal forests, including the northern hawk owl (*Surnia ulula*), the snowy owl (*Nyctea scandiaca*), the gyrfalcon (*Falco rusticolus*), and the white-winged crossbill (*Loxia leucoptera*), can occur intermittently farther south in southern Canada and the northern United States. Such eruptive movements push the edges of population ranges into locations not normally occupied.

More locally, seasonal eruptions or wandering movements occur with white-headed woodpeckers (*Picoides albolarvatus*), red crossbills, redpolls (*Carduelis hornemanni* and *C. flammea*), evening grosbeaks (*Coccothraustes vespertinus*), and other species. Irregular movements can help organisms fill vacant environments, thereby colonizing new habitats as founder populations (see chapter 9). For instance, Scott (1994) reported that unusual conditions of prevailing winds aided an irruptive dispersal of black-shouldered kites (*Elanus caeruleus*) over 80 km of open ocean to San Clemente Island in southern California. Such movements may be important for determining potential colonization rates of peripheral habitats and the conservation value of habitat in the periphery of some species ranges.

Stochastic Environments and Desired Conditions

Stochastic environments contain biotic and abiotic conditions that vary irregularly over time, usually substantially and unpredictably. All environments change, but not all populations and ecological communities are equally resilient or pliable to change. It is a matter of the intensity and duration of the change, coupled with the kinds of populations and ecological communities present, that determines whether a change produces long-lasting or permanent effects or whether populations or communities will return to a prior state. The degree to which populations are resilient to change or are pushed to new distributions or densities, or even to extirpation, is largely determined by the initial abundance and the vagility and habitat-specificity of the organisms, although certainly other factors play roles as well.

Populations can respond to a perturbation either functionally or numerically. The *functional response* of a population refers to changes in behavior of organisms, such as selecting different prey or using different substrates for resting or reproduction. Functional responses can also entail a temporary and localized increase in numbers resulting from immigration, or a decrease from emigration. The *numeric response* of a population refers to absolute changes in abundance of individuals through changes in recruitment.

A disturbance of a habitat might elicit one or both kinds of response. For example, a crown fire in subalpine forests of the northern Rocky Mountains might increase suitability of habitat for black-backed woodpeckers (*Picoides arcticus*) in several ways. The fire may induce a temporary influx of foraging woodpeckers into the forest (a shorter-term, functional response) as well as provide snag substrates for increased nesting density of woodpeckers in the area (a longer-term, numeric response).

It is important to distinguish between such responses to understand if management activities—especially intentional habitat restoration or enhancement activities—are truly serving to increase absolute population size or are simply redistributing organisms. In some cases, simple redistribution of organisms may be the goal, such as warding off foraging waterfowl from grain fields and agricultural lands and channeling them into nearby wetland refuges. In other cases, redistributions and local increases, such as displacement from disturbed or fragmented habitats, may belie an overall population decline.

We discuss disturbance dynamics of habitats further in chapter 9.

Evolutionary Adaptations and Selection Mechanisms

The evolution of individual behaviors of adaptive significance directly influences the overall distribution and abundance of populations. The degree of species specialization in habitat selection has importance for prioritizing habitat conservation activities and identifying at-risk species. Of prime interest to habitat ecologists and managers is evolution of habitat selection behaviors, including selection of prey and breeding, resting, and foraging habitats, for maximizing energy efficiency. For example, spotted owls in California may select roost sites as an adaptation to reducing heat stress (Barrows 1981).

Reproductive and hibernation ecology also has important bearing on habitat selection and conservation. For example, the stenotopic endangered Indiana bat (*Myotis sodalis*) selects natural, undisturbed caves as wintering hibernacula. Modification of cave entrances by humans has degraded the bat's winter habitat and altered its hibernation physiology (Richter et al. 1993). Caves with artificially modified (widened) entrances were found by Richter and colleagues to have a mean winter temperature 5.0°C higher than that of unmodified caves. Bats in modified caves entered hibernation at a 5 percent greater body mass and lost 42 percent more body mass, and small bats (<5.4 g) did not survive the winter. Remodification of the entrance to a more natural condition increased the colony from 2,000 to 13,000 bats over a decade. On the other hand, some congeners, such as the little brown myotis (*Myotis lucifugus*), have evolved to be more eurytopic. The little brown myotis uses a variety of substrates and locations for roosting, including buildings, snags, exfoliating bark of live trees, caves, rock outcrops and crevices, and mines (Marcot 1996), and thus probably is not in particular danger from cave modification.

Benkman (1993) studied the significance of crossbill (*Loxia*) adaptations to foraging on conifer seed cones. He found that four taxa or "types" (species or subspecies) of red crossbills in the Pacific Northwest have diversified morphologically in bill characters in response to alternative adaptive peaks in their foods of four conifer trees. Each adaptive peak corresponded to one conifer tree species whose seeds are produced regularly from year to year, held in cones through late winter when seed is most limited, and protected from depletion by potential noncrossbill competitors. Benkman concluded that (1) a critical feature of the ecology and evolution of crossbills is the reliability of seeds on key conifers during periods of food scarcity; (2) even in populations in highly variable environments, morphological traits have evolved to optimize food variability and availability; and (3) disruptive selection against intermediate phenotypes is likely serving to maintain, if not reinforce, the distinctiveness of types. Also, (4) the diversity of cone structure and seed size among key conifers is ultimately responsible for the diversification of crossbills. Further studies would determine the effect, if any, that selective foraging by crossbills, or by other spermivores that may take seeds of specific size or viability, would have on the evolution of conifer seed characteristics and conifer tree genetics and morphology.

Herbivory by ungulates and other species can have a major direct impact on vegetation conditions and habitat for other species. However, few studies have determined the specific mechanisms that induce such changes. To do this, Anderson and Briske (1995) conducted a controlled experiment with domestic herbivores in the southern true prairies in Texas. They concluded that selective herbivory of the late-seral dominant plant species, the perennial grass *Schizachyrium scoparium,* is the chief means by which plant species replacement occurs. This grass has a greater competitive ability and greater tolerance to herbivory pressure than the midseral grass species that make up the community. Thus, the herbivores modify the community because of differential, evolved tolerance to predation among different grass species. (Coevolution is also discussed in chapter 2.)

The evolution of species relations can have important implications for population persistence and management of habitat for populations. Many examples can be found in the literature. One pertains to obligate pollinator-host relations. Globally, insects—specifically Hymenoptera (bees and wasps), Lepidoptera (moths and butterflies), and Coleoptera (beetles)—are by far the most important pollinators of flowering plants. Recent declines in bees and wasps have sparked critical concern by some ecologists (Buchmann and Nabhan 1996). Upon loss of pollination vectors, plants may alter reproductive modes, suffer loss of genetic heterogeneity, or become locally extirpated (Washitani 1996). Aside from direct effects on vegetation and habitats as discussed above, herbivory also can have major indirect effects on pollination success and persistence of vegetation (Brody 1996).

Some vertebrate wildlife populations figure prominently in pollination ecology. Examples include many hummingbirds of Nearctic and Neotropical regions and their ecological vicariates, the sunbirds of Asia. Bats can play major roles in pollination of a variety of species in the tropics, although such roles likely are reduced or absent in many temperate and boreal ecosystems. Obligate pollinators and seed dispersers of "structural species" plants, such as some trees which serve as habitat or provide resources for a host of other species, were termed *mobile links* by Gilbert (1980). Terbourgh (1986) used the term *keystone mutualist* to refer to species whose extirpation would likely result in secondary extinctions of many associated obligate species. Habitat managers might want to identify mobile links and keystone mutualists— particularly important pollination vectors and dispersers of plant disseminules (plant seeds, fungi spores, lichen thalli, etc.)—to predict better the ecosystem effects of altering habitat on such species and to determine better the specific habitat requirements needed to provide for continuation of their ecological services. (However, see cautions in chapter 2 on use of these concepts in management.) Below, we further discuss the importance of species ecological functions to population conservation.

Exotic Species

Intrusion of natural environments by exotic species has become a major challenge for habitat conservation and restoration (Coblentz 1990; OTA 1993; Soule 1990). Exotic species occur in all taxonomic groups and all environments. Exotic plants can disturb native ungulate use of rangelands (Trammell and Butler 1995) and severely handicap management of natural conditions in parks (Wetman 1990; Tyser and Worley 1992). Exotic game bird and big game introductions can affect distribution of native plants and animals (OTA 1993). Unintentional invasions of exotic insects, plants, goats, pigs, and many other species in high-elevation parks in the Hawaiian Islands have caused a major disruption of native ecological communities and spurred intensive eradication management activities (Freed and Cann 1989; Vtorov 1993; Cole et al. 1995).

The Office of Technology Assessment (OTA 1993) used the term *nonindigenous species* to refer to alien or exotic species. The office concluded that, in a worst-case scenario, major economic losses could ensue with the continued spread of harmful nonindigenous species. OTA identified several major issues needing attention, including the need for a more stringent national policy in the United States, managing nonindigenous species for the spread of disease, and controlling the spread of weed plants and damage into natural areas.

However, it is not always clear which species are exotic (nonindigenous) and which are native. There are various cases of range expansions that may be natural, that may have been induced or enhanced by human alteration of environments, or that began as a minor introduction by humans. An example of a "natural" invader of North America is the cattle egret (*Bubulcus ibis*), which spread from Africa to South America about 1880, reached Florida and Texas in the 1940s and 1950s, and rapidly expanded north and west in North America (Ehrlich et al. 1988). Exotic escapee species that have spread throughout the continent include the European starling (*Sturnus vulgaris*); after two unsuccessful introductions, 60 birds were released into New York's Central Park in 1890, and within 60 years they had spread to the Pacific and have since outcompeted and threatened many other bird species (Ehrlich et al. 1988). Another example is brown-headed cowbirds (*Molothrus ater*), which have greatly expanded their range with the opening of forests and development of agriculture. On the other hand, the spotted dove (*Streptopelia chinensis*) was intentionally introduced in the Los Angeles area, and its spread has been limited to coastal southern California, despite its wide natural geographic range throughout Asia.

Along with moose (*Alces alces*), white-tailed deer (*Odocoileus virginianus*) are native to North America and have dramatically increased their geographic range northward and westward in the past several decades, including expansion into areas that were not modified much by humans (L. Irwin, pers. comm.). Their changes in distribution are not well understood but apparently are not all the result of human-caused changes in vegetation.

In recent years, barred owls have spread westward from subboreal forests in the northern United States and southern Canada and southward down the Cascades Mountains and into the Klamath Mountains of northwestern California, coming in secondary contact with the closely related and stenotopic northern spotted owl (Taylor and Forsman 1976; Hamer 1985). Reasons for the barred owl's recent and quick spread may have to do with its eurytopic habitat use or perhaps with the degree that clearcutting of conifer and mixed forests in North America has afforded habitats for foraging. Is the barred owl to be considered a desired native species of the West or an undesirable exotic from the East, indirectly introduced through major changes in forest conditions by human activities? Federal land management currently views the barred owl as a native species in the West, but ecological causes of its range expansion have not been determined.

Some exotic species are desired by people for their color or for sport. Such North American species that may be termed desired nonnatives include many introduced game birds, such as the ring-necked pheasant (*Phasianus colchicus*). Some populations of desired nonnatives— such as the nutria (*Myocaster coypus*), introduced into the United States for its pelt and once sought after—have proved to become harmful to other species or to wetlands or other sensitive ecosystems. The regulations that implement the National Forest Management Act (36 CFR 219) allow the USDA Forest Service to recognize "desired nonnative species" but specify neither how such designations should be made nor the ecological assessments required to determine effects of nonnative species on indigenous species. Clearly, in accord with OTA's (1993) recommendations, work remains to be done in better defining exotics and desired nonnatives and in projecting potential harm to native populations and ecosystems before introductions are made.

Human Disturbance Activities

Population introductions. Populations of some species have been translocated or introduced in order to recover species extirpated from an area or to provide opportunities for hunting or fishing of exotic species. Introductions, however, do not always succeed, particularly with threatened or endangered species recovery efforts. Introduction success has been related to initial population size (see box fig. 3.4B), degree of environmental stochasticity, subsequent poaching or exploitation pressures, and other factors.

Berger (1990) reported that, over 70 years of data with 122 populations of bighorn sheep (*Ovis canadensis*), 100 percent of the populations with fewer than 50 individuals went extinct within 50 years, whereas populations with more than 100 individuals persisted for up to 70 years. Berger concluded that causes of extinction were not food shortages, severe weather, predation, or interspecific competition. Likely candidates for causes of extinction were demographic stochasticity caused by small population size and lack of demographic rescue effects (adequate numbers of immigrants supplementing the population during critical population lows). However, Berger's study did not quantify food supply relations and focused only on transplant population size. Additional work could examine records of historical vegetation conditions. The vegetation has likely changed in many of the areas Berger evaluated, some because of livestock grazing and some because of fire suppression. Also, the dynamics of individual transplant populations, seasonal occurrence and food quality on local transplant ranges, and whether transplants are from different ecotypes need additional study.

Another major disturbance element affecting native wildlife populations is intentional or unintentional introduction of exotic competitors, parasites, predators, or pathogens, as discussed above. In some cases, cultivars or horticultural escapees of plants have wreaked ecological havoc on native habitats and biota. Classic examples include the widespread dispersal of kudzu (*Pueraria lobata*) in southeastern U.S. forests and the occurrence of dodder (*Cuscuta* spp.) on native flora and agricultural crops. Plants intentionally introduced for one reason may spread beyond their intended beneficial use. What seems amazing in hindsight, kudzu was intentionally introduced in the South for erosion control and forage in the 1940s because of its rapid growth, ease of propagation, and wide adaptability, but those very characteristics have caused it to stifle native vegetation well beyond its intended range (OTA 1993). Many other examples of introduced species gone madly awry were cited by OTA (1993).

Feral populations of domesticated goats (*Capra hircus*) and pigs (*Sus scrofa*), as well as introduced plants, have caused horrific ecological damage in isolated and fragile montane or island ecosystems such as the Channel Islands off southern California or many of the main Hawaiian Islands. Located on the upper slopes of Haleakala on Maui, Hawaii, is Hosmer Grove, a plantation of exotic trees created in the early part of the twentieth century to determine which tree species would do well in Hawaii for reforestation purposes. Curiously, the native koa (*Acacia koa*) and ohia (*Metrosideros spp.*) trees were not included in the experiment. Species selected for the test were exotic eucalyptus, pines, acacias, and others. Ironically, none of the species did well or were selected for reforestation use, although at least the pines have escaped from the grove and continue to encroach on the rare and declining, adjacent native plant communities of Waikamoi Preserve (The Nature Conservancy) (B. Marcot, pers. obs.). The preserve is one of the last remaining habitats for some of the rare and locally endemic native Hawaiian honeycreepers, particularly the 'akohekohe, or crested honeycreeper (*Palmeria dolei*), and the Maui parrotbill (*Pseudonestor xanthophrys*). Coupled with incursion into the preserve by feral goats and livestock, ecological pressure from exotic species continues to cause the native habitat to suffer. Nothing short of ongoing Herculean efforts by USDI National Park Service personnel to fence out feral ungulates and manually remove exotic organisms will provide the native plant and animal populations with much of a future.

Controlling populations of ecologically harmful exotic species is typically an expensive and too often a fruitless proposition. For example, Bock and Bock (1992) found that both exotic grass species (lovegrasses, *Eragrostis* spp., native to southern Africa) and local native grass species in an Arizona grassland were equally tolerant of fire, because both probably evolved in fire-prone systems. Thus, fire could not be used to eradicate the exotic species and recover the diverse native flora.

Hunting. The potential effects of hunting on wildlife are as varied as the species pursued. From the habitat perspective, Cook and Cable (1990) argued for the creation and conservation of windbreaks as an economic value for hunting. Some aspects of habitat conservation, particularly of summer range habitat for big game and wetlands for waterfowl, were intially protected largely for sport hunting. However, times have changed and user fees for hunting have become an economic polemic among hunters and managers alike. Fried et al. (1995) assessed the willingness to pay for elk-hunting opportunities in eastern Oregon and found that the mean willingness to pay was $287/trip for a virtually certain opportunity to shoot at an elk and an estimated mean value of $1,063 for a harvested elk (median of $333). Hammitt et al. (1990) listed a number of factors contributing to the hunting experience for white-tailed deer (*Odocoileus virginianus*).

In some cases, changing public attitudes toward hunting and the attitudes of hunters themselves (Heberlein 1991) has led to new legislation. For example, Loker and Decker (1995) reported that controversies over the hunting of black bears (*Ursus americanus*) in Colorado resulted in a referendum (Amendment 10) on the statewide ballot in November 1992 to prohibit specific hunting methods and the spring hunting of bears. The vote was swayed by concern for the animals' welfare largely from the nonhunting segment of the voting population. Presenting to the public the results of studies on the effects of hunting on population structure and trend, such as with black bears (Miller 1990), moose (Miller and Ballard 1992), and Dall sheep (*Ovis dalli*) (Murphey et al. 1990), had little to do with the referendum vote.

Should the economics of hunting change the fee system and procedure for funding habitat restoration and conservation activities on public lands (Williams and Mjelde 1994)? What should be the role of fee hunting on private lands in the United States (e.g., Smith et al. 1992)? The debate has yet to be settled.

Development. Effects of human developments and habitat alteration on wildlife populations are varied. Although much of the literature deals with the adverse effects of development on wildlife (e.g., Beatley 1994), there is also promise and potential for providing habitat conditions in urban environments for selected wildlife species more tolerant of human proximity (Adams 1994). The kinds of habitat alterations that are acceptable are largely a social issue. Depending on wildlife conservation goals, at least some wildlife in areas of human habitation can be provided for, and the choice is not always one of strict preservation versus total usurpation.

Blair (1996) studied bird community diversity along an urban gradient in southern California and found that bird species richness, Shannon diversity, and bird biomass peak at moderately disturbed sites, and species composition varies among all sites. Patterns related significantly to differences in habitat structures, particularly in the percentage of land covered by pavement, buildings, lawn, grasslands, and trees or shrubs. His work identifies manipulatable components in areas of more intense human habitation that would provide better for species tolerant to moderate disturbance. Urban park size may play a role in providing habitat for urban bird populations as well (Garvareski 1976). However, some native habitats, such as southern California chaparral (Soule et al. 1988), and associated native wildlife populations may be significantly more sensitive to urban encroachment and may require special conservation measures.

In some cases, urbanization can tip the balance between species interactions to the detriment of species more closely associated with undeveloped habitats. For example, Engels and Sexton (1994) verified their hypothesis that urban development in the Austin, Texas, area favored blue jays (*Cyanocitta cristata*) to the detriment of the endangered golden-cheeked warbler (*Dendroica chrysoparia*). The researchers conjectured that, although the two species show significant and inverse correlations to urban development, it is the foraging activity of jays in canopies of the warbler's juniper-oak woodland habitat that may have been the key factor. Jays may have deterred successful establishment of territories by male golden-cheeked warblers or successful attraction of warbler mates. But not all potential competitors produced adverse effects; the researchers did not find a negative correlation between the warblers and scrub jays (*Aphelocoma coerulescens*), which also occurred in the urban environments but in lower numbers. These are good examples of why it is important to unravel the causal factors influencing population size and trends, rather than focusing merely on habitats or environmental conditions.

Roads. The effect of roads on wildlife populations has not been well studied for many native species. The grizzly bear, wolverine (*Gulo gulo*), lynx (*Lynx lynx*), Florida panther, and other large carnivores are thought to be particularly sensitive to the presence of humans, including even sporadically used roads in otherwise rural or natural environments, although this theory needs empirical testing for some species. Some empirical evidence on grizzly bears (McLellan and Shackleton 1988) supports this, and road sensitivity is used in the cumulative effects modeling of grizzly bears in the western United States. Mitigation of the adverse effects of roads on Florida panthers has been somewhat successful by providing highway underpasses (Foster and Humphrey 1995), although it is difficult to determine the cumulative effect of development and highways on this endangered subspecies.

The effects of roads on small-bodied wildlife (e.g., Oxley et al. 1974) have not been well studied; the topic needs much work. In some cases, roads can serve as corridors by which weeds and undesired species spread into native environments (Tyser and Worley 1992). In Australia, Seabrook and Dettmann (1996) found that roads in forested and densely vegetated habitats act as activity corridors for the introduced, toxic cane toad (*Bufo marinus*), a species that has the potential to do much ecological harm to native wildlife communities. However, in montane temperate environments, seldom-used rural roads can usefully provide travel lanes for snowshoe hares and other species.

Some effects of roads on habitat conditions and ecosystem processes have been studied. Some rural roads can change landscape processes. Amaranthus et al. (1985) reported that logging and forest roads have contributed to debris slides in southwestern Oregon, which in turn have adversely altered riparian and aquatic stream habitats through increased sedimentation. Off-road vehicles have the potential for altering habitat conditions and directly disturbing some wildlife, although careful management of sites provided for such use can do much to mitigate adverse effects (Lacey et al. 1982).

Another aspect of potential disturbance of wildlife populations from human activities deals with noise (Fletcher and Busnel 1978). Noise seems inevitably to accompany most human activities, particularly noise associated with roads and recreation. Many wildlife species seem to accommodate to repetitive highway noise; egrets and herons commonly forage or even nest along median strips of busy highways in many coastal areas. However, noise sensitivity in wildlife has been poorly studied, and we should not be fooled by a few incidental and salient observations. Studies have shown declines in high-frequency auditory ability by some wildlife in the presence of noise (Viemeister 1983), but few studies have tested the effect of noise on wildlife dependent on high-frequency communication.

Pesticides. Much has been researched and published on the effects of pesticides on wildlife populations. In fact, it was the use of DDT and organochlorines and their biological magnification in higher trophic levels of food webs (e.g., Johnston 1975) that spurred the environmental movement of the 1960s. Effects of pesticides on wildlife populations has impelled development of risk analysis procedures in the Environmental Protection Agency of the United States (Tiebout and Brugger 1995) and guidelines for measuring and avoiding toxicity to wildlife (Hudson et al. 1984).

In one study, Moulding (1976) monitored the impact of the insecticide Sevin, used for controlling gypsy moths (*Porthetria dispar*), on a forest bird community in New Jersey. He reported a consistent, gradual decline in bird numbers, species richness, and diversity during eight weeks following spraying to 55 percent below control numbers of birds, and declines continued the following summer although spraying had ceased. Canopy-foraging birds were affected more than ground foragers, possibly because of the greater proportion of invertebrates in their diets. In a study of herbicide effects, Sullivan (1990) examined the effects of forest application of glyphosate herbicide on recruitment, growth, and survival in deer mice (*Peromyscus maniculatus*) and Oregon voles (*Microtus oregoni*) and found a temporary suppression of deer mice populations and no effect on vole populations. The glyphosate treatment had little or no direct effect on metabolic or general physiological processes in the young mice or voles, and any individual effect did not manifest itself as changes of population densities and recruitment. Secondary effects of herbicides on habitat alteration were studied by Morrison and Meslow (1983) in western forest plantations, where response of vegetation and bird assemblages to the use of glyphosate and 2,4-D was found to be short-lived.

Tracking Long-term Changes in Abundance and Distribution

In North America, long-term monitoring of population attributes can help to determine the contribution of habitat to species conservation. Standardized surveys of breeding and wintering birds—through U.S. Fish and Wildlife Service Breeding Bird Surveys (BBS) (Bystrak 1981) and wintering bird counts (WBC, or Audubon Christmas Bird Counts) (Bock and Root 1981), respectively—provide broad-scale information from which long-term trends may be inferred (Geissler and Noon 1981; Drennan 1981). Except as occasionally integrated into long-term ecological research (LTER) sites and other lengthy ecological studies, few if any of the established broad-based and long-term monitoring programs have been instituted for other wildlife taxonomic groups, although such studies on amphibians are beginning. Further discussions of using the BBS data for assessing and interpreting population trends, with an example of data on bobolinks (*Dolichonyx oryzivorus*), is presented in box 3.8.

BBS and other similar broad-scale survey and monitoring data by themselves do not explain the causes of population changes. For that, we need to marry geographic-specific evidence of population trends with information on changes of habitat and environmental conditions. One promising source is the Gap program databases that map habitat conditions interpreted for each species (Scott et al. 1993). There is great potential to integrate closely the BBS and WBC data with Gap and other similar data on habitat distribution, to verify the predictions of species distributions from the Gap models, and to determine the potential contributions of habitat changes to population trends. However, additional studies on factors other than habitat are essential as well. There is a major leap between identifying vegetation and environmental correlates of the distribution of species on the one hand and explaining the causal mechanisms of population change on the other.

Guidelines for Habitat Management for Metapopulations

Given the dynamics of metapopulations in heterogeneous environments, guidelines for habitat management might include the following:

1. Guidelines for species richness and overall biodiversity. Guidelines can be applied on three scales and can be derived from understanding how populations vary as a function of habitat complexity on each scale: (a) within habitat patches, or alpha diversity, such as with the observed correlations between foliage height diversity and bird species diversity presented earlier in this chapter; (b) between habitat patches, or beta diversity; and (c) among broader geographic areas, or gamma diversity. Beta diversity is often maintained by moderate disturbance regimes that provide for variations across space—typically within subbasins—in vegetation elements, substrates, and abiotic characteristics (including soils) of the environment. Beta diversity appears on different spatial scales for species with different body size, home range area, and vagility (Holling 1992). Gamma diversity is often controlled by climate, landform, geographic location, and broad-scale vegetation formation features.

Box 3.8. An example of long-term surveys aimed at determining population trends: a description of Breeding Bird Survey (BBS) methods and data

In this box, we explore in some detail the intent, methods, and results of the BBS survey program. We interpret results in terms of what sorts of bias might arise in estimating population trends and their causes and what kinds of conclusions should or should not be made from such survey data. This is a good example of how real-world complications need to be addressed when designing and implementing a trend study for wildlife populations. Our discussion here can pertain to all general count-type surveys of wildlife populations as well.

The BBS Program

Initiated in the mid-1960s by the USDI Fish and Wildlife Service (FWS), the BBS program aims at sampling regions and habitats throughout the United States and southern Canada for the purpose of indexing breeding bird population trends of a broad geographic extent (continentwide). BBS surveys are conducted annually at the height of the breeding season and currently use about 3700 survey routes, nearly 2900 of which are sampled every year.

The BBS Method

A BBS survey consists of a 39.4-km roadside route with 50 bird count points ("stops") spaced at 0.8-km intervals. At each stop, 3-minute counts are made of all birds seen or heard within 0.4 km of the count point. Routes are randomly located in order to sample habitats within each region.

Counts of each species are tallied per route and later pooled across all survey routes in the region. The average counts per route are linearly regressed against the year of survey under the assumption that changes in mean counts reflect trends in bird populations.

Trends of counts are summarized by U.S. state, Canadian province, physiographic stratum, BBS region (eastern, central, western), FWS administrative region (numbering six regions), the United States as a whole, Canada as a whole, and the entire survey region. To date, analyses have presented statistical tests of trends in counts over three time periods: 1966–1979, 1980–1994 (or current year with full data), and the overall period 1966–1994 (or current year). Results are typically presented as the mean trend of the linear regression (change in counts per year) and 95 percent confidence interval, with sample size (N, number of survey routes) and confidence (*P*) value of the regression coefficient.

Survey Biases

The BBS survey data have been criticized for containing too much sampling noise and other sources of error to provide an accurate picture of population trends. The intention in instituting the current BBS survey method was to reduce several potential sources of bias; the new method is intended to

- use a consistent technique for counting birds,
- ensure observer expertise,
- use the same routes and stops each year, and
- survey only during days of acceptable weather.

Additionally, the sample sizes of number of routes and number of count points per route were chosen to average out local variation and to help reduce overall statistical sampling error. This approach adheres to many of the criteria for a well-designed scientific study (see chapter 4).

However, several other sources of bias are not accounted for by the BBS method: The portions of species' ranges outside the BBS routes and survey regions are not sampled. Conditions alongside roads are represented but not those in unroaded environments. And some habitats are simply not represented well or at all in the surveys, including scarce habitats, small and irregularly distributed habitats (e.g., wetlands, talus and rock outcrops), and largely unroaded habitats (e.g., alpine tundra, upper elevation ice and rock fields, wildernesses).

(box continued on following page)

Box 3.8 Continued

Some Additional Considerations

The BBS data have provided useful information on broad-scale bird population trends over a period now exceeding three decades in North America. Data on breeding-season population trends of Neotropical migratory birds have proved to be particularly useful in identifying declining species and regions. Even if statistical changes in counts for a species are not significant, a decreasing trend can provide a tentative hypothesis of population change that can alert potential conservation action and be tested by more intensive sampling and local trend studies.

However, BBS surveys are designed to be quick and simple and merely to index numbers and trends. Thus, they do not estimate overall population sizes or the ecological density of species (numbers per unit area of habitat). Trends might belie the severity (or lack of significance) of a trend. For example, a significant declining trend might mean a great absolute number of birds are vanishing every year in a larger population than in a smaller population with the same trend slope. However, the smaller population might be at greater risk of extirpation, even with less severe declining trends. Thus, the BBS data—and survey information like the BBS data—need to be coupled with additional knowledge of trends in local habitats and trends and sizes of local populations.

Also, the time period over which population trend data provide at least somewhat statistically reliable information tends to span more than a decade. Small populations are likely to exhibit increasingly accelerated declines that would be masked by averaging trends over such time periods (see box 3.3). Thus, special monitoring studies are likely needed even for species sampled adequately by the BBS method but suspected of increasingly severe declines.

Nor do the BBS surveys account for differential detectability of species, as would be provided by other, more intensive survey methods, such as the variable circular plot technique (Ramsey and Scott 1979; Reynolds et al. 1980). Differential detectability can bias estimates of population size (not estimated in the BBS method) and possibly trend.

Using the same survey routes and stop points each year helps reduce interannual variation that would be caused if new routes or points were chosen each year. But it increases uncertainty of the representativeness of the originally chosen samples, particularly in sample-poor regions. An alternative approach would be to use a sampling frame that allows samples with partial replacement annually, as was developed for sampling habitats and the breeding status of northern spotted owls in the Pacific Northwest (Max et al. 1990). However, even this approach has its tradeoffs, because it may be much more difficult to manage.

BBS data do not give a good picture of the margins of a species' range, because rare occurrences of a species do not have a high likelihood of being included in a survey. However, the periphery of at least some species' ranges may be of particular interest in conservation, because this is where different selective pressures occur and novel genetic forms emerge (see text, table 3.5). Thus, the manager interested in range margins of breeding birds and population trends of birds along margins should also look elsewhere (particularly to local autecological studies) to supplement the BBS data. Also, at least one recent simulation study (NCASI 1996) suggested that errors in mapping the margins of species' ranges can result in large errors of identifying hot spots of biodiversity (species richness).

Finally, the BBS trend data do not provide information to explain the causes of population trends, especially important for interpreting declines. Many factors may contribute to declines (see text, table 3.3) which can be discerned only by additional local or autecological studies.

Species Poorly Sampled or Surveyed

Three general classes of bird species are undersampled by the BBS method and are not included in trend data analyses and summaries. These include:

1. Rarely encountered species. These are species detected on fewer than two BBS routes per year. They include high-latitude northern breeders (bird migration category L3, in table 3.4); Neotropical resident species that occur in the survey area only rarely (bird migration category PR for Neotropical species, in table 3.4); pelagic or coastal species; accidental species (including nomads, bird migration category NOM, in table 3.4), and some exotics.

(box continued on following page)

Box 3.8 Continued

2. Small-sample-size species. These are species with trend estimates <14 df (degrees of freedom = number of BBS routes minus number of states in which the species occurred). Some 20 percent of all species recorded on BBS routes have <14 df.
3. High-trend-variance species. These are species—some 22 percent of all species recorded on BBS routes—that might be counted frequently but that show great variation in counts and trends. They include species that have specialized habitats, very localized distributions, spruce-budworm association (irruptive), and that nest in colonies. Several herons, many ducks and waterbirds, a few species with far-northern-latitude breeding ranges, and several warblers, sparrows, blackbirds, and other species fall into this category.

An Example of Estimating Population Trends of Bobolinks

Bobolinks are Neotropical migrant passerines of the northern United States and southern Canada (see fig. 3.9 in text). They build open-cup nests on or close to the ground and favor tall grass, flooded meadows, farmlands, and clover fields with shrub cover for protection and perch sites. Many researchers have expressed great concern over the loss of native grasslands of North America and their associated native wildlife species (Samson and Knopf 1994). A focus on bobolinks from the BBS database may help determine trends for at least this grassland species and signal potential concern for others (particularly, grassland species not included in the BBS database).

The BBS data show the trend in counts of bobolinks per survey route over the period 1966–1994 (box fig. 3.8A). Visual inspection of this BBS-rangewide trend suggests an overall declining trend and a possible acceleration in declines after 1980.

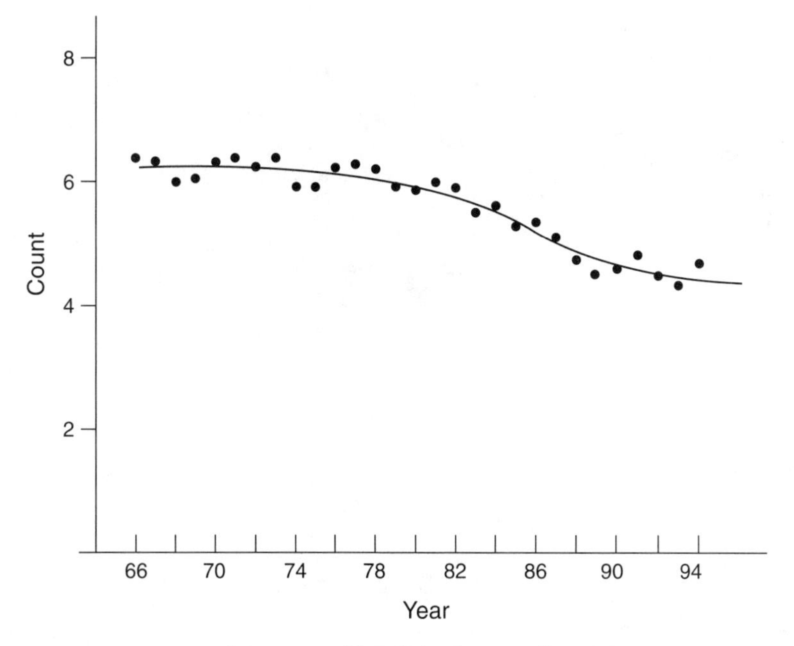

Box Figure 3.8A. Trends in counts of bobolinks along Breeding Bird Survey routes during 1966–1994 in North America. The overall trend shows a significant decline. (From Sauer et al. 1996)

(box continued on following page)

Box 3.8 Continued

If we count the total number of sampling strata (states, provinces, physiographic strata, regions, and countries) with positive trends, no trends, and negative trends, the BBS data suggest that populations are declining in most locations surveyed and indeed that overall declines are accelerating. Over the entire survey period of 1966–1994, some 19 survey strata had statistically significant ($P < 0.05$) declines, only 1 had a significant increase, and the rest (31) had no significant change. Evidence for accelerated declines is found by splitting the time periods. During the earlier period of 1966–1979, 12 survey strata had significant declines, 9 had significant increases, and the rest had no significant change, whereas during the more recent period of 1980–1994, there were 30 sample strata with significant declines, none with significant increases, and the rest with no significant change. Clearly, declines are becoming more pervasive throughout the sampling area.

Where are bobolinks declining? This is revealed in the BBS map of trends smoothed across survey routes (box fig. 3.8B). It appears from this map that bobolinks are declining in a large portion of their breeding range and are widely distributed within it. Local spots of stable or increasing population centers, however, seem to be suggested for a number of locations scattered widely throughout the range, such as those bordering the eastern and western fringes of the Great Lakes states. Interestingly, many of these stable or increasing centers occur along the periphery of the survey region. Whether these signal important population refuges, temporary sinks, permanent source centers, or other demographic phenomena (or perhaps sampling bias) awaits more local study. Initial modeling of metapopulation demographics, however, might help determine the overall stability of this emerging bobolink metapopulation should this trend continue and the centers become more isolated from one another.

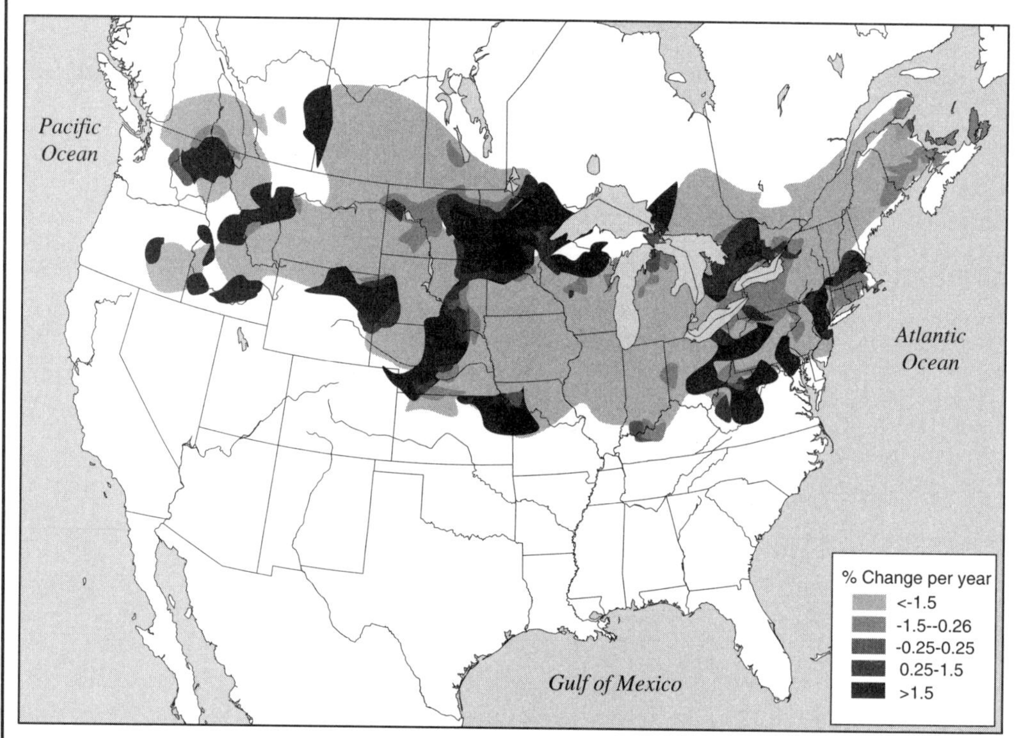

Box Figure 3.8B. Locations of changes in populations of bobolinks (based on changes in bird counts per Breeding Bird Survey route) during 1966–1994 in North America. (From Sauer et al. 1996)

(box continued on following page)

Box 3.8 Continued

Also, the trend map (box fig. 3.8B) should be inspected alongside the density map (fig. 3.9 in text) to determine if areas with the same trend pertain to vastly different population conditions, as discussed above. For example, the area of increasing trend (>1.5 percent/year) along the Kansas–Missouri border occurs in an area of lowest population density as indexed by the BBS route data (<1 mean count/route) and occurs along the periphery of the species' breeding range, whereas another area of the same increasing trend category along the North Dakota–Minnesota border occurs in an area of highest population density (30–100 mean counts/route) and is more centrally located in the species' overall breeding range. The demographic implications of maintaining suitable habitat conditions in these two areas having the same trends but different densities are likely to differ vastly. Such differences can be used to help prioritize grassland conservation measures at least for this species.

It is not immediately evident, however, which of these two sites would be more important for maintaining the species. The central-range location may provide a vital source of individuals (should further surveys suggest that this condition is real and not just a sampling artifact or temporary fluctuation), whereas the peripheral-range location could provide conditions just as important for potential adaptation of demes to new conditions there. Priorities for such conservation options need to be determined by inspecting likely future trends in habitats, as well as by considering the potential for habitat restoration, for maintenance of population sources in range centers, and for the emergence of populations with unique gene pools (see table 3.5). Modeling of metapopulation response to increasing isolation of these centers might help determine which sites contribute greater to overall near-term stability of the species (see chapter 10).

Additional Considerations for Interpreting Bobolink Trends

Of course, the BBS data do not tell the full ecological story. During the nineteenth century, increases in agricultural fields led to increases in bobolink populations and range extensions in the eastern part of the continent. This was followed by declines in bobolink populations in the twentieth century, particularly after 1940, when changes in agricultural practices (principally, the frequent mowing of hayfields) reduced the area of suitable grassland habitat for this species.

Bobolinks are one of the few North American songbirds that winter entirely south of the equator. Their winter biology is mostly unknown, and they could be experiencing declines on their wintering grounds from habitat loss, environmental degradation, or other causes of mortality. The BBS data do not by themselves explain the causes of declines or identify conditions, trends, or potential causes beyond the survey region.

2. Guidelines to maintain within-patch conditions. It may be desired to ensure that specific habitat patches remain viable environmental units. In this case, the size, as well as the topographic location, adjacency of other patches, and susceptibility to disturbances such as floods or fires, can influence within-patch conditions (see chapter 9). Within-patch conditions are usually primarily mediated by patch size and within-patch dynamics, but also by the context of the patch (i.e., the kinds of vegetation conditions in adjacent patches), the type and intensity of natural disturbances within the patch, and particularly the kind, frequency, and intensity of management activities within the patch.
3. Guidelines to maintain a desired occupancy rate of habitat patches. A collection of habitat patches may be occupied by organisms over time, depending on the size and quality of the patches, the type and quality of the intervening matrix environment, and the spatial juxtaposition of the patches. In dynamic environments such as subtidal marine substrates or high-elevation fire-adapted forests, the quiltwork of patches may change over time under disturbance regimes and presence of species that provide physical surfaces and substrates ("structural" species *sensu* Huston 1994).

Empirical studies and models of populations in patchy environments can help provide realistic expectations for occupancy rates within and among habitat patches. This has important implications for designing inventory and monitoring studies and interpreting results of such studies. Depending on the species and habitat conditions, not all patches may be occupied all the time. But seemingly unoccupied patches may still be important as "stepping stones," or key links, for distribution and dispersal of organisms throughout a landscape. Habitat patches can often be mapped and key links identified as part of a set of habitat features important to metapopulation dynamics (table 3.5). Stepping stones or habitat corridors do not necessarily have to consist of primary habitat (i.e., to provide for all life history needs to maximize realized fitness) in order to provide value to connect populations across a landscape.

Seemingly unoccupied patches may also serve as occasional sink habitat to house nonreproductive floaters not yet assimilated into the breeding segment of the population. Floaters are often important for maintaining populations and often occupy marginal habitats. Examples of studies of floaters are few but include gulls (Shugart et al. 1987), wrentits (Geupel et al. 1992), sparrows (Smith 1978), and chickadees (Smith 1984, 1989).

Other situations that affect desired occupancy rates are listed in table 3.5. By doing the following, habitat management guidelines can address these situations: ensure that habitat for naturally disjunct populations and peripheral parts of a species' range are not overlooked (it is in such places that interesting, new "natural experiments" in species lineages take place); restore or at least maintain habitat in bottlenecks, corridors, and key links to ensure connectivity of demes and subpopulations; and account for what constitutes primary habitat when engaging in habitat acquisition or restoration programs. In addition, monitoring programs and metapopulation modeling projects should take into account locations of dispersal barriers and filters, areas with unsuitable environments, and population isolates.

4. Guidelines for habitat configuration. Configuration of habitat patches may be directly manipulated by management activities, as an indirect result of other human land-altering activities such as growth of towns and roads, or as a result of stochastic and unpredictable natural disturbance events such as wildfire, mass wasting, and vulcanism, among other dynamics. Where it is possible to guide direct management activities and account for indirect effects and natural disturbances, to provide for long-term viability of populations, habitats should be configured so as to afford rescue effects of habitat colonization by organisms over time (see chapter 9 for further discussion of rescue effects). Habitat patches can be managed to remain reachable by dispersing and migrating individuals and by organisms moving within their home ranges. Feeding and resting habitats can be provided within daily dispersal distances or home range areas, such as has been included in habitat management guidelines for red-cockaded woodpeckers in forests of longleaf pine in the southeastern United States. Secondary or marginal habitats can be provided close to primary habitats to help serve as sinks for surplus or floater individuals in good reproductive years.

Additional ideas on guidelines for managing and monitoring habitats in landscapes are presented in chaper 9.

Table 3.5. Some elements of habitat distribution and pattern, important to maintaining metapopulations, that can be mapped and monitored

Elements that can be mapped	Description of element	Why the element is important to maintaining populations
Habitat isolates	Segments of a species' distribution or habitat distribution that is separated from all other segments by at least one standard deviation of average dispersal distance (to maintain genetic continuity), or twice the average dispersal distance (to maintain demographic continuity), of the species	Isolated or disjunct populations or reproductive individuals should be modeled separately. Small, isolated populations have higher susceptibility to decline or extirpation from a variety of factors. Subpopulations need to be connected both genetically and demographically to maintain continuity.
Habitat bottlenecks	A narrowing of habitat to accommodate only one connection between demes or populations across an area, rather than redundant connections in several directions	Severing of habitat bottlenecks may partly or fully isolate demes or populations. Can help identify areas of habitat conservation or restoration priority
Habitat corridors	Areas of more or less stable habitat serving to link population centers Corridors can facilitate actual movement of organisms between centers or provide for reproductive individuals so that centers are linked by transmission of genes.	Corridors serve to enlarge the effective (reproductive) size of populations, which increases the likelihood of persistence (viability); some, however, might serve as "predator traps" or avenues of dispersal for undesirable exotic species.
Areas of unsuitable conditions	Areas lacking habitat for a species	Can help identify priorities for restoration or habitat creation if needed for population size or connectivity Through empirical survey and modeling, can help set realistic expectations for presence of organisms
Habitat barriers	More than just areas of unsuitable conditions, barriers are absolute obstacles to movement and interaction of organisms.	Barriers demarcate the demographic boundary of populations and are important for devising realistic guidelines for population connectivity.
Key link habitat patches	Single patches that serve as critical links ("cut points") in the distribution of habitat in an area	Key links may be vital conduits for ensuring connectivity within a population.
Degrees of habitat suitability	In its simplest, two levels of suitability can be described: primary or source habitat with sufficient total area, patch size, food, water, and substrates that are required by the species and that serve to maintain the population; and secondary or sink habitat that can temporarily hold nonreproductive or floater individuals, in which populations likely are not self-maintaining.	Provides realistic expectations of reproductive performance of organisms within each habitat suitability class Identifies critical core habitats
Peripheral or marginal habitat	Habitat occurring at the edges of a species' distributional range. (Note that secondary or sink habitat, above, might also occur elsewhere in a species' range.)	Peripheral habitats can provide conditions (selection pressures) for the emergence of new subspecies, ecotypes, morphs, or eventually species.
Habitat patch density	The number of patches of habitat within a general area. The size of patches and the general area would vary by species body size, life history, and home range size.	Provides a basis for calculating crude and ecological density (Caughley 1977)

Summary: Understanding and Managing Populations in the Context of Habitats and Biodiversity

This chapter has presented a number of topics related to population conservation in the context of habitat ecology. It is important to study and understand population dynamics to inform us of species' potential responses to conditions and changes in habitats and environments and to understand the sometimes limited degree to which habitat management alone can provide for populations. Although much conservation work is done under the assumption that wildlife are a function of their habitats and that providing habitat provides for wildlife, the real world is far more complicated. It is a truism that habitat provides the basis for wildlife conservation, but it does not ensure it.

Considering Relations of Vegetation and Populations in Conservation

Population trends and viability are influenced not just by environmental conditions, but also by demographic and genetic conditions and variations and by the presence and roles of other organisms. Few wildlife species exist as ideal genetic populations; most occur as metapopulations with weakly interacting subpopulations or with isolated or disjunct population centers. This complicates the answers provided by simple models of minimum viable population sizes and suggests a need for assessing species individually rather than using a single golden rule for target population size. Small populations often can be at substantially greater risk of decline and extirpation.

Habitat conservation needs to attend to how populations occupy their distributional range and how organisms move through space and respond to stochastic environments. Ultimately, these factors—along with others, including the effects of exotic species and human disturbances—affect the evolution and expression of adaptive traits.

Long-term changes in populations can be provided by some ongoing surveys for breeding and wintering birds, but few such surveys have been instituted for other wildlife taxa. In the United States, state natural heritage programs provide sources of data on occurrence of rare plants and selected wildlife species but not specifically under a trend-monitoring program. Some broad-based programs can help provide coarse-grained information on habitat distribution with which population presence can be projected or integrated, but we cannot ignore the important contributions of factors other than habitat in affecting population density and trend.

Maintenance of habitats and wildlife populations is but one facet of maintaining overall biological diversity. We have provided a set of simple guidelines for maintaining biodiversity, species, and populations, particularly metapopulations in heterogeneous environments. These guidelines pertain to maintaining species richness and biodiversity, within-patch conditions, desired habitat occupancy rates, and habitat configuration for populations. Additional species-specific guidelines would be needed on other nonhabitat factors that influence population density and trend.

Toward a New Focus for Habitat and Population Management in an Ecological and Evolutionary Context

Many ecological entities, although poorly represented by accepted taxonomies, particularly of species and subspecies, may nonetheless warrant attention for conserving the evolutionary potential of species lineages. Such entities include: demes, subpopulations, disjunct populations, ecotypes, ecological functional groups of species, organisms at the periphery of their distributional range, organisms at the limits of their range of physiological tolerance, as well as organisms with ecological roles central to maintaining the richness of an ecological community (table 3.6). Groups of organisms in these conditions usually do not fit traditional Linnean taxonomy and are typically excluded from administrative conservation programs. They nonetheless constitute valid ecological categories deserving of understanding and potential conservation focus, as suggested by lessons and examples presented in this chapter on woodland caribou, rough-skinned newts, garter snakes, red-legged frogs, and many others. Ultimately, public policy makers and managers must decide if they are merely to maintain current conditions for the short term or attempt to manage evolutionary conditions for the long term.

Table 3.6. Conditions of organisms, populations, and species warranting particular management attention for conservation of biodiversity

Name of ecological component	Description	Why it is important to biodiversity	Examples
Species richness	Numbers of species of particular taxonomic group	Total representation of phylogenetic taxa	Total number of species by taxonomic class in an area
Functional species redundancy	Number of species having the same key ecological function	Provides continuation of ecological functions (at least at general level), should some species become extirpated	Ungulate herbivores in African savannas; primary cavity excavators; primary burrow excavators
Isolated or disjunct populations	Demes and populations separated from others of the same species by more than their median dispersal distance	Locally isolated populations can adapt to local conditions, as first stages of allopatric speciation or development of ecotypes	Western jumping mouse (*Zapus princeps*), on isolated mountain ranges of the Great Basin, western North America
Edges of distributional ranges of species	Peripheral parts of a species' range, often with suboptimal, sink habitat conditions	Character divergence of organisms or local adaptation to conditions different from those in the core of a species' range can occur at range margins	Coati (*Nasua nasua*), in open forests of southeastern Arizona
Coevolved species complexes	Species, with their environmental requirements, that have evolved in commensal or mutualistic relations	Changes in populations of one species can alter those of other species	Pollinator-plant relations, for example, bats and calabash tree (*Crescentia*) or saguaro cactus (*Carnegiea*) in southwest U.S. deserts

(*table continued on following page*)

Table 3.6 Continued

Name of ecological component	Description	Why it is important to biodiversity	Examples
Sibling or cryptic species complexes	Closely related species nearly identical in morphology but with different resource-use characteristics	The full range of resource and environmental requirements of all species of a complex is not represented by just one species of a complex	Dusky flycatcher (*Empidonax oberholseri*) in early and midstage forests and Hammond's flycatcher (*E. hammondii*) in old-growth forests in western North America
Mobile links (Gilbert 1980)	Organisms significant in the persistence of several plant species which support separate food webs	Supports the basis for several food webs in a community and numerous dependent species	Bees (Hymenoptera), butterflies and moths (Lepidoptera), flies (Diptera), and some beetles (Coleoptera) as obligate pollinators of many plant species in many parts of the world
Keystone mutualists (Gilbert 1980)	Organisms (typically plants) which provide critical support to large complexes of mobile links	Loss would trigger additional loss of mobile links, link-dependent plants, and decline in host-specialist insect diversity	Fig trees (*Ficus* spp.) in many New and Old World tropics
Locally endemic species	A taxonomic species or morphospecies complex found in only one small geographic location	Represents a novel genetic entity	Glacier Bay water shrew (*Sorex alaskanus*), endemic to southeast Alaska
Locally endemic subspecies, forms, and varieties	Taxonomic subspecies or characteristically unique phenotypes of a species found in only one small geographic location	Represents potential trend toward adaptation to local environmental conditions and possible emergence of new species lineage	Prince of Wales flying squirrel (*Glaucomys sabrinus griseifrons*), Prince of Wales Island
Ecotypes	Locally adapted genotypes that perform better in the native environments than in other parts of their parent species' range	Ecotypes signal evolution of local adaptations and contribute to overall genetic diversity of the species.	Mountain and northern ecotypes of woodland caribou (*Rangifer tarandus caribou*) in northern Idaho (see text)
Unique morphs and polymorphic populations	Local occurrences of organisms with unique phenotypic differences in meristics or coloration	Could signal adaptive significance, as with differential survival to predation or harsh environmental conditions, or emergence of new genetic entities	Color morphs of northern rough-skinned newt (*Taricha granulosa*) in western U.S. ponds (see box 3.5)

Note: Most of the components listed here would not be adequately represented simply by listing species presence, counting species in an area, or focusing conservation activity at the species level. However, these components may be of great interest in maintaining ecological processes of ecosystems and providing for evolutionary lineages in biogeographic settings fostering uniquely adapted organisms.

Literature Cited

Adams, L. W. 1994. *Urban wildlife habitats: A landscape perspective.* Minneapolis: University of Minnesota Press.

Akçakaya, H. R., M. A. McCarthy, and J. L. Pearce. 1995. Linking landscape data with population viability analysis: Management options for the helmeted honeyeater. *Biological Conservation* 73:169–76.

Aldrich, J. W., and C. S. Robbins. 1970. Changing abundance of migratory birds in North America. *Smithsonian Contributions to Zoology* 26:17–26.

Allen, S. H., and A. B. Sargeant. 1993. Dispersal patterns of red foxes relative to population density. *Journal of Wildlife Management* 57(3):526–33.

Amaranthus, M. P., R. M. Rice, N. R. Barr, and R. R. Ziemer. 1985. Logging and forest roads related to increased debris slides in southwestern Oregon. *Journal of Forestry* 83:229–33.

Anderson, V. J., and D. D. Briske. 1995. Herbivore-induced species replacement in grasslands: Is it driven by herbivory intolerance or avoidance? *Ecological Applications* 5(4):1014–24.

Arthur, S. M., T. F. Paragi, and W. B. Krohn. 1993. Dispersal of juvenile fishers in Maine. *Journal of Wildlife Management* 57(4):868–74.

Askins, R. A. 1995. Hostile landscapes and the decline of migratory songbirds. *Science* 267:1956–57.

Askins, R., J. F. Lynch, and R. Greenberg. 1990. Population declines in migratory birds in eastern North America. *Current Ornithology* 7:1–57.

Bahre, C. J. 1991. *A legacy of change: Historic human impact on vegetation of the Arizona borderlands.* Tucson: University of Arizona Press.

Barbour, M. G. 1996. American ecology and American culture in the 1950s: Who led whom? *Bulletin of the Ecological Society of America* 77:44–51.

Barrows, C. W. 1981. Roost selection by spotted owls: An adaptation to heat stress. *Condor* 83:302–9.

Batzli, G. O. 1994. Special feature: Mammal-plant interactions. *Journal of Mammalogy* 75:813–15.

Beatley, T. 1994. *Habitat conservation planning: Endangered species and urban growth.* Austin: University of Texas Press.

Beier, P. 1995. Dispersal of juvenile cougars in fragmented habitat. *Journal of Wildlife Management* 59(2):228–37.

Belthoff, J. R., and G. Ritchison. 1989. Natal dispersal of eastern screech-owls. *Condor* 91(2):254–65.

Benkman, C. W. 1993. Adaptation to single resources and the evolution of crossbill (*Loxia*) diversity. *Ecological Monographs* 63:305–25.

Berger, J. 1990. Persistence of different-sized populations: An empirical assessment of rapid extinctions in bighorn sheep. *Conservation Biology* 4:91–98.

Blair, R. B. 1996. Land use and avian species diversity along an urban gradient. *Ecological Applications* 6(2):506–19.

Block, W. M., M. L. Morrison, J. Verner, and P. H. Manley. 1994. Assessing wildlife-habitat relationships: A case study with California oak woodlands. *Wildlife Society Bulletin* 22:549–61.

Bock, J. H., and C. E. Bock. 1992. Vegetation responses to wildfire in native versus exotic Arizona grassland. *Journal of Vegetation Science* 3(4):439–46.

Bock, C. E., and T. L. Root. 1981. The Christmas Bird Count and avian ecology. *Studies in Avian Biology* 6:17–23.

Bokdam, J., and M. F. W. de Vries. 1992. Forage quality as a limiting factor for cattle grazing in isolated Dutch nature reserves. *Conservation Biology* 6:399–408.

Bonnell, M. L., and R. K. Selander. 1974. Elephant seals: Genetic variation and near extinction. *Science* 184:908–9.

Bonser, S. P., and R. J. Reader. 1995. Plant competition and herbivory in relation to vegetation biomass. *Ecology* 76:2176–83.

Boucher, M. H. 1985. Lotka-Volterra models of mutualism and positive density-dependence. *Ecological Modelling* 27:351–70.

Boyce, M. 1992. Population viability analysis. *Annual Review of Ecology and Systematics* 23:481–506.

Boyce, M. S., J. S. Meyer, and L. Irwin. 1994. Habitat-based PVA for the northern spotted owl. In *Otago Conference Series 2: Statistics in Ecology and Environmental Monitoring*, ed. D. J. Fletcher and B. J. F. Manley, 63–85. Dunedid, New Zealand: University of Otago Press.

Brody, A. K. 1996. Linking herbivory and pollination. *Bulletin of the Ecological Society of America* 77(2):103–5.

Brown, J. H., D. W. Mehlman, and G. C. Stevens. 1995. Spatial variation in abundance. *Ecology* 76(7):2028–43.

Brownlow, C. A. 1996. Molecular taxonomy and the conservation of the red wolf and other endangered carnivores. *Conservation Biology* 10(2):390–96.

Buchmann, S. L., and G. P. Nabhan. 1996. *The forgotten pollinators.* Covelo, Calif.: Island Press.

Buechner, M. 1987. A geometric model of vertebrate dispersal: Tests and implications. *Ecology* 68:310–18.

Bystrak, D. 1981. The North American Breeding Bird Survey. *Studies in Avian Biology* 6:34–41.

Caswell, H. 1989. Matrix population models: Construction, analysis, and interpretation. Sunderland, Mass.: Sinauer Associates, Inc.

Caughley, G. 1977. *Analysis of vertebrate populations.* New York: John Wiley and Sons.

Chambers, S. M., and J. W. Bayless. 1983. Systematics, conservation and the measurement of genetic diversity. *Genetics and conservation: a reference for managing wild animal and plant populations,* ed. C. M. Schonewald-Cox, S. M. Chambers, B. MacBryde, and L. Thomas, 349–63. Menlo Park, Calif.: Benjamin/Cummings.

Chaney, R. W. 1947. Tertiary centers and migration routes. *Ecological Monographs* 17:139–48.

Clements, F. E. 1904. *The development and structure of vegetation.* Studies in the Vegetation of the State, Botanical Survey of Nebraska 7, Lincoln, Neb.

Clements, F. E. 1916. *Plant succession.* Carnegie Institute of Washington, Publication 242, Washington, D.C.

Clements, F. E. 1920. *Plant indicators.* Carnegie Institute of Washington, Washington, D.C.

Coblentz, B. E. 1990. Exotic organisms: A dilemma for conservation biology. *Conservation Biology* 4:261–65.

Cole, F. R., L. L. Loope, A. C. Medeiros, J. A. Raikes, and C. S. Wood. 1995. Conservation implications of introduced game birds in high-elevation Hawaiian shrubland. *Conservation Biology* 9(2):306–13.

Colwell, R. K., and D. J. Futuyma. 1971. On the measurement of niche breadth and overlap. *Ecology* 52:567–76.

Conant, R. 1975. *A field guide to reptiles and amphibians of eastern and central North America.* Boston: Houghton Mifflin Co.

Cook, P. S., and T. T. Cable. 1990. The economic value of windbreaks for hunting. *Wildlife Society Bulletin* 18(3):337–42.

Crow, J. F., and M. Kimura. 1970. *An introduction to population genetics theory.* New York: Harper and Row.

Dasmann, R. F. 1964. *Wildlife biology.* New York: John Wiley and Sons.

Daubenmire, R. F. 1974. Plants and environment: A textbook on autecology. 3d ed. New York: John Wiley and Sons.

DeGraaf, R. M., and D. D. Rudis. 1990. Herpetofaunal species composition and relative abundance among three New England forest types. *Forest Ecology and Management* 32:155–65.

Diamond, J. M. 1984. "Normal" extinctions of isolated populations. In *Extinctions,* ed. M. H. Nitecki, 191–246. Chicago: University of Chicago Press.

Drennan, S. R. 1981. The Christmas Bird Count: An overlooked and underused sample. *Studies in Avian Biology* 6:24–29.

Ehrlich, P. R., D. S. Dobkin, and D. Wheye. 1988. *The birder's handbook.* New York: Simon and Shuster.

Engles, T. M., and C. W. Sexton. 1994. Negative correlation of blue jays and golden-cheeked warblers near an urbanizing area. *Conservation Biology* 8(1):286–90.

Falconer, D. S. 1989. *Introduction to quantitative genetics.* 3d ed. New York: Longman Scientific and Technical.

Feinsinger, P., and E. E. Spears. 1981. A simple measure of niche breadth. *Ecology* 62:27–32.

Finch, D. M. 1991. *Population ecology, habitat requirements, and conservation of Neotropical migratory birds.* USDA Forest Service, Rocky Mountain Forest and Range Experiment Station, General Technical Report RM-205. Fort Collins, Colo. 26 pp.

Finch, D. M., and P. W. Stangel, ed. 1992. *Status and management of Neotropical migratory birds.* USDA Forest Service, Rocky Mountain Forest and Range Experiment Station, General Technical Report RM-229. Fort Collins, Colo. 422 pp.

Fletcher, J. L., and R. G. Busnel. 1978. *Effects of noise on wildlife.* New York: Academic Press.

Foster, M. L., and S. R. Humphrey. 1995. Use of highway underpasses by Florida panthers and other wildlife. *Wildlife Society Bulletin* 23(1):95–100.

Frankel, O. H., and M. E. Soule. 1981. *Conservation and evolution.* Cambridge: Cambridge University Press.

Freed, L. A., and R. L. Cann. 1989. Integrated conservation strategy for Hawaiian forest birds. *BioScience* 39(7):475–79.

Fried, B. M., R. M. Adams, R. P. Berrens, and O. Bergland. 1995. Willingness to pay for a change in elk hunting quality. *Wildlife Society Bulletin* 23(4):680–86.

Futuyma, D. J. 1986. *Evolutionary biology.* 2d ed. Sunderland, Mass.: Sinauer Associates.

Garber, D. P., and C. E. Garber. 1978. A variant form of *Taricha granulosa* (Amphibia, Urodela, Salamandridae) from northwestern California. *Journal of Herpetology* 12:59–64.

Garvareski, C. A. 1976. Relationships of park size and vegetation to urban bird populations of Seattle, WA. *Condor* 78:375–82.

Geissler, P. H., and M. R. Fuller. 1985. Detecting and displaying the structure of an animal's home range. In *Proceedings of the American Statistical Association,* 378–83. Alexandria, Va.: American Statistical Association.

Geissler, P. H., and B. R. Noon. 1981. Estimates of avian population trends from the North American Breeding Bird Survey. *Studies in Avian Biology* 6:42–51.

Geupel, G. R., O. E. Williams, and N. Nur. 1992. Factors influencing the variation in abundance of "floaters" in a population of wrentits. Paper presentation and abstract in the 62d Annual Meeting of the Cooper Ornithological Society, 22–28 June 1992, Seattle, Wash.

Gilbert, L. E. 1980. Food web organization and conservation of neotropical diversity. In *Conservation biology: An evolutionary-ecological perspective,* ed. M. E. Soule and B. A. Wilcox, 11–34. Sunderland, Mass.: Sinauer Associates, Inc.

Gilpin, M. E., and I. Hanski, eds. 1991. *Metapopulation dynamics empirical and theoretical investigations.* Orlando: Academic Press.

Gilpin, M. E., and M. E. Soule. 1986. Minimum viable populations: Processes of species extinction. In *Conservation biology: The science of scarcity and diversity,* ed. M. E. Soule, 19–34 Sunderland, Mass.: Sinauer Associates.

Ginzburg, L. R., S. Ferson, and H. R. Akçakaya. 1990. Reconstructibility of density dependence and the conservative assessment of extinction risks. *Conservation Biology* 4:63–70.

Ginzburg, L. R., L. B. Slobodkin, K. Johnson, and A. G. Bindman. 1982. Quasiextinction probabilities as a measure of impact on population growth. *Risk Analysis* 2:171–81.

Gleason, H. A. 1939. The individualistic concept in plant succession. *American Midland Naturalist* 21:92–110.

Glenn-Lewin, D. C., R. K. Peet, and T. T. Veblen, eds. 1992. *Plant succession: Theory and prediction.* New York: Chapman and Hall.

Gliwicz, J. 1988. The role of dispersal in models of small rodent population dynamics. *Oikos* 52(2):219–21.

Gosz, J. R. 1993. Ecotone hierarchies. *Ecological Applications* 3:369–76.

Green, D. M. 1985. Differentiation in amount of centromeric heterochromatin between subspecies of the red-legged frog, *Rana aurora. Copeia* 1985(4):1071–74.

Green, D. M. 1986. Systematics and evolution of western North American frogs allied to *Rana aurora* and *Rana boylii:* Electrophoretic evidence. *Systematic Zoology* 35(3):283–96.

Greenwood, P. J., and P. H. Harvey. 1982. The natal and breeding dispersal of birds. *Annual Review of Ecology and Systematics* 13:1–21.

Guthery, F. S., and R. L. Bingham. 1992. On Leopold's principle of edge. *Wildlife Society Bulletin* 20:340–44.

Haas, C. A. 1995. Dispersal and use of corridors by birds in wooded patches on an agricultural landscape. *Conservation Biology* 9(4):845–54.

Hamer, T. 1985. Continued range expansion of the barred owl (*Strix varia*) in western North America from 1976–1985. Paper presentation and abstract in the Symposium on the Biology, Status, and Management of Owls, 9–10 Nov. 1985, Raptor Research Foundation, Sacramento, Calif.

Hammitt, W. E., C. D. McDonald, and M. E. Patterson. 1990. Determinants of multiple satisfaction for deer hunting. *Wildlife Society Bulletin* 18(3):331–37.

Harris, L. D. 1988. Landscape linkages: The dispersal corridor approach to wildlife conservation. *American Wildlife Natural Resources Conference* 53:595–607.

Hartl, D. L., and A. G. Clark. 1989. *Principles of population genetics.* 2d ed. Sunderland, Mass.: Sinauer Associates.

Hayes, M. P., and D. M. Krempels. 1986. Vocal sac variation among frogs of the genus *Rana* from western North America. *Copeia* 1986(4):927–36.

Hayes, M. P., and M. M. Miyamoto. 1984. Biochemical, behavioral and body size differences between *Rana aurora aurora* and *R. a. draytonii. Copeia* 1984(4):1018–22.

Heberlein, T. A. 1991. Changing attitudes and funding for wildlife—preserving the sport hunter. *Wildlife Society Bulletin* 19(4):528–34.

Hilden, O. 1965. Habitat selection in birds. *Annales Zoologici Fennici* 2:53–75.

Hill, C. J. 1995. Linear strips of rain forest vegetation as potential dispersal corridors for rain forest insects. *Conservation Biology* 9(6):1559–66.

Holdridge, L. R. 1947. Determination of world plant formations from simple climatic data. *Science* 105:367–68.

Holling, C. S. 1992. Cross-scale morphology, geometry, and dynamics of ecosystems. *Ecological Monographs* 62(4):447–502.

Hudson, R. H., R. K. Tucker, and M. A. Haegele. 1984. *Handbook of toxicity of pesticides to wildlife.* USDI Fish and Wildlife Service, Resource Publication 153. Washington, D.C. 90 pp.

Huey, R. B., C. R. Peterson, S. J. Arnold, and W. P. Porter. 1989. Hot rocks and not-so-hot rocks: Retreat-site selection by garter snakes and its thermal consequences. *Ecology* 70(4):931–44.

Huston, M. A. 1994. *Biological diversity: The coexistence of species on changing landscapes.* Cambridge: Press Syndicate of University of Cambridge.

Hutto, R. L. 1985. Habitat selection by nonbreeding, migratory land birds. In *Habitat selection in birds,* ed. M. L. Cody, 455–76. San Diego, Calif.: Academic Press.

Irwin, L. L. 1994. A process for improving wildlife habitat models for assessing forest ecosystem health. *Journal of Sustainable Forestry* 2(3-4):293–306.

James, F. C. 1971. Ordinations of habitat relationships among breeding birds. *Wilson Bulletin* 83:215–36.

Johnson, D. H. 1980. The comparison of usage and availability measurements for evaluating resource preference. *Ecology* 61:65–71.

Johnson, M. L. 1995. Reptiles of the state of Washington (1954). *Northwest Fauna* 3:3–80.

Johnston, D. W. 1975. Organochlorine pesticide residues in small migratory birds, 1964–73. *Pesticides Monitoring Journal* 9:79–88.

Jones, H. L., and J. M. Diamond. 1976. Short-time-base studies of turnover in breeding bird populations on the California Channel Islands. *Condor* 78:526–49.

Karr, J. R., and R. R. Roth. 1971. Vegetation structure and avian diversity in several New World areas. *American Naturalist* 105:423–35.

Kephart, D. G., and S. J. Arnold. 1982. Garter snake diets in a fluctuating environment: A seven-year study. *Ecology* 63(5):1232–36.

Klopfer, P. H. 1959. An analysis of learning in young Anatidae. *Ecology* 40(1):90–102.

Koenig, W., and P. Hooge. 1992. Philopatry, detectability, and the distribution of dispersal distances. Paper presentation and abstract in the 62d Annual Meeting of the Cooper Ornithological Society, 22–28 June 1992, Seattle, Wash.

Lacava, J., and J. Hughes. 1984. Determining minimum viable population levels. *Wildlife Society Bulletin* 12:370–76.

Lacey, R. M., R. S. Baran, H. E. Balbach, R. G. Goettel, and W. D. Severinghaus. 1982. Off-road vehicle site selection. *Journal of Environmental Systems* 12:113–40.

Laikre, L., and N. Ryman. 1991. Inbreeding depression in a captive wolf (*Canis lupus*) population. *Conservation Biology* 5(1):33–40.

Lamberson, R. H., B. R. Noon, C. Voss, and K. S. McKelvey. 1994. Reserve design for territorial species: The effects of patch size and spacing on the viability of the northern spotted owl. *Conservation Biology* 8(1):185–95.

Lande, R. 1987. Extinction thresholds in demographic models of territorial populations. *American Naturalist* 130(4):624–35.

Lande, R., and G. F. Barrowclough. 1987. Effective population size, genetic variation, and their use in population management. In *Viable populations,* ed. M. E. Soule, 87–123. New York: Cambridge University Press.

Larkin, R. P., and D. Halkin. 1994. Wildlife software: a review of software packages for estimating animal home ranges. *Wildlife Society Bulletin* 22(2):274–87.

Leopold, A. 1933. *Game management.* New York: Charles Scribner's Sons.

Loker, C. A., and D. J. Decker. 1995. Colorado black bear hunting referendum: What was behind the vote? *Wildlife Society Bulletin* 23(3):370–76.

Luken, J. O. 1990. *Directing ecological succession.* New York: Chapman and Hall.

MacArthur, R. H., and J. W. MacArthur. 1961. On bird species diversity. *Ecology* 42:594–98.

McLellan, B. N., and D. M. Shackleton. 1988. Grizzly bears and resource-extraction industries: Effects of roads on behaviour, habitat use and demography. *Journal of Applied Ecology* 25:451–60.

Maehr, D. S., and G. B. Caddick. 1995. Demographics and genetic introgression in the Florida panther. *Conservation Biology* 9(5):1295–98.

Marcot, B. G. 1985. Habitat relationships of birds and young-growth Douglas-fir in northwestern California. Ph.D. dissertation, Oregon State University, Corvallis.

Marcot, B. G. 1996. An ecosystem context for bat management: A case study of the interior Columbia River basin, U.S.A. In *Bats and forests symposium: October 19–21, 1995, Victoria, British Columbia, Canada,* ed. R. M. R. Barclay and R. M. Brigham, 19–36. Victoria, B.C.: B.C. Ministry of Forests.

Marcot, B. G., L. K. Croft, J. F. Lehmkuhl, R. H. Naney, C. G. Niwa, W. R. Owen, and R. E. Sandquist. 1998. *Macroecology, paleoecology, and ecological integrity of terrestrial species and communities of the interior Columbia River Basin and portions of the Klamath and Great Basins.* USDA Forest Service, General Technical Report PNW-GTR-410. Portland, Oreg.

Margalef, R. 1963. On certain unifying principles in ecology. *American Naturalist* 97:357–74.

Martin, T. E., and D. M. Finch, ed. 1995. *Ecology and management of Neotropical Migratory birds.* Oxford: Oxford University Press.

Maruyama, T. 1977. *Stochastic problems in population genetics.* Berlin: Springer-Verlag.

Marzluff, J. M., and R. P. Balda. 1989. Causes and consequences of female-biased dispersal in a flock-living bird, the pinyon jay. *Ecology* 70:316–28.

Max, T. A., R. A. Souter, and K. A. O'Halloran. 1990. Statistical estimators for monitoring spotted owls in Oregon and Washington in 1987. USDA Forest Service, Research Paper PNW-RP-420. Portland Oreg. 13 pp.

Menges, E. S. 1991. The application of minimum viable population theory to plants. In *Genetics and conservation of rare plants,* ed. D. A. Falk and K. E. Holsinger, 45–61. New York: Oxford University Press.

Merriam, C. H. 1898. *Life zones and crop zones.* USDA Biological Services Bulletin no. 10.

Miller, S. D. 1990. Impact of increased bear hunting on survivorship of young bears. *Wildlife Society Bulletin* 18(4):462–67.

Miller, S. D., and W. B. Ballard. 1992. Analysis of an effort to increase moose calf survivorship by increased hunting of brown bears in south-central Alaska. *Wildlife Society Bulletin* 20(4):445–54.

Mills, L. S. 1995. Edge effects and isolation: Red-backed voles on forest remnants. *Conservation Biology* 9:395–403.

Mills, L. S., and P. E. Smouse. 1994. Demographic consequences of inbreeding in remnant populations. *American Naturalist* 144(3):412–31.

Morrison, M. L., and E. C. Meslow. 1983. Impacts of forest herbicides on wildlife: Toxicity and habitat alteration. *Transactions of the North American Wildlife and Natural Resources Conference* 48:175–85.

Morton, M. L. 1992. Effects of sex and birth date on premigration biology, migration schedules, return rates and natal dispersal in the mountain white-crowned sparrow. *Condor* 94:117–33.

Moulding, J. D. 1976. Effects of a low-persistence insecticide on forest bird populations. *Auk* 93:692–708.

Murphy, E. C., F. J. Singer, and L. Nichols. 1990. Effects of hunting on survival and productivity of Dall sheep. *Journal of Wildlife Management* 54(2):284–90.

Murray, J. D. 1988. Spatial dispersal of species. *Trends in Ecology and Evolution* 3(11):307–9.

Nagorsen, D. W., and R. M. Brigham. 1993. *Bats of British Columbia.* Vancouver, B.C.: University of British Columbia Press.

NCASI. 1996. *The National Gap Analysis Program: Ecological assumptions and sensitivity to uncertainty.* National Council of the Paper Industry for Air and Stream Improvement, Inc., Technical Bulletin No. 720. Research Triangle Park, N.C. 56 pp. + 16 pp. appendix.

Nussbaum, R. A., E. D. Brodie, Jr., and R. M. Storm. 1983. *Amphibians and reptiles of the Pacific Northwest.* Moscow, Idaho: University of Idaho Press.

Odum, E. P. 1969. The strategy of ecosystem development. *Science* 164:262–70.

Odum, E. P. 1971. *Fundamentals of ecology.* 3d ed. Philadelphia: Saunders.

OTA (Office of Technology and Assessment). 1993. *Harmful non-indigenous species in the United States.* OTA-F-565, 2 vols. U.S. Congress, Office of Technology Assessment. 391 pp. + 57 pp. summary.

Oxley, D. J., M. B. Fenton, and G. R. Carmody. 1974. The effects of roads on populations of small mammals. *Journal of Applied Ecology* 11:51–59.

Paton, P. W. C. 1994. The effect of edge on avian nest success: how strong is the evidence? *Conservation Biology* 8:17–26.

Peterson, C. R. 1987. Daily variation in the body temperatures of free-ranging garter snakes. *Ecology* 68(1):160–69.

Petraitis, P. S. 1981. Algebraic and graphical relationships among niche breadth measures. *Ecology* 62:545–48.

Pollard, E., K. H. Lakhani, and P. Rothery. 1987. The detection of density-dependence from a series of annual censuses. *Ecology* 68:2046–55.

Possingham, H. P., I. Davies, I. R. Noble, and T. W. Norton. 1992. A metapopulation simulation model for assessing the likelihood of plant and animal extinctions. *Mathematics and Computers in Simulation* 33:367–72.

Quinlan, J. R. 1986. Induction of decision trees. *Machine Learning* 1(1):81–106.

Ramsey, F. L., and J. M. Scott. 1979. Estimating population densities from variable circular plot surveys. In *Sampling biological populations,* ed. R. M. Cormack, G. P. Patil, and D. S. Robson, 155–81. Statistical Ecology Series Vol. 5. Fairland, Md.: International Cooperative Publishing House.

Raphael, M. 1988. Long-term trends in abundance of amphibians, reptiles, and mammals in Douglas-fir forests of northwest California. *Management of amphibians, reptiles, and small mammals in North America,* ed. R. C. Szaro, K. E. Severson, and D. R. Patton, 23–31. USDA Forest Service General Technical Report RM-166. Rocky Mountain Forest and Range Experiment Station, Flagstaff, Ariz.

Reed, J. M., P. D. Doerr, and J. R. Walters. 1988. Minimum viable population size of the red-cockaded woodpecker. *Journal of Wildlife Management* 52:385–91.

Reynolds, R. T., J. M. Scott, and R. A. Nussbaum. 1980. A variable circular-plot method for estimating bird numbers. *Condor* 82:309–13.

Richter, A. R., S. R. Humphrey, J. B. Cope, and V. Brack, Jr. 1993. Modified cave entrances: thermal effect on body mass and resulting decline of endangered Indiana bats (*Myotis sodalis*). *Conservation Biology* 7(2):407–15.

Robinson, S. K., F. R. Thompson III, T. M. Donovan, D. R. Whitehead, and J. Faaborg. 1995. Regional forest fragmentation and the nesting success of migratory birds. *Science* 267:1987–90.

Robinson, W. L., and E. G. Bolen. 1984. *Wildlife ecology and management.* New York: Macmillian Publishing Co.

Rodgers, A. R., and W. E. Klenner. 1990. Competition and the geometric model of dispersal in vertebrates. *Ecology* 71:818–22.

Rotenberry, J. T. 1985. The role of habitation avian community composition: Physiognomy or floristics? *Oecologia* 67:213–17.

Roth, R. R. 1976. Spatial heterogeneity and bird species diversity. *Ecology* 57:773–82.

Roy, M. S., E. Geffen, D. Smith, E. A. Ostrander, and R. K. Wayne. 1994. Patterns of differentiation and hybridization in North American wolflike canids, revealed by analysis of microsatellite loci. *Molecular Biology and Evolution* 11:553–70.

Ruggiero, L. F., R. S. Holthausen, B. G. Marcot, K. B. Aubry, J. W. Thomas, and E. C. Meslow. 1988. Ecological dependency: The concept and its implications for research and management. *North American Wildlife Natural Resources Conference* 53:115–26.

Samson, F., and F. Knopf. 1994. Prairie conservation in North Ameria. *BioScience* 44(6):418–21.

Samuel, M. D., and E. O. Garton. 1985. Home range: a weighted normal estimate and tests of underlying assumptions. *Journal of Wildlife Management* 49:513–19.

Sauer, J. R., S. Schwartz, B. G. Peterjohn, and J. E. Hines. 1996. The North American Breeding Bird Survey home page. Version 95.1. Patuxent Wildlife Research Center, Laurel, Md.

Savory, A. 1988. *Holistic resource management.* Washington, D.C.: Island Press.

Scott, J. M., S. Mountainspring, F. L. Ramsey, and C. B. Kepler. 1986. *Forest bird communities of the Hawaiian Islands: Their dynamics, ecology, and conservation.* Studies in Avian Biology 9. Los Angeles: Cooper Ornithological Society.

Scott, J. M., F. Davis, B. Csuti, R. Noss, B. Butterfield, C. Groves, H. Anderson, S. Caicco, F. D'erchia, T. C. Edwards, Jr., J. Ulliman, and R. G. Wright. 1993. Gap analysis: A geographic approach to protection of biological diversity. *Wildlife Monographs* 123:1–41.

Scott, T. A. 1994. Irruptive dispersal of black-shouldered kites to a coastal island. *Condor* 96:197–200.

Seabrook, W. A., and E. B. Dettmann. 1996. Roads as activity corridors for cane toads in Australia. *Journal of Wildlife Management* 60(2):363–68.

Senner, J. W. 1980. Inbreeding depression and the survival of zoo populations. In *Conservation biology: An evolutionary-ecological perspective,* ed. M. E. Soule and B. A. Wilcox, 209–24. Sunderland, Mass.: Sinauer Associates.

Shaffer, M. L. 1983. Determining minimum viable population sizes for the grizzly bear. *International Conference on Bear Research and Management* 5:133–39.

Shapiro, A. D. 1987. *Structured induction in expert systems.* Reading, Mass.: Addison-Wesley.

Shugart, G. W., M. A. Fitch, and G. A. Fox. 1987. Female floaters and nonbreeding secondary females in herring gulls. *Condor* 89(4):902–6.

Sibley, C. G., and B. L. Monroe, Jr. 1991. *Distribution and taxonomy of the birds of the world.* New Haven, Conn.: Yale University Press.

Smith, J. L. D., A. H. Berner, F. J. Cuthbert, and J. A. Kitts. 1992. Interest in fee hunting by Minnesota small-game hunters. *Wildlife Society Bulletin* 20(1):20–26.

Smith, S. M. 1978. The "underworld" in a territorial sparrow: Adaptive strategy for floaters. *American Naturalist* 112:571–82.

Smith, S. M. 1984. Flock switching in chickadees: Why be a winter floater? *American Naturalist* 123:81–98.

Smith, S. M. 1989. Black-capped chickadee summer floaters. *Wilson Bulletin* 101(2):344–49.

Snyder, M. A. 1993. Interactions between Abert's squirrel and ponderosa pine: The relationship between selective herbivory and host plant fitness. *American Naturalist* 141:866–79.

Soule, M. E. 1980. Thresholds for survival: maintaining fitness and evolutionary potential. In *Conservation biology: An evolutionary-ecological perspective,* ed. M. E. Soule and B. A. Wilcox, 151–70. Sunderland, Mass.: Sinauer Associates.

Soule, M. E. 1987. *Viable populations for conservation.* Cambridge: Cambridge University Press.

Soule, M. E. 1990. The onslaught of alien species, and other challenges in the coming decades. *Conservation Biology* 4:233–39.

Soule, M. E., D. T. Boulger, A. C. Alberts, R. Sauvajot, J. Wright, M. Sorice, and S. Hill. 1988. Reconstructed dynamics of rapid extinctions of chaparral-requiring birds in urban habitat islands. *Conservation Biology* 2:75–92.

Spencer, S. R., G. N. Cameron, and R. K. Swihart. 1990. Operationally defining home range: temporal dependence exhibited by hispid cotton rats. *Ecology* 71:1817–22.

Stebbins, R. C. 1966. *A field guide to western reptiles and amphibians.* Boston: Houghton Mifflin Co.

Stiles, F. G., and D. A. Clark. 1989. Conservation of tropical rain forest birds: a case study from Costa Rica. *American Birds* (Fall):420–28.

Storm, R. M., and W. P. Leonard, eds. 1995. *Reptiles of Washington and Oregon.* Seattle: Seattle Audubon Society.

Sullivan, T. P. 1990. Influence of forest herbicide on deer mouse and Oregon vole population dynamics. *Journal of Wildlife Management* 54:566–76.

Tansley, A. G. 1935. The use and abuse of vegetational concepts and terms. *Ecology* 16:284–307.

Taylor, A. D. 1990. Metapopulations, dispersal, and predator-prey dynamics: An overview. *Ecology* 71:429–33.

Taylor, A. L., and E. D. Forsman. 1976. Recent range extensions of the barred owl in western North America, including the first records for Oregon. *Condor* 78(4): 560–61.

Terborgh, J. 1986. Keystone plant resources in the tropical forest. In *Conservation biology: The science of scarcity and diversity,* ed. M. E. Soule, 330–44. Sunderland, Mass.: Sinauer Associates.

Terborgh, J. 1989. *Where have all the birds gone?* Princeton, N.J.: Princeton University Press.

Thomas, C. D. 1990. What do real population dynamics tell us about minimum viable populations sizes? *Conservation Biology* 4:324–27.

Thomas, J. W., R. J. Miller, C. Maser, R. G. Anderson, and B. E. Carter. 1979. Plant communities and successional stages. In *Wildlife habitats in managed forests: The Blue Mountains of Oregon and Washington,* ed. J. W. Thomas, 22–39. USDA Forest Service, Agricultural Handbook No. 553.

Tiebout, H. M., III, and K. E. Brugger. 1995. Ecological risk assessment of pesticides for terrestrial vertebrates: evaluation and application of the U.S. Environmental Protection Agency's quotient model. *Conservation Biology* 9(6):1605–18.

Trammel, M. A., and J. L. Butler. 1995. Effects of exotic plants on native ungulate use of habitat. *Journal of Wildlife Management* 59(4):808–16.

Turesson, G. 1922. The genotypical response of the plant species to habitat. *Hereditas* 3:211–350.

Tyser, R. W., and C. A. Worley. 1992. Alien flora in grasslands adjacent to road and trail corridors in Glacier National Park, Montana (U.S.A.). *Conservation Biology* 6(2):253–62.

Van Hulst, R. 1992. From population dynamics to community dynamics: Modelling succession as a species replacement process. *Plant succession: Theory and prediction,* ed. D. C. Glenn-Lewin, R. K. Peet, and T. T. Veblen, 188–214. New York: Chapman and Hall.

Viemeister, N. F. 1983. Auditory intensity discrimination at high frequencies in the presence of noise. *Science* 221:1206–7.

Vtorov, I. P. 1993. Feral pig removal: Effects on soil microarthropods in a Hawaiian rain forest. *Journal of Wildlife Management* 57(4):875–80.

Warren, C. D., J. M. Peek, G. L. Servheen, and P. Zager. 1996. Habitat use and movements of two ecotypes of translocated caribou in Idaho and British Columbia. *Conservation Biology* 10(2):547–53.

Washitani, I. 1996. Predicted genetic consequences of strong fertility selection due to pollinator loss in an isolated population of Primula sieboldii. *Conservation Biology* 10(1):59–64.

Wayne, R. K., and S. M. Jenks. 1991. Mitochondrial DNA analysis implying extensive hybridization of the endangered red wolf *Canis rufus. Nature* 351:565–68.

Westman, W. E. 1990. Park management of exotic plant species: Problems and issues. *Conservation Biology* 4:251–60.
Whittaker, R. H. 1975. *Communities and ecosystems.* 2d ed. New York: Macmillian.
Wiens, J. A. 1969. *An approach to the study of ecological relationships among terrestrial birds.* Ornithological Monographs no. 8. Washington, D.C.: American Ornithologists, Union.
Williams, C. F., and J. W. Mjelde. 1994. Conducting a financial analysis of quail hunting within the Conservation Reserve Program. *Wildlife Society Bulletin* 22(2):233–41.
Willson, M. F. 1974. Avian community organization and habitat structure. *Ecology* 55:1017–29.
Wolf, C. M., B. Griffith, C. Reed, and S. A. Temple. 1996. Avian and mammalian translocations: Update and reanalysis of 1987 survey data. *Conservation Biology* 10(4):1142–54.
Worton, B. J. 1995. Using Monte Carlo simulation to evaluate Kernel-based home range estimators. *Journal of Wildlife Management* 59(4):794–800.

PART 2

The Measurement of Wildlife-Habitat Relationships

4 The Experimental Approach in Wildlife Science

One way to approach definition is to consider science as a process of questioning and answering. The questions are, by definition, scientific if they are about relationships among observed phenomena. The proposed answers must, again by definition, be in natural terms and testable in some material way. On that basis, a definition of science as a whole would be: Science is an exploration of the material universe that seeks natural, orderly relationships among observed phenomena and that is self-testing [that is, our understanding of the universe can be tested by the same kinds of observations from which they arose].

—G. G. SIMPSON, *THIS VIEW OF LIFE*

Introduction

Study of the relationships between animals and their habitats is no different from any other scientific endeavor in that the "scientific method" must be used. In the broadest sense, "scientific thinking" is the logical thought process one might use to discover, for example, why a car will not start. The activities of scientists striving to understand how the universe works, however, are collectively called the scientific method. This method appears relatively easy to apply when described in its simplest form, yet its application, particularly in ecology and related fields, has generated considerable discussion. The discussions have focused on which activities are the most useful or important (e.g., Romesburg 1981; Peters 1991) and how to perform them correctly (e.g., Murphy 1990; Nudds and Morrison 1991; Drew 1994). Thus, telling someone to follow the scientific method when designing and carrying out a study, although good advice, is inadequate as a set of instructions for what actually to do.

This chapter provides a brief description of the activities associated with the scientific method and their relation to the study of animals and habitats. We also present some issues about the scientific method that have generated discussions among ecologists and wildlife biologists. Our purpose is not to resolve the controversies but to identify the kinds of information needed by those who study wildlife-habitat relationships. We also discuss the design and application of experiments in wildlife science and some of the problems inherent in experiments in field and laboratory situations. We focus on the experimental approach because it is a powerful tool that has been underutilized by wildlife biologists. We do not review statistical models, even though the design of an experiment is intimately linked to a statistical model. Such a review is beyond the scope of this chapter, and several texts on statistics and experimental design are available (e.g., Skalski and Robson 1992; Scheiner and Gurevitch 1993; Kuehl 1994; Sokal and Rohlf 1995). We end the chapter with a discussion of when to use field and laboratory experiments and what to do when tightly controlled experiments cannot be done because of practical constraints.

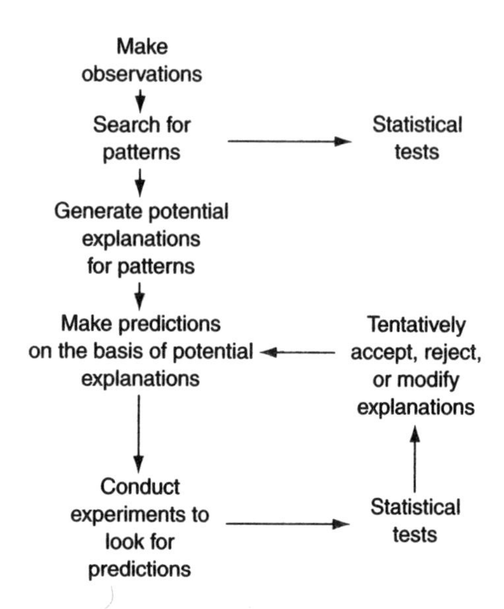

Figure 4.1. Activities in the scientific method and common places in the method where statistical tests are employed

The Scientific Method

Carey (1994:5) defined the scientific method as "a rigorous process whereby new ideas about how some part of the natural world works are put to the test." Descriptions of the scientific method often include three to six "steps," or activities, depending on how they are combined: (1) making observations; (2) searching for patterns among observations; (3) generating a potential explanation for the patterns detected; (4) deducing predictions from the proposed explanation; (5) testing the potential explanation by looking for the predicted phenomena, often after some experimental manipulation; and (6) tentatively accepting, rejecting, or modifying the proposed explanation, depending on how well predictions were met (fig. 4.1).

Observations are the starting point of science (Simpson 1964:88). Some philosophers argue about whether what we perceive is real, but most ecologists are willing to accept that our senses provide us with a good approximation of the existing world (Simpson 1964). That is not to say that our observations are always free of error, but numerous observations by different observers from different points in space and time generally help separate observational errors from reality (Matter and Mannan 1989).

Observations tell us little about the natural world when made independently; their value becomes apparent, however, when patterns can be recognized among them. If, for example, we consistently were to observe individuals of a given species in a particular type of vegetation and recognize the association, then some level of prediction might be possible. A simple prediction from this pattern is that we can expect to see individuals of this species when we visit different sites within the same vegetation type. A more complex prediction is that the number of individuals of this species will increase if we increase the amount of the vegetation type through some management activity. Several factors, such as whether the species successfully survives and breeds in the vegetation type, will influence whether the second prediction is accurate. The point, however, is that predictions of this kind become possible only after patterns of association are recognized.

The process by which we come to recognize patterns is called *induction* (Hanson 1965), or reasoning from particulars to generals. The "particulars" are the individual observations we make about the natural world, and the "generals" are the associations or patterns we detect among them. Humans have the ability to recognize patterns partly because our brains can catalog and retain information for long periods of time, but recognition of subtle patterns requires keen observational skills and often immersive experience. Skills associated with observation almost certainly vary among individuals, but they can be enhanced for most with practice and knowledge.

Scientists usually search for patterns in nature in a more formal manner than casual observers. Scientific observations often are made systematically or come from random "samples" so that they provide a picture of the natural world that is as accurate and unbiased as possible. If a relationship between a species and a particular vegetation type is suspected, then a series of surveys for the species might be conducted throughout the year, over several years, in a random sample of areas within the vegetation type to determine if the relationship is real. Furthermore, scientists might employ statistics to help them detect patterns or correlations among variables that otherwise would be difficult to discern (fig. 4.1).

It is important to remember that a correlation between two variables says nothing concrete about the nature of the relationship between the variables. A positive correlation between the abundances of two species—say, mosquitoes and deer ticks—in an area does not indicate that an increase in the abundance of mosquitoes *caused* or somehow influenced the increase in deer ticks, or vice versa. The two species simply could be responding to the effects of increased rainfall on the different resources of each and be increasing independently of each other. Conversely, if the abundances of two species are positively correlated and one species (e.g., a flycatcher) depends on another (e.g., a woodpecker, to excavate its next cavities), then there could be a mechanism relating the covariance of the two species. Patterns alone, however, tell us little about these mechanisms.

Once correlations or patterns among variables have been identified and confirmed, some people are stimulated to wonder what processes or mechanisms might have created the patterns; that is, they begin to ask why. This speculation can result in the development of potential explanations of observed patterns. Such "speculative explanations" are called research hypotheses (Romesburg 1981:295), and they articulate a mechanism or process that might be responsible for the pattern of interest.

The term sometimes used to describe the process by which we generate research hypotheses, *induction* (Platt 1964; Popper 1981a; Medawar 1984), is the same one used to describe how we recognize patterns. But the mental activities associated with hypothesis generation, although they involve pattern recognition, probably require more information, creativity, and insightfulness than pattern recognition alone. Neither process, however, is well understood or completely under our control (Matter and Mannan 1989). Thus, the word *method* in the term *scientific method* is somewhat misleading if it is interpreted to mean that all people will be able to perform all steps or activities equally if given enough instruction. We emphasize that anyone can improve his or her chances for significant participation in the creative aspects of science by being well informed in the area of his or her research and understanding how science works.

Potential explanations of patterns or phenomena in nature are constrained by what is already known about the world. They cannot, for example, contradict the laws of physics or umbrella theories, such as evolution. But even with these constraints, it is possible (and even likely) that more than one research hypothesis will be proposed for a given pattern or phenomenon. Chamberlin (1897) advocated working with multiple research hypotheses, because he felt that a single hypothesis is more likely to become accepted without having undergone some form of test. We do not think that the scientific method *requires* multiple explanations, but it does require that research hypotheses be viewed as *candidate* explanations until they are tested. Prior to testing, scientists must remain objective about their hypotheses, no matter how convincing they sound. If more than one potential explanation for a given pattern exists, tests should help determine which is valid. Deciding which of several competing hypotheses to test first depends, in part, on which best fits existing theories and empirical information.

The approach for testing research hypotheses, and either gaining or losing confidence in their validity, is called the hypothetico-deductive (H-D) method (Romesburg 1981). In this activity, scientists devise tests that put their hypotheses at risk; that is, they perform experiments or propose novel observations that potentially refute or cast doubt on their candidate explanations. Romesburg (1981) noted that the causal process or mechanism described in a research hypothesis might be difficult to observe directly. Thus, "a research hypothesis must be tested indirectly because it *embodies a process,* and experiments can only give facts entailed by a process" (Romesburg 1981:295; emphasis added). A general model of an H-D test might be: If my hypothesis is true (i.e., if the mechanistic process I envision is actually ongoing), and if I manipulate a critical element associated with the proposed process while simultaneously controlling other critical elements, certain things must happen, or, conversely, certain things cannot happen.

The "things" in this model are predictions—observable events or patterns—derived from what should happen after a manipulation, given that the tentative explanation being tested is valid. *Deduction* is the word which describes the process by which predictions are generated in an H-D test. Deduction, by definition, is the derivation of a conclusion by reasoning or logic, where the conclusion follows necessarily from the premises. The predictions of an H-D test, then, are logically deduced on the basis of the premise of the hypothesis (hence the term *hypothetico-deductive*), and they should be specific to the hypothesis being tested. If the predictions generated in an H-D test could come about by processes other than the one being tested (e.g., by a competing hypothesis), then the test would not be a good one.

The best tests of research hypotheses, according to Popper (1981b), are falsification tests—those in which all elements are controlled except the element critical to the proposed explanation, and those that involve attempts to find results that are logically prohibited from occurring if the hypothesized explanation is true. If experiments produce logically prohibited consequences, the hypothesis is falsified and will need to be modified or discarded. Failure to find the prohibited consequences does not necessarily "prove" the hypothesis to be true but does increase our confidence in it (Simpson 1964). A high level of confidence in the validity of a given hypothesis should come only after it has been tested repeatedly in a variety of ways.

Tests of the kind Popper described are possible only if we can envision events or patterns that will refute the hypothesis under consideration. If hypotheses are vague or general, they sometimes can be interpreted to explain nearly any related observation in nature or any outcome of a pertinent manipulation. Hypotheses of this kind are considered unfalsifiable, that is, not amenable to tests. Unfalsifiable hypotheses sometimes can be improved by describing more explicitly the mechanism or process being tested or by making them more specific. Identifying critical elements to manipulate in an experiment and making what Popper called "risky predictions" often are easier if the details of the proposed process are explicitly described.

Predictions that come from hypotheses about ecological phenomena often involve the responses of individual animals or plants. These responses are not likely to be identical among individuals, no matter how carefully a test is controlled, at least partly because each individual has a unique genotype and environmental history. For example, providing the same amount of an important nutrient to individual plants, even under highly controlled conditions, will not result in the same amount of growth in each plant. This kind of variation often requires that statistical tests be used to assess whether the predictions of an H-D test are met. In this example, a scientist might have hypothesized that the nutrient in question plays an important function in the growth of the plant and articulates a mechanism through which the nutrient aids plant growth. A prediction of the test might be that plants which receive the nutrient will grow more rapidly than those which do not. A statistical test would be used to compare the average growth of plants in the two groups. The null "statistical hypothesis" in this example typically would be that there is no difference between the mean growth rates of the two groups. Rejecting or failing to reject the null statistical hypothesis determines whether the predictions of the test have been met (fig. 4.1). Thus, the assessment of patterns with statistics can play an important role in the validation of a research hypothesis, but this role is most powerful when the statistical test is embedded in the broader framework of an H-D test.

Many discussions about the scientific method among ecologists and wildlife biologists have focused on formulating and testing "hypotheses." What constitutes a research hypothesis is one question that sometimes causes confusion among students and professional biologists alike, because the term is used in a variety of ways. Romesburg (1981) recognized this problem and clearly differentiated between research hypotheses (i.e., those that include a process and explain a pattern) and statistical hypotheses, which he defined as conjectures about classes of facts. Carey (1994:9) noted, "In the jargon of the scientist, just about any claim that may require testing before it is accepted will be called a hypothesis." Thus, a question about whether a species is more abundant in one mountain range than in another might be framed as a hypothesis. But Carey (1994) was clear that statements of this kind are not research hypotheses because they *do not include an explanation.* We suggest that questions of this nature, which address patterns, are best framed as statistical hypotheses, and that the

H-D method is best suited for testing hypotheses that include explanations.

Another issue that has generated considerable discussion among ecologists is the use of "null models" as tests of the influence of ecological processes on the structure of communities (Conner and Simberloff 1979; Strong et al. 1979; Diamond and Gilpin 1982; Harvey et al. 1983). Null model "tests" typically involve identifying a pattern that is presumed to have been caused primarily by some ecological process, such as competition. The observed pattern is then compared with a model of what is expected if the variable involved in the pattern is distributed by some "randomization procedure" (Scheiner 1993) or by a set of "random-process community assemblage rules" (Pleasants 1990). For example, one could compare the composition of species on the islands of an archipelago, presumed to have been determined largely by competition, with the composition that would occur if the presence and absence of species were determined by chance (see Harvey et al. 1983 for a review of this and other examples). The idea underlying this approach is that the random or null model represents what would occur in the community in the absence of the hypothesized ecological process (Harvey et al. 1983). The value of the approach is that, if the patterns observed in ecological communities potentially could come about by a series of chance events, then the ecological processes under consideration need not be invoked. Alternatively, if the patterns are different from those that are expected by chance, they may lend support to the idea that the hypothesized process has played a role in structuring the community.

Concern about the use of null models as tests focuses on at least two issues. The first is whether true null or random models can be created from the data that generated the pattern of interest. Harvey et al. (1983:190) noted that, "if some rearranged or reshuffled version of the data is used in constructing the null model, one cannot be sure that some biological interactions that may have shaped the data are not woven into the model." This problem is especially prevalent in null models in which constraints are placed on what constitutes chance. In the example presented above regarding species composition on islands of an archipelago, the null model might keep the *number* of species on the islands the same as in the original pattern (Harvey et al. 1983). The idea would be to account for effects, such as vegetation complexity or distance to sources of colonists, that partly determine species richness. But if competition also played a role in determining species richness, the effects of the very process that is supposed to be excluded in the null model may still be present.

A second issue about the use of null models is whether they actually determine, with a high level of confidence, if a hypothesized process has played a significant role in structuring a community. The design of a null model test clearly is different from that of an H-D test in that no critical element is manipulated and other potential processes are not completely excluded or controlled. Thus, null models may not determine whether a proposed process did or did not act in a given community, because alternative hypotheses for the observed patterns cannot be excluded (Keddy 1989). Manipulative experiments, however, are not likely to be feasible when dealing with patterns that span archipelagos and processes that occurred over long periods of time. Thus, approaches other than H-D tests may sometimes be the only recourse for understanding how the natural world works. The level of confidence in explanations evaluated with null models, however, should be tempered with the understanding that other explanations may still be feasible.

Application of the Experimental Approach

In the past, wildlife scientists did not often participate in experimental studies, partly because the questions they asked were generally about patterns of natural phenomena, not the processes that caused them. Species studied frequently were those that were of interest from a management perspective, and questions addressed patterns of relative abundance, habitat use, survival, and productivity. Wildlife science focused on describing these patterns because the information they provided generally was sufficient for the development of successful management strategies. Studies that document patterns in nature often require intensive and extensive surveys of animals or painstaking observations of animal behavior (see chapter 7), but usually not experiments.

Wildlife scientists were not alone in the past in ignoring experiments. Hairston (1989:x) noted that experiments were used infrequently by ecologists in general in the 1950s and 1960s, despite the pleas of advocates of the experimental approach. Of interest is that some of the early field and laboratory experiments performed by ecologists related directly to habitat relationships. For example, Harris (1952) attempted to test the idea of habitat selection experimentally by presenting individual prairie deer mice (*Peromyscus maniculatus bairdi*) and woodland deer mice (*P. m. gracilis*) with the choice of artificial woods or artificial fields. He found that test animals preferred the artificial habitat that most closely resembled the natural environment of their own subspecies and concluded that mice were reacting to visual cues provided by the artificial vegetation. Wecker (1963) later attempted to determine whether early experience (i.e., learning) plays a role in habitat recognition and selection in deer mice. He found that the behaviors associated with habitat selection are primarily controlled by what Mayr (1974) called closed genetic programs; that is, they are not greatly affected by early experience. Similar experiments conducted by Klopfer (1963), however, revealed that habitat selection by chipping sparrows (*Spizella passerina*) can be modified to some extent by early experience.

Experiments are a potentially powerful tool for increasing understanding of ecological processes and phenomena, and their use in ecological research, including wildlife science, has increased dramatically in recent years. However, the role of experiments in the scientific method is still sometimes misunderstood. Therefore, students hoping to pursue careers as scientists in wildlife ecology or related fields should familiarize themselves with the experimental approach. Knowledge of experiments should not diminish the value of wildlife studies designed to identify and elucidate patterns, but a more complete understanding of the purpose and design of experiments will facilitate their use when they are needed.

Purposes of Experiments

The primary purpose of an experiment, stated simply, is to test an idea about how the natural world works. In terms presented in the previous section, the purpose of an experiment is to test the validity of a research hypothesis.

A secondary purpose of experiments is to corroborate or describe the existence or nature of a pattern (i.e., for "fact-finding"; Kneller 1978:116). The design of fact-finding experiments is the same as or similar to those used to test research hypotheses, but there is no hypothesis being tested and there are no predictions made because the purpose is to find out what happens when, through manipulation, we mimic some natural event.

Considerations of Experimental Design

The model of a hypothetico-deductive test, presented in the previous section, forms the conceptual foundation for designing an H-D experiment, but there is no simple formula or methodology that can provide experimental details. Each experiment is an exercise in creativity because the experimenter must be able to identify an element that plays an important role in the hypothesis being tested, manipulate it without changing other important elements, and logically deduce (predict) what will happen after the manipulation.

The strength of experimental results either to support or to reject a research hypothesis depends on whether the experiment is properly designed and executed. Consideration of several rules and concepts about experimental design should help strengthen experimental results. Below, we briefly review several factors that should be considered during the design of experiments. Our review focuses on field experiments because they are more common in wildlife science than experiments conducted in laboratories.

One important rule associated with experimentation is that the experiment must be repeatable. This means that other scientists must be able to duplicate the methods used in an experiment so that they can duplicate the experiment, if they wish. The methods, therefore, must not include any subjective assessments of important variables, and they must be carefully and accurately described, usually in a written report or publication. In fact, field experiments in ecology and wildlife science rarely are repeated, but that is another matter, the implications of which are discussed below.

Another important consideration when designing an H-D experiment is that the experimental units (e.g., individual animals or plants, or plots of land) involved must accurately represent the domain to which the hypothesis is supposed to apply. Thus, the *sampling frame* (Skalski and Robson 1992), or pool from which samples will be drawn, should match the target population if the experiment is to have relevance to the target population. For example, if the hypothesis being tested is about a process that may be operating within a given vegetation type, the plots used in the experiment ideally should be selected at random from the entire distribution of the vegetation type. In reality, experimental units used in field experiments often are constrained by lack of access to private lands or by lack of travel funds. And, in some situations, experimental units are purposely chosen so that they are similar to each other in order to reduce variation (see below). Thus, for several reasons, the experimental units used in an experiment might come from a subset of the "target population," and the inferences that can be drawn from the experiment are then limited (see Skalski and Robson 1992).

Manipulation of an element that is central to the hypothesis being tested is a critical component of an H-D experiment. Manipulations in experiments usually are called *treatments*. Ideally, important biotic and abiotic factors not being manipulated are controlled by the experimenter; thus, changes in the variable of interest after the manipulation can be attributed to the treatment. In field experiments, these other factors usually are not under control, but the effects their changes have on experimental results must be assessed. This is done in two ways. The first is with information about the variable of interest before and after treatment. There are no set criteria to determine the period of time over which these data should be collected, but Hairston (1989:24) noted that lack of adequate baseline (before-treatment) data was a common failing among ecological experiments. Measurements after treatment should extend at least through the period during which the variable of interest is expected to change.

The second way to assess the effects of uncontrolled factors on field experiments is with the use of *controls* (Hairston 1989:24–26). Controls are experimental units that are not manipulated but are assessed in the same manner as units that are. Assessing both the changes in the variable of interest before and after manipulation and the changes in the unmanipulated control units provides a measure of how much the variable of interest would have changed in the absence of the manipulation. For example, Franzblau and Collins (1980) hypothesized that food availability is a primary part of the process that determines the size of territories in birds. They predicted that, if their hypothesis were true, adding food to territories of rufous-sided towhees (*Pipilo erythrophthalmus*) should result in a decrease in the size of the birds' territories. The first steps in their experiment were to identify a sample of territories of rufous-sided towhees and measure the size of each territory. Food was then added to some of the territories, the *treatment* or *experimental territories,* and other territories were left untreated as *controls.* The size of each territory was remeasured after the addition of food, and changes in the size of treatment territories were assessed relative to changes in the controls. The value of the controls becomes apparent if, for example, the size of all territories had decreased because of some change in the environment other than the addition of food or because of some effect of the observer. If this had happened, there would have been no way to evaluate changes in the treatment territories without comparing them with changes in the controls.

Application of treatments and selection of experimental units are associated with two concepts, *randomization* and *replication* (Fisher 1947), which are integral to the design of experiments. Randomization is one way to protect the experiment from unknown sources of experimental bias (Skalski and Robson 1992). When experimental units are selected, there inevitably will be variation among them, and these differences could influence the effects of a treatment. Therefore, designation of which experimental units receive treatment should be made randomly (or randomly with some constraints) (Cox 1958) so that "a treatment is not consistently favored or impaired by extraneous or unexpected sources of variation" (Skalski and Robson 1992:16). For example, Franzblau and Collins (1980) provided food to randomly selected territories within their study area so that differences between the territories would not uniformly bias the effects of their treatment. If, instead, they had arbitrarily provided food to territories on the north half of their study area, some environmental factor in that portion of the study area could have either ameliorated or enhanced the effects of additional food on territory size and compromised their experimental results.

Responses of experimental units to treatments will not be identical, no matter how carefully a test is controlled. Therefore, the responses of several experimental units must be assessed to increase the precision of the estimate of the variable of interest and thus help determine if the predictions of the experiment were met. This approach is called *replication,* and each set of treatment and control units is called a *replicate.* The number of replicates needed in an experiment is positively related to the amount of variation or error in the experiment that will mask the effects of the treatment. A host of factors can influence experimental error (see Skalski and Robson 1992 for a review), but it can be estimated from other similar experiments or preliminary surveys. Error in an experiment can be reduced through appropriate design. For example, if experimental units are to be selected along an environmental gradient that will influence the effects of the treatment, one way to reduce error is to group experimental units along the gradient and assess the effects of the treatment on groups. This technique, called *blocking,* is an attempt to remove experimental error by assigning as much of it as possible to differences between groups or blocks (Skalski and Robson 1992).

Another consideration related to selection of replicates or experimental units is that they should be *independent* of each other. Independent experimental units "are interspersed in time and space so that the response [to a treatment] of any one unit has no influence on the response of any other unit" (Smith 1996:17). Selecting independent replicates in field experiments is sometimes difficult because of interactions between animals and the movements of animals, water, or other materials between experimental units. For example, if the movements or behavior of towhees in territories receiving food had influenced the size of adjacent territories, or vice versa, the replicates used by Franzblau and Collins (1980) would not have been independent. Another way that independence sometimes is violated in experiments is by sampling repeatedly from one site (e.g., an area treated with prescribed fire) and considering each sample a replicate. The samples in this situation are not independent of each other because they all come from a single application of the treatment. Analyzing samples of this kind as if they were independent replicates is one form of *pseudoreplication* (Hurlbert 1984).

Rejecting or failing to reject the null statistical hypothesis in an H-D test either lends some support to the research hypothesis under consideration or falsifies it. It is important, therefore, to understand the kinds of errors that can be made when deciding the outcome of a statistical test. A Type I error is made if the null hypothesis is rejected when it is true, and a Type II error is made if the null hypothesis is not rejected when it is false. The probability of a Type I error (α) in an experiment is set by the investigator, and convention has established the value of $\alpha = 0.05$. The probability of a Type II error in an experiment is denoted by β. The reciprocal of β, $(1 - \beta)$, is called the *statistical power* of a test and is the probability of correctly rejecting a null hypothesis that is false.

Committing a Type II error would, in the typical design of a statistical test, indicate that no pattern exists when in fact one does. In an H-D experiment, a Type II error could lead to an inappropriate rejection of the research hypothesis. Analysis of the statistical power of an experiment during its design can help avoid Type II errors. Statistical power in a test increases with increasing sample size, *n,* and effect size. Effect size is the degree to which application of treatment causes a change and generally is estimated as the minimum change that is biologically meaningful (Steidl et al. 1997). If effect size can be estimated, analysis of statistical power during the design of an experiment can help determine the number of samples needed to detect the predetermined level of response (Peterman 1990: Steidl et al. 1997).

Field Experiments

Field experiments are powerful tools, in part, because researchers can bring about specific conditions in study plots that, without manipulation, might never occur or might take decades to occur. Ecologists often favor field experiments over laboratory experiments (Hairston 1989) because they are conducted in natural settings, and hence the results are generally considered more believable. The price of increased realism, however, is loss of control over most environmental variables. This loss of control may preclude testing some research hypotheses because the critical elements may be difficult or impossible to manipulate in field situations.

The dependence of field experiments on *controls* and the spatial scale at which many field experiments must be conducted make them vulnerable to several practical problems (Hilborn and Ludwig 1993). Finding sites or populations to serve as controls may be difficult, especially if the experiment concerns endangered species, limited environments, or mechanisms that operate on broad spatial scales. Furthermore, as the spatial scale of an experiment increases, experimental error also is likely to increase because of an increase in heterogeneity within and among experimental units. Increasing sample size to deal with this error may be impractical or impossible because of cost and availability. The small sample sizes often associated with field experiments also make randomization and proper interspersion of treatments difficult to achieve. Thus, field experiments are difficult to design and execute, and the difficulty usually increases as the spatial scale of the experiment increases. Unfortunately, mechanisms that operate at broad spatial scales sometimes are among the most important for understanding the causes of ecological patterns (Lawton 1996a).

Another problem inherent in field experiments (and to some degree in all ecological experiments) is that patterns and phenomena in ecology often have multiple and interacting causes (Peters 1991). Experimental designs exist that examine the effects and interactions of more than one treatment at a time, but they require more replicates than simpler experiments, and thus the practical problems outlined above are exacerbated.

Field experiments rarely are repeated (Hairston 1989). This is not surprising, given the cost and difficulties associated with designing and executing them and the fact that many take years to complete. Furthermore, most scientists probably would prefer to test a new idea than repeat someone else's work. However, confidence in a research hypothesis should come only after other scientists have tested it in different ways and found it to hold in a variety of places or situations (Kneller 1978:117). The lack of repetition of field experiments may mean that some (many?) hypotheses in ecology are accepted prematurely.

The difficulty of designing field experiments and the cost and time needed for their execution also may discourage experiments beyond initial, relatively crude levels of explanation. For example, an experiment could be done to test whether application of prescribed fire causes an increase in the abundance of an endangered species. The experiment might involve comparing the abundance of the species in treatment and control plots before and after burning. The results of such an experiment may support the idea that a causal relationship exists between fire and abundance of the species, but additional experiments would be necessary if biologists were interested in knowing, for example, whether fire increases the abundance of the species by increasing food, or by improving the structure of vegetation relative to the needs of the species, or both. Biologists also would want to know whether the increase in abundance is due to a functional or a numeric response and whether abundance is positively related to survival and productivity. The frequent need to explain patterns in nature at multiple levels is one of the concerns that Peters (1991) expressed about the utility of ecological experiments (also see Gavin 1991).

The level of explanation needed for management purposes depends on the situation. If, in the example above, prescribed fire was consistently effective in increasing the abundance of the endangered species, and if patterns of landownership and other social constraints permitted burning, then the level of explanation provided by the hypothetical experiment might be sufficient. If, on the other hand, burning was not feasible throughout the range of the species, it would be important to know what features of the habitat were changed by fire so that alternative manipulations could be devised to produce the same effect.

Laboratory Experiments

The design of ecological experiments in laboratories and other highly controlled environments is conceptually no different from those in field situations. The primary difference between laboratory and field experiments is the degree of control over experimental conditions. In a laboratory, nearly all important variables can be controlled. This level of control conceptually makes designing experiments simpler (although not necessarily easy), because the investigator can include and manipulate the environmental elements that are needed to test the idea under investigation and can exclude or control everything else. The questions that can be asked in laboratory experiments are thus more precise than those that can be asked in field experiments (Hairston 1989). Control of nearly all variables in an experiment also facilitates replication of trials within an experiment and duplication of the experiment itself, because conditions do not change over time and presumably can be re-created by other scientists.

Significant advances in understanding the natural world have been made in laboratory experiments in many scientific fields, including biology. However, there is reluctance by some, perhaps many, ecologists to accept the findings of laboratory experiments in ecology (Mertz and McCauley 1980) primarily because "its artificiality may simply swamp processes of ecological relevance" (Peters 1991:137).

Lawton (1996b) listed and discussed the primary arguments against experiments in the Ecotron, a set of highly controlled environmental chambers designed to replicate miniature terrestrial ecosystems (sometimes called bottle communities). Not surprisingly, the arguments against Lawton's (1996b) "bottle experiments" are the same arguments leveled against many other laboratory experiments in ecology. The arguments focus on the lack of applicability of the experiments to actual wildland environments and include concerns about the choice of species composition in experiments and the lack of major perturbations, immigration, emigration, seasons, appropriate scale, and important biological processes (Lawton 1996b). A major concern is that manipulating a single factor in a setting that does not include the full array of ecological interactions may produce results that could either unrealistically emphasize or diminish the importance of that factor (Peters 1991). Because of these concerns, results of laboratory experiments in ecology have been most widely accepted when they have addressed questions about physiological processes, bioenergetics, or the causes of relatively simple animal behaviors (Hairston 1989).

Improvement in the design of laboratory settings may help reduce some of the concerns about the applicability of results that come from them. The "big bottle" experiments of

Lawton (1996b) represent an effort to make an experimental setting for study of miniature terrestrial ecosystems more realistic, despite the expressed concerns. Lawton (1996b:668) noted, "Many ecologists seem to understand the need for simple experiments in pots and greenhouses; and yet a minority of colleagues become highly critical and concerned when an attempt is made to make artificial, controlled environments more rather than less realistic!"

One specific concern about artificiality in laboratory experiments is that the experimental settings (e.g., cages or aquaria) can frustrate or prevent behaviors that animals would perform in natural situations (Peters 1991). This is an obvious problem because many important ecological interactions, such as competition and predator-prey relations, directly involve animal behavior. Clearly, laboratory settings must be designed (e.g., be large enough) to allow animals to respond naturally to experimental conditions if the results are to be more acceptable (see Matter et al. 1989; Glickman and Caldwell 1994).

Animal-habitat relationships obviously involve the behavior of habitat selection. Animals seeking a place to live can respond to a specific set of environmental conditions in at least two ways—stay and try to establish themselves or leave. Experimental enclosures with no exits are most likely adequate settings for assessing the preference of animals for a given physical variable. For example, if fish are placed in a tank of water with a gradient of temperatures, most generally will stay in the part of the tank where the temperature is most acceptable to them (Warren 1971:186). However, enclosures are not adequate for assessing whether conditions, as a whole, inside the enclosure represent a suitable environment for an animal, because the animal cannot leave.

Enclosures with exits that allow animals to leave have been used to study a variety of subjects including social behavior and population size (Butler 1980), habitat relationships (Wilzbach 1985), and emigration and population regulation (McMahon and Tash 1988). Allowing animals to leave enclosures may represent a significant improvement in the design of ecological experiments in controlled settings (Matter et al. 1989), because the animals' responses to the conditions inside the enclosure indicate whether they are suitable.

How to Proceed

Romesburg (1981) felt that many of the ideas upon which the science of wildlife management has been founded are hypotheses that have not been tested. He argued convincingly that wildlife biologists do a good job of making observations, searching for patterns, and formulating hypotheses, but that attempts to apply the H-D method to verify hypotheses are rare. He encouraged wildlife biologists to participate more actively in the search for causal mechanisms through experimentation. More recently, Peters (1991) suggested that we have overemphasized the search for causal mechanisms in ecology (rather fruitlessly in his opinion), and that we should put more effort into the search for predictive relationships to help us solve, or at least deal with, ecological problems.

Given the problems inherent in both laboratory and field experiments and the somewhat conflicting advice in the literature about both the scientific method and its application, a legitimate question for those studying wildlife-habitat relationships is how to proceed with investigations. The answer, in our opinion, is dependent upon the kind of investigation that is undertaken and the resources available to the investigator. In wildlife ecology, the questions at the heart of many investigations are "species driven." That is, studies are done primarily because information about some aspect of the life history of a species is needed for management purposes. Often even basic information about these species, such as the types of vegetation they use, is lacking, so studies that document patterns of this kind are necessary. If an investigation is about identifying or verifying *patterns* in nature, then H-D tests are not necessary or applicable.

Studies of pattern are critical in science and management and should not be viewed as second-class activities (Weiner 1995; Lawton 1996a). In science, patterns stimulate the development of explanations and therefore are a vital part of the scientific method. Patterns may even help assess, with weak inference, the validity of research hypotheses (e.g., null models). In management, decisions frequently need to be made quickly and almost always without all the necessary information; thus, inferences from patterns alone may be sufficient for management action. For example, if we know that a species is declining in abundance and that the habitat upon which it depends also is declining, then some form of habitat conservation probably should be initiated, even though a causal relationship between the two patterns has not been established definitively.

If studies of pattern are undertaken, however, they should be conducted at several spatial scales over relatively long periods of time to ensure that the true nature of the patterns under investigation are revealed. Also, "attention should be paid to details such as using unbiased sampling techniques, collecting adequate numbers of samples, and employing appropriate statistics" (Nudds and Morrison 1991:759). Following these suggestions will help ensure that the patterns detected are real and that they are accurately described. Inaccurate descriptions of patterns can lead to inappropriate management actions and will hinder development of realistic explanations (Kodric-Brown and Brown 1993).

When the purpose of an investigation is to discover the nature of a process or mechanism that underlies a pattern in nature (i.e., when the study is "question driven"), an H-D test is needed. The species or ecological system to be studied in an H-D test might be chosen on the basis of how amenable the species or system is to control and manipulation. Neither laboratory nor field experiments are perfect for answering questions about ecological processes, yet both have value. Their relative values can perhaps be assessed by thinking of each as a tool designed for a specific function. Like most tools, each will do "some things well, some things badly, and other things not at all" (Lawton 1996b:669). We suggest that, when possible, investigators make use of both field and laboratory experiments to gain the highest level of confidence possible in their hypotheses.

Murphy (1990) advocated applying the H-D method to single, large-scale field manipulations, such as the creation of refuge boundaries, and treating these manipulations as experiments. He proposed that we could make predictions about what will happen after the manipulation is made, wait to see what in fact does happen, and then make inferences about the processes that cause the outcomes. Sinclair (1991) echoed these ideas, and Nudds and Morrison (1991:758–59) noted, "Hypothetico-deductive research is not characterized by whether it is experimental, because hypotheses can be tested with data not collected by experiment."

There are almost certainly some situations where hypotheses can be falsified conclusively by a single observation or a few observations

Box 4.1. Alternative approaches to classical statistical study design

What should be done when sample sizes are small, as with unreplicable landscapes or with threatened species with tiny populations? What should be done when controls do not exist, when baseline conditions cannot be established, and when unforeseen disturbances wreck the experiment? These and other nightmares haunt many real-world wildlife studies where funding levels and the pace of human activities and natural perturbations seldom allow for perfect experimental designs.

One answer may lie in the use of Bayesian statistics. Although their utility is controversial, they may provide a useful complement to other traditional approaches.

Bayesian approaches entail first describing *a priori* probabilities of outcomes given specific conditions, such as the specific environmental states, *E*, present with populations of specific sizes, *S*. Priors are thus denoted as the conditional probability of the environmental condition given a specific population size, or $P(E \mid S)$. For various population sizes, *S*, a *likelihood function* of priors can then be plotted. We will return to likelihood functions in a moment.

For a particular study area, the unconditional probabilities (overall frequency distribution) of population sizes $P(S)$ and of environmental conditions $P(E)$ are additional factors. Then, the posterior probability $P(S \mid E)$ of predicting a population size given an environmental state can be calculated by using Bayes theorem: $P(S \mid E) = P(E \mid S) \cdot P(S)/P(E)$. A graph can then be plotted showing posterior probabilities of population size for various environmental states.

Another way of expressing Bayes theorem (Reckhow 1990:2056) more explicitly displays the role of null hypotheses, H_o, and competing alternative hypotheses, H_A, in relation to data *x*:

$$P(H_o|x_1, \ldots, x_n) = \frac{P(x_1, \ldots, x_n| H_o) \quad P(H_o)}{P(x_1, \ldots, x_n) \quad P(H_o) + P(x_1, \ldots, x_n| H_A)\, P(H_A)}$$

In this formulation, the odds for H_o against H_A can be calculated as the ratio of the likelihood function of conditions *x* given the null hypothesis to the likelihood function of conditions *x* given the alternative hypothesis. This odds ratio is roughly analogous to the *P* value in classical statistics. The basic formula also can be extended to accommodate more than one alternative hypothesis.

Major advantages of the Bayesian approach are: (1) it makes use of existing knowledge or expert judgment in the estimation of the prior probabilities; and (2) it produces a useful formula that predicts outcomes in terms of likelihoods or odds.

Major complaints about the Bayesian approach are: (1) prior probabilities can be biased when based on best guesses rather than on empirical research; (2) prior probability values often greatly influence the posterior probabilities so that even minor bias or inaccuracy will change outcomes; and (3) the environmental states must be depicted in only a few, oversimplified categories.

In more complex Bayesian approaches, the "independent variable" *E* can be partitioned into multiple components, and the unconditional and conditional likelihoods for each component can be evaluated separately. Other variants to the approach provide for sequential estimation of the posteriors so that biases can be reduced by successive approximations or by continuous data collection. Some authors have revised the formulae for calculating posterior probabilities under various assumptions of statistical distributions of the prior probabilities.

not derived from experiments, but the level of confidence in accepting or rejecting an explanation under investigation is dependent upon the number of alternative explanations that also could account for observed outcomes. Ecosystems and the ecological processes that drive them are sufficiently complex that alternative explanations for observed phenomena usually are abundant (Peters 1991). Hence, large-scale manipulations, which usually are not replicated or randomly applied and during which there is no control or assessment of most environmental variables, are not likely to be exclusionary tests of underlying processes. Drew (1994:597) concluded, "The assertion that any question can be made scientifically rigorous by forcing it into [the H-D method] is false. Demographic and environmental indicators can be useful management tools, but when attempts are made to raise management targets to the level of scientific hypotheses—and exclusionary hypotheses at that—we only confuse the issue and give managers a false sense of assuredness."

Manipulating the environment, monitoring what happens, and changing the manipulation if we do not get the results we expect is called adaptive management (see chapters 11 and 12). This approach is very useful and provides a wealth of information about what happens when we change the environment under specific conditions at given points in time. It also may provide hints about the processes that cause the results we observe. Caution should be used, however, if adaptive management exercises are treated as H-D tests, because most are not designed as conventional experiments and cannot effectively be analyzed as such.

A concern of many ecologists is that field experiments, particularly those dealing with large spatial or temporal scales or small populations, will continue to be plagued by the lack of adequate controls and replication. Does this mean that understanding of broad-scale ecological processes will continue to be based on weak inferences? Bayesian statistics (box 4.1) may provide an alternative and more effective way to analyze the results of manipulations that lack classic experimental designs. This method of analysis is not a substitute for careful experimentation, but it may be useful when experimentation is not possible (Hilborn and Ludwig 1993).

Approaches used to understand how the world works vary according to circumstances and subject matter. Some subjects and circumstances may not lend themselves to experimentation. Understanding in these situations may need to be based on relatively weak inferences derived from patterns or creative analyses of manipulations without replication or randomization. However, properly designed experiments embedded in H-D tests are perhaps the best tool available for understanding ecological processes and, when feasible, should be considered.

Conclusion

It is critical that individuals participating in the study of wildlife-habitat relationships have a thorough understanding of all activities involved in the scientific method. This understanding does not mean that each individual will perform all the activities in every study he or she undertakes. Such an understanding does, however, help ensure that the approach taken in a study will be appropriate for the kind of question being asked. The benefit of the scientific method is its power in helping humans understand how nature works. Thus, all scientific activities—observing, recognizing patterns, and formulating and testing hypotheses—should be carried out so as to move toward understanding.

Literature Cited

Butler, R. G. 1980. Population size, social behavior, and dispersal in house mice: A quantitative investigation. *Animal Behavior* 28:78–85.

Carey, S. S. 1994. A beginner's guide to scientific method. Belmont, Calif.: Wadsworth Publishing Company.

Chamberlin, T. C. 1897. Studies for students: The method of multiple working hypotheses. *Journal of Geology* 5:837–48.

Conner, E. F., and D. Simberloff. 1979. Assembly of species communities: Chance or competition? *Ecology* 60:1132–40.

Cox, D. R. 1958. *Planning experiments.* New York: John Wiley and Sons.

Diamond, J. M., and M. E. Gilpin. 1982. Examination of the "null" model of Conner and Simberloff for species co-occurrences in islands. *Oecologia* 52:64–74.

Drew, G. S. 1994. The scientific method revisited. *Conservation Biology* 8:596–97.

Fisher, R. A. 1947. *The design of experiments.* 8th ed. London: Oliver and Boyd.

Franzblau, M. A., and J. P. Collins. 1980. Test of a hypothesis of territory regulation in an insectivorous bird by experimentally increasing prey abundance. *Oecologia* 46:164–70.

Gavin, T. A. 1991. Why ask "why": The importance of evolutionary biology in wildlife science. *Journal of Wildlife Management* 55:760–66.

Glickman, S. E., and G. S. Caldwell. 1994. Studying behavior in artificial environments: The problem of "salient elements." In *Naturalistic environments in captivity for animal behavior research,* ed. E. F. Gibbons Jr., E. J. Wyers, E. Walters, and E. W. Menzel Jr., 197–216. Albany: State University of New York Press.

Hairston, N. G., Sr. 1989. *Ecological experiments: Purpose, design, and execution.* Cambridge: Cambridge University Press.

Hanson, N. R. 1965. *Patterns of discovery.* Cambridge: Cambridge University Press.

Harris, V. T. 1952. *An experimental study of habitat selection by prairie and forest races of the deer mouse,* Peroyscus maniculatus. Contributions from the Laboratory of Vertebrate Biology No. 56. University of Michigan, Ann Arbor.

Harvey, P. H., R. K. Colwell, J. W. Silverton, and R. M. May. 1983. Null models in ecology. *Annual Review of Ecological Systems* 14:189–211.

Hilborn, R., and D. Ludwig. 1993. The limits of applied ecological research. *Ecological Applications* 3:550–52.

Hurlbert, S. H. 1984. Pseudoreplication and the design of ecological field experiments. *Ecological Monographs* 54:187–211.

Keddy, P. A. 1989. *Competition.* New York: Chapman and Hall.

Klopfer, P. H. 1963. Behavioral aspects of habitat selection: The role of early experience. *Wilson Bulletin* 75:15–22.

Kneller, G. F. 1978. *Science as a human endeavor.* New York: Columbia University Press. 333 pp.

Kodric-Brown, A., and J. H. Brown. 1993. Incomplete data sets in community ecology and biogeography: A cautionary tale. *Ecological Applications* 3:736–42.

Kuehl, R. O. 1994. *Statistical principles of research design and analysis.* Belmont, Calif.: Duxbury Press.

Lawton, J. H. 1996a. Patterns in ecology. *Oikos* 75:145–47.

Lawton, J. H. 1996b. The ecotron facility at Silwood Park: The value of "big bottle" experiments. *Ecology* 77:665–69.

McMahon, T. E., and J. C. Tash. 1988. Experimental analysis of the role of emigration in population regulation of desert pupfish. *Ecology* 69:1871–83.

Matter, W. J., and R. W. Mannan. 1989. More on gaining reliable knowledge: A comment. *Journal of Wildlife Management* 53:1172–76.

Matter, W. J., R. W. Mannan, E. W. Bianchi, T. E. McMahon, J. H. Menke, and J. C. Tash. 1989. A laboratory approach for studying emigration. *Ecology* 70:1543–46.

Mayr, E. 1974. Behavior programs and evolutionary strategies. *American Scientist* 62:650–59.

Mayr, E. 1982. *The growth of biological thought.* Cambridge, Mass.: Belknap Press of Harvard University Press.

Medawar, P. 1984. *Pluto's republic.* New York: Oxford University Press.

Mertz, D. B., and D. E. McCauley. 1980. The domain of laboratory ecology. In *Conceptual issues in ecology,* ed. E. Saarinen, 229–44. Dordrecht: D. Reidl.

Murphy, D. D. 1990. Conservation biology and scientific method. *Conservation Biology* 4:203–4.

Nudds, T. D., and M. L. Morrison. 1991. Ten years after "reliable knowledge": Are we gaining? *Journal of Wildlife Management* 55:757–60.

Peterman, R. M. 1990. Statistical power analysis can improve fisheries research and management. *Canadian Journal of Fisheries and Aquatic Science* 47:2–15.

Peters, R. H. 1991. A critique for ecology. Cambridge: Cambridge University Press.

Platt, J. R. 1964. Strong inference. *Science* 146:347–53.

Pleasants, J. M. 1990. Null-model tests for competitive displacement: the fallacy of not focusing on the whole community. *Ecology* 71:1078–84.

Popper, K. 1981a. The myth of inductive hypothesis generation. In *On scientific thinking*, ed. R. D. Tweney, M. E. Doherty, and C. R. Mynatt, 72–76. New York: Columbia University Press.

Popper, K. 1981b. Science, pseudo-science, and falsification. In *On scientific thinking*, ed. R. D. Tweney, M. E. Doherty, and C. R. Mynatt, 92–99. New York: Columbia University Press.

Reckhow, R. L. 1990. Bayesian inference in non-replicated ecological studies. *Ecology* 71:2053–59.

Romesburg, H. C. 1981. Wildlife science: Gaining reliable knowledge. *Journal of Wildlife Management* 45:293–313.

Scheiner, S. M. 1993. Introduction: Theories, hypotheses, and statistics. In *Design and analysis of ecological experiments*, ed. S. M. Scheiner and J. Gurevitch, 1–13. New York: Chapman and Hall.

Scheiner, S. M., and J. Gurevitch, eds. 1993. *Design and analysis of ecological experiments.* New York: Chapman and Hall.

Simpson, G. G. 1964. *This view of life.* New York: Harcourt, Brace, and World.

Sinclair, A. R. E. 1991. Science and the practice of wildlife management. *Journal of Wildlife Management* 55:767–73.

Skalski, J. R., and D. S. Robson. 1992. *Techniques for wildlife investigations: Design and analysis of capture data.* New York: Academic Press.

Smith, R. L. 1996. *Ecology and field biology.* 5th ed. New York: HarperCollins College Publishers.

Sokal, R. R., and F. J. Rohlf. 1995. *Biometry.* San Francisco: W. H. Freeman and Company.

Steidl, R. J., J. P. Hayes, and E. Schauber. 1997. Statistical power analysis in wildlife research. *Journal of Wildlife Management* 61:270–79.

Strong, D. R. J., L. A. Szyska, and D. S. Simberloff. 1979. Tests of community-wide character displacement against null hypotheses. *Evolution* 33:897–913.

Warren, C. E. 1971. *Biology and water pollution control.* Philadelphia: W. B. Saunders Company.

Wecker, S. C. 1963. The role of early experience in habitat selection by the prairie deer mouse, *Peromyscus maniculatus bairdi. Ecological Monographs* 33:307–25.

Weiner, J. 1995. On the practice of ecology. *Journal of Ecology* 83:153–58.

Wilzbach, M. A. 1985. Relative roles of food abundance and cover in determining the habitat distribution of stream-dwelling cutthroat trout (*Salmo clarki*). *Canadian Journal of Fisheries and Aquatic Science* 42:1668–72.

5 Measurement of Wildlife Habitat: What to Measure and How to Measure It

Introduction

We first review two of the major aspects of measuring wildlife habitat: what to measure and how to measure it. Implicit in our discussion is a careful evaluation of the niche gestalt of the animal. As discussed elsewhere, we must find ways to identify what the animal perceives as important to its survival: the *what* question. That is, we must try to see through the eyes of the animal and understand its sensory perceptions. We review literature on niche descriptions and on resource partitioning and allocation to identify factors of importance in a study.

We then discuss common methods used in measuring wildlife habitat: the *how* question. We identify major techniques, discuss their relationship to analysis of community structure, and outline their pros and cons. Techniques include the use of organism-centered plots versus various randomization procedures, how vegetation and other environmental features have been measured, and measurements of habitat diversity and heterogeneity.

"When" to measure, sample-size requirements, observer biases, and related topics are covered in the next chapter. "Whom" to measure—an equally important part of wildlife-habitat analysis—is covered in chapter 4. A severe weakness in any one of these aspects of designing a study, or a series of weaknesses across several of them, will place severe limitations on the scientific credibility of results and the applicability of the results to management situations. Wiens (1989:307–12) previously developed a categorization of measuring habitat similar to ours.

This and the following chapter can perhaps be best described as an exercise in "pattern seeking." That is, in most cases we are seeking to describe patterns of habitat use rather than testing specific hypotheses about the causes of the relationships we observe in nature. Although one can almost always develop a hypothesis, there is no reason to do so just for the sake of developing one. A hypothesis such as "there is no difference between used versus available habitat" is trivial on several fronts. Most notably, if the hypothesis were not rejected, we would have to conclude that an animal is randomly distributed, a situation we know simply does not occur. In addition, we can always find a difference between used and available habitat, depending on how finely we divide our measurement variables. Pattern seeking is a necessary early step in analysis of an animal's habitat, and it falls into the natural history tradition that has marked the history of animal ecology. Pattern seeking can most readily be contrasted with various experimental methodologies, as developed in chapter 4.

What to Measure

Green (1979:10) asked, "What criteria should be applied to choice of a system and variables in it for use in applied environmental studies?" In answering his question, he named four criteria:

1. spatial and temporal variability in biotic and environmental conditions that would be used to describe or predict impact effects;
2. feasibility of sampling with precision at a reasonable cost;
3. relevance to the impact effects and a sensitivity of response to them; and
4. some economic or aesthetic value, if possible.

These criteria apply either in descriptive studies or in analyses of impact effects (e.g., the effects of a chemical spill or of forest harvesting). Understanding the variability inherent in the system of interest is critical in designing a study; the variability must be generally understood, and the sampling must be designed to capture it. This variability includes natural, stochastic, or systematic change, as well as measurement and sampling error. There is always a cost-benefit analysis that researchers conduct, either formally or informally, when choosing variables for measurement: one must explicitly identify the precision necessary to reach project goals and then match all sampling to this needed precision. "Mismatching" of precisions is a common mistake (i.e., a waste of time) in wildlife studies. For example, there is no reason to determine the density of small mammals to two or three significant figures (e.g., 3.25 animals/ha) if these data are to be correlated with ocular measurements of vegetative cover that can be honestly estimated only into broad categories (e.g., grass cover = 10–20 percent).

A classic "waste of time" inherent in habitat studies is to measure everything possible for the duration of the study, and then run correlations to try to identify variables important in describing the distribution of the animal of interest. As noted above by Green, the variables measured should have relevance and sensitivity to the question at hand. Researchers often fail to follow the lessons learned in other studies, and insist instead on reinventing the proverbial wheel at the start of each new study. The fear of missing something important certainly drives such decisions and as such is somewhat understandable. However, given the limited time and budgets that most studies operate on, more thought should go into variable selection before field studies are started. If questions remain regarding the necessity of including certain variables, then preliminary analyses of data should be used to further reduce the variables being collected in the field. This provides more time to increase both the precision of the remaining variables, as necessary, and the sample size.

Finally, Green notes that the variables selected could have some economic or aesthetic value. This criterion may or may not be a factor, depending on the purpose of the study. Naturally, variables of economic interest might be included in a study if the intended audience for the results has specific interests. For example, foresters would usually be more interested in the relationship between birds and an economically valuable tree species than in a relationship between birds and shrub cover.

Spatial Scale

Correct determination of the scale of analysis is the cornerstone of habitat analysis and model development. That is, we should match the scale of analysis with the scale that we wish to use in applying our results to management purposes. For example, there is likely little reason to spend the time collecting "microhabitat" variables if one is interested only in describing the distribution of an animal across general vegetation types. Green (1979) asserted that it is not the degree to which a model meets perfection that renders it valid, but rather its adequacy in fulfilling a prescribed purpose. There is no reason for high precision if the biological change to be detected and managed for is large.

The scales at which wildlife-habitat relationships can be examined fall along a continuum that is not unlike how an animal selects "appropriate" habitat (chapter 2). The way an animal perceives its environment and the way we relate these perceptions to an organized method of study have the utmost importance in model development. We can examine habitat use at its broadest, or biogeographic, scale, passing through successively finer-scale evaluations until we reach the level of the individual. It is important to note, however, that these fine-scale models almost always vary between locations and time periods and certainly between populations. The magnitude of these variations determines the generality of the model (*generality* refers to the applicability of a model developed at one time and place to other times and places). Much of the wildlife-habitat literature has been criticized because of its time and place specificity (e.g., Irwin and Cook 1985). This criticism is misplaced and shows a general lack of understanding of the relationship between the precision of the variables measured and the scale of application possible. We should add that researchers conducting most wildlife-habitat-relationship studies seldom acknowledge the generality of their data and models and, thus, the locations and conditions in which managers should use them. Therefore, the decision to develop relatively "extensive," or broad-scale, models or more "intensive," or fine-scale, models should be based on the objectives of the particular study. The more extensive approach typically cannot tell us such things as how an animal reacts to change in litter depth, local density of trees by species, or the occurrence of a predator in a specific patch of vegetation; such approaches are likely necessary for management of localized populations of animals. Once the level of specificity is determined, the researcher can determine the type, resolution, and geographic extent of data collection required.

Van Horne (1983) outlined a hierarchic approach to viewing wildlife-habitat relationships (fig. 5.1). Her level 1 applies to intensive, site-specific studies of individual species; level 2 uses more generalizable variables and likely allows application (or relatively easy adaptation) of a

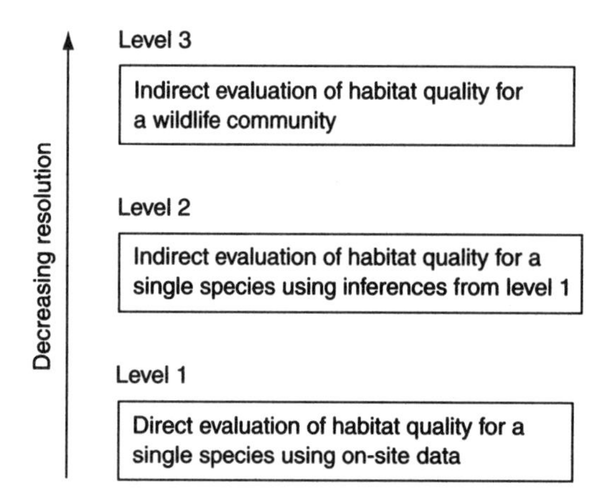

Figure 5.1. Hierarchic description of habitat quality assessments. (Reproduced from Van Horne 1983 [fig. 1], by permission of The Wildlife Society)

model to other locations; and level 3 is the most extensive approach that develops relationships for a host of species. Her scheme could easily be divided into many more levels, but it serves to indicate how habitat relationships can be studied, depicted, modeled, and applied along a continuum of resolution. Figure 5.2 depicts the major scales on which ecological field studies are typically conducted.

Wildlife managers' frequent frustration with the failure of most models to work as intended in their specific location is understandable. Yet, we have a dilemma: Models based on broad measurements of vegetation can seldom be applied to local situations (e.g., a small management unit or refuge), whereas models devel-

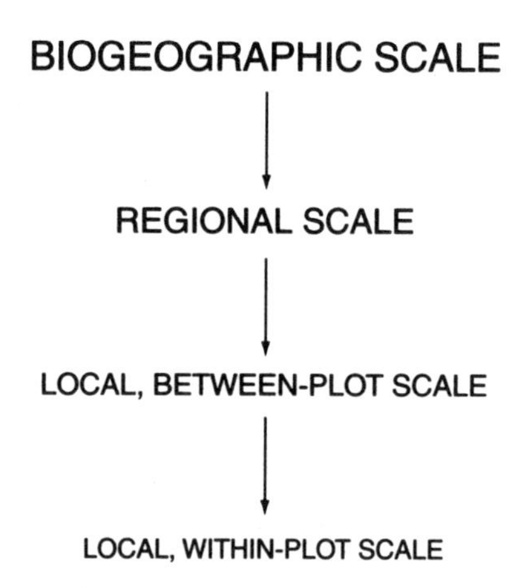

Figure 5.2. An outline of the spatial scales at which ecological field studies are conducted

oped on a fine scale for a specific location can seldom be adequately applied (generalized) to other locations (Block et al. 1994). The patterns of habitat use that emerge from our studies are sensitive to the scale of the comparison, because different relationships may exist in different subsets of the samples being compared. As shown by Wiens and Rotenberry (1981; see also Wiens 1989:56–57), a species will exhibit different patterns of habitat use depending on which portion of its distribution across the landscape is sampled by the observer (see fig. 5.3). In figure 5.3, sampling across the entire habitat gradient depicted would result in yet another ("averaged") pattern of habitat use. None of the depictions would be wrong per se; their "correctness" would depend on the scale of the question asked.

How can this dilemma be solved? It is possible in many situations to adapt a model developed for use in one location to the conditions that exist in another location. Again, we need not repeatedly reinvent the wheel. As discussed later in this chapter, both the structure and species composition of vegetation function in habitat selection. Obviously, models dominated by structural variables (e.g., vegetative cover, tree height) will be easier to adapt to different locations than models dominated by specific plant species (which may not occur in the new location). However, little work has been conducted in adapting site-specific models to different locations.

As summarized by Wiens (1989:239), studies across broad geographic areas are likely to overlook important details that account for the dynamics of local populations. Far from being idiosyncratic "noise," the variations within or among local populations may contain important mechanistic information about the factors causing the organisms' response. These variations tend to disappear on broader scales because of the consequences of averaging, unless the study is designed to determine such variation, as by stratified subsample or blocking designs.

We cannot specify here the level of detail required to develop proper habitat relationships for every situation. What we are trying to develop is the thought process that researchers must use on a study-specific basis. We will, however, present and discuss several examples of the types of variables collected by researchers attempting to develop predictive relationships. The list of potential studies is long, and our examples should not be taken as indicating only the best or even the most common techniques. Rather, we want researchers to see the range of variables being collected by workers in the field.

Measurements of the Animal

We have discussed the question of how refined our estimates of the numbers of an animal population need be for model development. For example, must the estimate be at the level of presence or absence of a species, index of abundance (e.g., numbers/count), or absolute density (numbers/unit area)? It is beyond the scope of this book to review the methods available for estimating animal abundance. In fact, many excellent works have been published on this subject (e.g., Cooperrider et al. 1986; Ralph et al. 1993; Bookhout 1994; Heyer et al. 1994). Here we wish to touch briefly on a method of evaluating wildlife-habitat relationships that centers on counting animals and then relating these numbers to environmental features.

Although we are not covering this topic in detail, the reliability of our estimates of the number of animals present is crucial. That is, are our estimates of absolute or even relative abundance a fair reflection of the number of

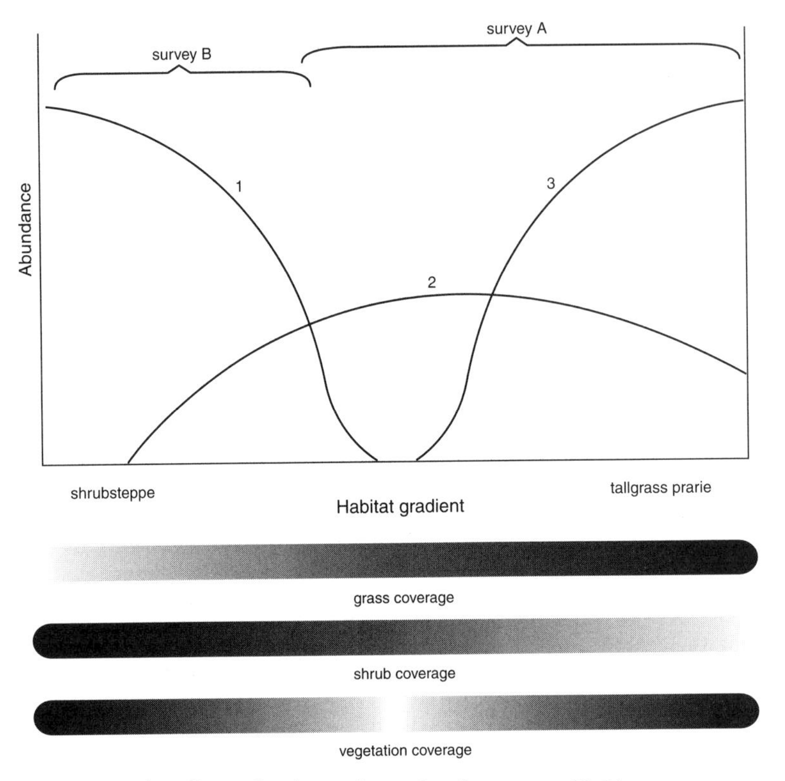

Figure 5.3. The effects of scale on the study of patterns of habitat association. In this schematic diagram, species 1–3 have characteristic distributions on a gradient from shrubsteppe through short- and mixed-grass prairies to tallgrass prairie. Grass coverage, shrub coverage, and overall vegetation height change on the gradient as shown below the graph. In survey A, a large portion of the gradient is sampled, but extreme shrubsteppe sites are omitted, whereas in survey B only shrubsteppe and a few grass-shrub sites are studied. The species will exhibit different patterns of habitat association in the two surveys. Species 1, for example, will exhibit a strong negative association with grass coverage and a positive association with shrub coverage in survey A but may fail to show either association in survey B. (Reproduced from Wiens and Rotenberry 1981 [fig. 3], by permission of the Ecological Society of America)

animals actually present? If we use multiple observers, are their estimates comparable? Can estimates derived by one observer of a group of observers in one study area be used to validate a model developed by other people at a different location? Throughout this book we discuss the varying effects of observer error, errors encountered because of hidden biases, and other factors that work against the development of habitat models. Here we are interested in errors associated with estimating numbers of animals. We wish to remind the reader of the many problems associated with counting animals, and we reiterate that poor estimates of animal numbers will negate conclusions about habitat relationships drawn on the basis of even the most carefully collected environmental variables.

An entire class of habitat models is based on correlating animal numbers to some features of the animal's biotic and abiotic environment. The purpose of these models is to develop adequate prediction of the presence, abundance, or density of the species on the basis of environmental features (see later in this chapter; also chapter 10). The methods vary according to available data: techniques for handling presence or absence data are not necessarily the same as those suitable for density data. Researchers and managers can use these models to predict the changes that will occur in animal abundance given changes in variables in the model.

A tremendous amount of research has been devoted to developing density estimates for vertebrates. Much of the theory in ecology, as well as applications in resource management and conservation, depend on reliable expression and comparison of numeric abundance (Smallwood and Schonewald 1996). Additionally, the development of many wildlife-habitat models is based, in part, on estimations of abundance (e.g., regression analysis; see below).

However, the expression of density (or an index thereof) can vary widely depending on the spatial scale of the study. That is, simply converting the number of animals to some standard area (e.g., extrapolating birds in a 1.5-ha study site to birds per 40 ha [100 acres]) in no way standardizes estimates for comparisons with other studies. Density for any one population is not likely to remain constant across spatial scales. For example, Smallwood and Schonewald (1996) found that $\log_{10}$ population-density estimates consistently decrease linearly with the $\log_{10}$ spatial extent of the study area for species of terrestrial Carnivora. They showed that the size of the study area accounts for most of the variation in population estimates.

Therefore, if density is to be related to ecological variables, then it should be estimated from a representative spatial scale; otherwise, the pattern will be masked by high variability in density among scales. The choice of spatial scale must be based on the species' relationship with the landscape, not on the spatial resolution of the technology that is used for observation or some artificial or convenient area (Addicott et al. 1987; Wiens 1989; Smallwood and Schonewald 1996). Determination of the effect of study-area size on abundance values should be incorporated into the preliminary sampling phase of all studies. Unfortunately, virtually no researcher takes the time to do so; this would be a very fruitful area for further research (see chapter 12).

Most studies of habitat use occur where the researcher has previous knowledge that the species of interest is in adequate abundance; this ensures accumulation of an adequate sample size. The home range or a finer-scale activity location within the home range (e.g., a foraging or nest site) is usually the focus of study. Results of these studies are often then extrapolated (often not by the original researcher) to a much larger area than was used in collecting the data. However, such results can be reliably extrapolated only if animals and their habitat are uniformly distributed across the landscape. The distribution of animals is, of course, aggregated in some manner.

Working with Swainson's hawks (*Buteo swainsoni*), for example, Smallwood (1995) showed that most of the density variation among study sites was due to the size of the study area chosen by the researcher. Further, studies conducted on larger spatial scales yielded habitat associations different from those yielded at more conventional, home-range levels. This relationship holds for other species (as reviewed above; Smallwood and Schonewald 1996). These studies clearly indicate that researchers must give close attention to the size of the study area used: Most such decisions are made either from convenience or justified on the basis of the size of the species' home range. Additionally, managers must use caution when trying to extrapolate abundances derived from small, relatively high-density areas.

Measurements of Environmental Features

Conceptual Framework

Next we describe some common techniques used to measure habitat. As discussed throughout this book, habitat is viewed along a continuum of spatial scale—the key word here being *continuum.* That is, all scales of habitat that we have invented or will invent, be they microhabitat, macrohabitat, or mesohabitat, are meaningful only in the context of specific spatial and temporal extents. Thus, it is important that researchers clearly elucidate their conceptual basis for making the measurements that they report.

Two basic and obvious aspects of vegetation can be distinguished: the structure, or physiognomy; and the taxa of the plants, or floristics. Many authors had initially concluded that vegetation structure and "habitat configuration" (size, shape, and distribution of vegetation in an area), rather than particular plant taxonomic composition, are most influential in determining patterns of habitat occupancy by animals, especially birds (see Hilden 1965; Wiens 1969; James 1971; Anderson and Shugart 1974; Willson 1974; James and Wamer 1982; Rotenberry 1985). As Rotenberry observed, however, more recent studies have shown that plant species composition plays a much greater role in determining patterns of habitat occupancy than previously thought. As detailed below, the relative usefulness of structural versus floristic measures is foremost a function of the spatial scale of analysis.

Many earlier researchers failed to place their studies adequately on a specific spatial scale, thereby obscuring the relative (and proper) roles that structure and floristics can play in predicting habitat relationships on different scales (Levin 1992). As further noted by Rotenberry, the same species that appears to respond to the physical configuration of the environment at the continental level may show little correlation with physiognomy at the regional or local level. Thus, many animals may be differentiating between gross vegetative types on the basis of physiognomy (i.e., they occupy a general area that is "appropriate" in its structural configuration), with further refinement of the distribution (and thus, abundance) within a local area based on plant taxonomic considerations. Note that this scenario of Rotenberry's relates closely to Hutto's (1985) hypothesized mode for the process of habitat "selection" in animals (see chapter 2). Rotenberry quantified his ideas using data from the shrubsteppe studies of Wiens and his co-workers. Rotenberry's analysis of these data indicated that, when the correlation between physiognomy and flora is statistically separated, a significant relationship remains between bird abundance and flora, but not between birds and physiognomy (fig. 5.4).

Thus, we return again to the important theme that the variables measured and the level of model refinement required should be based on the scale of interest. Simple presence or absence studies of animals on regional or broader scales likely do not require floristic

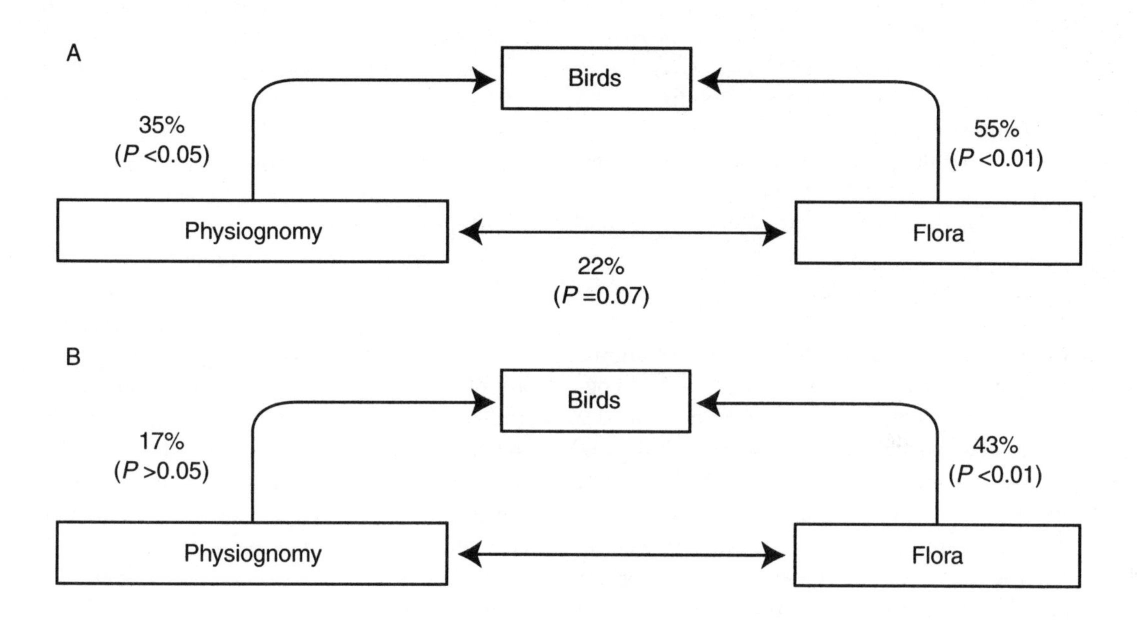

Figure 5.4. A. Coefficients of determination ($r^2 \times 100$) between similarity and distance matrices based on avian, floristic, and physiognomic composition of eight grassland study sites. Significance levels of association are given in parentheses. B. Partial coefficients of determination, as above. Correlation between physiognomy and flora has been partialed out. (From Rotenberry 1985: fig. 1)

analysis of vegetation. Broad categorization by physiognomy—probably including differentiation no lower than life form, such as deciduous and evergreen, or by general ecological classes or vegetative types—is likely to be adequate. Plant taxonomy becomes increasingly important, however, as our studies become increasingly site specific. As a general rule, then, one would always be safe in collecting data on as fine a scale as time and budget allow, using common sense not to unduly overdo it; lumping *a posteriori* is always possible. A better approach is to begin a study with a preliminary evaluation of the variables and sampling methods necessary to achieve the desired level of refinement; necessary sample sizes can also be determined from such preliminary work.

Macrohabitat

With the rise in interest of studying animal diversity, researchers have developed various measures that seek to relate the numbers and kinds of animals to the gross structure of the vegetation. Most famous are the foliage height diversity–bird species diversity (FHD-BSD) constructs of MacArthur and MacArthur (1961). The diversity of birds rises as vegetation becomes increasingly stratified (see fig. 3.1). A plethora of studies followed the early work of the MacArthurs, with most showing similar results. Wiens (1989:129–34) provided a review, with cautions, of BSD-FHD and related issues.

In vertically simple vegetation, such as brushlands and grasslands, FHD would not be expected to provide a good indicator of animal diversity (at least for most vertebrates). Recognizing this problem, Roth (1976) developed a method by which the dispersion of clumps of vegetation—here, shrubs—forms the basis for a measure of habitat "heterogeneity." Roth was able to relate BSD to this measure of habitat heterogeneity, or patchiness. Other workers also built relationships between measures of horizontal vegetation development and animal communities (see Wiens 1969; Morrison and Meslow 1983; see also chapter 8).

Returning to figure 3.1, note that there is considerable scatter around the regression line. Thus, the usefulness of this general principle as a site-specific predictor decreases as the scale of application becomes increasingly fine. Karr and Roth (1971) suggested that the scatter in a FHD-BSD relationship likely results from important but unmeasured variables influencing the avian community. Measures of diversity sacrifice complexity for simplicity; this is why they are useful primarily on larger spatial scales. These indices collapse detailed information on plants—species composition, foliage condition (vigor), arthropod abundance—into a single number.

The FHD-BSD relationships do indicate, however, that species respond to complexity in their environment. But to what specifically do they respond? The observer can increase FHD artificially by simply changing the number of categories used in the calculations. Thus, while FHD gives some indication of foliage strata and the response of animals to such vegetation development, it is much too arbitrary and general to result in more than a gross and relative examination of wildlife-habitat relationships in different areas.

As discussed above, indirect measures of foliage structure (e.g., FHD) are extremely gross and provide only limited predictive power. However, measures of canopy closure are being used as centerpieces of management guidelines for many species of wildlife, including the federally endangered northern spotted owl (*Strix occidentalis caurina*) and the northern goshawk (*Accipter gentilis*). Thus, if management guidelines are to be interpreted and applied adequately, we must know that our methodologies truly reflect something of ecological causation.

A more useful measure of foliage structure for many species may be foliage volume, the actual surface area of foliage available for consumption or as substrates for insects and other prey. It is extremely tedious, however, to measure actual foliage volume. Plant ecologists have done so by cutting down trees and then measuring and counting their leaves, needles, and other parts. Such data are necessary for accurate predictions of photosynthetic rates and measurements of tree growth (see Carbon et al. 1979). After ecologists have completed such work, they can develop statistical models that relate some more easily measured aspect of the plant to foliage volume (VanDeusen and Biging 1984). Plant ecologists also use dry volume of plant material as a measure of volume.

Researchers have developed even more indirect measures of foliage volume. In a comparison of habitas of the chestnut-backed chickadee

(*Parus rufescens*) and the black-capped chickadee (*P. atricapillus*), Sturman (1968) used equations for shapes that approximate the structure of conifers and hardwoods. Methods such as Sturman's (see also Mawson et al. 1976) can be viewed as compromises between labor-intensive techniques that involve sampling whole trees and the qualitative estimates of FHD.

Many other methods of estimating foliage development have been devised (see review by Ganey and Block 1994). A popular technique involves the use of a spherical densiometer. Most easily described as a round, concave mirror onto which a grid has been scribed, this device is used to estimate canopy closure directly over the device. Ganey and Block (1994) tested the relative results between the spherical densiometer and the sighting tube. A sighting tube is simply a piece of PVC pipe (or similar material) through which the observer estimates canopy closure. Ganey and Block found that the densiometer resulted in higher estimates of canopy closure than the sighting tube: 57 percent of all estimates showed >50 percent cover for the densiometer, whereas only 38 percent did so for the sighting tube. They could not, of course, determine which method was more accurate. Their results do show, however, that the method used can have a significant impact on the conclusions regarding canopy cover. And, in turn, this could have substantial ecological ramifications.

Many of the currently used habitat models operate on the macrohabitat scale, including most statewide wildlife-habitat relationship, or "WHR," constructs (Block et al. 1994), Gap models (Scott et al. 1993), and habitat suitability index (HSI) models (Fish and Wildlife Service 1981). Most of these models use broad-scale categorizations of vegetation types (often mislabeled as habitat types) as a predictor of animal presence or abundance. However, many developers of these models substantially mismatch scales in the variables used to develop their models; this is especially evident in HSI and WHR models. Such mismatching (e.g., entering micro- and macrohabitat variables into the same analysis) ignores current theories concerning the hierarchic nature of habitat selection and makes it difficult to interpret model output. Models developed at the macrohabitat level are very useful in our understanding of broad habitat relationships, but they should be limited to application on a broad scale.

Quite obviously, how we categorize vegetation is not a trivial matter if we expect our macrohabitat models to perform adequately. Unfortunately, many workers categorize vegetation types on the basis of the qualitative judgments of observers, which makes them difficult to replicate in other studies. The lack of any testing before most of these models are used to make management decisions further confounds our ability to evaluate their appropriateness (Block et al. 1994).

Microhabitat

To determine the relevant variables to collect and analyze, we must determine what aspects of the environment that organisms recognize as relevant. As noted by Krebs (1978:40), "We must be careful here to define the perceptual world of the animal in question before we begin to postulate the mechanisms of habitat selection." Krebs (1978:42) identified two basic factors that must be kept separate when discussing habitat selection: evolutionary factors, which ultimately confer reproductive fitness and survival value on habitat selection; and behavioral factors, which provide the proximate mechanisms by which animals select habitat. The proper behavioral factors that have direct ties to the survival of the animal must be identified. Thus, we are interested in behaviors that result from stimuli from the landscape; breeding, display, and feeding sites; food; and other animals (Hilden 1965; Krebs 1978:42).

Dueser and Shugart (1978) listed four criteria that can help guide selection of microhabitat variables for measurement:

1. Each variable should provide a measure of the structure of the environment that is either known or reasonably suspected to influence the distribution and local abundance of the species.
2. Each variable should be quickly and precisely measurable with nondestructive sampling procedures.
3. Each variable should have intraseasonal variation that is small relative to interseasonal variation.
4. Each variable should describe the environment in the immediate vicinity of the animal.

Dueser and Shugart noted that the final criterion reflects their concern for describing the environment in sufficient detail to detect subtle differences between microhabitats that appeared to be grossly similar. Their first criterion tells us that previous natural history information, plus a good deal of biological common sense, will help narrow the choice of variables. We must be careful, however, not to follow past mistakes or misinformation regarding the natural history of the animal and, further, not to let our preconceived notions and biases eliminate potentially important variables. It is often worthwhile to have your list of variables reviewed by a biologist who is familiar with the area but not with the species; for instance, a plant ecologist can probably offer valuable advice to a wildlife biologist in planning a study. Dueser and Shugart were apparently concerned with nondestructive sampling because of their desire to make repeated observations at a site. And, it would be difficult to repeat a study in the same area if the site has been severely disturbed (assuming that the disturbance is not part of the planned study). Their concern regarding sampling variation is both a biological and a statistical issue. They indicated that measurements should be sufficiently precise that variations within relatively short periods of time (within seasons, or intraseasonal) are not obscured by the interseasonal variations, which are likely to be much larger. In studies of microhabitat, we usually wish to determine how these often subtle measures of habitat vary; thus, a high degree of precision is required. Their final point is intuitive: Microhabitat should be measured in close proximity to the animal. It is not a mistake to measure variables from many scales; it may be a mistake to include them in the same analysis (unless a clear biological and statistical justification is provided).

Whitmore (1981) also outlined general criteria for selecting variables, listing three categories that involve "practical aspects" of variable selection:

1. Variables should be measurable at the desired level of precision.
2. Variables should be biologically meaningful.
3. Variables should be relevant to the species in question.

In reference to his second criterion, Whitmore asked what possible direct meaning measurements of tree roots would have on a canopy-feeding warbler. Regarding his third criterion, he asked how directly important the percentage of grass cover in the ratio of grasses to forbs could be to a bark-foraging bird. It can be argued that these variables may have some indirect relationship to a variable of more obvious importance to an animal; for example, root condition could relate to canopy volume or condition, which in turn might relate to insect density and thus the behavior of the warbler. However, such indirect measures increase the variance (weaken the relationship) between the warbler and the truly relevant variable. Many indirect measurements in the same analysis thus greatly compound the error in the results, making for weak conclusions.

A Major Weakness: Indirect Measurements of Habitat

We should note here that most of our measurements are only indirect reflections of what the animal is probably responding to in making decisions about habitat selection. Few would argue that insectivorous birds must have an adequate supply of the proper prey. Yet, few studies of such birds quantify prey because of the extreme difficulty in doing so (Morrison et al. 1990). Thus, we sometimes make indirect measurements of prey by assuming that more foliage likely results in more prey. Yet indirect measures do not end here. Measuring foliage volume also is extremely difficult. Thus, as described above, we usually use some indirect measure of foliage volume. What we have, then, is an indirect measure of foliage volume being used to index insect abundance indirectly; errors are certainly grossly compounded, and it is difficult to know the biological meaning of our results in terms of specific causes. Such problems are not restricted to bird studies: big-game biologists often make indirect estimates of food requirements, including micronutrient needs, through various indirect estimates of forage abundance (e.g., ocular estimates or line intercepts of shrub cover); herpetologists use numerous measures of ground, shrub, and canopy cover to describe indirectly the microsite conditions (e.g., soil moisture) available to salamanders. The prevalence of indirect measures—and indirect measures to measure other variables indirectly—is probably one of the primary weaknesses in the area of wildlife-habitat-relationship research and modeling. In fact, it will be difficult to increase further the accuracy of our models and thus our management decisions until we begin to make more direct measurements of the factors that influence the distribution, abundance, and behaviors of wildlife. We have done a lot of the relatively easy work; substantial advances await implementation of the much more difficult measurements. We are not calling for abandonment of indirect measures. Rather, we are saying that such measures have limited predictive power in many situations and that we should openly acknowledge the limits such measures place on our results.

Focal-Animal Approach

Most studies of microhabitat selection are variations of the "focal-animal" approach (see chapter 7 for focal-animal sampling in behavioral research). These methods use the presence of an animal as an indication of the habitat being used by the species. No correlation between abundance and the environment is involved. Rather, the location of individual animals is used to demark an area from which environmental variables are measured. As detailed in the following section, an animal's specific location might serve as the center of a sampling plot. Or a series of observations of an individual might be used to delineate an area from which samples are then made (e.g., see Wenny et al. 1993). In either case, the major assumption of this approach is that measurements indicate habitat preferences of the animal.

Animals do not, however, spend equal amounts of time at each activity and location during a day. Feeding, drinking, calling, resting, grooming, and other activities each consume different amounts of time and energy and often take place in different locations within the home range (e.g., calling from exposed locations or resting in the shade). The amount of time spent in any one activity may not indicate the importance of the activity to the animal: drinking may take only a few minutes each day, but without water the animal is unlikely to survive. Again, the researcher must be aware of these behaviors when designing a study based on the focal-animal approach; chapter 7 provides a more detailed description of behavioral sampling. This adds strength to the argument made above regarding the importance of considering time-activity budgets in analysis of habitat relationships.

For example, many studies have used the location of a singing male bird or a foraging individual as the center of plots describing the habitat of the species (e.g., James 1971; Holmes 1981; Morrison 1984; VanderWerf 1993). But how well do such plots indicate the species' habitat preference or even the individual's territory preference? Collins (1981) examined habitat data collected at perch and nest sites for several species of warblers. He found that 29 percent of the nest sites had vegetation structures that significantly differed from the corresponding perch sites within the territories. Not surprisingly, he found that basing habitat only on perch sites overestimated the tree component of habitat. This does not mean that a study based only on perches or only on nests is flawed. Rather, it means that such studies describe only "perch habitat" or "nest-site habitat" and, as such, describe only part of what one might call the breeding habitat of the species. Similar examples can be found in the literature for all major groups of wildlife: for example, bedding sites of deer (Ockenfels and Brooks 1994) and trapping locations of small mammals (e.g., Morrison and Anthony 1989; Kelt et al. 1994).

Use versus Availability: Basic Designs

As discussed throughout this book, the comparison of the use of an item or habitat element (food, habitat characteristic) relative to its abundance or availability is a cornerstone of wildlife-habitat analyses. Thomas and Taylor (1990) reviewed 54 papers published in the *Journal of Wildlife Management* that analyzed use and availability of food items or habitat characteristics (we ignore the conceptual differences between measureable "abundance" and what an organism perceives as "availability" for this discussion). Thomas and Taylor concluded that use-availability studies could be categorized into three basic designs:

Design 1: The availability and use of all items are estimated for all animals, but organisms are not individually identified, and only the item used is identified. Availability is assumed to be equal for all individuals. Here, they found that habitat studies often compare the relative number of animals or their sign of presence in each vegetation type with the proportion of that type in the study area. About 28 percent of the studies reviewed fell into this category.

Design 2: Individual animals are identified, and the use of each item is estimated for each animal. As for Design 1, availabilities are assumed equal for all individuals and are measured or estimated for the entire study area. Studies that compare the relative number of relocations of marked animals in each vegetation type with the proportion of that type in the area fall into this category; 46 percent of the studies reviewed were placed in this category.

Design 3: This design is the same as Design 2, except that the availability of the items is also estimated for each individual animal. Studies in this category often estimate the home range or territory for an individual and compare use and availabilities of items within that area; 26 percent of the studies were in this category.

Thomas and Taylor also provided a good review of studies that fit each of these categories, as well as guidelines for sample sizes necessary to conduct such analyses. Virtually all classes of statistical techniques have been used to analyze use-availability (or use or non-use) data, depending upon the objectives of the researcher, the structure of the data, and adherence to statistical assumptions (i.e., univariate parametric or nonparametric univariate comparisons, multivariate analyses, Bayesian statistics, and various indices). These techniques have been well reviewed (e.g., Johnson 1980; Alldredge and Ratti 1986, 1992; Thomas and Taylor 1990; Aebischer et al. 1993). Compositional analysis, only recently applied to habitat analysis, should be considered for use in these studies (Aebischer et al. 1993).

In all these designs, however, composition of available habitat is measured only once and then compared with observations of habitat used by the animals; the latter are usually collected over a period of time (e.g., a "season") to gather an adequate sample size. Habitat used by animals is, in essence, an average use over a defined period. Of course, habitat availability is not constant, and changes in various components of it could substantially influence habitat use, the ramifications of which depending largely on the goals of the investigation.

To rectify this potential problem in use-availability studies, Arthur et al. (1996) developed a method of estimating habitat use involving multiple observations of habitat availability. Their method, which is too detailed to develop here, allows for quantitative comparisons between habitat categories and is not affected by arbitrary decisions about which categories to include in a study (see also Manly et al. 1993).

Case Studies

As we showed earlier in this book, the distribution of plants is tied closely to regimes of temperature, soil, and moisture. Animals, in turn, are often linked in some fashion to plants for shelter, for food (directly or indirectly), or for both. But to what specific aspect of vegetation are animals responding? What are the stimuli causing the behaviors that we call resource use? Notwithstanding our comments above on indirect measurements, the fact remains that most studies have used, and will continue to use, indirect measures of habitat selection.

To answer these questions, we now turn to specific examples of variables collected by researchers seeking to describe the habitat use patterns of animals. The examples we use come from widely read and cited papers. We offer these few examples as good starting points for students planning similar evaluations of habitat use by a particular species. We strongly recommend that researchers concentrating on a particular taxon not restrict themselves to literature relating to that group only. For example, a researcher designing a study to examine habitats of ground-foraging birds would likely gain valuable information on types of variables and sampling design by reviewing papers on small mammals and reptiles. We will also give brief summaries of the objectives of each paper;

remember that the variables should follow closely from the objectives.

Birds have received by far the most attention with regard to the analysis of habitat use patterns. This is likely a reflection of the conspicuousness of birds: most are active during the day, most give at least some vocalizations during all parts of the year, and they are inexpensive to observe (one needs only binoculars and a notebook).

James (1971) conducted one of the first and most-cited studies quantifying bird-habitat relationships. As discussed earlier (chapter 2), she based her study on the conceptual framework of the niche and a bird's niche gestalt. She used 15 measures of vegetation structure to describe the multidimensional "habitat space" of a bird community in Arkansas. She followed closely the methods she developed with Herman Shugart, Jr. (James and Shugart 1970); these methods are described in the next section ("How to Measure").

The conceptual framework on which James based her work (niche gestalt) and the general analytical techniques she used (multivariate analysis of focal-bird observations) have led to a plethora of studies that have expanded upon her basic ideas. Her strategy and methods are still in wide use. Recently, for example, Murray and Stauffer (1995) based their vegetation sampling on the James and Shugart (1970) methodology.

Dueser and Shugart (1978) were among the first workers to quantify microhabitat use patterns of small mammals in a multivariate sense. The variables they selected, however, would apply regardless of the analytical techniques used. They had as their goal the description of microhabitat differences between the small mammal species of an upland forest in eastern Tennessee. Their specific objectives were to characterize and compare microhabitats of species within the forest and to examine the relationships of species abundance and distributions to the relative availability of selected microhabitats. This study is a good example of the pattern-seeking nature of most habitat descriptions. Although one could devise a null hypothesis here, it would be trivial: for example, test the null hypothesis of no difference in habitat use between species.

Dueser and Shugart gathered information for vertical strata at each capture site of a small mammal: overstory, understory, shrub level, forest floor, and litter-soil level. Table 5.1 lists the variables they collected. Note that they did not collect species-specific information on plants beyond designations of "woodiness," "evergreenness," and the like. This can be considered an unfortunate omission for a microhabitat analysis, the ramifications of which to the final results are unknown. They did, however, record the number of woody and herbaceous species. They paid special attention to features of the forest floor, such as litter-soil compactability, fallen log density, and short herbaceous stem density. They found that certain of these soil variables played a significant role in describing the differences in microhabitats of the species studied. Except for the lack of detailed information on plant taxonomy, we consider Dueser and Shugart's study a good example of a very detailed set of variables used to differentiate between concurring species of animals.

Reinert (1984) sought to differentiate microhabitats of timber rattlesnakes (*Crotalus horridus*) and northern copperheads (*Agkistrodon contortrix*), which occur sympatrically in temperate deciduous forests of eastern North America. His approach was a multivariate habitat description consistent with the Hutchinsonian definition of the niche (Green 1971) and was based on the concept of the niche gestalt

Table 5.1. Designation, descriptions, and sampling methods for variables measuring forest habitat structure, used by Dueser and Shugart

Variable	Method
1. Percentage of canopy closure	Percentage of points with overstory vegetation, from 21 vertical ocular tube sightings along the center lines of two perpendicular 20-m^2 transects centered on trap
2. Thickness of woody vegetation	Average number of shoulder-height contacts (trees and shrubs), from two perpendicular 20-m^2 transects centered on trap
3. Shrub cover	Same as (1), for presence of shrub-level vegetation
4. Overstory tree size	Average diameter (in cm) of nearest overstory tree, in quarters around trap
5. Overstory tree dispersion	Average distance (m) from trap to nearest overstory tree, in quarters
6. Understory tree size	Average diameter (cm) of nearest understory tree, in quarters around trap
7. Understory tree dispersion	Average distance (m) from trap to nearest understory tree, in quarters
8. Woody stem density	Live woody stem count at ground level within a 1.00-m^2 ring centered on trap
9. Short woody stem density	Live woody stem count within a 1.00-m^2 ring centered on trap (stems $\leq$0.40 m in height)
10. Woody foliage profile density	Average number of live woody stem contacts with an 0.80-cm-diameter metal rod rotated 360°, describing a 1.00-m^2 ring centered on the trap and parallel to the ground at heights of 0.05, 0.10, 0.20, 0.40, 0.60, . . ., 2.00 m above ground level
11. Number of woody species	Woody species count within a 1.00-m^2 ring centered on trap
12. Herbaceous stem density	Live herbaceous stem count at ground level within a 1.00-m^2 ring centered on trap
13. Short herbaceous stem density	Live herbaceous stem count within a 1.00-m^2 ring centered on trap (stems $\leq$0.40 m in height)
14. Herbaceous foliage profile density	Same as (10), for live herbaceous stem contacts
15. Number of herbaceous species	Herbaceous species count within a 1.00-m^2 ring centered on trap
16. Evergreenness of overstory	Same as (1), for presence of evergreen canopy vegetation
17. Evergreenness of shrubs	Same as (1), for presence of evergreen shrub-level vegetation
18. Evergreenness of herb stratum	Percentage of points with evergreen herbaceous vegetation, from 21 step-point samples along the center lines of two perpendicular 20-m^2 transects centered on trap
19. Tree stump density	Average number of tree stumps $\geq$7.50 cm in diameter, per quarter
20. Tree stump size	Average diameter (cm) of nearest tree stump $\geq$7.50 cm in diameter, in quarters around trap
21. Tree stump dispersion	Average distance (m) to nearest tree stump $\geq$7.50 cm in diameter, in quarters around trap
22. Fallen log density	Average number of fallen logs $\geq$7.50 cm in diameter, per quarter
23. Fallen log size	Average diameter (cm) of nearest fallen log $\geq$7.50 cm in diameter, in quarters around trap
24. Fallen log dispersion	Average distance (m) from trap to nearest fallen log $\geq$7.50 cm in diameter, in quarters around trap
25. Fallen log abundance	Average total length (>0.50 m) of fallen logs $\geq$7.50 cm in diameter, per quarter
26. Litter-soil depth	Depth of penetration (<10.00 cm) into litter-soil material of a hand-held core sampler with 2.00-cm diameter barrel
27. Litter-soil compactability	Percentage of compaction of litter-soil core sample (26)
28. Litter-soil density	Dry weight density (g/cm^2) of litter-soil core sample (26), after oven-drying at 45°C for 48 hr
29. Soil surface exposure	Same as (18), for percentage of points with bare soil or rock

Source: Dueser and Shugart 1978: appendix; reproduced by permission of the Ecological Society of America

(James 1971). Reinert was interested not only in features of the ground but also in the weather conditions immediately surrounding the snake. He measured temperature and humidity at several locations near an animal, as well as the structure of the surrounding vegetation. Here again, however, no information on plant taxa was included (table 5.2).

Morrison et al. (1995) used time-constrained surveys to describe the microhabitats of amphibians and reptiles in southeastern Arizona mountains. Observers walked slowly, searching the ground and tree trunks and turning over moveable rocks, logs, and litter to examine protected locations while a stopwatch ran. With the survey time stopped, a 5-m-diameter plot was then centered on the animal's location, serving as the location where microhabitat conditions were measured. They measured substrate temperature and various descriptors of the vegetation and other habitat characteristics.

Ockenfels and Brooks (1994) radio-tracked 22 Coues white-tailed deer to describe diurnal bedding sites. The proper diurnal sites are critical to this deer in the very hot summer temperatures in the southern Arizona study area. Using 40-m^2 circular plots, Ockenfels and Brooks measured topography, temperature, vegetative characteristics, and percentage of shade at bedding sites and similar, unused sites.

Welsh and Lind (1995) analyzed the habitat affinities of the Del Norte salamander (*Pleth-*

Table 5.2. Structural and climatic variables used by Reinert

Mnemonic	Variable	Sampling method
ROCK	Rock cover	Coverage (%) within 1-m^2 quadrant centered on snake location
LEAF	Leaf litter cover	Same as ROCK
VEG	Vegetation cover	Same as ROCK
LOG	Fallen log cover	Same as ROCK
WSD	Woody stem density	Total number of woody stems within 1-m^2 quadrant
WSH	Woody stem height	Height (cm) of tallest woody stem within 1-m^2 quadrant
MDR	Distance to rocks	Mean distance (m) to nearest rocks (>10 cm max. length) in each quarter
MLR	Length of rocks	Mean max. length (cm) of rocks used to calculate MDR
DNL	Distance to log	Distance (m) to nearest log (≥7.5 cm max. diameter)
DINL	Diameter of log	Max. diameter (cm) of nearest log
DNOV	Distance to overstory tree	Distance (m) to nearest tree (≥7.5 cm dbh [diameter at breast height])
DBHOV	Dbh of overstory tree	Mean dbh (cm) of nearest overstory tree within each quarter
DNUN	Distance to understory tree	Same as DNOV (trees <7.5 cm dbh >2.0-m height)
CAN	Canopy closure	Canopy closure (%) within 45° cone with ocular tube
SOILT	Soil temperature	Temp (°C) at 5-cm depth within 10 cm of snake
SURFT	Surface temperature	Temp (°C) of substrate within 10 cm of snake
IMT	Ambient temperature	Temp (°C) of air at 1 m above snake
SURFRH	Surface relative humidity	Relative humidity (%) at substrate within 10 cm of snake
IMRH	Ambient relative humidity	Relative humidity (%) 1 m above snake

Source: Reinert 1984; table 1; reproduced by permission of the Ecological Society of America

odon elongatus) in relation to landscape, macrohabitat, and microhabitat scales. They presented a detailed rationale for the selection of methods, including choice of analytical techniques, data screening, and interpretation of output. The variables they measured, separated by spatial scale, are shown in table 5.3.

How to Measure

Sampling Principles

In this section we review some of the common methods used to measure wildlife habitat. We do not present a survey of all literature available for all taxa; Cooperrider et al. (1986) provided a thorough review of basic sampling techniques for all major taxa of wildlife (see also Bookhout 1994).

Recall from chapter 3 our discussion of how vegetation traditionally forms the template for how we view wildlife-habitat selection (see also discussion of key environmental correlates in chapter 11). As such, it is not surprising that we turn to plant ecologists for advice on many of our fundamental sampling methods. Indeed, even a cursory review of the methods sections in wildlife publications shows a reliance on standard, classical methods of quantifying the structure and floristics of vegetation: point quarter, circular plots and nested circular plots, sampling squares, line intercepts, and so on. These methods are used for good reason: they have been developed and extensively tested by plant ecologists in a multitude of environmental situations for decades. This is not to say we should not be innovative. Standard methods do, however, provide both an established starting point from which wildlife biologists can adapt specific methods as needed and easy comparability between studies. There are many fine books available that review sampling methods in vegetation ecology (e.g., Daubenmire 1968; Mueller-Dombois and Ellenberg 1974; Greig-Smith 1983; Cook and Stubbendieck 1986; Bonham 1989; Schreuder et al. 1993).

For two reasons we must be careful, of course, not simply to accept as fact all statements concerning habitat use patterns given in the literature: First—and this is more likely to occur in books than in the primary literature (i.e., original publications in scientific journals)—authors tend to repeat what previous authors have written; these citations are known as secondary, tertiary, and so on, depending upon how far removed they are from the original paper. Carefully question how vegetation structure was categorized in earlier studies; we must avoid inserting old biases into new studies (see the following subsection on preliminary sampling).

Second, we should realize that most studies directly pertain only to the time and place in which the study was conducted. Such studies do serve as a fine starting point for development of a new study; and as aforementioned, we should avoid repeating the mistakes of others and avoid reinventing the wheel over and over again. However, although repeatability and corroboration (or, alternatively, falsification) are cornerstones of scientific understanding, there is no reason to keep repeating what others have already done simply because "it hasn't been done *here* before." In such cases, a rather brief study designed to see if the earlier work can be corroborated might be indicated. Unfortunately, though, in real-world management the reverse is more often the case: Studies are often applied to environments and locations far different from those intended. In such cases, local corroborative studies are useful for determining how pertinent or applicable findings are to local situations.

Table 5.3. Hierarchic arrangement[a] of ecological components represented by 43 measurements of the forest environment taken in conjunction with sampling for the Del Norte salamander (*Plethodon elongatus*)

Hierarchic scale
- Variable category
  - Variables[b]

- II. Landscape scale
  - A. Geographic relationships
    - Latitude (degrees)
    - Longitude (degrees)
    - Elevation (m)
    - Slope (%)
    - Aspect (degrees)
- III. Macrohabitat or stand scale
  - A. Trees: density by size[c]
    - Small conifers (C)
    - Small hardwoods (C)
    - Large conifers (C)
    - Large hardwoods (C)
    - Forest age (in years)
  - B. Dead and down wood: surface area and counts
    - Stumps (B)
    - All logs-decayed (C)
    - Small logs-sound (C)
    - Sound log area (L)
    - Conifer log-decay area
    - Hardwood log-decay area (L)
  - C. Shrub and understory composition (>0.5 m)
    - Understory conifer (L)
    - Understory hardwoods (L)
    - Large shrub (L)
    - Small shrub (L)
    - Bole (L)
    - Height II–ground vegetation (B) (0.5–2 m)
  - D. Ground-level vegetation (<0.5 m)
    - Fern (L)
    - Herb (L)
    - Grass (B)
    - Height I –ground vegetation (B) (0–0.5 m)
- III. Macrohabitat or stand scale (continued)
  - E. Ground cover
    - Moss (L)
    - Lichen (B)
    - Leaf (B)
    - Exposed soil (B)
    - Litter depth (cm)
    - Dominant rock (B)
    - Codominant rock (B)
  - F. Forest climate
    - Air temperature (°C)
    - Soil temperature (°C)
    - Solar index
    - % canopy closed
    - Soil pH
    - Soil relative humidity (%)
    - Relative humidity (%)
- IV. Microhabitat scale
  - A. Substrate composition
    - Pebble (P) (% of 32–64-mm-diameter rock)
    - Cobble (P) (% of 64–256-mm-diameter rock)
    - Cemented (P) (% of rock cover embedded in soil/litter matrix)

Source: Reproduced from Welsh and Lind 1995 (table 1), by permission of the Department of Zoology, Ohio University
Note: Level I relationships (the biogeographic scale) were not analyzed because all sampling occurred within the range
[a]Spatial scales arranged in descending order from coarse to fine resolution (Weins 1989)
[b]The abbreviations used for the variables are:
C = count variables (number per hectare)
B = Braun-Blanquet variables (the percentage of cover in 1/10-ha circle)
L = line transect variables (the percentage of 50-m line transects)
P = percentage within 49-m^2 salamander search area
[c]Small trees = 12–53 cm dbh; large trees = >53 cm dbh

Preliminary Sampling

We must remember that the variables we measur and the means by which we measure them will themselves play a substantial role in determining the results of our analyses. For example, if we do not record vegetation data by taxonomic classification, then floristics can play no role in our analysis or in subsequent management applications. Likewise, if vegetation structure is categorized into 2-m height intervals, then you are making the implicit assumption that vegetation so profiled has meaning to the animal(s) under study. As we cautioned above, when designing a study, researchers must walk the fine line between simply repeating what others have done and inventing new methods, and between economically measuring only the most pertinent variables and more finely dividing variables to test for new associations.

Unfortunately, the methods sections in most papers provide little if any information about *why* the author chose the methods that he or she used. Usually the methods used are simply stated; for example: Vegetation was placed in 2-m-height intervals; or trees were categorized into the following diameter classes. A common method is to record vegetation in 11.3-m-radius plots. The relatively young researcher must wonder how such an odd measure was chosen. This measure is a holdover from the old English measurement system: an 11.3-m radius results in a 0.04-ha plot, which corresponds to the 0.1-acre plot that has been the standard for vegetation measurements. Modern researchers use this radius because it is "standard" in the literature. However, we know of no studies that have shown that this radius holds any more ecological validity than any other radius; yet, its use continues. It probably does afford some degree of sampling efficiency. We guess it is as good a radius as any other, given the lack of studies showing the contrary; either we will all be correct or we will all be wrong!

Clearly, there is the need for preliminary sampling to establish the most predictive measurements of an animal's habitat. Although we certainly understand the limits that time and money place on the intensity and duration of a study, it makes little sense to spend one's effort in the field primarily on the basis of established dogma and untested techniques. We can safely state that our most popular publications (as measured by the number of reprint requests and comments from colleagues) are consistently those that report on evaluations of methods and analytic techniques, rather than those that center on descriptions of an animal's use of habitat in a particular area.

Preliminary sampling allows one both to test the predictability of the field methods and to ensure that adequate sample sizes are being accumulated (sample size analysis is discussed in the following chapter). Referring to our previous discussion of variable selection, few of the variables listed in table 5.1, for example, played any significant statistical role in describing the habitat of the species under study. Yet, all these variables were collected for the duration of the study. Recording such data is time consuming. Would not a more efficient procedure involve collecting initial samples and then conducting preliminary analyses to determine which variables are duplicative (highly intercorrelated) and which have little or no predictive power? Box 5.1 presents the details of such an analysis.

Sampling Methods

The most popular methods of measuring microhabitat originated with a protocol developed by Frances C. James and Henry Herman Shugart, Jr. (1970). They developed a quantitative method of obtaining vegetation data "in a

Box 5.1. Preliminary data collection and analysis

Ecologists have the tendency to want to collect everything possible when in the field. After all, we are out there, so should we not spend our time fruitfully? The environment is complex, with many interacting factors. As such, does it not follow that we are setting ourselves up for failure by restricting the types of data collected when in the field? Once that sampling period is over—be it the summer of 1996 or the winter of 1996–1997—it will never occur again.

Although these sentiments are understandable on the surface, they fail when put to both logic and reality. First, it is simply not logical to try to collect "everything." Many, many variables are intercorrelated, essentially measuring the same phenomenon. Second, thousands of wildlife-habitat studies have been conducted; might we not expect that knowledge has been furthered to some extent? Let us not try to reinvent the wheel with every study. From a realistic standpoint, time is simply not available to collect an adequate number of samples on every variable that we think might be important.

Although the preceding comments seem trivial on the surface, the fact is that both new graduate students and seasoned professionals fall into the "collect everything" trap. One need only note that most multivariate papers spend some time in methods explaining how the data set was reduced; analysis of principal components and methods of stepwise inclusion of variables are examples. Most multivariate models include only a few independent variables in the final model; however, most multivariate studies of wildlife-habitat relationships collect 20 or 30 or even over 100 variables.

Another issue concerns the intensity of sampling conducted in terms of both numbers of samples and intensity with which each sampling method is applied. That is, if you are using a point intercept, should sampling points be 1 m or 2 m apart along the line? If you are using a circular plot to record tree density, should the plot be 5 m or 10 m in radius? There are few studies available to help guide such decisions. Most "standards" became such through repeated use of a convenient metric (e.g., the 0.1-acre plot).

As we discuss in this chapter, preliminary data collection and analysis can prevent wasted effort. Some specific steps in such a design are provided below; they can easily be generalized to most situations.

Step 1: First conduct a thorough literature review of similar studies. Do not restrict this review to your own taxonomic family; studies on mammals in oak woodlands provide insight into environmental conditions that could affect herps or birds in oaks. Make a list of both the variables collected and the ones that were found to be important predictors. Make note of the methods used.

Step 2: Conduct preliminary sampling.

- Sample as intensively as the literature indicates could be necessary, and sample as many variables as the literature indicates might be important (the goal is to have a manageable list of variables, say 10–20). If you are not sure if the plot should be a 10-m or 15-m radius, then record your data such that they can later be analyzed by both radii.
- Run correlation analyses to determine the redundancy that exists among your independent variables.
- Analyze the preliminary data to determine what independent variables are the best predictors of your dependent variable.
- Make sure to include sample size analyses.

Preliminary sampling can be conducted during the first part of your first field season. If you are conducting a study during different seasons, then this procedure should be repeated at the start of each different season.

Step 3: Conduct the study. By following these or similar procedures, you can use your field time to collect a larger sample size or to include another sex or age group or even other species, rather than to collect data that will be excluded from later analysis.

simple and regular manner." Their original intent was to provide a method that could augment the data on bird populations being gathered in the National Audubon Society's Breeding-Bird Censuses and the Winter Bird–Population Studies throughout the United States. But as noted earlier, their strategy has found extremely wide applicability throughout the ecological community. As we detail later in this chapter, they started by evaluating the relative efficiency of various sampling methods. Their methods gathered data on the density, basal area, and frequency of trees, canopy height, shrub density, percentage of ground cover, and percentage of canopy cover. They established 0.1-acre (0.04-ha) plots to estimate tree density and frequency. To estimate shrub density they made two transects at right angles to one another across the 0.1-acre plots, counting the number of woody stems intercepted by their outstretched arms. An ocular tube was used to estimate vegetation cover. They also provided details on how the sampling equipment could be constructed and examples of data sheets.

We have previously discussed the importance of James's (1971) paper to our conceptualization of how animals perceive their environment—the niche gestalt. She felt that the size of plot used would give an adequate description of the vegetation within an individual bird's territory; we will see below that she actually tested this assumption by comparing several methods (preliminary sampling!). She also acknowledged a potential bias in concentrating on song perches, and assumed that song-perch habitat reflects the fuller array of habitat elements for each bird species. As we have seen (Collins 1981; as discussed previously), this assumption may not hold. But regardless of the problems inherent in using only one behavior as the basis for habitat evaluation, the methods used by James have had a positive and pronounced influence on most analyses of wildlife habitat that followed.

James and Shugart (1970) compared four of the standard methods recommended by plant ecologists for making quantitative estimates of vegetation. These are the plotless methods, namely, the quarter method (Cottam and Curtis 1956; Phillips 1959) and the wandering quarter method (Catana 1963); and the areal methods, namely, arm-length transects (Rice and Penfound 1955; Penfound and Rice 1957) and circular plots (Lindsey et al. 1958). James and Shugart compared the average amount of work accomplished in 30 minutes of field effort by one observer using each of these four sampling methods in turn, assuming that he or she was familiar with the method and species of plants in the study area (see table 5.4). Of course, the amount of work accomplished will vary with the terrain, the density of the vegetation, and the actual number of observers involved. They found that results from the two plotless methods (quarter and wandering quarter) tended to overestimate the total tree density and underestimate tree density by species. Results from the two areal methods they used—1/100-acre rectangles and 1/10-acre circles—gave fairly accurate estimates of total density and density by species.

Table 5.4. Average work accomplished in 30 minutes of field effort, recording the species and diameters of trees in an upland Ozark forest in Arkansas

Sampling method	Number of units	Number of trees identified and measured
Quarter method	12 quarters	48
Wandering quarter method	40 trees	40
1/10-acre circles	2 circles	57
1/100-acre rectangles	6 rectangles	19

Source: James and Shugart 1970: table 1

Circular plots are easy to establish, mark, measure, and relocate, and estimates of animal numbers within such plots can be statistically related to vegetation data in a straightforward manner. Plots provide for the sampling of vegetation and animals at specific locations in space and time. Thus it is also easy to pinpoint plots using geographic positioning systems (GPS) and subsequently to input their data into geographic information systems (GIS). If plots can be considered independent data points (a function of the sampling design and behavior of the animals), then one's sample size is equal to the number of plots sampled. Or if the plots are used to sample from a single study area, then plots sampled can be averaged and the associated measures of variance can be calculated. Noon (1981) presented a useful description and example of both the transect and areal plot sampling systems. The problem with transects is that they cover relatively large areas, and thus make it difficult to relate specific animal observations (or abundances) to specific sections of the transect. Transects are, however, widely used to provide an overall description of the vegetation of entire study areas.

In summary to this point, fixed-area plots and transects can be used to provide site-specific, detailed analyses of wildlife-habitat relationships. The majority of sampling methods used since the 1970s to develop wildlife-habitat relationships, for subsequent multivariate analyses, have used fixed-area plots (usually circular) as the basis for development of a sampling scheme that may then incorporate subplots, sampling squares, and transects. Next we describe examples of some of the more widely used methods.

Dueser and Shugart (1978) developed a detailed sampling scheme that combined plots of various sizes and shapes, as well as short transects (see fig. 5.5). Although designed for analysis of small mammal habitat, the techniques can be easily adapted for most terrestrial vertebrates. Dueser and Shugart established three independent sampling units, centered on each trap: a 1.0-m^2 ring, two perpendicular 20-m^2 arm-length transects, and a 10-m-radius circular plot. The 1.0-m^2 ring provided a profile of vertical foliage from the ground through 2 m height, for both herbaceous and woody vegetation. Also, four replicate core-sample estimates of litter-soil depth, compactability, and dry weight density were made on the perimeter of this central ring. The two arm-length transects provided measures of cover type, surface characteristics, and density and evergreenness of the four strata of vegetation. Data recorded for each quarter of the 10-m-radius plot included the species, dbh, and distance from the trap to the nearest understory and overstory trees, numbers of stumps and fallen logs, basal diameter and distance of nearest stump and fallen log, and total length of fallen logs.

We consider Dueser and Shugart's sampling scheme an excellent starting point for any study being established in forested areas; the basic sampling strategy could also be adapted for other vegetation types. Although all the variables they measured (see table 5.1) relate to "microhabitat," some relate to the specific trap location (e.g., litter samples, soil compactability), whereas others relate more to the conditions surrounding the trap (e.g., data collected along the arm-length transects, distance to trees). Thus, to a minor degree, the authors were mixing scales of measurement within the general construct of measuring microhabitat. As discussed previously, *microhabitat* is a general term that should be clearly defined for all research applications. Authors should explain how each variable fits within the overall concept of spatial scale and thus how it meets study objectives.

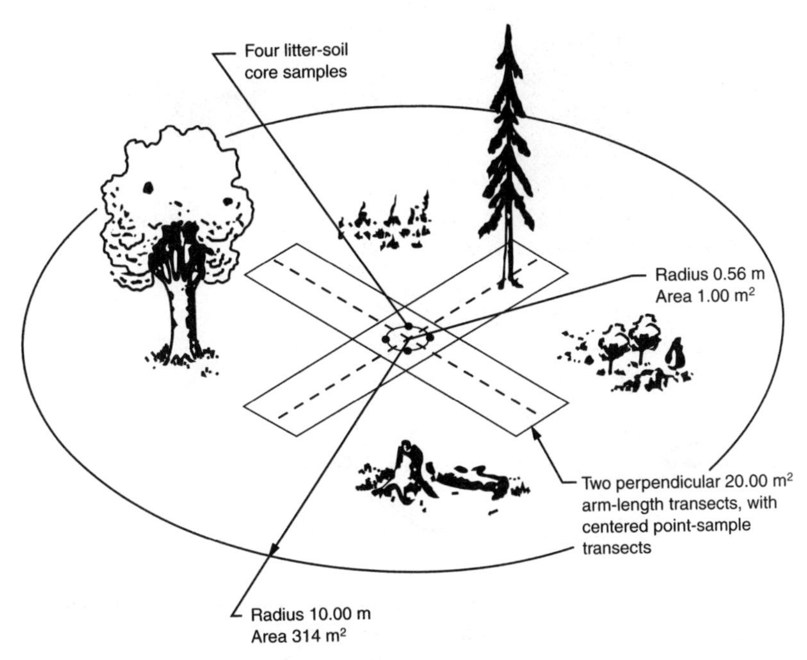

Figure 5.5. Habitat variable sampling configuration used by Dueser and Shugart in their study of small mammal habitat use. (Reproduced from Dueser and Shugart 1978 [fig. 1], by permission of the Ecological Society of America)

In his analysis of snake populations, Reinert (1984) adopted techniques similar to those used in the bird study by James (1971) and the small mammal study by Dueser and Shugart (1978). That is, Reinert applied the basic conceptual framework used by the earlier authors—the niche gestalt and multivariate representation of the niche—in developing the rationale for his methods. Here again, we see the commonality in methods running across studies of wildlife-habitat relationships.

Reinert made several modifications to the sampling methods used by James and by Dueser and Shugart. Notably, he used a 35-mm camera equipped with a 28-mm wide-angle lens to photograph 1-m^2 plots from directly above the location of a snake. He then determined the various surface cover percentages by superimposing each slide on a 10 × 10–square grid. Reinert, then, more rigorously quantified his measure of cover values than most workers, who usually use ocular estimates. As discussed in the following chapter, much error can be entered into a data set when ocular estimates are used to measure plant cover (Block et al. 1987). The specific sampling scheme used by Reinert is summarized in figure 5.6, and his variable list was previously presented in table 5.2. Note the similarity between Reinert's design and that of Dueser and Shugart, including the minor mixing of spatial scales. Reinert added several environmental variables that measured air, surface, and soil temperature and humidity. The values of these variables are

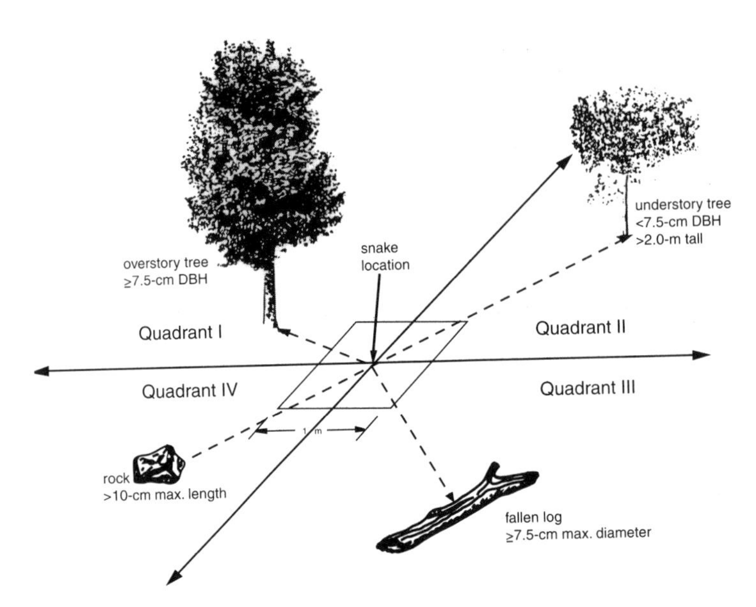

Figure 5.6. Sampling arrangement for snake locations. (Reproduced from Reinert 1984 [fig. 1], by permission of the Ecological Society of America)

obviously dependent upon the time of day and the general weather conditions at the time of measurement; such constraints do not influence (are not correlated with) the other variables measured. In addition, temperature data take on a much different statistical distribution from that of most vegetation variables. Such mixing of variable types in the same analysis should be undertaken with caution and, in fact, should probably be avoided (unless partitioned out in a stepwise analysis).

Literature Cited

Addicott, J. F., J. M. Aho, M. F. Antolin, D. K. Padilla, J. S. Richardson, and D. A. Soluk. 1987. Ecological neighborhoods: Scaling environmental patterns. *Oikos* 49:340–46.

Aebischer, N. J., P. A. Robertson, and R. E. Kenward. 1993. Compositional analysis of habitat use from animal radio-tracking data. *Ecology* 74:1313–25.

Alldredge, R. J., and J. T. Ratti. 1986. Comparison of some statistical techniques for analysis of resource selection. *Journal of Wildlife Management* 50:157–65.

Alldredge, R. J., and J. T. Ratti. 1992. Further comparison of some statistical techniques for analysis of resource selection. *Journal of Wildlife Management* 56:1–9.

Anderson, S. H., and H. H. Shugart, Jr. 1974. Habitat selection of breeding birds in an east Tennessee deciduous forest. *Ecology* 55:828–37.

Arthur, S. M., B. F. J. Manly, L. L. McDonald, and G. W. Garner. 1996. Assessing habitat selection when availability changes. *Ecology* 77:215–27.

Block, W. M., K. A. With, and M. L. Morrison. 1987. On measuring bird habitat: Influence of observer variability and sample size. *Condor* 72:182–89.

Block, W. M., M. L. Morrison, J. Verner, and P. N. Manley. 1994. Assessing wildlife-habitat-relationships models: A case study with California oak woodlands. *Wildlife Society Bulletin* 22:549–61.

Bonham, C. D. 1989. *Measurements for terrestrial vegetation.* New York: John Wiley and Sons.

Bookhout, T. A., ed. 1994. *Research and management techniques for wildlife and habitats.* 5th ed. Bethesda, Md.: The Wildlife Society.

Carbon, B. A., G. A. Bartle, and A. M. Murray. 1979. A method for visual estimation of leaf area. *Forest Science* 25:53–58.

Catana, A. J., Jr. 1963. The wandering quarter method of estimating population density. *Ecology* 44:349–60.

Cody, M. L. 1983. Continental diversity patterns and convergent evolution in bird communities. In *Mediterranean-type ecosystems,* ed. F. J. Kruger, D. T. Mitchell, and J. U. M. Jarvis, 357–402. Berlin: Springer-Verlag.

Collins, S. L. 1981. A comparison of nest-site and perch-site vegetation structure for seven species of warblers. *Wilson Bulletin* 93: 542–47.

Cook, C. W., and J. Stubbendieck, eds. 1986. *Range research: Basic problems and techinques.* Denver: Society for Range Management.

Cooperrider, A. Y., R. J. Boyd. and H. R. Stuart, eds. 1986. *Inventory and monitoring of wildlife habitat.* USDI Bureau of Land Management Service Center, Denver, Colo.

Cottam, G., and J. T. Curtis. 1956. The use of distance measures in phytosociological sampling. *Ecology* 37:451–60.

Daubenmire, R. 1968. *Plant communities: A textbook of plant synecology.* New York: Harper and Row.

Dueser, R. D., and H. H. Shugart, Jr. 1978. Microhabitats in a forest-floor small mammal fauna. *Ecology* 59:89–98.

Fish and Wildlife Service. 1981. *Standards for the development of suitability index models.* USDI Fish and Wildlife Service, Ecological Services Manual 103. Washington, D.C.: Government Printing Office. 68 pp.

Ganey, J. L., and W. M. Block. 1994. A comparison of two techniques for measuring canopy closure. *Western Journal of Applied Forestry* 9:21–23.

Green, R. H. 1971. A multivariate statistical approach to the Hutchinsonian niche: Bivalve mollusks of central Canada. *Ecology* 52:543–56.

Green, R. H. 1979. *Sampling design and statistical methods for environmental biologists.* New York: John Wiley and Sons.

Greig-Smith, P. 1983. *Quantitative plant ecology.* 3d ed. Berkeley: University of California Press.

Heyer, W. R., M. A. Donnelly, R. W. McDiarmid, L. C. Hayek, and M. S. Foster. 1994. *Measuring and monitoring biological diversity: Standard methods for amphibians.* Washington, D.C.: Smithsonian Institution Press.

Hilden, O. 1965. Habitat selection in birds. *Annales Zoologici Fennici* 2:53–75.

Holmes, R. T. 1981. Theoretical aspects of habitat use by birds. In *The use of multivariate statistics in studies of wildlife habitat,* ed. D. E. Capen, 33–37. USDA Forest Service General Technical Report RM-87.

Hutto, R. L. 1985. Habitat selection by nonbreeding, migratory land birds. In *Habitat selection in birds,* ed. M. L. Cody, 455–76. New York: Academic Press.

Irwin, L. L., and J. G. Cook. 1985. Determining appropriate variables for a habitat suitability model for pronghorns. *Wildlife Society Bulletin* 13:434–40.

James, F. C. 1971. Ordinations of habitat relationships among breeding birds. *Wilson Bulletin* 83:215–36.

James, F. C., and H. H. Shugart, Jr. 1970. A quantitative method of habitat description. *Audubon Field Notes* 24:727–36.

James, F. C., and N. D. Wamer. 1982. Relationships between temperate forest bird communities and vegetation structure. *Ecology* 63:159–71.

Johnson, D. H. 1980. The comparison of usage and availability measurements for evaluating resource preference. *Ecology* 61:65–71.

Karr, J. R., and R. R. Roth. 1971. Vegetation structure and avian diversity in several New World areas. *American Naturalist* 105:423–35.

Kelt, D. A., P. L. Meserve, and B. K. Lang. 1994. Quantitative habitat associations of small mammals in a temperate rainforest in southern Chile: Empirical patterns and the importance of ecological scale. *Journal of Mammalogy* 75:890–904.

Krebs, C. J. 1978. *Ecology: The experimental analysis of distribution and abundance.* New York: Harper and Row.

Levin, S. A. 1992. The problem of pattern and scale in ecology. *Ecology* 73:1943–67.

Lindsey, A. A., J. D. Barton, and S. R. Miles. 1958. Field efficiencies of forest sampling methods. *Ecology* 39:428–44.

MacArthur, R. H., and J. W. MacArthur. 1961. On bird species diversity. *Ecology* 42:594–98.

Manly, B. F. J., L. L. McDonald, and D. L. Thomas. 1993. *Resource selection by animals: statistical design and analysis for field studies.* London: Chapman and Hall.

Mawson, J. C., J. W. Thomas, and R. M. DeGraaf. 1976. *Program HTVOL: The determination of tree crown volume by layers.* USDA Forest Service Research Paper NE-354.

Morrison, M. L. 1984. Influence of sample size on discriminant function analysis of habitat use by birds. *Journal of Field Ornithology 55:330–35.*

Morrison, M. L., and R. G. Anthony. 1989. Habitat use by small mammals on early-growth clear-cuttings in western Oregon. *Canadian Journal of Zoology* 67:805–11.

Morrison, M. L., and E. C. Meslow. 1983. Bird community structure on early-growth clearcuts in western Oregon. *American Midland Naturalist* 110:129–37.

Morrison, M. L., B. G. Marcot, and R. W. Mannan. 1992. *Wildlife-habitat relationships: Concepts and applications.* 1st ed. Madison: University of Wisconsin Press.

Morrison, M. L., W. M. Block, L. S. Hall, and H. S. Stone. 1995. Habitat characteristics and monitoring of amphibians and reptiles in the Huachuca Mountains, Arizona. *Southwestern Naturalist* 40:185–92.

Morrison, M. L., C. J. Ralph, J. Verner, and J. R. Jehl, Jr. 1990. *Avian foraging: Theory, methodology, and applications.* Studies in Avian Biology no. 13.

Mueller-Dombois, D., and H. Ellenberg. 1974. *Aims and methods of vegetation ecology.* New York: John Wiley and Sons.

Murray, N. L., and D. F. Stauffer. 1995. Nongame bird use of habitat in central Appalachian riparian forests. *Journal of Wildlife Management* 59:78–88.

Noon, B. R. 1981. Techniques for sampling avian habitats. In *The use of multivariate statistics in studies of wildlife habitat,* ed. D. E. Capen, 42–52. USDA Forest Service General Technical Report RM-87.

Ockenfels, R. A., and D. E. Brooks. 1994. Summer diurnal bed sites of Coues white-tailed deer. *Journal of Wildlife Management* 58:70–75.

Penfound, W. T., and E. L. Rice. 1957. An evaluation of the arms-length rectangle method in forest sampling. *Ecology* 38:660–61.

Phillips, E. A. 1959. *Methods of vegetation study.* New York: Holt, Rinehart, and Winston.

Ralph, C. J., G. R. Geupel, P. Pyle, T. E. Martin, and D. F. DeSante. 1993. *Handbook for field methods for monitoring landbirds.* USDA Forest Service General Technical Report PSW-144. 41 pp.

Reiner, H. K. 1984. Habitat separation between sympatric snake populations. *Ecology* 65:478–86.

Rice, E. L., and W. T. Penfound. 1955. An evaluation of the variable-radius and paired-tree methods in the blackjack-post oak forest. *Ecology* 36:315–20.

Rotenberry, J. T. 1985. The role of habitat in avian community composition: Physiognomy or floristics? *Oecologia* 67:213–17.

Roth, R. R. 1976. Spatial heterogeneity and bird species diversity. *Ecology* 57:773–82.

Schreuder, H. T., T. G. Gregoire, and G. B. Wood. 1993. *Sampling methods for multiresource forest inventory.* New York: John Wiley and Sons.

Scott, J. M., F. Davis, B. Csuti, R. Noss, B. Butterfield, C. Groves, H. Anderson, S. Caicco, F. D'Erchia, T. C. Edwards, Jr., J. Ulliman, and R. G. Wright. 1993. *Gap analysis: A geographic approach to protection of biological diversity.* Wildlife Monograph 123.

Smallwood, K. S. 1995. Scaling Swainson's hawk population density for assessing habitat use across an agricultural landscape. *Journal of Raptor Research* 29:172–78.

Smallwood, K. S., and C. Schonewald. 1996. Scaling population density and spatial pattern for terrestrial, mammalian carnivores. *Oecologia* 105:329–335.

Sturman, W. A. 1968. Description and analysis of breeding habitats of the chickadees, *Parus atricapillus* and *P. rufescens. Ecology* 49:418–31.

Thomas, D. L., and E. Y. Taylor. 1990. Study designs and tests for comparing resource use and availability. *Journal of Wildlife Management* 54:322–30.

VanderWerf, E. A. 1993. Scales of habitat selection by foraging 'elepaio in undisturbed and human-altered forests in Hawaii. *Condor* 95:980–89.

VanDeusen, P. C., and G. S. Biging. 1984. *Crown volume and dimension models for mixed conifers of the Sierra Nevada.* Northern California Forest Yield Cooperative Research Note no. 9, Department of Forestry and Resource Management, University of California, Berkeley.

Van Horne, B. 1983. Density as a misleading indicator of habitat quality. *Journal of Wildlife Management* 47:894–901.

Welsh, H. H., Jr., and A. J. Lind. 1995. Habitat correlates of Del Norte salamander, *Plethodon elongatus* (Caudata: Plethodontidae), in northwestern California. *Journal of Herpetology* 29:198–210.

Wenny, D. G., R. L. Clawson, J. Faaborg, and S. L. Sheriff. 1993. Population density, habitat selection and minimum area requirements of three forest-interior warblers in central Missouri. *Condor* 95:968–79.

Whitmore, R. C. 1981. Applied aspects of choosing variables in studies of bird habitats. In *The use of multivariate statistics in studies of wildlife habitat,* ed. D. E. Capen, 38–41. USDA Forest Service General Technical Report RM-87.

Wiens, J. A. 1969. *An approach to the study of ecological relationships among terrestrial birds.* Ornithological Monographs no. 8.

Wiens, J. A. 1989. *The ecology of bird communities.* Vol. 1: *Foundations and patterns.* Cambridge: Cambridge University Press.

Wiens, J. A., and J. T. Rotenberry. 1981. Habitat associations and community structure of birds in shrubsteppe environments. *Ecological Monographs* 51:21–41.

Willson, M. F. 1974. Avian community organization and habitat structure. *Ecology* 55:1017–29.

6 Measurement of Wildlife Habitat: When to Measure and How to Analyze

Introduction

In this chapter we first develop the concept of when to measure wildlife habitat. *When* to measure involves both between-season and within-season analyses. It has been clearly shown that resource requirements of animals vary considerably between seasons; the use of different winter and summer ranges by big game is one example. However, species that are permanent residents in an area frequently switch to different foraging substrates and food sources as seasons change. Further, within a given time period, resource use and activity patterns often vary with time of day, stage of the breeding cycle, and other divisions of an animal's life cycle. Thus, it is critical that one understands the temporal aspects of resource use prior to designing a study of habitat.

The distribution and abundance of animals vary with variations in food supply, weather conditions, predator activity, and many other biotic and abiotic factors. As we show throughout this book, these factors must drive how we design and then interpret our studies of wildlife habitat. Few studies, however, have explicitly recognized these variations in species distributions and abundances, but instead have based their conclusions on time- and site-species analyses.

In the final sections of this chapter we discuss the use of multivariate statistics in analysis of wildlife habitat. We first examine the rationale for using multivariate statistics, relating these techniques to our conceptualization of wildlife-habitat studies. We then review the all-important assumptions associated with these statistical techniques. We follow with a classification and discussion of several of the more commonly used methods and their applications to wildlife studies. Sample-size requirements and a review of statistical computer packages are also included. A formal course in multivariate statistics is not required for understanding this chapter. Many fine multivariate texts are available and should be consulted for a general overview of this class of analyses (e.g., Morrison 1967; Cooley and Lohnes 1971; Pimentel 1979; Afifi and Clark 1984; Dillon and Goldstein 1984; Digby and Kempton 1987; Krzanowski 1990).

When to Measure

As is commonly known, the behavior, location, and needs of animals change, often substantially, throughout the year. Too many researchers, however, ignore such temporal variations in habitat use, in at least two major ways: While often acknowledging that temporal variations do occur, they sample from such a narrow time (usually only during spring and/or summer) that their results are too time and location-specific for more than minimal application to other situations. Second, they may sample from across some broad time period (like the summer season) and then "average-out" the relationships over the period. We are not faulting the first strategy with regard to the detail of the data collected, if the samples are adequate; applicability is the issue. Without knowledge of an animal's total requirements, management recommendations have limited and perhaps faulty implications.

Few would argue, of course, that the preferred study design is repetition of study for every appropriate biological period (e.g., prebreeding, breeding). It is easy to conceptualize what "averaging" may mean to resulting descriptions of habitat relationships: In figure 6.1 we see that the average use over time by two hypothetical species is not, in fact, a close approximation of their actual behavior. Rather, the average implies that the animals use this tree species basically identically. Here we see, however, that the more finely we stratify our sampling, the greater the number of levels of resolution we have available for subsequent analysis. The question concerning how finely to measure is a study-specific problem, and further, within a particular study, the level of resolution will probably vary with time (e.g., between seasons). Hypothetical scenarios such as that depicted in figure 6.1 should concern researchers. Such concerns can be evaluated early in a study by making sure that sampling is sufficiently intensive that such relationships can be identified. Subsequent sampling can be adjusted after such preliminary evaluations of data are completed.

Schooley (1994) reviewed 43 papers that examined habitat use of terrestrial vertebrates and were published in the *Journal of Wildlife Management* between 1988 and 1991. He found that most studies pooled data on habitat use over a number of years, evidently without testing for annual variation. Using the black bear (*Ursus americanus*) as an example, he illustrated the misleading inferences that can result from pooling data across years. Patterns of annual variation in habitat use were lost when data were pooled (see fig. 6.2.)

No distinct separation actually exists between sampling wildlife populations within or

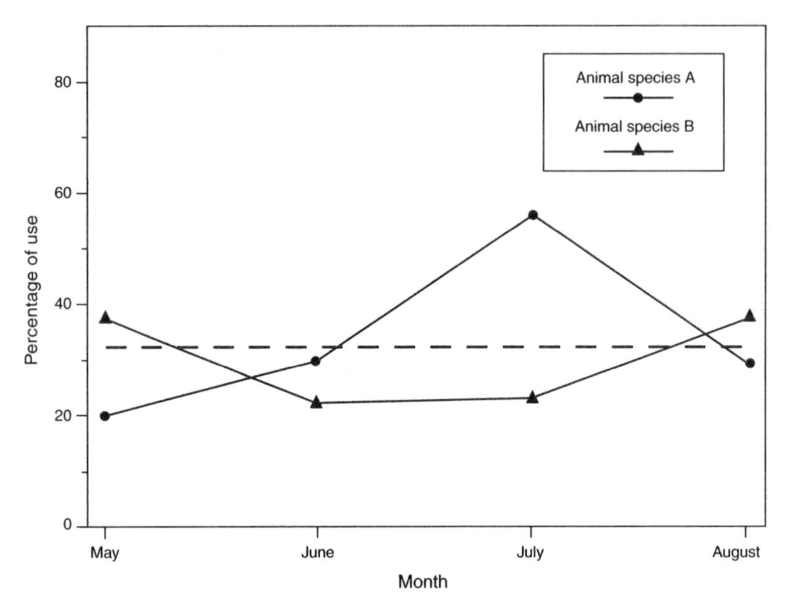

Figure 6.1. Use of a species of tree by two hypothetical animal species during "summer." The dashed horizontal line represents the approximate average of use values for both species (calculated separately) across time.

between seasons (intra- or interseasonally, respectively). Further, even within relatively small areas, the seasons can differ widely for species both within and especially between vertebrate groups. In most regions, for example, the breeding season for many rodents can start as early as February, months before many breeding bird species even arrive in the area. Moreover, within the birds, many resident species begin breeding several months before nonresidents. Few studies acknowledge these sampling problems; or if they do, they sample from the "middle" of the season (which, for logistical reasons, usually conforms to the summer recess of most universities).

Because of limited funding, availability of personnel, cold weather, and other problems related to management of studies, most researchers have concentrated their studies during the summer. But it is now well known that the survival of the animal may depend on the nonbreeding, particularly the wintering, period. Intuitively, we would expect that the fall and winter periods—when populations are at their greatest numbers (because of offspring), resources are declining (trees and arthropods are dying or going dormant), animals are physiologically stressed by dispersal or migration movement, and the weather is becoming harsher—would be the most difficult times for an animal. Further, because of these changes in environmental conditions, the habitat itself changes. Large-scale migration by numerous species of animals provides a good indication of

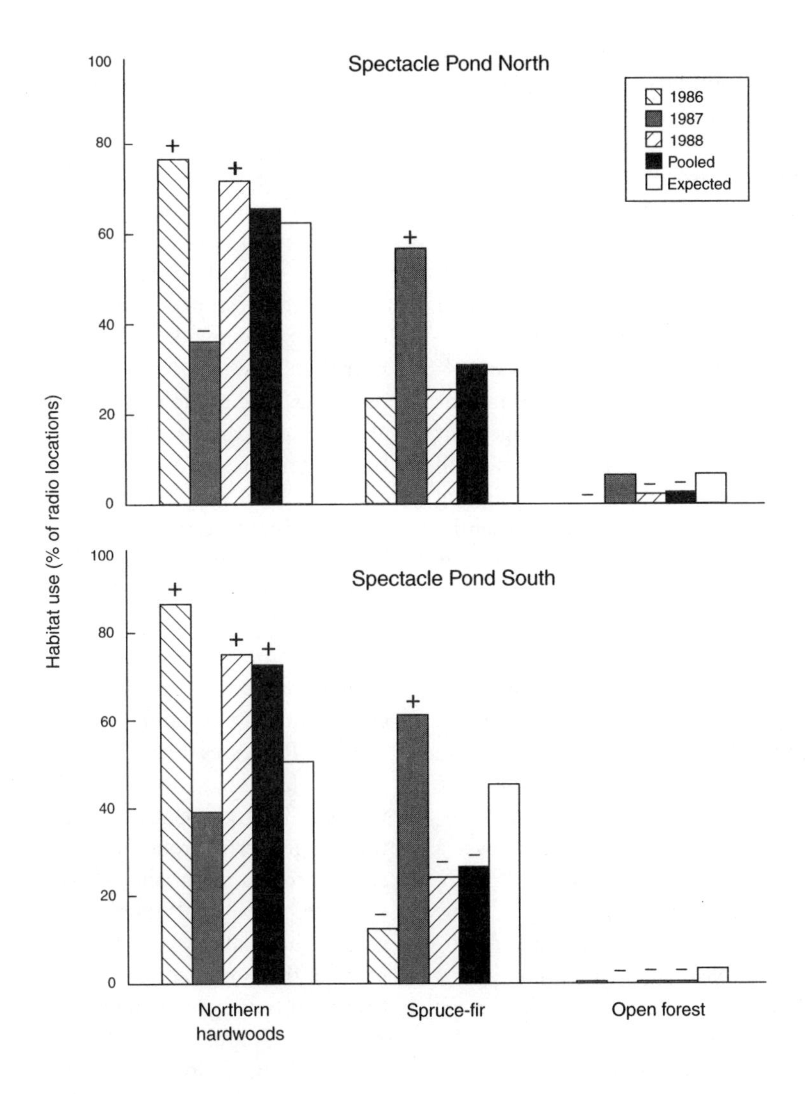

Figure 6.2. Habitat selection by female black bears at two study areas in northern Maine during fall (one September den entry), 1986–1988. Habitat use is presented for individual years and for all years pooled. Expected use is estimated from availability of habitats. Habitat availability was estimated separately for each year, but only 1987 data are presented, because availability differed little between years. A + indicates use was greater than availability, a − indicates use was less than availability, and no symbol indicates use was equal to availability. Selection was based on Bonferroni-adjusted 90% CIs. (Reproduced Schooley 1994 [fig. 1], by permission of The Wildlife Society)

the lengths to which animals must go to find favorable living conditions. The importance of nonbreeding periods to animals was popularized by Fretwell (1972), although he was certainly not the first worker to recognize this importance. If an animal changes its pattern of habitat use or if a change occurs in the environment itself, then a summer-based study cannot be used with any confidence to predict the subsequent responses of the animal.

At a minimum, researchers should acknowledge these sampling limitations and the ramifications of such weaknesses on management recommendations. Ideally, researchers will avoid many of these problems by designing studies that actually determine if such potential temporal variations are influencing their results in major ways.

Long-Term Temporal Changes

We have just discussed the need to evaluate intra- and interseasonal variations in habitat use. The magnitude of the temporal variation, however, is a question of scale, and the scale can run across as well as within years. Thus, relationships developed within a year or over a period of a few years may not hold for a different period of years. We all know how environmental conditions can change drastically across years: "Normal" weather conditions seldom occur, but are in reality a crude average of extremes in rain, temperature, and wind. An animal's reaction to such conditions over time will provide the most important data with which one can evaluate habitat relationships.

As reviewed by others (e.g., Likens 1983; Wiens 1984; Strayer et al. 1986; Leigh and Johnston 1994), a tradition of pursuing short-term studies has developed over the past several decades, especially among North American scientists. This trend arose from constraints imposed by short funding duration, the need to finish graduate programs within short periods of time (actually a noble goal), and the pressure placed on researchers to publish. But by restricting the duration of our investigations, we adopt a snapshot approach to studying nature. Our wildlife studies usually run from one to three years and, at best, give us only a partial view of most ecological situations; at worst, they give us a false interpretation.

Long-term studies are especially suited to exploring four major classes of ecological phenomena: slow processes, rare events, subtle processes, and complex phenomena (Strayer et al. 1986; also discussed in chapter 9 on disturbance dynamics). Forest succession, invasion of exotic species, and vertebrate population cycles are prominent examples of slow processes that have obvious importance in formulating management decisions. The substantial impact of El Niño occilations, catastrophic natural events (e.g., floods, disease outbreaks, fires), and population eruptions are examples of rare events that are certainly missed by short-term studies. Processes that change over time in a regular fashion, but where the year-to-year variance is large relative to the magnitude of the longer-term trend, are examples of the subtle processes that short-term studies cannot evaluate. Finally, although scientists certainly realize that nature is complex, seldom do we provide the time necessary for the phenomena to reveal enough of their characteristics to allow meaningful interpretation.

How long *is* long enough? Strayer et al. (1986) provided two rather different definitions for the concept of long-term. Their first definition considers the length of the study in terms of natural processes. Here a study is long-term if it continues for as long as the generation time of the dominant organism or long enough to include examples of the important processes that structure the ecosystem under study; the length of study is measured against the dynamic speed of the system being studied. Obviously, such a criterion demands that the researcher has a good understanding of the system of interest. Strayer and his colleagues' other approach is to view the length of study in a relative fashion, with long-term studies being those that have continued for a longer time than most other such studies. Here they are accepting the constraints applied by human institutions, not the rate of natural processes.

Of course, not all studies need be long term to provide useful results. Descriptive studies of essentially static patterns (e.g., morphology, genetic characteristics of species), of processes at the individual level (e.g., growth, behavior, physiology), or of patterns of adaptation do not necessarily require long time periods. Ecological studies conducted on broad spatial scales (e.g., relating presence or absence to vegetation types) also are usually relatively static. The principal disadvantages of long-term studies are, of course, practical rather than ecological.

There are four classes of short-term studies that can potentially provide insight into long-term relationships: (1) retrospective studies, (2) substitution of space for time, (3) use of systems with fast dynamics as analogs for systems with slow dynamics, and (4) modeling (Strayer et al. 1986). Further, a series of short-term studies can be incorporated into a longer-term plan for research; such a plan works especially well in a professor–graduate student relationship. Substitution of space for time and modeling are commonly used for longer-term work. In substitution studies, sites with different characteristics are used instead of following a few sites for an extended period (a chronosequence). An example is examining the relationship between birds and plant succession over 1 year using sites of different ages (e.g., 1, 5, 15, 30, or more years postharvest or postburn). To provide valid results, such a design requires that all the sites have similar histories and characteristics. Naturally, a large number of replicates enhances the reliability of such a study. Of course, these studies cannot capture the historical events that shaped each site; they can only swamp the effect through a large sample size; ideally, this will yield adequate results. An increase in experimental manipulations of vegetation, food, competitors, predators, and other parameters is needed in association with demographic studies.

Can short-term studies, then, tell us much about animals? Wiens (1984:202–3) thought not, answering, ". . . a short-term approach is likely to produce incomplete or incorrect perceptions of a complex reality . . . perhaps obtaining results that are superficial and quite possibly incorrect." Short-term studies can provide useful information if they have a specific focus that is not likely to be obscured by the background variation in habitat relationships. However, available evidence clearly indicates the necessity of the implementation of longer-term studies if the development of wildlife-habitat relationships is to advance. We should not be surprised that the relatively "easy" studies have already been conducted; the task presented to new students should not be expected to ease with time. Graduate students cannot be expected, of course, to conduct the long-term work in their thesis projects. It is the duty of graduate advisors to help direct students toward projects whose results are not likely to be swamped by unknown long-term variance. Long-term studies done in parallel with carefully matched, short-term substitution studies are needed. Wiens (1989:174–96) provided additional review material on the usefulness of long-term ecological studies.

Sample Size Requirements

Regardless of the care taken in designing a study, all is for naught if an insufficient number of observations is made. To most, this must seem an obvious statement. However, very little attention has been paid in the scientific literature to this fundamental question of study design and statistical analysis. We are not sure why this is so and can only advance the suggestion that researchers have generally tried to collect the largest sample size their budgets and time would allow. All of us have heard this, and several of us must admit to having followed this "strategy" earlier. If researchers must limit their sampling because of time or monetary constraints, then they would be better advised to limit the scope of study and produce one good result rather than many weak ones.

Most general statistics texts discuss determination of the sample size necessary to perform certain analyses. The specific methods vary according to the type of data being collected. In general, the sample size needed depends upon four factors: (1) the variance of the population, (2) the size of the difference between sampling units to be detected, (3) the statistical probability (Type I error), and (4) the assurance one desires in detecting the difference (Type II error). The researcher can establish the final three factors; obviously, any two of these factors will determine the third (Zar 1984). Note that an initial estimate of the population variance is required. These variances can be estimated either from the literature or from data collected during a preliminary analysis of the population of interest. See box 6.1 for details on calculating sample size.

Another method of estimating necessary sample size involves continually evaluating the data as increasing sample sizes are obtained. Such "sequential sampling" is especially useful when one has no idea of the true population variance, and especially when one is gathering data on numerous variables (as is often the case in habitat studies). Further, continual evaluation of data gives the added benefit of providing in-depth familiarity with the data set, familiarity that may identify unexpected or interesting facts about the species in time to modify the sampling protocol. Such a procedure fits as part of the preliminary sampling strategy that we previously discussed. Mueller-Dombois and Ellenberg (1974) suggested plotting the standard error as a function of the number of samples and denoting an adequate sample size when the curve decreased to less than a 5–10 percent gain in precision for a 10 percent increase in sample size.

A major problem encountered by many wildlife researchers is the adequate sampling of rare populations. Two major issues are involved here: First is the issue of being forced to sample the same individual animals repeatedly. Such repeated sampling usually violates statistical independence, which prevents each observation from being treated as a "sample." Although beyond the scope of this book, statistical methods such as time-series analysis and repeated-measures analysis of variance are available to handle such situations (see Raphael 1990). In addition, decent biological rationales can often be used to minimize repeated sampling (e.g., sampling an individual animal only once per sampling session). True independence is never achieved, but compromises are often necessary (and should be fully explained and defended in any publication).

Small mammal trapping presents a special problem with repeated sampling: the same individuals will certainly be captured within and also between trapping sessions. Researchers have handled this dilemma in various ways. Kelt et al. (1994) concluded that the repeated captures of the same individuals at trap stations could introduce a bias, so they used only first-time captures in their analyses. In contrast, Morrison and Anthony (1989) considered each capture of a new individual or recapture at a new location a single sample. For example, if two *Microtus* individuals were captured at the same trap, the habitat characteristics associated

Box 6.1. How many samples are needed?

In wildlife research—directly manipulative experiments as well as observational studies—it is always better to plan ahead for sampling design than to struggle with incomplete data after the fact. Correctly designing a study typically entails asking the following kinds of questions:

- What and how many conditions (control and treatment groups) am I trying to test?
- How variable are the groups?
- How confident do I need to be in detecting differences?
- How precise should the test be?
- How many samples do I need?

Much has been written on sampling designs in response to these and related questions (see text for discussion and references). Here, we focus on the last question, that of sample size.

Determining sample size needed to detect differences between means in a multigroup study is a relatively simple procedure, but one that is seldom conducted in wildlife monitoring or research studies. As a result, many investigators (including managers interested in data on monitoring of wildlife populations or habitats) struggle with samples inadequate for statistically answering questions about differences and must resort to pooling samples, which is often undesirable.

To illustrate the calculation, let's work with an example. Let us assume we want to detect differences in the mean number of large snags (dead standing trees) in four successional stages of forest development. The question is then, How many (what sample size of) vegetation sampling plots (or forest stands) are needed per successional stage? Our successional stages comprise sample groups. Comparing means between more than two groups, in this case four groups, entails use of analysis of variance (ANOVA) techniques and use of Student's *t* statistics to determine mean differences. We will assume a balanced ANOVA design with equal sample sizes in each group. Also, making the calculation entails using an initial sampling from which preliminary estimates of sample standard deviations and coefficients of variation are derived.

First, we need to specify the level of certainty of detecting a difference and the percentage of difference between means to detect. In our hypothetical study, we can specify, for example, an 80 percent certainty of detecting a 5 percent difference between two of the four group means at a confidence level of 0.01. This is a reasonable expectation, although a confidence level of 0.05 is also typically used in such studies.

Since $a = 4$, the ANOVA error MS (mean square) has $v = a(n - 1) = 4\ (n - 1)$ degrees of freedom. The sample size n is thus:

$$n \geq 2(\sigma/\delta)^2(t_{\alpha(v)} + t_{2(1-P)(v)})^2$$

where

σ = population standard deviation
δ = smallest true difference that is desired to be detected
t = values of Student's t (found in statistical tables)
v = degrees of freedom (df) of the sample standard deviation with a groups each with n replications in a balanced ANOVA
α = significance level (here, 0.01)
P = desired probability (here, 80 percent)

Note that only the ratio σ/δ is needed, not the individual factors; this ratio is specified in our example here at 5 percent.

Next, assume that a sample resulted in a coefficient of variation (CV = σ/y, where y = sample mean) among the groups at 6 percent. We proceed by making an initial guess at n, say $n = 20$ (always a safe starting point unless samples are extremely variable, in which case begin with $n = 40$. Then, $v = 4(20 - 1) = 76$ df. Since CV = 6 percent, then $s = 6y/100$, where s = the sample standard deviation. We wish δ to be 5 percent of the mean; that is, $\delta = 5y/100$. Using s as an estimate of σ, which is obtained from initial exploratory sampling, we obtain $\sigma/\delta = (6y/100)(5y/100) = 6/5$, so that

$$\begin{aligned} n &\geq 2(6/5)^2(t_{0.01(76)}+t_{2(1-0.80)(76)})^2 \\ &= 2(6/5)^2(2.64+0.847)^2 \\ &= 2(1.44)(12.16) \\ &= 35.0 \end{aligned}$$

Next, we try another value for n, say, $n = 35$. So, $v = 4(35 - 1) = 136$; and $n \geq 2(1.44)(2.61 + 0.845)^2 = 34.4$, or $n \approx 35$ again. Thus, is seems that 35 samples per population (or forest plots per successional stage) are necessary.

Our example depicts sampling forest vegetation plots. The approach can be used to determine the number of samples for monitoring programs, such as those needed to detect differences between treatments and controls in adaptive management studies (see chapter 12).

with the trap were entered into the analysis twice. Recaptures of the same individual at the same trap were not analyzed. They followed this procedure to weight the analysis toward favored trap sites and their associated habitat characteristics. They rationalized that, if that procedure were not used, trap sites with one individual of a species and those with multiple individuals would likely show false similarity in habitat use. Their procedure was thus an attempt to weight habitat use by animal abundance. We recommend, however, that researchers avoid the problem of independence and design studies to fit within repeated-measures designs. Or if this is not possible (e.g., an adequate sample of recaptures cannot be obtained), then base trapping grid habitat analysis on the area used by each individual within defined sampling seasons. Here, a crude "used area" delineated by trap locations serves as the sampling area for vegetation analysis. After all, the use of a trap—into which an animal has been attracted with baits—is unlikely to represent an unbiased sample of habitat.

Second, detecting the presence of rare animals typically cannot be accomplished through the use of standard sampling methods. The abundance of rare animals usually approximates a Poisson distribution. Such distributions are characterized by many zeroes and only an occasional location of the animal of interest. The sample size necessary for detecting species in such situations has been developed by Green and Young (1993); a generalized view of sampling efforts needed is provided in figure 6.3. As noted by Green and Young, "rarity" is a relative

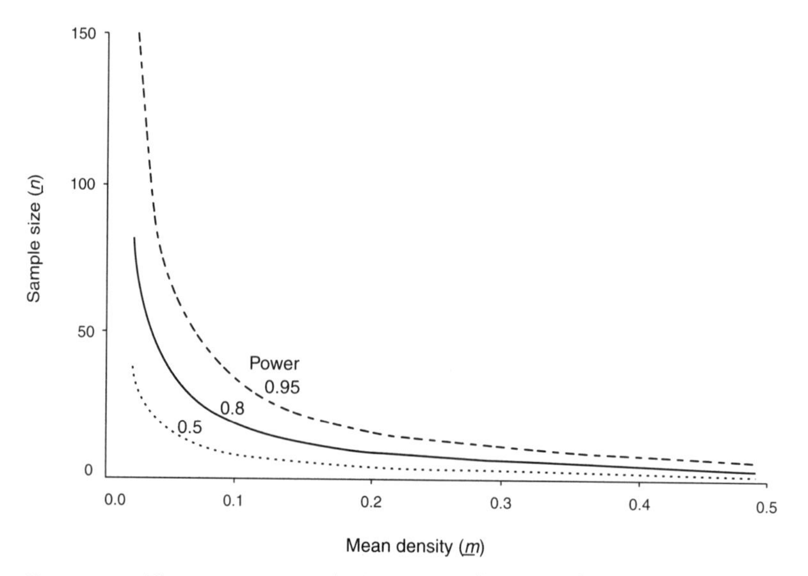

Figure 6.3. The necessary sample size, *n*, as a function of mean density, *m*, for various degrees of power, $1-\beta$, when sampling the Poisson distribution. (Reproduced from Green and Young 1993 [fig. 3], by permission of the Ecological Society of America)

concept; they define anything that has a density of <0.1 per sampling unit size (e.g., quadrat size) as rare. They present formulae that are applicable to specific sampling situations.

Block et al. (1987) studied the sample size necessary in making estimates of vegetation characteristics in the mixed-conifer zone of the western Sierra Nevada. Locations of foraging birds (although the results of their study are applicable to any forest animal) served as the center of 0.04-ha circular plots in which structural and floristic variables were quantified. Measurements were taken using both visual (ocular) estimates and more objective techniques involving standard measuring devices (diameter tape for three dbh, clinometer for tree heights, measuring tape for distances). For each variable, they randomly selected, with replacement, 10 subsamples of 5, 10, 15, 20, 30, 40, 50, and 60 plots. They then calculated the mean for each sample and then the mean of the means for each subset. Stability of the estimates (i.e., an adequate sample size) was defined as the point where the estimate of the mean remained within one standard deviation of the subsequent estimates and there was little variation in the magnitude of the confidence interval (CI).

Block and his colleagues found that the type of variable collected, whether visually estimated or measured, affected the sample size required. Minimum sample size requirements for measured variables ranged from 20 (for average tree dbh) to 50 (for average tree height), and those for estimated variables ranged from 20 (for average tree dbh) to more than 75 (for average height to the first live tree branch). More important, the final point estimates of variables often differed between estimates and measurements (fig. 6.4.)

The Block et al. (1987) study shows that researchers must be concerned not only with the size of the sample gathered but also with the way in which variables are estimated. Note again (fig. 6.4) that the two techniques of making observations often result in different values. Many such variables entered into a multivariate analysis can alter the resulting conclusions regarding habitat use by an animal, especially if accompanied by small sample sizes (Morrison 1984b). These researchers used "standard" methods of recording data, which raises questions regarding the conclusions drawn by other workers using similar techniques. In fact, a review of studies of wildlife-habitat relationships indicates that extremely small sample sizes were used in many instances (Morrison 1984a,b; Morrison et al. 1990). Students are encouraged to examine carefully the sample sizes used and the techniques used to gather data in the publications they read in support of their studies.

Observer Bias

The scientific method is designed, in part, to help us avoid allowing our preconceived notions to determine the outcome of our studies. We all carry numerous biases into the design of any study: our training predisposes us to prefer certain sampling methods; we might believe that animals respond in specific ways to environmental conditions, which tends to narrow our sampling focus; time and monetary constraints limit our ability to sample in the way we would prefer; and so forth. Biases not only influence the way we design a study, but additional ones are also inserted into field sampling when multiple observers—likely all with different training and experience—are used to gather data. As noted by Gotfryd and Hansell (1985:224), "Ignoring observer-based variability may lead to conclusions being precariously balanced on artifacts, spurious relations, or irreproducible trends." This is a strong

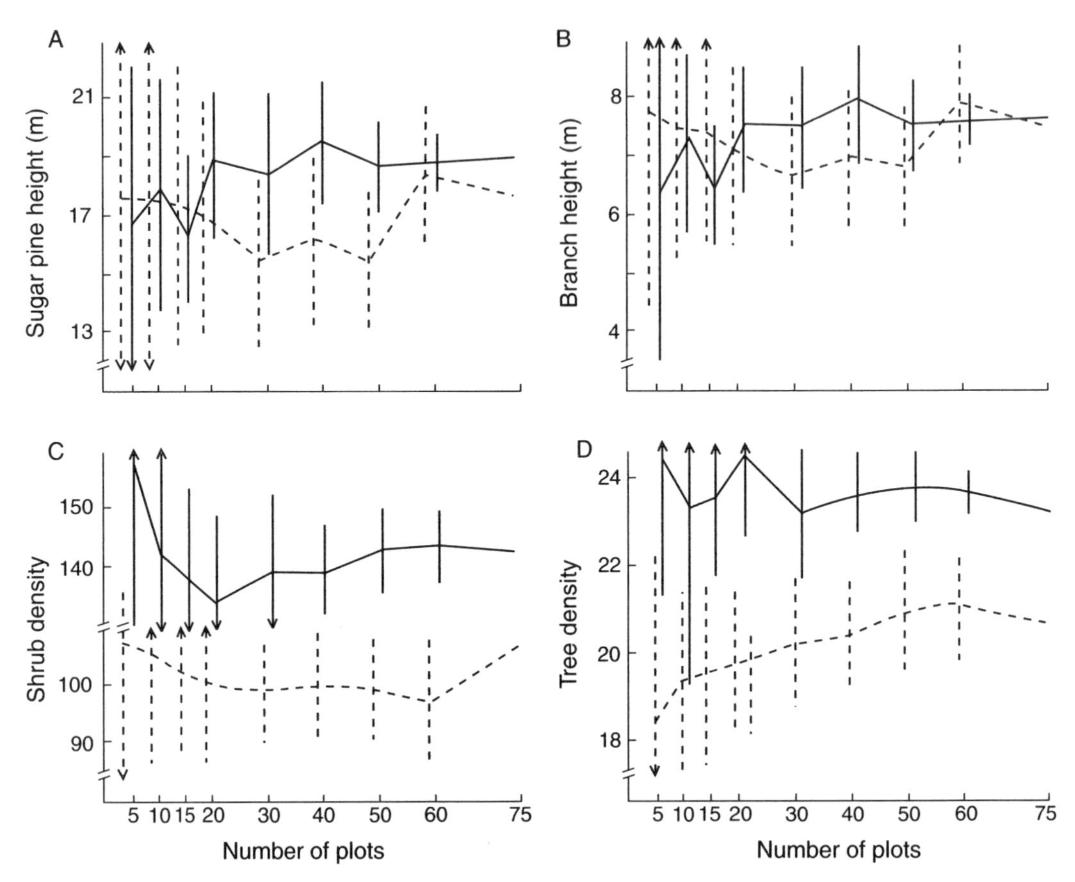

Figure 6.4. Influence of sample size on the stability of estimates (dashed horizontal lines) and measurements (solid horizontal lines) of bird-habitat characteristics. Dashed or solid vertical lines represent 1 SD from point estimates for estimates and measurements, respectively. Variables shown are A, average height of sugar pine; B, average height to the first live branch of sugar pine; C, average number of shrubs within sample plots; and D, average number of trees within sample plots. (Reproduced from Block et al. 1987 [fig. 2], by permission of Hastings Natural History Reservation)

statement that is, unfortunately, ignored in the design and implementation of many studies of wildlife and their habitats. Here we will concentrate on the evaluation and subsequent reduction of bias in field sampling.

Gotfryd and Hansell (1985) used four observers to sample independently eight plots located within an oak-maple forest near Toronto, Ontario. They followed methods detailed by James and Shugart (1970) and measured the variables given in table 6.1. They found that observers differed significantly in their measurements for 18 of the 20 vegetation variables measured. As these authors noted, their study

Table 6.1. Vegetational habitat variables and their mnemonics, used by Gotfryd and Hansell

Mnemonic	Variable
TRSP	Number of tree species
SHRSP	Number of shrub species
*SDEN	Density of woody stems <7.6-cm diameter at breast height
*CC	Canopy cover
*GC	Ground cover
BAA	Basal area (BA) of trees 7.6–15.2-cm
BAB	BA of trees 15.2–23-cm
*BA1	BAA + BAB
*BA2	BA of trees 23–53-cm
*BA3	BA of trees >53-cm
CH1	Maximum canopy height
CH4	Maximum canopy height in quadrant having lowest canopy
*CHAV	Average of canopy height maxima by quadrant
*CHRNG	CH1 to CH4
*CHCV	CHRNG/CHAV
DTR1	Distance to nearest tree >15.2-cm
DTR4	Maximum of by-quadrant nearest tree distances
*DAV	Average of by-quadrant nearest tree distances
*DRNG	DTR1 to DTR4
*DCV	DRNG/DAV

Source: Reproduced from A. Gotfryd and R. I. C. Hansell, "The Impact of Observer Bias on Multivariate Analyses of Vegetation Structure," *Oikos*, vol. 45 © 1985 Oikos
*Variables used in multivariate analyses

addressed only the precision of estimates between observers; no measure of the accuracy of their results was conducted.

Block et al. (1987) used several univariate and multivariate analyses to test for differences between three observers in estimating plant structure and floristics. They found that ocular estimates by the three observers differed significantly for 31 of the 49 variables they measured. Perhaps the most confounding aspect of using multiple observers they found was the unpredictable nature of variation among observers. Multiple comparisons of estimates for the 31 significant variables resulted in all possible combinations of observers. Thus, when samples from different observers are pooled, sampling bias can increase. As we show later in this chapter, this variability has, especially when combined with low sample sizes, a particularly profound influence on multivariate analyses.

Ganey and Block (1994) used three observers to sample plots for canopy closure employing two different estimation techniques (spherical densiometer and sighting tube; results described above). They found significant variation among observers in estimates of canopy cover for both methods. Results were, however, relatively more consistent for the sighting tube.

The biases associated with estimations of animal abundances should also be considered carefully in habitat studies. This is because many of our analytic procedures correlate animal numbers with features of the environment. Obviously, a study that has low bias among habitat characteristics can be ruined by biased count data, and vice versa. For example, Dodd and Murphy (1995) evaluated the accuracy and precision of nine techniques used to count heron nests. Although they found rather high error rates among the techniques, observer bias was low for most methods. Interestingly, the highest observer bias was found for their point counting technique, a result apparently due to the observers' varying choices of the optimum vantage point from which to count nests at colonies.

Researchers can reduce interobserver variability by following closely a set of well-defined criteria for selecting and training observers. Although designed for bird censusing, the steps outlined by Kepler and Scott (1981) for bird counting methods can be applied generally to most types of sampling: Carefully screen applicants initially to eliminate the more obvious visual, aural, and psychological factors that increase observer variability. In addition, researchers should organize a rigorous observer training program, which will further reduce inherent variation but will not eliminate it. In a field experiment, Scott et al. (1981) found that observers, after training, could estimate their distance from a singing bird with an accuracy of a 10–15 percent variation from the true distance. Also, periodic training sessions with observers should be conducted to counter their "drift" and thus recalibrate their recording to standard and known values (see Block et al. 1987). However, even though observers can be trained to develop more precise ocular estimates of vegetation conditions, such training does not necessarily reduce bias.

Plant ecologists have long recognized differences between data collection techniques (see Cooper 1957; Lindsey et al. 1958; Schultz et al. 1961; Cook and Stubbendieck 1986; Hatton et al. 1986; Ludwig and Reynolds 1988; Kent and Coker 1994). Although the cost of measuring plant structure and floristics in an adequate number of plots usually is great in both time and money, the ramifications of not following a rigorous sampling design are severe. Again, it is better to limit the scope of a study so that the data that are collected are done so properly; preliminary sampling and analysis help identify potential problems.

Multivariate Assessment of Wildlife Habitat

In this section we first examine the rationale for using multivariate techniques, relating these methods of analysis to our conceptualization of wildlife-habitat studies. We then review the all-important assumptions associated with these methods. We follow with a classification and brief discussion of some of the current techniques and their applications to wildlife studies, concentrating on the frequently used multiple regression and discriminant analysis procedures. Our intent here is to introduce briefly the application of multivariate techniques to the analysis of wildlife habitat data, concentrating on problems encountered in their use. A formal course in multivariate statistics is not a prerequisite for understanding this chapter. We do, however, strongly recommend such a course in the study plan of all graduate students. Such a course assists with your ability to evaluate the literature, even if you never conduct a multivariate analysis of your own data. Many good multivariate texts are available and should be consulted for details not included herein (e.g., Cooley and Lohnes 1971; Pimentel 1979; Dillon and Goldstein 1984; Johnson 1992; Manly 1994); Dillon and Goldstein's text is an especially readable work.

Conceptual Framework

Multivariate analysis is a branch of statistics used to analyze multiple measurements that have been made on one or more samples of individuals. Multivariate analysis is distinguished from other forms of statistical procedures in that multiple variables are considered in combination as a system of measurement. Because the variables are typically interdependent, we cannot separate them and examine each individually (Cooley and Lohnes 1971:3).

Multivariate statistical techniques were not originally designed for analysis of wildlife habitat and behavioral data. Indeed, multivariate techniques have been used since the late 1880s, with a progression of new, more sophisticated methods following throughout the 1900s (Cooley and Lohnes 1971:4). Only since the 1960s have researchers placed emphasis on quantitative analyses of wildlife habitat.

The application of multivariate analysis of wildlife data is the product of a synthesis beginning in the early 1970s that united several lines of ecological research. As outlined by Shugart (1981), this synthesis involves linking two analytic tools with three ecological concepts—niche theory, microhabitat, and individual response (fig. 6.5). The availability of high-speed computers along with prewritten multivariate computer programs (i.e., "canned" statistical packages) allows easy access to this general field of analysis. Hutchinson's reformulation of the niche concept (1957) in terms of an "*n*-dimensional hypervolume" caused ecologists to alter their view of wildlife habitat and the ways in which they analyze data.

Green (1971) was one of the first ecologists to apply multivariate analyses formally to Hutchinson's concept. The *n*-dimensional concept of the niche and the *n*-dimensional sample space of multivariate analysis are analogous in many ways, and this similarity has led to an obvious application of multivariate methods to ecological data (Shugart 1981).

Assumptions

Four common assumptions are associated with parametric multivariate analyses: multivariate normality, equality of the variance-covariance matrices (group dispersions), linearity, and independence of the error terms (residuals). Violation of any of these assumptions can bias or taint analysis results and conclusions derived from them. Unfortunately, many published papers in the wildlife literature have failed to discuss these assumptions and their ramifications on results. As we discuss below, low sample size is often a major factor in such violations.

Normality

The assumption of multivariate normality is more than simply an assumption that each variable is itself normally distributed, as in the univariate sense. Unfortunately, tests of this assumption are cumbersome and yield only approximations. Variable-by-variable examination of normality, however, will certainly help identify variables that depart greatly from normality. Standard univariate transformations (e.g., log, square-root transformations) can be applied when appropriate.

Variance

In ecology, however, it is well known that the distribution and behavior of animals vary along gradients of environmental variables (e.g., soil moisture, canopy cover, air temperature). As noted by Pimentel (1979:177), however, biologists seem too concerned with the mean responses of animals to environmental gradients, rather than with the distribution of animals along such gradients. Concentrating on mean responses certainly lowers the ability of research results from one location (and thus one point on the gradient) to be applied to another geographic location (and thus likely

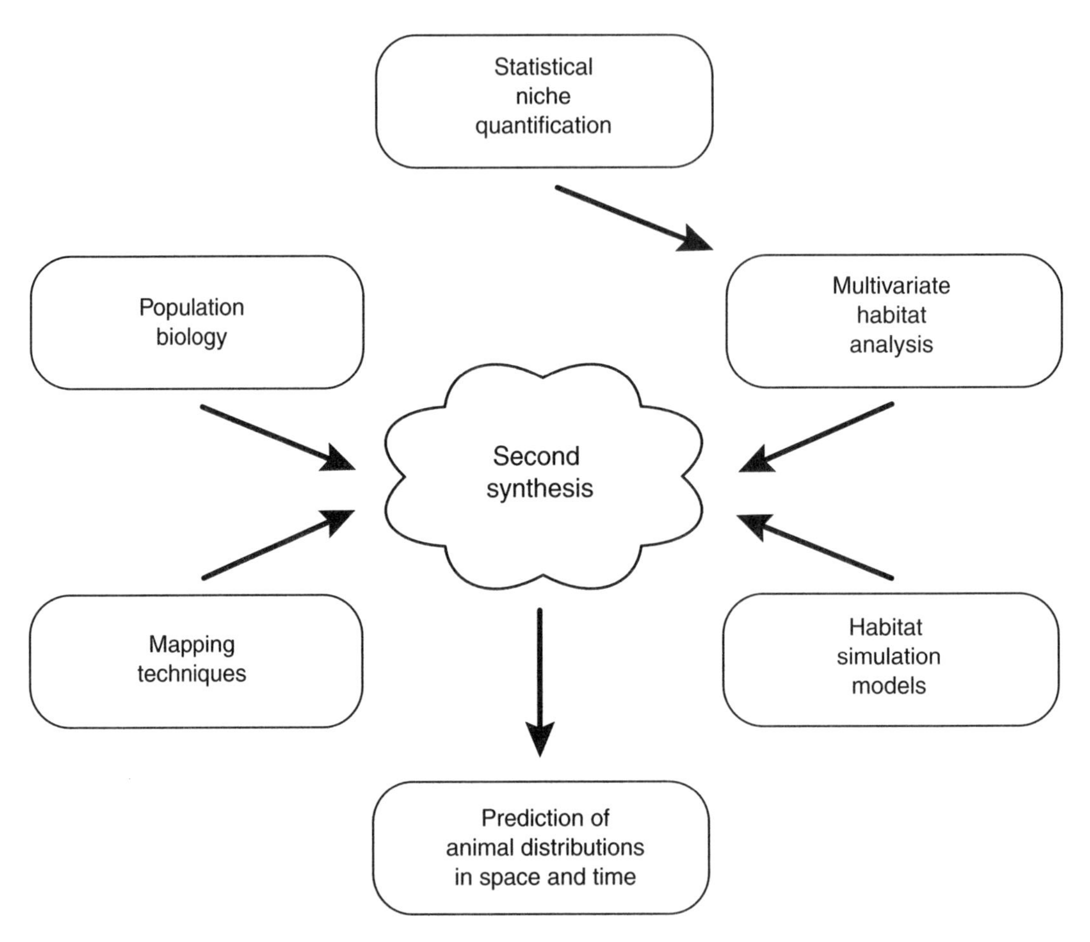

Figure 6.5. Schematic diagram of the scientific research elements that combine in a synthesis to produce multivariate habitat analysis. (From Shugart 1981: fig. 1)

from another point on the gradient). If populations have unequal dispersions, then the populations are different even if their central tendencies (means) are the same. Thus, trying to force normality to meet the formal assumption of equality of dispersions is biologically unsound. In fact, the magnitude of this dispersion can be used as a multivariate measure of niche breadth (see Carnes and Slade 1982). Nothing prevents one from comparing populations having unequal dispersions. Fortunately, research indicates that tests of equality of group centroids are rather insensitive to moderate departures of multivariate normality and homoscedasticity. If sample sizes are large and equal between groups, then the inequality of dispersions has no real effect on interpretation of results. Application of this rationale requires careful planning to ensure that large and equal samples are collected for all groups under study—a caveat that has been ignored in most studies of wildlife-habitat relationships. As noted by Wiens (1989:66), a considerable portion of the variance that is so easily discarded in statistical analyses may contain important insights into the dynamics of the system under study.

Linearity

Linearity is important in multivariate analyses in two main ways. First, most models are based on a linear relationship. Second, the correlation coefficient, which forms the basis of most multivariate calculations, is sensitive only to the linear component of the relationships between two variables. Also, assuming linear relations is a more parsimonious approach than the use of nonlinear multivariate models, nonlinear models should be used only if linear ones prove to be poor fits. Fortunately, many relationships can be approximated using linear models even though nonlinear components may exist. The data transformations noted above may help to linearize a nonlinear relationship. However, nonlinearity changes the probabilities in tests of significance; for example, one is likely to fail to reject null hypotheses of equality of group centroids but to succeed in appropriately rejecting null hypotheses of equality of group dispersions (Pimentel 1979:178–79. Researchers would be well served by first examining the linearity of their data, variable by variable, before plunging them into the black box of a canned multivariate statistical package. Highly nonlinear data should not be combined into a single analysis with linear data.

Thus, the selection of linear models is usually more a matter of statistical convenience than a decision based on ecological reality. As noted above, we should expect biological entities often to have nonlinear and nonnormal distributions. Linear analyses are, however, relatively straightforward; their statistical properties are well understood. Further, a model has little utility if other researchers and especially managers find it difficult to understand. Thus, it can be argued that linear models should be used unless the data are *highly* nonlinear.

There are two basic methods for applying nonlinear analyses to statistical models. The first is to transform variables by various orders of polynomial transformations. Linear models are, specifically, first-order polynomial models. Second-order, or quadratic, models are obtained by squaring a variable (X^2), resulting in a U-shaped relationship. Third-order, or cubic, models are obtained by cubing a variable (X^3), resulting in a curving relationship. Higher-order models are possible, but they become increasingly difficult to visualize and are unlikely to be applicable. An interesting example is provided by Burger et al. (1994), who regressed the arcsine-transformed percentage of predation on nests onto a natural log-transformed area (ln[size]) and the square of log transformed area (ln[size]^2). This polynomial regression explained 77 percent of the variation in predation rates among study tracts of varying sizes (fig. 6.6). However, it is exceedingly difficult to interpret these transformations in terms of real-world natural history.

Because it is relatively difficult to interpret high-order models, you should first try to linearize data through appropriate transformations. It is, after all, difficult to explain that a population's abundance is a function of cubed foliage volume, squared snag density, and shrub cover. If a linear model can meet your objectives, then use it. However, if a linear model does not provide an adequate prediction after

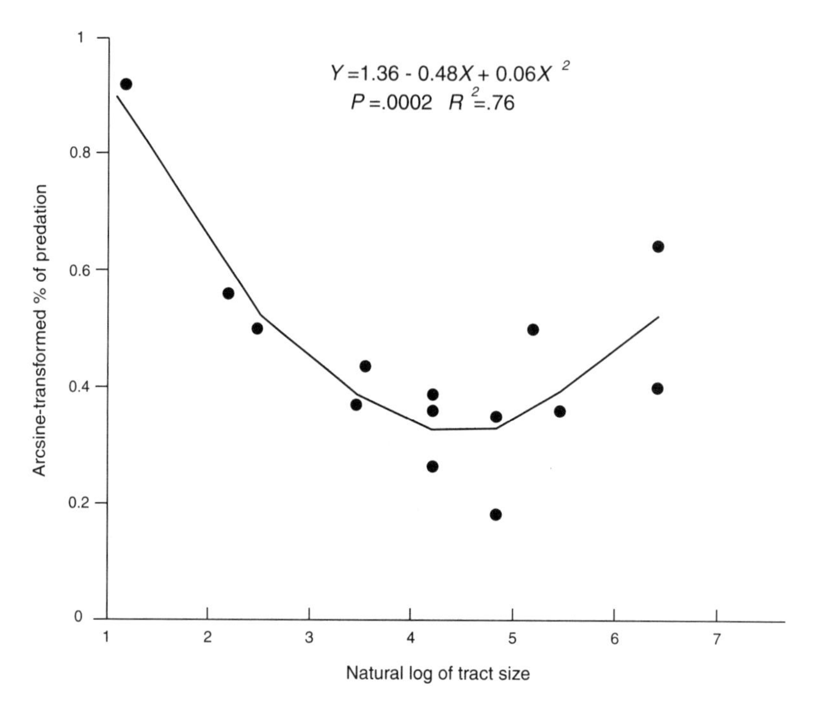

Figure 6.6. Arcsine-transformed predation on artificial nests (n = 540) in 15 prairie fragments regressed onto a natural log-transformed tract area (*ln* [size]) and (*ln* [size]^2). (Reproduced from Burger et al. 1994 [fig. 1], by permission of The Wildlife Society)

appropriate testing (see below), then nonlinear alternatives should be used before declaring that the linear model did not reveal significant environmental correlations (see also Nadeau et al. 1995).

Meents et al. (1983) evaluated the use of first-, second-, and third-order independent variables for predictions of bird abundance using multiple regression. They found that the polynomial variables resulted in significant regressions in many cases in which linear variables did not do so. They also showed that, while linear relationships dominated during certain parts of the year, nonlinear relationships were important during other periods of time. This explains, in part, why models often must be season-specific. In addition, they showed why nonlinear relationships can have important management implications. For example, woodpecker abundance might be related to snag density in a curvilinear fashion, where relatively low or high densities of snags result in a lower bird abundance than that found for moderate snag densities. Thus, woodpeckers may disappear when snags decline below some critical nonzero threshold. In this simple example, linear models could mask important biological relationships, resulting in faulty management decisions. Other examples are apparent (e.g., soil-moisture tolerance among salamanders, micronutrient content of herbivore forage, grass density for rodents). Other examples of nonlinear analysis of wildlife-habitat data are included below with specific analytic techniques (e.g., logistic regression).

Purely nonlinear models exist in a form other than the standard linear least squares model (with polynomial transformations). Draper and Smith (1981: chapter 10) gave examples of such nonlinear models, and Seber (1989) and Ratkowsky (1990) provided thorough development of nonlinear modeling as applied to regression analysis. On computers, nonlinear models are solved as approximations through relatively complicated iterative numeric calculations. The form of the equations and the iterative methods are beyond the scope of this book. Nonlinear models are, however, being incorporated into most of the larger statistical software packages.

Independence

Independence and random sampling are cornerstones of all biological investigations, whether they are later analyzed by uni- or by multivariate methods. In studies of biological populations in which the population of interest is hard to define and sampling procedures may affect the probability of an individual being included in the sample, truly random sampling is difficult to achieve. Behavioral responses may vary between sexes and among age classes, locations, and the like, rendering the possibility of gathering a truly random sample highly unlikely. Thus, we should tightly restrict the definition of the population sampled. For example, populations designated by sex, age, time, and location should provide reliable interpretations. When these factors are ignored, unequal representation of the factors between samples might imply differences that do not exist between populations (Pimentel 1979:176). Unfortunately, few published studies specifically or adequately define their statistical, sampled population. Failure to define the population has clear and adverse management implications: users will be unable to apply results appropriately to the proper time and place, thus extending results beyond the proper target population (Tacha et al. 1982).

Exploratory versus Confirmatory Analyses

Scientists often collect wildlife data in observational studies not specifically designed to test a statistical hypothesis. Here, the researcher does not have sufficient information to state and test *a priori* a null hypothesis. Analysis of such data is often called exploratory analysis. In contrast, if a researcher has some prior information regarding the theoretic structure of the data, such as that gathered from an earlier exploratory study, then "corroborative" studies and analyses are performed. If data fail to meet the assumptions associated with the particular method(s) used, however, then results of these studies are also stated in terms of "exploratory" or "descriptive" analyses. That is, because formal tests of null hypotheses require that all assumptions are met (or at least not grossly violated), one cannot technically test these hypotheses under violations of assumptions. Unfortunately, many researchers use this rationale—calling their results descriptive—as an excuse for not gathering an adequate number of samples in the first place. Authors must clearly state the severity of the violations of assumptions, provide details on the specific form of these violations, and interpret how these violations could bias results, interpretations, and conclusions.

Classifications of Multivariate Techniques

We can divide multivariate techniques into two general categories: one group, data reduction or ordination procedures, also called dependence models; and two or more group classification procedures, also called interdependence models. Both categories, while containing primarily parametric procedures, also include nonparametric methods. Multivariate techniques can also be subdivided into linear and nonlinear techniques.

A chart of many of the most common statistical methods (table 6.2) reveals the interrelated nature of multivariate methods. Methods are broadly classified by the number of dependent and independent variables of interest and the goal of the researcher analyzing the data. That is, does the researcher wish to examine the structure, or interdependence, of variables (PCA); examine the relationship (correlation) between variables (multiple *R*); separate groups (MANOVA); or develop predictive equations (DA)? Dillon and Goldstein (1984: fig. 1.5–1) and Harris (1985: table 1.1) gave similar classifications of multivariate techniques. Below we briefly outline some of the specific methods found within these categories and provide examples of their applications to wildlife-habitat data. Because most studies of wildlife-habitat relationships have concentrated on principal components analysis (PCA), multiple regression (MR), and discriminant analysis (DA), we will concentrate our discussion on how these methods are used and interpreted. There are many additional analyses, such as detrended correspondence analysis (DCA), nonmetric multidimensional scaling (NMDS), and reciprocal averaging (RA), that could be applied to habitat analysis. Capen (1981) and Verner et al. (1986) should be consulted for many more specific examples.

Table 6.2. Classification of common statistical techniques

Major research question	Number of dependent variables	Number of independent variables	Analytical method[a]
Degree of relationship between variables	One	One	Bivariate *r*
		Multiple	Multiple *R*
	Multiple	Multiple	Canonical *R*
Significance of group differences	One	One	One-way ANOVA; *t*-test
		Multiple	Factorial ANOVA
	Multiple	One	One-way MANOVA; T^2
		Multiple	Factorial MANOVA
Prediction of group membership	One	Multiple	One-way DA; logistic regression
	Multiple	Multiple	Factorial DA
Structure	Multiple	Multiple	PCA; FA; metric and nonmetric multidimensional scaling; correspondence analysis

Source: B. G. Tabachnick and L. S. Fidell *Using Multivariate Statistics,* 3d ed. Copyright © 1996 by HarperCollins Publishers Inc.; reprinted by permission of Addison-Wesley Educational Publishers Inc.

[a]Analysis of variance (ANOVA); multivariate analysis of variance (MANOVA); Hotelling's T^2; discriminant analysis (DA); principal components analysis (PCA); factor analysis (FA).

Data Structure: Ordination and Clustering

Methods within this broad category of analyses seek to reduce a complex data set (i.e., many variables) to a few number of dimensions (axes) that are internally correlated but unique (not correlated) with regard to other derived dimensions. Ordination and clustering are similar in that no groups are assumed to exist prior to analysis. Because wildlife biologists usually collect data on numerous, usually intercorrelated variables, we need a method of reducing the number of variables to a manageable level. A common technique involves a two-step procedure: First, variables that are highly intercorrelated (e.g., an r of >0.7) are identified. Second, for each correlated pair identified, the one with the least power of separating groups is removed from the analysis (as identified through a t- or F-test). Unfortunately, unless the excluded variable had a perfect correlation with the included variable, this method results in some degree of loss of information from the data set. The lower the correlation (r), the greater the loss. Thus, "data reduction" techniques that retain all the information in the variables are desirable.

Principal components analysis is a method that identifies new sets of orthogonal (mutually perpendicular and thus not correlated) axes in the direction of greater variance among observations. The first axis is the line in a direction through the observations that is oriented such that the projections of the observations onto the axis have maximum variance. The second axis is in the direction of the greatest variance perpendicular ("orthogonal") to the first axis; that is, this axis does not duplicate the variance "explained" in the first axis. Additional axes are derived until all the explainable variance is accounted for.

Kelt et al. (1994) used multivariate analyses of distributions of small mammals to identify important habitat associations in a temperate rain forest in Chile. They centered their vegetation sampling at each trap station and, during winter and summer, collected detailed data on canopy, shrub, and ground cover (table 6.3). They then used principal components analysis (PCA) to identify vegetation variables that helped distinguish the plots. In summary, variables for ground cover were important descriptors for both seasons, and shrub and tree variables were relatively less important in describing the vegetation. The relative importance of characteristics of ground cover appeared greater in winter than in summer, when other variables were of generally equal importance (table 6.3). Eigenvalues greater than 1.0 were retained for ecological evaluation. They then plotted captures of rodents onto the PCA-reduced habitat dimensions and showed that the species tended to overlap considerably in their use or habitat characteristics (figure 6.7). This study serves as a good example of how PCA can be used to assist in data reduction (i.e., to simplify interpretation of the many correlated vegetation variables), then help to understand how the community of small mammals separates along the PCA axes, and then to interpret the principal causes of such separation in terms of the original variables.

General Interpretations

It is often difficult to determine which variables are involved with what principal component and to what degree. By comparing component loadings, we can determine those variables most related to a component. Dillon and Goldstein (1984:69) observed that it is easier to adopt heuristics for the purpose of interpreting the pattern. The procedure given by Dillon and Goldstein can simplify interpretation

Table 6.3. Eigenvectors for the significant components resulting from a principal components analysis conducted on summer and winter vegetative parameters

Season Parameter	PC1	PC2	PC3	PC4	PC5	PC6	PC7
			Summer				
Eigenvalues	3.570	2.790	2.270	1.640	1.330	1.160	1.020
Canopy cover	−0.166	0.180	0.003	−0.189	0.018	0.408*	−0.331*
Number of logs	0.206	−0.144	−0.212	0.274*	−0.285*	0.250*	0.007
Distance to nearest shrub	−0.015	0.197	−0.023	0.491*	0.163	−0.294*	−0.084
Width of nearest shrub	0.016	0.243	0.457*	0.208	−0.004	0.306*	0.146
Height of nearest shrub	0.128	0.212	0.415*	0.213	0.095	0.373*	−0.110
Number of species of shrubs	−0.092	0.063	0.226	−0.004	−0.199	0.086	0.754*
Shrub cover	0.273*	−0.116	0.251*	−0.192	−0.009	0.112	−0.293*
Number of trees	−0.283*	0.154	−0.204	−0.253*	0.292*	0.132	0.049
Number of species of trees	−0.174	0.177	−0.228	−0.179	0.459*	0.177	0.233
Bare ground cover at 2 m	0.204	0.367*	0.012	−0.254	−0.204	−0.167	0.074
Herbaceous ground cover at 2 m	−0.395*	−0.217	0.203	0.147	0.015	−0.068	−0.168
Tree plus trunk ground cover at 2 m	0.276*	−0.117	−0.300*	0.117	0.169	0.254*	0.143
Bare ground cover at 4 m	0.260*	0.366*	0.076	−0.283*	0.023	−0.118	−0.086
Herbaceous ground cover at 4 m	−0.421*	−0.215	0.135	0.030	−0.121	−0.045	0.018
Tree plus trunk ground cover at 4 m	0.256*	−0.135	−0.215	0.278*	0.227	0.207	0.109
Soil hardness	−0.175	0.154	0.136	0.172	0.409*	−0.035	−0.010
Foliage density at 15 cm	−0.024	−0.360*	0.096	−0.228	0.043	0.321*	−0.074
Foliage density at 30 cm	0.120	−0.328*	0.185	−0.217	0.153	0.042	0.175
Foliage density at 50 cm	0.187	−0.242	0.245	−0.189	0.241	−0.262*	0.159
Foliage density at 100 cm	0.235	−0.139	0.206	0.068	0.400*	−0.230	−0.064
			Winter				
Eigenvalues	3.380	2.360	1.530	1.040			
Canopy cover	−0.060	−0.117	0.299*	−0.162			
Logs	0.196	0.303	0.020	0.254*			
Distance to nearest shrub	0.122	−0.177	−0.132	0.746*			
Shrub cover	−0.020	0.392*	−0.121	−0.493*			
Number of species of shrubs	0.045	0.071	0.097	−0.035			
Number of trees	−0.102	−0.390*	0.472*	−0.073			
Number of species of trees	0.010	−0.390*	0.446*	−0.085			
Bare ground cover at 2 m	0.407*	−0.299*	−0.202*	−0.180			
Herbaceous ground cover at 2 m	−0.494*	0.059	−0.025	0.112			
Tree plus trunk ground cover at 2 m	0.185	0.394*	0.385*	0.089			
Bare ground cover at 4 m	0.448*	−0.163	−0.169	−0.140			
Herbaceous ground cover at 4 m	−0.502*	−0.020	−0.110	0.038			
Tree plus ground cover at 4 m	0.184	0.352*	0.465*	0.151			

*Denotes loadings ≥0.25

Source: Kelt et al. 1994: table 3

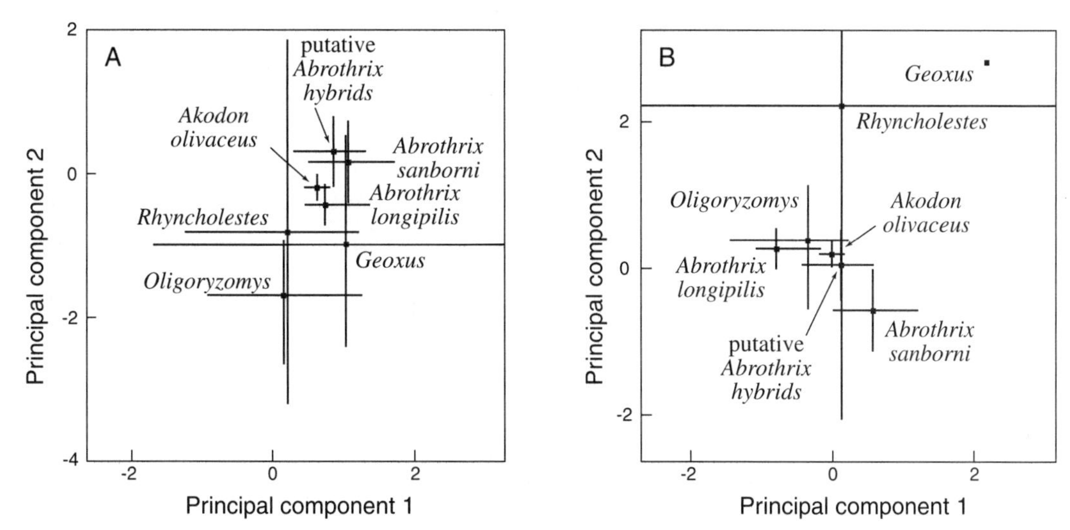

Figure 6.7. Results of principal components analysis of habitat variables associated with capture sites of small mammals at La Picada, Chile. (A, mean principal component scores during summer 1983–1984; B, mean principal component scores during winter 1984). Projections are 95 percent CI. (From Kelt et al. 1994: fig. 2)

considerably. The results in tables 6.4 and 6.5 serve as examples.

1. Starting with the first variable and first component and moving horizontally from left to right across the components, circle the variable loading with the largest absolute value (e.g., in table 6.5 the "litter" variable has the highest loading with PC1). Then find the second highest variable on any component and circle it; continue this procedure for each variable.
2. After evaluating all the variables, examine each loading for "significance." This assessment can be made on the basis of the statistical significance (see table 6.5) of the correlation coefficient (loading) or on the basis of the practical significance (table 6.4). For statistical significance, in most instances with sample sizes less than 100, the smallest loading would have to be greater than 0.3 in order to be considered significant. Practical significance involves some reasonable or practical rule for the minimum amount of a variable's variance that must be accounted for by a component. This is especially important in ecological studies in which biological significance may require a much higher loading than indicated by the associated statistical significance (the latter of which is largely based on sample size). Significant loadings should be underlined.
3. Examine this pattern matrix to identify the variables that have not been underlined and therefore do not "load" on any component (e.g., possibly "percentage of shrub cover" in table 6.4 and "litter" in

Table 6.4. Summary of the first four principal components

Variable	Mnemonic	PC1	PC2	PC3	PC4
Eigenvalue		3.72	1.79	1.56	1.29
Percentage of variance		28.6	13.7	12.0	9.9
Percentage of ground cover	GC	0.111	0.577	0.261	−0.051
Percentage of shrub cover	SC	−0.532	−0.314	0.160	0.551
Percentage of conifers	CO	0.238	0.709	−0.423	0.258
Canopy height	CH	−0.739	0.466	−0.281	0.111
Number of tree species	SPT	0.056	−0.510	−0.016	0.508
Trees 7.5–15 cm dbh	T1	0.677	0.019	0.264	0.299
Trees 15.1–23 cm dbh	T2	0.876	0.205	−0.154	0.160
Trees 23.1–30 cm dbh	T3	0.818	0.059	−0.368	−0.032
Trees 30.1–38 cm dbh	T4	0.267	−0.196	−0.549	−0.601
Trees 38.1–53 cm dbh	T5	−0.397	−0.076	−0.711	0.002
Trees 53.1–68 cm dbh	T6	−0.657	0.208	−0.167	0.270
Trees >68-cm dbh	T7	−0.511	0.169	0.166	0.175

Source: Collins 1983: table 2; reproduced by permission of the American Ornithologists' Union

Table 6.5. Results of principal components analysis using weighted averages of eight habitat variables for 34 bird species

Component	PC1	PC2	PC3
Variation explained (%)	67.46	23.47	5.50
Cumulative variation (%)	67.46	90.93	96.43
Variable		Correlations with orginal variables	
Litter	0.85**	0.23	0.42**
Slash	−0.38*	0.86	0.32
Herbaceous Vegetation	−0.93**	−0.29	0.07
$\overline{X}$ height of herbaceous vegetation	−0.96**	−0.01	−0.04
Canopy layers	0.98**	−0.14	0.01
Maximum canopy height	0.90**	−0.41*	0.08
Canopy cover	0.95**	0.11	−0.20
Trees <12.7 cm	0.27	0.90**	−0.33*

*$P < 0.05$
**$P < 0.01$
Source: Maurer et al. 1981: table 3

table 6.5). The researcher then must decide whether to rest the analysis only on those variables with significant loadings or to evaluate critically each variable with regard to the research objective and biological knowledge.

4. On the basis of the results of the previous steps, attempt to assign a biologically meaningful interpretation to the pattern of component loadings. Variables with higher loadings have greater influence; variables with negative loadings have in-

verse influence. Assign a name that reflects, to the extent possible, the combined meaning of the variables that load on each such component. In table 6.4, for example, PC1 could be interpreted as a forest-height component separating areas with large trees (T6, T7), tall canopies (CH), and a low amount of shrub cover (SC) from areas characterized by smaller trees (T1–T3) (see Collins 1983). Note that both the magnitude and direction (+ or −) of the loading were used to infer this relationship. In practice, having many variables with moderate-sized loadings complicates this step.

Various other parametric and nonparametric methods fall within this category of multivariate techniques, including factor, principal coordinates, and correspondence analyses, nonmetric multidimensional scaling, cluster analysis, and their relatives. Miles (1990), for example, compared results using several of these methods. The nonparametric techniques are, of course, designed for situations where data are highly nonlinear or when sample sizes are too small for normality-linearity to be determined adequately. No method is without drawbacks, and further, few people have much practical experience with the analysis of data using nonparametric techniques (i.e., even many of the more popular statistical packages include only a few of these methods). Thus, we recommend that students proceed cautiously when selecting an analytic technique. Unless your data departs substantially from parametric assumptions, it might be better to stick with the more common and well understood techniques. Because of the qualitative nature of the interpretation of the output from all these multivariate methods, one must fully understand how a technique operates in order to assign meaningful biological interpretation to the results. The greatest problem involves not so much the specific method used as the absence of any follow-up confirmatory or validation studies (e.g., Marcot et al. 1983; Raphael and Marcot 1986; Fielding and Haworth 1995). Even if multiple techniques are used to "confirm" study results, this does not justify inadequate sample sizes, biased sampling methods, or gross violations of statistical assumptions. It is best to remember that the components derived from these techniques cannot be properly considered to be "niche dimensions" or "habitat dimensions" except by means of an arbitrary, operational decision by the researcher (Wiens 1989:65).

Assessing Relationships: Multiple Regression Analysis

Regression analysis is the most widely used method of data analysis. Regression provides three general types of results: First, regression can predict or estimate one response variable from one or more predictor variables. The estimated variable is called a predicted, criterion, or (most commonly) dependent variable. The one or more variables that estimate a dependent variable are termed predictors, covariates, or (most commonly) independent variables. Second, regression analysis can determine the best formula (on the basis of available data) for predicting some relationship. Third, the success (precision) of a regression analysis can be ascertained, usually through use of correlation coefficients (Pimentel 1979:33).

Although popular, regression analysis has many problems associated with its use that can substantially bias biological interpretations of data. Pimentel (1979) provides an especially sobering review of regression analysis. As in all multivariate techniques, the multivariate extension of simple linear regression magnifies these problems.

Draper and Smith (1981: chapter 8) provided a flow diagram of the steps necessary to ensure proper development of predictive models, a frequent goal of biologists using multiple regression (MR) (fig. 6.8). Their diagram identifies three primary stages in such a plan: planning the analysis, developing the models, and verifying (validating or testing) the initial model outputs.

Regression analysis allows us to identify how much of the observed variation in the dependent variable is explained by the independent variables and how much is not (the error term, *e*). A measure of the relative importance of each of these sources of variation (independent variables) is termed the coefficient of multiple

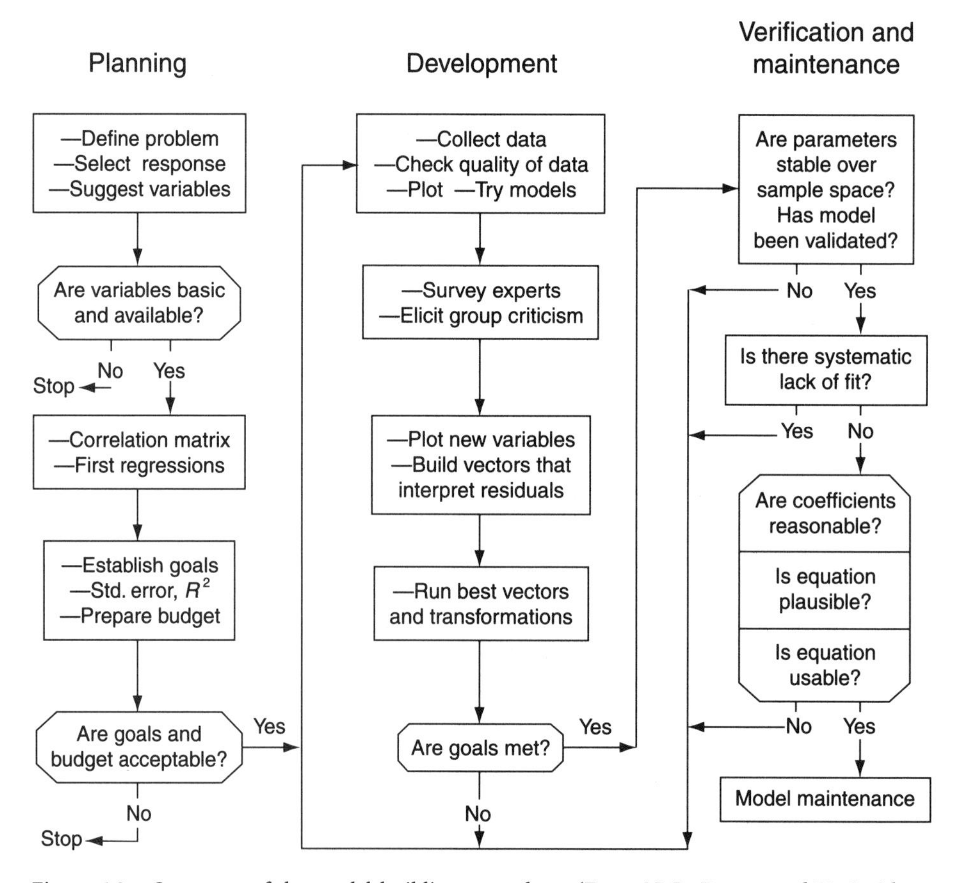

Figure 6.8. Summary of the model-building procedure. (From N. R. Draper and H. Smith, *Applied Regression Analysis*, 2d ed. [New York: John Wiley and Sons © 1981]; reprinted by permission of John Wiley and Sons, Inc.)

determination, R^2. R^2 ranges from 0 for no linear relationship to 1 for a perfectly linear relationship. The value of R^2 is thus a measure of the explanatory value of the linear relationship (Wesolowsky 1976:43).

However, when sample sizes are small in relation to the number of parameters fitted (number of independent variables), it is possible to get a high R^2 even when no linear relationship exists. Conversely, a low value does not necessarily indicate a "bad" relationship. Here, one or more individual coefficients may be significant, and the corresponding parameters rather than the overall regression model may be of primary interest. In addition, a low R^2 may simply show insufficient variation in mean values. Here again we see the overriding role that spatial scale plays in determining the results of our studies. In figure 6.9A, we see that fitting a line through all the points results in a good model for predicting the abundance of a bird species in relation to tree density. However, when our interest becomes more site specific (fig. 6.9B), we see that tree density alone is a poor predictor of bird abundance. This is because we have moved more into the realm of microhabitat analysis, in which increasingly fine, site-specific measures of an animal's habitat must be measured to explain its abundance. This is a simple, univariate example, but illustrates the relationship between sampling scale and results of regression analysis. You can almost always find a significant regression result by sampling across a wide enough range of conditions, for example, by including young-growth forest in an analysis of the relationship between numbers of canopy-dwelling birds and forest structure.

Classification: Discriminant Analysis

Discriminant analysis (DA) is widely applied throughout the scientific disciplines, including wildlife science. DA refers to a general group of methods, each of which has slightly different objectives. The overall goal of DA, however, is the classification of individuals into specific groups (such as species, vegetation types). Researchers can use methods of DA to evaluate similarities and differences between sites or individual samples; it thus resembles PCA in its ordination capabilities. Unlike PCA, however, DA starts with sets of groups (two or more) and a sample from each group. Thus, while the goals of PCA and some applications of DA are similar, the experimental designs for collecting data differ markedly.

Dillon and Goldstein (1984: chapter 10) provided an excellent description of how DA works, which we will summarize here, along with material from Pimentel (1979: chapter 10) and Neff and Marcus (1980). Under DA, researchers evaluate one categorical dependent variable and a set of independent variables. Although there is no requirement that these independent variables be continuous in nature, DA often performs poorly when independent variables are categorical. The categorical dependent variable is a grouping factor that places each observation in one and only one predefined group.

For example, we might be interested in examining differences between species or study sites on the basis of a series of environmental characteristics. After all individuals are assigned to these groups, we further wish to "discriminate" bewteen the groups on the basis of the values of the independent (predictor) variables (the habitat characteristics). DA is thus a method of separating groups on the basis of measured characteristics and determining the degree of dissimilarity of observations and groups and the specific contribution of each independent variable to this dissimilarity (as in the variable loadings described for PCA).

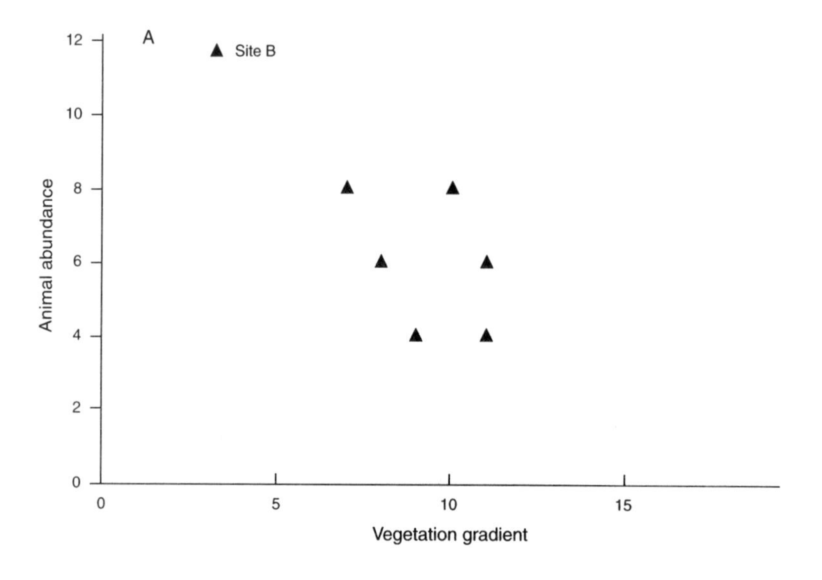

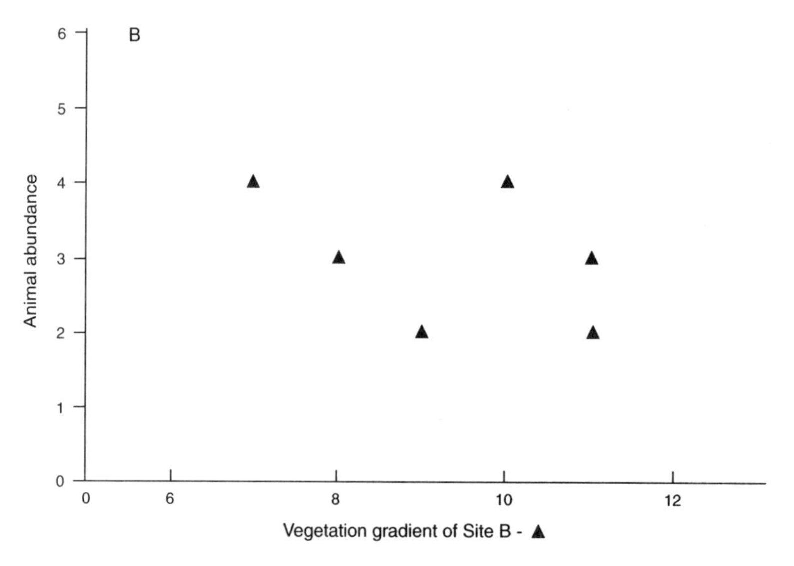

Figure 6.9. Examples of how the extent, or range, of an environmental gradient from which samples are taken can influence conclusions. A. Sampling from an arbitrary, relatively wide vegetation gradient shows a positive relationship between animal abundance and vegetation. B. Sampling from only a subset of the gradient ("Site B") results in a negative abundance-vegetation relationship.

Table 6.6. Classification matrix derived from a discriminant function program showing actual and predicted species (group) membership for singing male warblers based on habitat use on the deciduous tree and nondeciduous tree sites

	Predicted group membership (in percentages)[a]		
Actual group	Orange-crowned	MacGillivray's	Wilson's
		Deciduous Tree Sites	
Orange-crowned warbler	*13.0*	47.8	39.1
MacGillivray's warbler	9.1	*87.9*	3.0
Wilson's warbler	25.9	14.8	*59.3*
		Nondeciduous Tree Sites	
Orange-crowned warbler	*16.1*	58.1	25.8
MacGillivray's warbler	16.1	*74.2*	9.7
Wilson's warbler	26.1	13.0	*60.9*

Source: Morrison 1981: table 5; reproduced by permission of the American Ornithologists' Union

[a]Italicized numbers denote percentages correctly classified

After developing the linear discriminant functions, the researcher can then use these functions to identify or classify "unknowns" into the group predicted by the discriminant analysis. Such a classification analysis is often used to determine how well the DA can identify members of the groups used; table 6.6 presents a simple example. In table 6.6, species not well separated from each other by the DA will show high classifications (misclassifications) for a different species; here, many orange-crowned warblers were misclassified as MacGillivray's warblers, indicating high overlap between the habitats used by two of the species included in this study.

Interpretation

The use of discriminant loadings will give the most straightforwrad and useful interpretation of a discriminant analysis. Similar to principal component loadings, discriminant loadings give the simple correlation of an independent variable with a discriminant function. Most statistical packages will produce discriminant loadings.

A stepwise selection procedure is commonly used to determine the set of independent variables that best separate the groups being considered. Stepwise procedures function similarly for all multivariate techniques, and they are in fact independent of the actual multivariate analysis. That is, they "screen" variables by a set of statistical criteria to determine if the variables should be allowed to enter into the actual multivariate calculations. In DA, an F-test is usually used to order each variable by its ability to separate the groups. Depending on the F-value (or corresponding P-value) used, variables are entered into the analysis until some user-defined cut-off is reached (e.g., $P < 0.05$ or 0.1). There are many modifications of the stepwise procedure and several other less-used but potentially more appropriate methods for variable entry (e.g., all possible subsets). Researchers should review Dillon and Goldstein (1984:234–42) regarding specific steps to take when using variable selection procedures.

Table 6.7 shows a typical presentation of results of a discriminant analysis. For the "all

Table 6.7. Discriminant function analysis of small mammal habitat use, western Oregon

Group discriminant function	Relative variance (%)	Cumulative variance (%)	Wilk's	X^2	df	p	Box's M (p)
All species							
DF I	75.1	75.1	0.700	93.45	12	<0.0001	
DF II	24.9	100.0	0.910	24.77	5	0.0002	<0.0001
Rodents only (excluding *Zapus* spp.)[a]							
DF 1	98.0	98.0	0.716	68.76	6	<0.0001	<0.0001
Sorex spp. only							
DF I	100.0	100.0	0.789	9.00	2	0.0111	<0.0001

Source: Reproduced from M. L. Morrison and R. G. Anthony, "Habitat Use by Small Mammals on Early-Growth Clear Cuttings in Western Oregon," *Canadian Journal of Zoology* 67 (1989): table 4

[a]DF II accounted for 2 percent of the variance and was not significant at $P = 0.45$

species" analysis, note that two signnificant functions were derived. Box's M is the test of the equality of the variance-covariance matrices between species—the multivariate equivalent of the univariate test of equality of variances. Here, Box's M was significant for all comparisons, meaning that the species did have different variances (in multivariate habitat use). A significant Box's M means that the formal test of the null hypothesis of equality of group centroids (i.e., are the species different in multivariate habitat use?) is technically invalid. As noted earlier, we should expect that species differ in the way they use habitat characteristics, in terms of both their mean use and the distribution of values about the mean. Thus, transforming data to meet normality results in loss of ecological information about the animals. We are thus left with a dilemma: Should we try to meet formal statistical assumptions, or should we let the data speak for themselves? In the face of such a situation, most biologists present their results as "descriptions" of the habitat use of the species rather than as formal tests of null hypotheses. The important points are to make sure that you have an adequate sample size (so you know that the distribution of samples is accurate) and that you fully discuss violations of assumptions.

As noted by Wiens (1989:65), just because a discriminant analysis derives a primary multivariate dimension that is statistically significant and thus provides separation in habitat use between the species measured, it does not necessarily follow that this dimension represents the primary means of habitat separation in the community. Wiens (1989: chapter 9) provides additional examples and cautions concerning the use of discriminant analysis and other multivariate procedures as applied to wildlife studies.

Findley and Black (1983; see also Findley 1993) summarized their conception of how a Zambian insectivorous bat community would appear in multivariate space. Their drawing (fig. 6.10) provides a good example of how projections of species can be interpreted; this interpretation applies to most methods. Each sphere represents the morpho- or ecospace occupied by one species. The volume of the sphere equals the total niche volume, the diameter equals the amount of intraspecific variation, and the center is the species' centroid (i.e., the mean value of all its morphological or

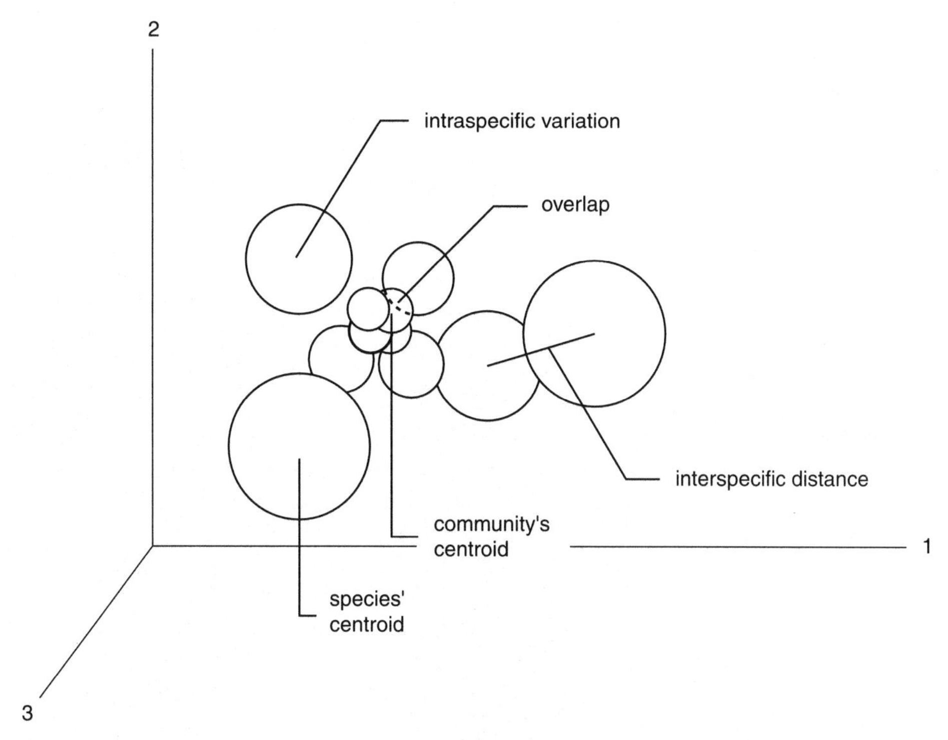

Figure 6.10. Tentative model of a community in attribute space. The attributes could be morphology or diet, or perhaps other physiological, behavioral, or ecological parameters. The community is composed of some closely packed, rather invariable species, and some more distant, more variable kinds. (Reproduced from Findley and Black 1983 [fig. 1], by permission of the Ecological Society of America)

ecological variables). Overlap is shown if the spheres intersect each other; overlap is a function of both interspecific distance and intraspecific variability. In fig. 6.10, the region of the community centroid is occupied by a number of closely packed species with low intraspecific variability (i.e., narrow niches).

Rexstad et al. (1988) conducted a simulation study to examine, in detail, the ramifications of violating the assumptions of discriminant analysis. Although some of their conclusions should be tempered (see Taylor 1990; also Rexstad et al. 1990), the message contained in their paper remains: There is little published evidence to suggest the widely held beliefs that discriminant analysis is robust to violations of the variance-covariance matrix, or that using the analysis for "descriptive" purposes results in only ecologically meaningful analyses. Because of these shortcomings, many people are now using logistic regression analysis as a replacement for discriminant analysis; logistic regression is the topic of our next section.

Logistic Regression Analysis

Logistic regression analysis (LR) has been gaining in use in recent years as a nonparametric, nonlinear alternative to two-group discriminant analysis. LR can be used to analyze independent variables that are true categorical data (such as coat color) and those that have been summarized into categories (such as height intervals); it can also analyze continuous data. LR can be used to develop predictive models as well.

Nadeau et al. (1995) used logistic regression to create a habitat model based on muskrat (*Ondatra zibethicus*) presence along wetlands of James Bay, Quebec. They developed models based on the presence of burrows only and another using the presence of muskrat feeding signs and droppings as dependent variables. Independent variables were a mix of continuous (e.g., water depth) and categorical (e.g., bank slope) variables (table 6.8). Chandler et al. (1995) used LR to predict the probability of bald eagle use of a shoreline segment on the basis of a series of habitat variables. Brennan et al. (1986) showed that LR could develop more rigorous habitat models than other linear techniques under certain conditions.

LR is potentially more robust than similar parametric procedures to deviations from multivariate normality and equal covariation. Further, various studies have found that LR provides better between-group separations and classification success than discriminant analysis does. Moreover, the nonlinear tendencies of ecological data are usually distorted by linear analyses. Thus, the nonparametric procedures should provide a more meaningful interpretation of ecological phenomena when linear models fail to explain variation adequately (see Efron 1975; Press and Wilson 1978; Brennan et al. 1986).

LR does have several limitations. First, software with LR routines has not been as widely available as that with parametric procedures,

Table 6.8. Results of univariate tests on habitat variables measured along 60-m shore sections of various wetlands in James Bay, Quebec, and their significance (*P*) in regard to the presence/absence of muskrat burrows, feeding signs, and droppings

	Burrows			Feeding signs and droppings		
Variables	*n*	Test	*P*	*n*	Test	*P*
Water depth 2 m from shore	200	Z = −0.35	0.723	234	Z = −0.25	0.804
Distance to 50-cm water depth	199	Z = 0.67	0.503	233	Z = −0.57	0.566
Distance to 50-cm elevation	200	Z = −2.42	0.015	234	Z = −2.17	0.030
Submerged plant cover (%)	200	Z = 2.55	0.010	235	Z = 3.53	<0.001
Floating plant cover (%)	203	Z = 2.11	0.035	235	Z = 1.25	0.212
Emergent plant cover (%)	200	Z = −0.56	0.558	235	Z = 0.11	0.911
Cover of hydrophytes (%)	203	Z = 2.64	0.008	235	Z = 2.89	0.004
Surface covered by hydrophytes	202	Z = 2.97	0.003	235	Z = 3.04	0.002
Distance to hydrophyte community	200	Z= 2.41	0.016	235	Z = 2.42	0.015
Water velocity	200	Fisher	0.137	235	Fisher	0.258
Protection from wind	200	Fisher	0.322	233	Fisher	0.157
Bank slope	200	Fisher	0.013	235	Fisher	0.078
Width of shore herbaceous belt	200	χ^2 = 7.58	0.023	235	χ^2 = 10.88	0.004
Dominance of clay-loam soil	200	Fisher	0.003	235	Fisher	<0.001
Dominance of sand-gravel soil	200	Fisher	0.062	235	Fisher	0.034
Dominance of peat soil	200	Fisher	0.137	235	Fisher	0.130
Wetland type	200	χ^2 = 32.81	<0.001	235	χ^2 = 32.10	<0.001

Source: Nadeau et al. 1995: table 2, reproduced by permission of The Wildlife Society

including discriminant analysis and multiple regression. Thus, relatively few people have direct experience with running and interpreting LR, including resource managers. This limitation is easing, however, as LR becomes more commonplace. Second, LR is limited to the two-group situation. Thus, even within a single study the researcher often needs to run both two-group (e.g., use vs. nonuse) and greater than two-group (e.g., comparison of community separation) analyses. Clearly, mixing analyses makes comparison of results difficult. The choices of method should be carefully evaluated and then clearly elucidated in all research papers.

Cautions about Mulivariate Analyses

As discussed in our opening comments in this chapter, and in more detail in chapters 5 and 8, there are important limitations to the pattern-seeking, descriptive methods of habitat use typically found in multivariate analyses to which the researcher need attend. Kaufman and Kaufman (1989) aptly summarized the cautions that should be applied to results of descriptive studies:

1. Statistical separation of species on a habitat axis does not prove that the animals recognize and respond to the characteristics measured. At best, this provides testable hypotheses to evaluate with experimentation.
2. Even though a series of species occupy different average positions along habitat axes, interspecific overlap may be considerable and must be explained (this also applies to intraspecific, sex-age comparisons).
3. Most studies are not replicated in time or space, so their generality is unknown.
4. Differences in resource use do not necessarily elucidate the mechanisms that are ultimately responsible for the patterns observed.
5. Understanding cause and effect of observed patterns of habitat use requires experimental manipulations, not just additional studies using refined descriptions of habitat use (chapter 5).

It is critical that students carefully examine the magnitude of differences for each independent variable between groups in a multivariate analysis. Remember that the variable(s) entered and the order of entry are primarily a statistical decision based on some selection criterion (e.g., *F*-value). A variable with a small difference in means and a small variance could have a higher *F*-value and be entered into an analysis before a variable with a large difference in means but also a relatively large variance. As discussed earlier, biologists tend to concentrate on statistical significance rather than on an ecological interpretation of means *and* variances of variables prior to entry into an analysis. Thus, the ecological interpretation made after an analysis in such situations can be misleading. Multivariate methods worsen this tendency because of the "black box" that field data disappear into and the myriad of statistical parameters that return on the printout. The high natural variance usually seen in biological populations makes application of models to other places and times difficult (see Fielding and Haworth 1995 for a detailed analysis and discussion).

Sample Size Analyses

Inadequate sample size is a chronic problem found in many papers using multivariate methods. Regardless of the refinement of study design and the care taken in recording data, proper ecological interpretation of multivariate analyses is difficult at best; interpretations based on inadequate samples are a waste of time and potentially misleading. Earlier in this chapter we identified methods of determining proper sample sizes in the univariate case; those methods apply here on a variable-by-variable basis. In multivariate analyses, however, our problems are increased by orders of magnitude given our desire to interpret ecological phenomena in many dimensions for many species simultaneously.

Johnson (1981b) outlined general guidelines for determining adequate sample sizes in multivariate studies, noting that more observations are needed when the number of independent variables is large. Many published studies, however, have only slightly more observations than variables; some even have fewer. Johnson thought that an appropriate minimum sample size for multivariate analyses is 20 observations, plus 3–5 additional observations for each variable in the analysis. We suggest that an additional 5–10 observations for each variable would provide a more conservative target for the sample size. Larger sample sizes do not, however, provide an answer for poorly designed studies or biased data. As Johnson (1981b:56) noted, "Calls for larger samples are the 'knee-jerk' reaction when variability is excessive."

In a study of habitat use by birds in Oregon, Morrison (1984b) found that a minimum of 35 plots was necessary to obtain stable results; stability was determined when means and variances did not change with an increase in sample size. Morrison's review of the wildlife literature showed, however, that very few studies met the minimum criteria established by Johnson (1981b). In addition, few researchers even discussed the issue of sample size. Claiming that you have collected all the data that you could serve as no excuse for publishing results based on inadequate sample sizes. A similar study by Block et al. (1987) found that an even larger number of plots—up to 75 or more—as needed in their bird-habitat analysis.

The minimum size of the sample needed is a study-specific question; the papers reviewed above give approximations of the range of samples you can expect to need. Remember, however, that these minimum sample sizes apply to each biological period being considered (e.g., winter, fall), and the appropriate n might vary for different periods. All studies should include a justification of the amount of data used in the analyses.

Computer Statistical Packages

Previously available only on mainframe computers, numerous, powerful statistical packages are now available also for microcomputers. Unfortunately, the availability and ease of use of these packages likely have led to an increasing number of misuses of statistical methods. Novice users of these packages should be aware that the default settings for the specific analytic methods often must be adjusted for each application. Further, each package has a set of options and statistics that must be specifically requested. Many university computer centers offer short courses in the use of statistical packages; an increasing number of statistics departments are offering more in-depth courses on the use and interpretation of such software.

The user manuals for some of the packages offer descriptions of the analytic methods being used and step-by-step instructions on how to interpret results. This is especially true of the widely used SPSS (Statistical Package for the Social Sciences) software, which offers an excellent user's manual and both basic and advanced manuals that explain most statistical procedures. White and Clark (1994), in their review of microcomputing for wildlife, preferred the SAS (Statistical Analytical System) software, primarily because of its useful programing capabilities, a capability lacking in SPSS. Although the ability to program within SAS certainly offers long-term advantages to many users, it is an option that most wildlife biologists will seldom, if ever, use. We find that SPSS offers what most researchers require, and the user's manuals make effective teaching tools. We have heard researchers express the view that SPSS is actually too easy to use, equating "ease of use" with "ease of misuse." There is truth in this opinion, and we caution that the ability to run your data through a computer does not mean that you have necessarily done the analysis correctly.

Several statistical and ecological textbooks also include examples using output from one or more of the statistical packages; some even provide a disk containing specialized programs for the analysis of ecological data (Berenson et al. 1983; Tabachnick and Fidell 1983; Afifi and Clark 1984; Harris 1985; Ludwig and Reynolds 1988; Kent and Coker 1994).

Literature Cited

Afifi, A. A., and V. Clark. 1984. *Computer-aided multivariate analysis.* Belmont, Calif.: Lifetime Learning Publications.

Berenson, M. L., D. M. Levine, and M. Goldstein. 1983. *Intermediate statistical methods and applications: A computer package approach.* Englewood Cliffs, N.J.: Prentice-Hall.

Block, W. M., K. A. With, and M. L. Morrison. 1987. On measuring bird habitat: Influence of observer variability and sample size. *Condor* 72:182–89.

Brennan, L. A., W. M. Block, and R. J. Gutierrez. 1986. The use of multivariate statistics for developing habitat suitability index models. In *Wildlife 2000: Modeling habitat relationships of terrestrial vertebrates;* ed. J. Verner, M. L. Morrison, and C. J. Ralph, 177–82. Madison: University of Wisconsin Press.

Burger, L. D., L. W. Burger, Jr., and J. Faaborg. 1994. Effects of prairie fragmentation on predation on artificial nests. *Journal of Wildlife Management* 58:249–54.

Capen, D. E., ed. 1981. *The use of multivariate statistics in studies of wildlife habitat.* USDA Forest Service General Technical Report RM–87.

Carnes, B. A., and N. A. Slade. 1982. Some comments on niche analysis in canonical space. *Ecology* 63:888–93.

Chandler, S. K., J. D. Fraser, D. A. Buehler, and J. K. D. Seegar. 1995. Perch trees and shoreline development as predictors of bald eagle distribution on Chesapeake Bay. *Journal of Wildlife Management* 59:325–32.

Cody, M. L. 1983. Continental diversity patterns and convergent evolution in bird communities. In *Mediterranean-type ecosystems,* ed. F. J. Kruger, D. T. Mitchell, and J. M. Jarvis, 357–402. New York: Springer-Verlag.

Collins, S. L. 1983. Geographic variation in habitat structure of the black-throated green warbler (*Dendroica virens*). *Auk* 100:382–89.

Cook, C. W., and J. Stubbendieck, eds. 1986. *Range research: Basic problems and techniques.* Denver: Society for Range Management.

Cooley, W. W., and P. R. Lohnes. 1971. *Multivariate data analysis.* New York: John Wiley and Sons.

Cooper, C. F. 1957. The variable plot method for estimating shrub density. *Journal of Range Management* 10:111–15.

Digby, P. G. N., and R. A. Kempton. 1987. *Multivariate analysis of ecological communities.* London: Chapman and Hall.

Dillon, W. R., and M. Goldstein. 1984. *Multivariate analysis: Methods and applications.* New York: John Wiley and Sons.

Dodd, M. G., and T. M. Murphy. 1995. Accuracy and precision of techniques for counting great blue heron nests. *Journal of Wildlife Management* 59:667–73.

Draper, N. R., and H. Smith. 1981. *Applied regression analysis.* 2d ed. New York: John Wiley and Sons.

Efron, B. 1975. The efficiency of logistic regression compared to normal discriminant analysis. *Journal of the American Statistical Association* 70:892–98.

Fielding, A. H., and P. F. Haworth. 1995. Testing the generality of bird-habitat models. *Conservation Biology* 9:1466–81.

Findley, J. S. 1993. *Bats: A community perspective.* Cambridge: Cambridge University Press.

Findley, J. S., and H. Black. 1983. Morphological and dietary structuring of a Zambian insectivorous bat community. *Ecology* 64:625–30.

Fretwell, S. D. 1972. *Populations in a seasonal environment.* Princeton: Princeton University Press.

Ganey, J. L., and W. M. Block. 1994. A comparison of two techniques for measuring canopy closure. *Western Journal of Applied Forestry* 9:21–23.

Gotfryd, A., and R. I. C. Hansell. 1985. The impact of observer bias on multivariate analyses of vegetation structure. *Oikos* 45:223–34.

Green, R. H. 1971. A multivariate statistical approach to the Hutchinsonian niche: Bivalve molluscs of central Canada. *Ecology* 52:543–56.

Green, R. H., and R. C. Young. 1993. Sampling to detect rare species. *Ecological Applications* 3:351–56.

Harris, R. J. 1985. *A primer on multivariate statistics.* 2d ed. Orlando, Fla.: Academic Press.

Hatton, T. J., N. E. West, and P. S. Johnson. 1986. Relationships of error associated with ocular estimation and actual cover. *Journal of Range Management* 39:91–92.

Hutchinson, G. E. 1957. Concluding remarks. *Cold Spring Harbor Symposium on Quantitative Biology* 22:415–27.

James, F. C., and H. H. Shugart, Jr. 1970. A quantitative method of habitat description. *Audubon Field Notes* 24:727–36.

Johnson, D. H. 1981a. The use and misuse of statistics in wildlife habitat studies. In *The use of multivariate statistics in studies of wildlife habitat,* ed. D. E. Capen, 11–19. USDA Forest Service General Technical Report RM–87.

Johnson, D. H. 1981b. How to measure habitat—a statistical perspective. In *The use of multivariate statistics in studies of wildlife habitat,* ed. D. E. Capen, 53–57. USDA Forest Service General Technical Report RM –87.

Johnson, R. A. 1992. *Applied multivariate statistical analysis.* 3d ed. Englewood Cliffs, N.J.: Prentice-Hall.

Kaufman, D. W., and G. A. Kaufman. 1989. Population biology. In *Advances in the study of Peromyscus* (Rodentia), eds. G. L. Kirkland, Jr., and J. N. Layneeds, 233–70. Lubbock: Texas Tech University Press.

Kelt, D. A., P. L. Meserve, and B. K. Lang. 1994. Quantitative habitat associations of small mammals in a temperate rainforest in southern Chile: Empirical patterns and the importance of ecological scale. *Journal of Mammalogy* 75:890–904.

Kent, M., and P. Coker. 1994. *Vegetation description and analysis: A practical approach.* New York: John Wiley and Sons.

Kepler, C. B., and J. M. Scott. 1981. Reducing count variability by training observers. *Studies in Avian Biology* 6:366–71.

Krzanowski, W. J. 1990. *Principles of multivariate analysis: A user's perspective.* Oxford: Oxford University Press.

Leigh, R. A., and A. E. Johnston, eds. 1994. *Long-term experiments in agricultural and ecological sciences.* Oxford: CAB International.

Likens, G. E. 1983. A priority for ecological research. *Bulletin of the Ecological Society of America* 64:234–43.

Lindsey, A. A., J. D. Barton, and S. R. Miles. 1958. Field efficiencies of forest sampling methods. *Ecology* 39:428–44.

Ludwig, J. A., and J. F. Reynolds. 1988. *Statistical ecology: A primer on methods and computing.* New York: John Wiley and Sons.

Manly, B. F. J. 1994. *Multivariate statistical methods: A primer.* 2d ed. New York: Chapman and Hall.

Marcott, B. G., and M. G. Raphael, and K. H. Berry. 1983. Monitoring wildlife habitat and validation of wildlife-habitat relationships models. *Transactions of the North American Wildlife and Natural Resources Conference* 48:315–29.

Maurer, B. A., L. B. MacArthur and R. C. Whitmore. 1981. Habitat associations of breeding birds in clearcut deciduous forests in West Virginia. In *The use of multivariate statistics in studies of wildlife habitat,* ed. D. E. Capen, 167–72. USDA Forest Service General Technical Report RM–87.

Meents, J. K., J. Rice, B. W. Anderson, and R. D. Ohmart. 1983. Nonlinear relationships between birds and vegetation. *Ecology* 64:1022–27.

Miles, D. B. 1990. A comparison of three multivariate statistical techniques for the analysis of avian foraging data. *Studies in Avian Biology* 13:295–308.

Morrison, D. F. 1967. *Multivariate statistical methods.* 2d ed. New York: McGraw-Hill.

Morrison, M. L. 1981. The structure of western warbler assemblages: Analysis of foraging behavior and habitat selection in Oregon. *Auk* 98:578–88.

Morrison, M. L. 1984a. Influence of sample size and sampling design on analysis of avian foraging behavior. *Condor* 86:146–50.

Morrison, M. L. 1984b. Influence of sample size on discriminant function analysis of habitat use by birds. *Journal of Field Ornithology* 55:330–35.

Morrison, M. L., and R. G. Anthony. 1989. Habitat use by small mammals on early-growth clear-cuttings in western Oregon. *Canadian Journal of Zoology* 67:805–11.

Morrison, M. L., B. G. Marcot, and R. W. Mannan. 1992. *Wildlife habitat relationships: Concepts and applications.* 1st ed. Madison: University of Wisconsin Press.

Morrison, M. L., C. J. Ralph, J. Verner, and J. R. Jehl, Jr. 1990. *Avian foraging: theory, methodology, and applications.* Studies in Avian Biology no. 13.

Mueller-Dombois, D., and H. Ellenberg. 1974. *Aims and methods of vegetation ecology.* New York: John Wiley and Sons.

Nadeau, S., R. Decarie, D. Lamber, and M. St-Georges. 1995. Nonlinear modeling of muskrat use of habitat. *Journal of Wildlife Management* 59:110–17.

Neff, N. A., and L. F. Marcus. 1980. *A survey of multivariate methods for systematics.* New York: privately published. Printed by the American Museum of Natural History, New York.

Pimentel, R. A. 1979. *Morphometrics.* Dubuque, Iowa: Kendall/Hunt Publishing Co.

Press, S. J., and S. Wilson. 1978. Choosing between logistic regression and discriminant analysis. *Journal of the American Statistical Association* 73:699–705.

Raphael, M. G. 1990. Use of Markov chains in analyses of foraging behavior. *Studies in Avian Biology* 13:288–94.

Raphael, M. G., and B. G. Marcot. 1986. Validation of a wildlife-habitat-relationships model: Vertebrates in a Douglas-fir sere. In *Wildlife 2000: modeling habitat relationships of terrestrial vertebrates,* ed. J. Verner, M. L. Morrison, and C. J. Ralph, 129–38. Madison: University of Wisconsin Press.

Ratkowsky, D. A. 1990. *Handbook of nonlinear regression models.* New York: Marcel Dekker.

Rexstad, E. A., D. D. Miller, C. H. Flather, E. M. Anderson, J. W. Hupp, and D. R. Anderson. 1988. Questionable multivariate statistical inferences in wildlife habitat and community studies. *Journal of Wildlife Management* 52:794–98.

Rexstad, E. A., D. D. Miller, C. H. Flather, E. M. Anderson, J. W. Hupp, and D. R. Anderson. 1990. Questionable multivariate statistical inferences in wildlife habitat and community studies: A reply. *Journal of Wildlife Management* 54:189–93.

Schooley, R. L. 1994. Annual variation in habitat selection: Patterns concealed by pooled data. *Journal of Wildlife Management* 58:367–74.

Schultz, A. M., R. P. Gibbens, and L. DeBano. 1961. Artificial populations for teaching and testing range techniques. *Journal of Range Management* 14:236–42.

Scott, J. M., F. L. Ramsey, and C. P. Kepler. 1981. Distance estimation as a variable in estimating bird numbers from vocalizations. *Studies in Avian Biology* 6:334–40.

Seber, G. A. F. 1989. *Nonlinear regression.* New York: John Wiley and Sons.

Shugart, H. H., Jr. 1981. An overview of multivariate methods and their application to studies of wildlife habitat. In *The use of multivariate statistics in studies of wildlife habitat,* ed. D. E. Capen, 4–10. USDA Forest Service General Technical Report RM–87.

Strayer, D., J. S. Glitzenstein, C. G. Jones, J. Kolasa, G. E. Likens, M. J. McDonnell, G. G. Parker, and S. T. A. Pickett. 1986. *Long-term ecological studies: An illustrated account of their design, operation, and importance to ecology.* Institute of Ecosystem Studies, Occasional Paper No. 2. 38 pp.

Stubbendieck, J. L. 1986. *Range research: Basic problems and techniques.* Denver, Colo.: Society for Range Management.

Tabachnick, B. G., and L. S. Fidell. 1983. *Using multivariate statistics.* New York: Harper and Row.

Tacha, T. C., W. D. Warde, and K. P. Burnham. 1982. Use and interpretation of statistics in wildlife journals. *Wildlife Society Bulletin* 10:355–62.

Taylor, J. 1990. Questionable multivariate statistical inferences in wildlife habitat and community studies: A comment. *Journal of Wildlife Management* 54:186–89.

Verner, J., M. L. Morrison, and C. J. Ralph, eds. 1986. *Wildlife 2000: Modeling habitat relationships of terrestrial vertebrates.* Madison: University of Wisconsin Press.

Wesolowsky, G. O. 1976. *Multiple regression and analysis of variance.* New York: John Wiley and Sons.

White, G. C., and W. R. Clark. 1994. Microcomputer applications in wildlife management and research. In *Research and management techniques for wildlife and habitats,* ed. T. A. Bookhout, 75–95. Bethesda, Md.: The Wildlife Society.

Wiens, J. A. 1984. The place of long-term studies in ornithology. *Auk* 101:202–3.

Wiens, J. A. 1989. *The ecology of bird communities.* Vol. 1: *Foundations and patterns.* Cambridge: Cambridge Univerisity Press.

Wiens, J. A., and J.T. Rotenberry. 1981. Habitat associations and community structure of birds in shrubsteppe environments. *Ecological Monographs* 51:21–41.

Zar, J. H. 1984. *Biostatistical analysis.* 2d ed. Englewood Cliffs, N.J.: Prentice-Hall.

7 Measuring Behavior

Introduction

Many studies of habitat use have focused on behavior because of its obvious importance in understanding the distribution, abundance, and needs of animals. Analysis of behavior shows how animals *actively* use their environment. In contrast, presence/absence and relative abundance studies rely on indirect correlations between the animal and features of the area it is found in to infer preference for habitat components, substrates, or foods.

Although they are usually more informative than purely correlational studies, behavioral observations by themselves are fraught with uncertainties. In particular, how are the behaviors we observe related to the survival and, ultimately, to the reproductive success of the individual? It should be evident that the study of behavior must begin with the development of a sound theoretical framework that elucidates how the animal perceives and then uses its environment. Determination of an animal's perception of its environment is difficult to achieve through behavioral observations. Thus, three general areas of study have contributed to our understanding of how animals perceive their environment: development of theoretical models, laboratory experiments, and field studies. Each of these areas of study is incorporated into this chapter.

Regardless of the theoretical basis upon which we construct our behavioral studies, we must employ proper methods of data collection and analysis. Such considerations include: specific methods by which animals are located and visually observed, biases associated with observation methods, the application of statistical procedures, and an adequate sample size. Unfortunately, scant attention has historically been given these concerns altogether in any given behavioral study.

In this chapter we will review the general theoretical framework upon which studies of wildlife behavior have been conducted. The principal methods used to observe animal behavior and to assess resource abundance and use are then discussed. Because of the obvious importance that the gathering of food plays in the life of an animal, methods of studying foraging behavior and diet are also covered. Finally, we discuss how studies of energetics can be used to advance our knowledge of the habitat-use patterns of wildlife.

Theoretical Framework

We should remember that the study of wildlife habitat is essentially the study of animal behavior in the broadest sense. How organisms select habitat reflects their evolutionary history as well as the current ecological conditions of the environment and the influence of other animals. In the end, all behavior—whether for foraging, dispersal, migration, reproduction, or predator avoidance—is modified and guided by the selective advantage of increasing fitness of individuals and vitality of populations.

The theoretical framework we developed in chapter 2 on the perception of habitat by an animal—its niche gestalt, stimulus summation, and so forth—sets the background by which we can observe and then interpret animal behavior. How an animal views its environment—namely, as a series of different spatial scales, from the general vegetation type down to specific features of the microsite—is reflected in an individual's behavior. By observing this behavior in a systematic manner, we can learn much about why animals succeed or fail to succeed in a particular environment.

Three terms often used in behavioral studies are *use, selection,* and *requirement.* In this chapter we define *use* as the demonstrated presence of a particular item in an animal's behavioral repertoire, for example, a den, perch, or foraging site, or a prey item in the diet. *Selection,* in contrast, has typically been defined as use coupled with evidence that the frequency of the item's occurrence in an animal's behavioral repertoire is significantly greater (statistically and biologically) than its frequency in the animal's environment. More recently, Hall et al. (1997) suggested that *selection* should be used in reference to the process by which innate and learned behavioral decisions are made. Further, they defined *preference* as the consequence of this selection process, which is measured as the disproportional use of some resources over others. There is much confusion in the literature regarding these terms, and we suggest that the standardizations suggested by Hall et al. (1997) be adopted. We have tried to avoid confusion in the use of terms without altering the original intent of the researchers.

Finally, *requirement* is the presence of a resource in an animal's environment— particularly, the minimum amount that the animal must obtain so it can live and reproduce. Unfortunately, few studies have identified requirements. Observational studies in the wild usually cannot be designed so that the researcher can conclude whether an animal selects a resource because of a behavioral or physiological requirement. As such, most studies of wildlife behavior report the use of various resources, although they may inappropriately label their data as showing preference (*sensu* Hall et al., 1997). However, if many studies conducted across different time periods and locations consistently show preference of a particular resource or behavior, then one can likely infer that the species is exhibiting a behavior of adaptive significance; this implies a requirement (Ruggiero et al. 1988).

Measuring Behavior

Introduction

Martin and Bateson (1993:19–23) outlined the general steps to follow in designing and implementing a behavioral study. Here we are specifically interested in the choice of variables and recording methods, the accumulation of an adequate number of samples, and data analysis. More thorough reviews of research methods in behavioral studies have been given in many publications (see especially Altmann 1974; Kamil and Sargent 1981; Hazlett 1977; Colgan 1978; Kamil et al. 1987; Gottman and Roy 1990; Morrison et al. 1990; Sommer and Sommer 1991; Lehner 1996). Every issue of the behavioral journals (e.g., *Journal of Animal Behaviour; Behaviour; Behavioral Ecology*), including those devoted to the study of specific taxa such as primates (e.g., *International Journal of Primatology; Primates*) presents many papers that contain examples of behavioral methods and analyses.

Most behavioral observations can be broken down into three general categories: structure, consequence, and spatial relation (Martin and Bateson 1993:57–58). Researchers must recognize these categories before defining specific variables to record; otherwise, problems will likely arise in the analysis and interpretation of data. *Structure* describes the appearance, physical form, or temporal pattern of the animal's behavior in terms of posture and movements. For example, saying that a deer "bends down and removes a leaf from a bush" describes the structure of the behavior. *Consequence,* in contrast with structure, describes the effects of the animal's behavior. Here, behavior is described without reference to how the effects are achieved (see also Dewsbury 1992). Saying that the deer is "browsing" describes the consequences of the behavior but says nothing about how the browsing takes place (i.e., the structure). *Spatial relation* describes behavior in terms of the animal's spatial proximity to features of the environment, including other animals. For example, "approaching a bush" describes the relation of the animal to a potential foraging substrate. Recording the relation or orientation of the animal adds substantial information to the behavioral description.

A common mistake is for behavior to be described in terms of presumed consequences, rather than in neutral terms that do not impose a meaning that is likely to be tainted by observer biases. For example, labeling a behavior as stalking implies a consequence that may be unwarranted with further, more objective study (see Martin and Bateson 1993:57–58).

Several commonly used terms in behavioral studies must first be defined before observations can be recorded (Martin and Bateson 1993:62–66). *Latency* is the time from some specified event (such as the beginning of the recording session) to the onset of the first occurrence of the behavior of interest. *Frequency* is the number of occurrences of the behavior pattern per unit of time (the total number of occurrences should not be used as frequency unless accompanied by a time unit). *Duration* is the length of time that a single occurrence of the behavior pattern lasts. *Intensity* is widely used but has no specific definition. According to Martin and Bateson (1993:65), it is best viewed as a measure of the amplitude or magnitude of a behavior (e.g., how loud a call was, or how high a jump was). They use the term *local rate* as an index of intensity, where local rate is the number of component acts per unit of time spent performing the activity (e.g., the number of bites taken within a specific period of time).

Selection of Variables

In chapter 5 we discussed selection of variables in habitat analysis. Much of that discussion applies, in general, to analysis of animal behavior. Whereas habitat analysis focuses on the description of the environment surrounding the animal, the behavioral descriptions concentrate on the specific actions of the animal within its habitat.

Animals perform a myriad of activities during a day. They sleep, groom, engage in intra- and interspecific interactions, feed, and so on. To quantify these behaviors in any practical way requires that we devise some form of record keeping. One could first compile a catalog of behaviors that describes the behavioral repertoire of the species. Known as ethograms (from *etho-* as in the biological study of behavior, or "ethology"; Martin and Bateson 1993:6), such descriptions are useful starting points for the design of a behavioral study. However, ethograms have been published for only a few species. Developing an ethogram is especially appropriate during preliminary sampling. Although it is too detailed to describe here, we urge readers to review Schleidt et al. 1984 on the development of ethograms.

Slater (1978) provided five basic points for classifying behaviors. Slater's first point is that behavioral categories must be discrete. This means that acts within a category must have clear points of similarity that they do not share with acts outside the category. Second, behavioral acts must be homogeneous. All the acts in a category should be similar in form so that there is little danger of massing two different behaviors in the same category. Third, it is better to split than to lump. Within reason, two similar behaviors with possibly different consequences should be placed into separate categories; they can always be lumped later. Fourth, names of categories should avoid implying a causal or functional role; this was discussed above in our initial discussion of consequence. Rather, names should clinically describe the behavior in terms of actions, not outcomes or motives. Last, the number of categories must be manageable. Too many categories will reduce an observer's reaction time and lower accuracy. Automated recording devices, including computer-aided data-entry programs, can help overcome this problem to a degree (see Raphael 1988; Martin and Bateson 1993: chapter 7; Paterson et al. 1994; Samuel and Fuller 1994).

Recording Methods

Once researchers have developed behavioral categories, they must rigorously define methods for recording these behaviors. Martin and Bateson (1993: chapter 6) provided an excellent framework for recording behavioral observations by defining sets of sampling and recording rules (fig. 7.1). Below we summarize these rules and provide brief examples of their applications to wildlife-habitat studies.

Sampling Rules

Before collecting data, you must objectively decide which individuals to observe (sample) and when to observe them. In chapter 4 we discussed methods for the selection of study species; and in chapter 6 (and elsewhere) we discussed the importance of temporal and spatial stratification of data collection. Here we concentrate on observations of individual animals within the chosen species and within the context of a proper spatial and temporal study design.

Focal-Animal Sampling. Focal-animal sampling, or just focal sampling, involves observing one individual, even if it is located within a group of animals, for a specified period of time. This method has the clear advantage of allowing adequate records to be collected on various classes of a species, for example, data by sex and age. Further, if animals are individually tagged or show unique markings (e.g., coat colors and patterns of wild horses, size and shape of antlers or horns), then you can accumulate information on individual variation rather than simply lumping information for all individuals in a group. For example, Carranza

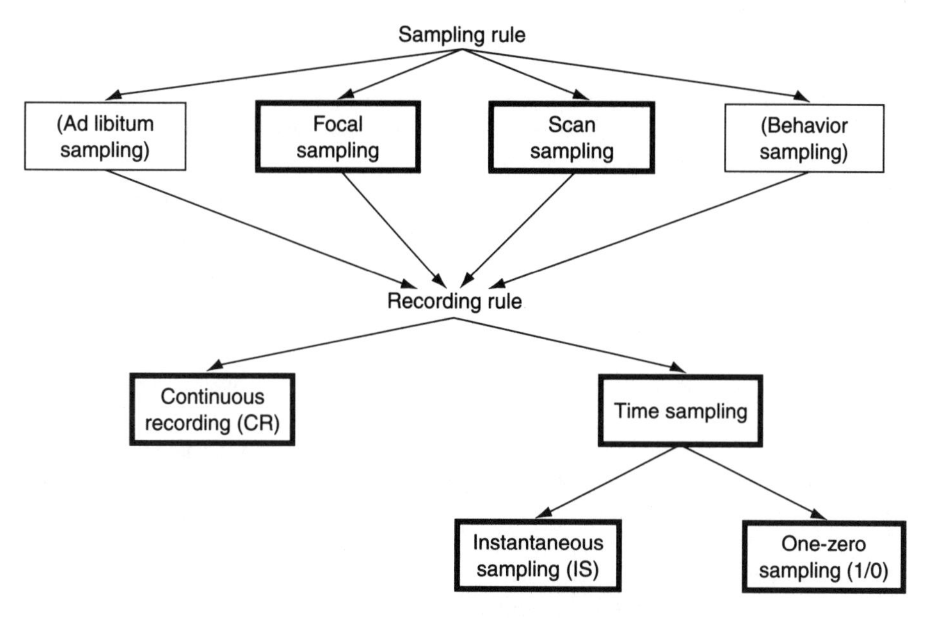

Figure 7.1. The hierarchy of *sampling rules* (determining who is watched and when) and *recording rules* (determining how their behavior is recorded) (From P. Martin and P. Bateson, *Measuring Behaviour,* 2d ed. [Cambridge: Cambridge University Press © 1993], 88, fig. 6.1; reprinted with the permission of Cambridge University Press)

(1995) used antler shape and branching pattern to identify red deer (*Cervus elaphus*) individually. Stamps and Krishnan (1994) chose one individual as a focal individual from a dyad of lizards (*Anolis aeneus*) to determine the outcome of social interactions. Several other examples involving focal sampling are provided below.

A problem with focal sampling involves the conspicuousness of individual animals. By focusing on individuals, researchers may produce records of behavior that vary considerably in length. For example, foraging birds are obvious when they are near branch tips and lower in a tree, but they usually disappear when they move upward and inward in the foraging substrate. Likewise, the response of ungulates to disturbance is easy to observe and quantify (e.g., number of steps or bounds, head movements) when they are at the edge of an opening, but observations of their behavior become difficult when they move into the trees. Further, our attempts to follow animals and observe their activities undoubtedly influence their behavior in ways we are largely unaware of (more on this below). In addition, visibility bias can be introduced into the data when trying to follow animals in dense vegetation.

Thus, sampling periods must be short and are therefore biased toward those individuals who are most conspicuous. Unfortunately, no sampling method employing only the observer's eyes can overcome this problem, although many researchers have made attempts to do so.

For example, students of bird foraging behavior often observe a bird for a short period of time (usually a few seconds) before beginning data recording. The rationale here is that they are allowing the bird to move into a less conspicuous position than the one in which it was initially observed. In addition, singing birds are often ignored because such individuals are usually in a high, conspicuous position (Hejl et al. 1990).

The advent of small radio transmitters promises to solve this problem in some measure. Although one still cannot see specifically what the out-of-sight animal is doing, at least the general location and movement of the animal can be approximated without causing undue disturbance. Many radios can also be fitted with mercury switches or vital rate sensors to detect body position, heart rate, body temperature, and other activities. Event recorders can automate much of the recording of these data (e.g., Berdoy and Evans 1990). Further, miniaturized video cameras are being developed so that behavior can be observed from the animal's point of view. Clearly, technology is rapidly advancing our ability to observe behavior in increasingly unbiased ways. Williams (1990) provided a review of the use of telemetry in studies of bird foraging behavior, and many publications and books are available on the use of radio-tagging wildlife (e.g., Mech 1983; Kenward 1987; White and Garrott 1990; Samuel and Fuller 1994).

Losito et al. (1989) developed a modification of focal sampling to address the problem of individuals disappearing from view, a problem especially evident in areas with limited visibility (e.g., dense brush). Termed focal-switch sampling, their method allows switching to a new focal individual (nearest neighbor) if the original focal animal goes out of view. They found that focal switching is more efficient than standard focal sampling because it saved 24 percent of samples from premature termination while also reducing recording biases in several analyses.

Scan Sampling. Using scan sampling, observers collect data on an entire group of animals (a flock, pack, herd) at regular intervals, recording the behavior of each individual; individual identification is usually not made, however. The behavior of each individual scanned is intended to be an instantaneous sample. A scan sample usually restricts the observer to recording only a few, simple categories of behavior. Scan sampling takes only a few seconds to several minutes, depending upon the amount of information being recorded, the group size, and the activity level of the group (e.g., a grazing herd versus a stampeding one).

Bias in scan sampling is influenced largely by the period of time that elapses between scans, or samples. Rare or inconspicuous behaviors will likely be observed at a lower frequency than they actually occur. For example, Harcourt and Stewart (1984) showed that studies using scan sampling of foraging gorillas resulted in underestimates of actual foraging time. Focal-animal sampling resulted in a more accurate estimate of the actual feeding time because the researchers could follow individuals continuously. Biases that can occur using scan sampling include (1) differential visibility between activity categories and seasons, (2) poor correlation between group composition and the representation of age-sex classes in scans, (3) diurnal variation in sample sizes, and (4) increased habituation to the observers with duration of the study (Newton 1992).

When rare events are the activity of interest, however, a generalized *behavior sampling* (or ad libitum) methodology is preferred. Such a methodology involves watching the entire group for long periods of time, recording each instance of the behavior of interest; such sampling can be incorporated as a subpart of either focal or scan sampling (Martin and Bateson 1993:87). Masataka and Thierry (1993) used ad libitum sampling in their study of macaques.

Scan sampling can also be combined with focal sampling during the same recording session. Here, researchers record in detail the behavior of focal individuals, but at specific intervals they also scan sample the group for simpler behavioral categories. Workers have shown, for flocking birds and herding mammals, that the time an individual spends foraging is related to the number and proximity of group members; for example, an individual must watch more vigilantly for predators as group size decreases (Hamilton 1971; Pulliam 1973; Caraco et al. 1980; Kildaw 1995). Using both methods in this manner thus allows the researcher to place focal-animal observations in an overall context of group activities.

For example, L'Heureux et al. (1995) used scan and focal sampling to study mother-yearling associations in bighorn sheep (*Ovis canadensis*) in Alberta, Canada. Scan sampling, at 15-minute intervals, was used to record the distance between a mother and yearling. Focal sampling was also used by observing sheep for at least 40 minutes, until they were out of sight, or for a maximum of 2 hours. Carranza (1995) also used both scan and focal sampling to study male-female interactions in red deer—the focal method to record interactions between territorial males, and scan samples at 1-minute intervals to record the number of females in a territory.

Recording Rules

The means of actually recording behavior, after the sampling rule has been chosen, are called recording rules. Figure 7.1 depicts the relationship between sampling rules and recording rules. We can divide recording rules into two basic categories: continuous recording and time sampling. Although we would prefer to be consistent and use the term *time "recording"* to clearly identify its classification as a recording rule, we will follow the terminology used by Martin and Bateson (1993); changing terms would only further confusion. Note that focal sampling (a sampling rule) is not synonymous with continuous recording (a recording rule); and scan sampling (a sampling rule) is not synonymous with instantaneous sampling (a recording rule).

Continuous Recording. In continuous recording, researchers record each occurrence of a behavioral act, as well as the time of the activity and pertinent environmental information. Continuous recording gives true frequencies, latencies, and durations of behavior *if* an animal can, indeed, be watched continuously for a sufficient period of time ("sufficient" must be determined through sample-size analyses). Termination of a recording session will usually result in an unreliable estimation of the duration of certain behaviors. Continuous recording is necessary when the sequence of behaviors is of interest. The method is, however, tedious, and in practice only a few categories can be measured reliably.

In studies of foraging behavior, observers often record the behavior of an animal continuously for a specific period of time. After some time has elapsed, the worker observes the same or a different individual for the same period of time. Within continuous recording periods, the duration of each behavior is timed. For example, Block (1990) watched foraging white-breasted nuthatches (*Sitta carolinensis*) for 10–15 seconds at a time, noting the duration of time spent searching for and procuring prey. Because a foraging bird can seldom be continuously observed for more than a few seconds to a few minutes, accurate estimates of latency and especially duration are difficult to obtain. Loughry (1993) used 5-minute continuous focal-animal sampling to study the behavior of black-tailed prairie dogs (*Cynomys ludovicianus*). In their study of foraging in the wapiti (*Cervus elaphus*), Wilmshurst et al. (1995) defined a cropping sequence as beginning when the animal put its head down to graze and ending when it lifted its head. The cropping rate was calculated as the number of bites taken during the cropping sequence divided by time. Weckerly (1994) watched foraging black-tailed deer (*Odocoileus hemionus columbianus*) for periods of 7–10 minutes.

Time Sampling. Time sampling is a general category that involves recording behavior periodically; each observation session is divided into successive short periods of time called sample intervals. However, time sampling has been criticized because there is no standardized time period used among scientists. Thus, the data on frequencies (of behavioral acts) lack any universal interpretation (Quera 1990). Therefore, time sampling is reported as relative durations or prevalence of the acts.

There are two basic types of time sampling. *Instantaneous sampling,* sometimes referred to as point sampling, records the instant (or point) at which the activity of an individual occurs. The structure of the data obtained depends largely upon the length of the sample interval. If the interval is long relative to the average duration of the behavior, then one can obtain a measure of the proportion (not frequency) of all sample points at which the behavior occurred. If the sample interval is short relative to the average duration of the behavior, however, then instantaneous sampling can approximate the results of continuous recording. In a detailed analysis of instantaneous sampling, Poysa (1991) concluded that averages of a great number of individuals give reliable estimates of true time budgets even for behavioral sequences of short duration. However, Poysa showed that if duration of a particular act is very short compared with other acts under study, then rate measures should be used instead of instantaneous sampling. It is also critical that samples from each individual under observation are long and roughly equal.

Instantaneous sampling is a common method used to record wildlife behavioral data. Typically, the observer follows an animal while it remains in view (or places a limit on the observation period) and records specific behavioral acts at set intervals. For example, Stamps and Krishnan (1994) recorded the location of every marked lizard every 20 minutes in a study of territoriality. Students of bird foraging commonly use instantaneous sampling (Hejl et al. 1990). Paterson et al. (1994) recommended focal-animal sampling, instantaneous scan sampling, and focal-time sampling (see Baulu and Redmond 1978) as the most flexible and statistically meaningful techniques.

One-zero sampling is similar to instantaneous sampling in that the recording session is subdivided into short sample intervals. Here, however, the observer merely records whether or not a particular behavior occurred during the sample period, recording no information on the frequency or duration of the act. Unlike instantaneous sampling, one-zero sampling consistently overestimates duration, because the behavior is recorded as though it occurred throughout the sample interval when it need have occurred only once. In fact, Altmann (1974) argued against the use of one-zero sampling in all applications. In contrast, Martin and Bateson (1993:97–98) provided a strong rationale for using one-zero sampling in certain situations, and included examples of how this method should be applied. Bernstein (1991) concluded that one-zero sampling may be preferred when acts are clustered and the goal is to predict the probability of at least one such act occurring in a given time interval. Clearly, caution should be employed when using any method, but especially when using one that involves such controversy. One-zero sampling is seldom used in behavioral studies.

Sampling Concerns

Included in our discussions above on sampling and recording rules were various cautionary notes on biases associated with specific methods. In the following section we discuss sampling issues that are common to all behavioral observations, including independence, observer bias, and sample size requirements.

Independence

Most statistical analyses carry the assumption that the data represent random samples from populations and that each datum is statistically independent from other data points (Zar 1984). Thus, the number of *individuals* for which data are taken characterizes or defines one's total sample. Machlis et al. (1985) noted that the objective of research is to obtain measurements from an adequate number of individuals, not to obtain large samples of measurements; they termed this the "pooling fallacy." Although pooling might not be a fatal error in certain situations (see below), studies of behavior are replete with examples of pooling errors (see also Hurlbert 1984).

In behavioral sampling, researchers commonly collect a series of instantaneous records on a single individual and then consider each point sample or each interval as an independent sample. That is, recording six instantaneous samples every so often (be it 1 sec, 1 min, or 1 hr) on the same individual results in a sample size of one. One procedure, termed aggregation, would be to average these six samples, resulting in $n = 1$ individual, where n is the sample size. However, this procedure can lead to false conclusions if the distribution of the behavior is bimodal (or, more generally, nonnormal), in which case the mean score for an individual will be the one that rarely occurs (Leger and Didrichsons 1994). Another sample interval taken a short time later, likewise, would not represent another sample. Another procedure would be to randomly select *a posteriori*, one sample from the repeated samples originally taken on an individual.

The problem, then, becomes one of determining when samples become independent. Martin and Bateson (1993:86) noted that samples "must be adequately spaced out over time." Defining *adequate* is difficult, however, and is related to the goal of the study. That is, if your objective is to describe behavior at the population level (i.e., individual variation within a population), then sampling new individuals across the population is indicated. Resampling an individual should be avoided so that results are not biased toward a certain segment of the population; independence is thus assured. This statement must be tempered, of course, in cases where the population is localized and rare. Likewise, if the goal is to examine a small segment of a larger population, independence *and* an adequate sample size will be difficult to achieve. Time-series analyses and repeated-measures designs are clearly indicated in such situations (described below).

Leger and Didrichsons (1994) noted that pooling is especially common in field studies of endangered species or other small populations, because multiple samples are taken from each individual to accumulate large sample sizes.

They noted that the central question about pooling is whether the population can be represented in an unbiased manner by repeated sampling of the same individuals. If, for instance, all individuals in a theoretical population have the same mean and variance in the behavior of interest, then it would not matter if one obtains 100 data points by sampling 100 individuals once each, 50 individuals twice each, or even 1 individual 100 times.

Leger and Didrichsons (1994) hypothesized that pooling is a reliable procedure *if* intrasubject variance exceeds intersubject variance. They evaluated several data sets and concluded that pooling can provide estimates of population means and variances that are at least as reliable as those provided by single sampling (of individuals) and aggregation, provided that samples sizes are about equal among individuals, or that intrasubject variance is higher than intersubject variance. When intrasubject variance exceeds intersubject variance, then unequal sample sizes among individuals become problematic.

McNay et al. (1994) examined independence of movements using telemetry data with black-tailed deer. They found that, even with six-week intervals between samples (eight samples/year), observations were still dependent for over 50 percent of the deer tested. Because most animal location data sets are likely to have a skewed distribution of data points, McNay and colleagues recommended placing emphasis on sampling animal locations systematically through time rather than trying to determine a time interval that would provide independent location samples. In other words, do not try to achieve independence of samples at the expense of gathering an adequate understanding of animal behavior. As noted by McNay and colleagues, time intervals between samples should be chosen with the understanding that, with the intervals' increase, potential gains in behavioral information are decreased. Swihart and Slade (1985) and White and Garrott (1990:148) provide additional discussion of this topic.

Intrasubject variation is an important, although seldom analyzed, aspect of behavioral research. Researchers have tended to treat populations as homogeneous units, thus largely ignoring individual differences in behavior. Implicit in such an attitude is that variance about a mean (behavior) is irrelevant and not of biological interest. In a monograph on this subject, Lomnicki (1988) contended that further advances in population ecology will require consideration of individual differences, such as unequal access to resources. Thus, repeatedly collecting samples on known individuals provides us with important ecological data. For example, changes in foraging strategies are common during ontogenetic size development, a phenomenon especially evident in reptiles (see Webb and Shine 1993 on blindsnakes [Typholopidae]; and Wikelski and Trillmich 1994 on iguanas [*Amblyrhynchus*]). Likewise, changes in energy stores can influence the trade-off between foraging time and vigilance in Belding's ground squirrels (*Spermophilus beldingi*) (Bachman 1993).

Thus, it should be clear that some advanced planning will avert most problems involving data independence. We can define some recording rules that seek to maximize independence of observations, while realizing that complete independence will not be achieved in all cases. A few such rules are: Only one individual within a group (flock, herd) should be recorded per sampling session. Observers should systematically cover the study area (which is as large as possible, especially if site-specific data are not needed), seeking out new individuals or groups; that is, avoid resampling the same group. Sampling sessions should be stratified by time period, both within and between days. Sessions should be distributed throughout the identified biological period (e.g., breeding) to avoid grouping samples within short periods of time. In small populations, an attempt should be made to identify animals individually.

Observer Bias

Avoiding bias is virtually impossible. Observer bias takes various forms: the influence of observers on an animal, intra- and interobserver consistency in recording data, preconceived notions regarding how an animal "should" behave, differences in observer abilities, and so forth. Each bias is serious; when combined, they can make interpretation of results difficult.

Researchers conducting behavioral studies should be aware that their presence and activities likely influence an animal's behavior. Researchers often state that their presence did not markedly alter an individual's behavior, or that they waited until the animal returned to normal activities before recording data. This seems to be an exercise in self-delusion. Wild animals are constantly vigilant for predators and competitors; your presence likely heightens their awareness. Such high awareness or responsiveness is termed *sensitization*. Further, it is likely that the animal knows you are there long before you ever see it. The waning of responsiveness is termed *habituation*, and is considered by most to be a form of learning (Immelmann and Beer 1989). Animal species vary widely in their ability to learn. However, research has shown that birds and mammals have the ability to perform both temporal and numeric operations in parallel (Roberts and Mitchell 1994). For example, many corvids can remember the location of food caches for months and also which caches they visited previously. Shettleworth (1993) reviewed the topic of learning in animals. Rosenthal (1976) presented a detailed analysis of the effects of the researcher in behavioral studies.

Animals that appear to become habituated to the presence of observers have thus adopted a modified pattern of behavior that allows them to keep the observer under surveillance. Animals adjust their behavior according to the costs and benefits associated with different courses of action available to them (e.g., hiding, fleeing). Further, detection of a potential predator (or human observer) may precede the observable response by a significant period of time. For example, Roberts and Evans (1993) showed that sanderlings (*Calidris alba*), in not tolerating any close approaches by a human, acted to minimize both the number of flights they made and the distance of each flight.

Intraobserver reliability is a measure of the ability of a specific observer to obtain the same data when measuring the same behavior on different occasions (Martin and Bateson 1993:32–34). This measure thus indicates the ability of an observer to be precise in his or her measurements. *Precision* describes the repeatability of a measurement and is not synonymous with accuracy. Because we seldom know what the actual behavioral pattern is—that is, the "true" pattern—we cannot directly measure an observer's accuracy. Assessing intraobserver reliability in field-based studies is difficult; animal seldom repeat their behavior in exactly the same fashion. One test is to videotape animals and then repeatedly to present (in some random fashion) individual sequences to the observer. Researchers can use the results of such trials to estimate the degree of observer reliability.

Interobserver reliability measures the ability of two or more observers to obtain the same results on the same occasion (Martin and Bateson

1993:117). To what extent is interobserver reliability a problem in field studies? Ford et al. (1990), for example, found that comparisons of foraging behaviors of individuals of the same species in different areas or years, recorded by different observers, needed to be treated cautiously. Problems were particularly evident when observers had not previously agreed on standard methods of observation or classification of terms. Differences in experience between observers apparently accounted for much of the interobserver variability noted.

Intra- and interobserver reliabilities have been substantially improved by careful, rigorous, and repeated training. Each new observer is taught how data should be recorded, initially working with others in the field, comparing data and discussing reasons for decisions; videotape, as noted above, or captive animals can also be used. Observer reliability increases, in our experience, when observers become informed about and comfortable with why a particular behavior is categorized in a certain manner. Training should continue throughout the study, with frequent sessions to "recalibrate" the observers. Each behavior should be carefully defined in writing. It usually helps to define a behavior by its structure (e.g., *probe* means "insert bill beneath surface of substrate"). Commonly, in protracted studies definitions and criteria tend to "drift" with the passage of time, as observers become more familiar with behaviors and possibly lazier in their evaluations. Careful and repeated training will help solve this problem. Further, efforts have been made to standardize terminology in various disciplines. For example, Remsen and Robinson (1990) developed a standardized terminology and definitions for avian foraging studies. The methodology of Schleidt et al. (1984) for a standard ethogram is a useful starting point for behavioral studies.

Sample Size Requirements

Researchers *must* incorporate evaluation of sample size requirements into the design phase of each study. Such planning guides the collection of data, avoiding both under- and overkill of sampling efforts. Further, it immediately informs you if you are trying to accomplish too much for the time and resources available. In our experience with graduate students, we have found that our first job when reviewing research proposals is to prevent the well-meaning student from attempting too much. It is, indeed, far better to do one thing very well than to do a bunch of things poorly. Many, many workers have been forced to combine data across seasons, years, ages, and sexes because they did not plan properly.

Throughout this book we have discussed how temporal and sex differences affect differential habitat use by animals. Studies of behavior show that, even within what we consider a "season," combining data over just a few months can obscure important patterns of resource use. In designing studies, researchers must carefully determine the number and types of variables for which they will have time to collect data.

For example, Brennan and Morrison (1990) found that significant variation in the use of tree species by foraging chestnut-backed chickadees (*Parus rufescens*) occurs throughout the year (fig. 7.2). Using some of the same data, Morrison (1988) showed how lumping of data can result in inappropriate interpretations: lumping tends to "average out" many possibly important ecological relationships (fig. 6.1). Likewise, Sakai and Noon (1990) showed that Pacific-slope flycatchers (*Empidonax difficilis*) significantly alter their foraging behavior within the breeding period.

These results are not surprising. Animals must respond to changes in resource availabil-

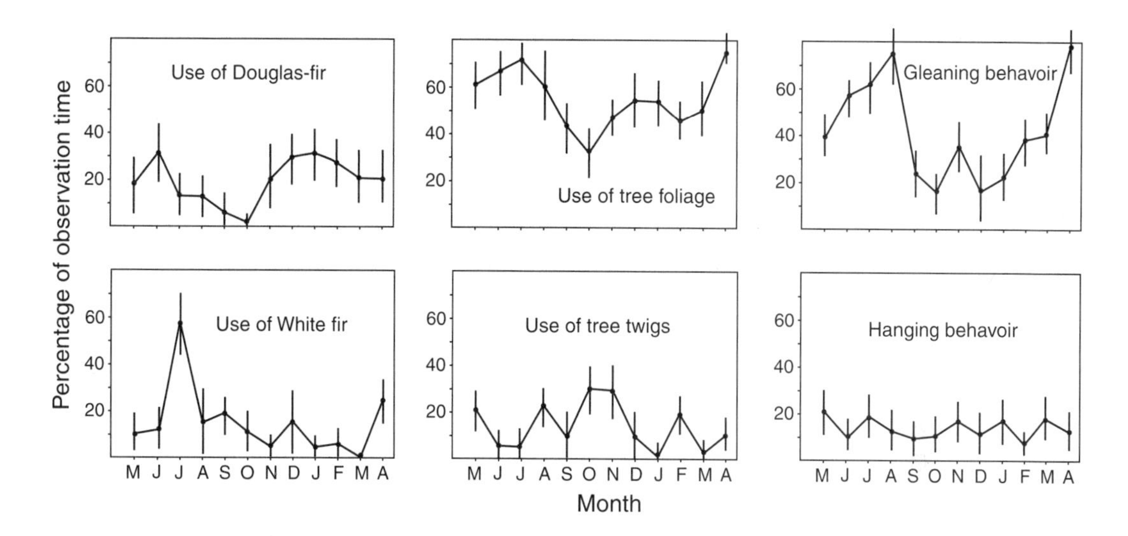

Figure 7.2. Seasonal variation in the use of tree species, the use of substrates, and foraging modes by chestnut-backed chickadees, using 1-month intervals. Dots represent mean values, and vertical bars represent one standard deviation. (Reproduced from Brennan and Morrison 1990 [fig. 4], by permission of the Department of Biology, University of California at Riverside)

ity and abundance and the demands placed upon them by both abiotic and biotic factors. Many of the equivocal results obtained in behavioral studies likely result from the lumping of data. Thus, not only must researchers evaluate sample sizes needed for reliable interpretations; they must also carefully evaluate these samples over relevant periods of time—periods that are usually shorter than those presented in most papers.

Activity Budgets and Energetics

Natural historians have long been interested in quantifying how animals allocate their time among the requirements for foraging, sleeping, moving, breeding, and so on. Developing these *activity budgets* (or time budgets) is the first step in developing an understanding of the relationship between time allocation and survivorship, reproductive success, body condition, and other aspects of natural history. A simple time budget for the loggerhead shrike (*Lanius ludovicianus*) is given in table 7.1. Defler (1995) presented a good example of time budgets in monkeys (see also Newton 1992). As noted by Weathers et al. (1984), however, quantifying time allocation is easy relative to assessing the energy an animal expends while conducting the activities. In this section we present a brief review of time-energy budgets as they relate to advancing our understanding of animal behavior and, ultimately, survival and fitness.

Assessments of animal energetics commonly assume that the goal of an animal is to maximize its net energy balance. Many theoretical and empirical studies have examined how the

Table 7.1. Time budgets of loggerhead shrikes, in hours spent per activity

		Perching							
Bird number	Run number	Night	Day	Eating	Preening	Flying	Hopping	Other	Total
218	1	11.96	10.79	0.41	0.22	0.10	0.03	0.66	24.17
220	2	12.54	9.44	0.49	0.11	0.07	0.06	0.28	22.99
232	3	11.47	10.70	0.05	0.12	0.08	0.02	0.38	22.82
221	4	11.67	10.40	0.22	0.04	0.20	0.04	0.09	22.66
230	5	11.43	9.42	0.70	0.10	0.10	0.03	1.88	23.66
225	6	12.28	6.73	0.01	0.01	0.19	0.05	2.74	22.01
218	7	11.97	10.30	0.30	0.35	0.08	0.11	0.39	23.50
232	8	12.17	10.49	0.00	0.71	0.05	0.00	0.58	24.00
Mean		11.94	9.78	0.27	0.21	0.11	0.04	0.88	23.23
(SD)		(0.39)	(1.34)	(0.25)	(0.23)	(0.06)	(0.03)	(0.93)	(0.73)

Source: Weathers et al. 1984: table 1; reproduced by permission of the American Ornithologists' Union

animal achieves this balance. Thus, the neurological and physiological capabilities of the animal itself link it to its environment. One should think of the environment in its functional relation to the animal, rather than merely as the geographic area and physical structure of the habitat in which the animal lives, in order to understand the dynamic relationships between an animal and its environment (Moen 1973:21).

Thermal energy is exchanged between the animal and its environment by radiation, conduction, convection, and evaporation. Each of the methods of thermal energy exchange changes relative to one another as the animal's environment alters: Rain, wind, and other abiotic factors affect energy exchange over the short term. Changes in such factors as plant cover, for example, influence energy exchange over longer periods of time.

A wild animal has an ecological metabolic rate that is an expression of the energy "cost of living" for the purpose of daily activities and other physiological processes. Ecological metabolism varies from one activity to another and from one species to another. There are various direct and indirect methods for determining metabolic rates of animals under varying environmental conditions, both in the laboratory and in the field.

Studies of physiology set in an ecological context, or physiological ecology, are relatively rare in the field of wildlife management. As developed throughout this book, most wildlife studies are based on indirect measures of an animal's requirements, for example, the recording of foraging location rather than foods eaten, or the recording of rates of movement rather than energy expended. It seems intuitive, however, that tying energy expenditures with observed behaviors will go a long way toward furthering our understanding of why animals behave, and ultimately survive and reproduce, in the manner that we observe.

The time-energy budget (TEB) method is a commonly used technique for estimating total daily energy expenditure in animals. TEB has two parts: First, you develop an activity budget as described above. Second, you convert these activity data to energetic equivalents from estimates of energy costs for each activity as determined from controlled studies or the literature (Haufler and Servello 1994). Numerous workers have conducted laboratory studies in order to determine the relationship between activity and energy expenditure; thus, for the species you are studying, approximations can usually be made on the basis of general allometric equations. Naturally, these values are only approximations, but they do provide a general understanding of activity-energy relationships (described below). Weathers et al. (1984) showed up to a 40 percent error rate for time-budget estimates that assign to behaviors energy equivalents that have been derived from the literature rather than empirically.

Finch (1984) determined the daily activity budget of Abert's towhees (*Pipilo aberti*) by quantifying the duration of four activities: perching, ground foraging, flying, and nest attendance. These data were transformed into percentages of the observation periods and the activity day. Daily energy expenditure was then determined using published estimates of basal and thermostatic requirements and estimates of the energy requirements for the four activities. Hobbs (1989) developed a model of energy balance in the mule deer (*Odocoileus hemionus*) that predicted changes in body condition of does and fawns and predicted the relationship of those changes to rates of mortality. He used literature values to develop a detailed model of energy balance based on animal activity. Another good example of the development of an activity budget and subsequent estimation of energy expenditure was provided by Dasilva (1992) for a colobine monkey (*Colobus polykomos*).

A more direct method of determining energy expenditures is the doubly labeled water technique. This technique involves the injection (labeling) of oxygen (oxygen 18) and hydrogen (tritium or deuterium) isotopes into an animal prior to its release and calculating the rate of CO_2 production, which can be equated to metabolic rate from the relative turnover rates of isotopes measured upon resampling (recapture) of the animal; this value is termed the field metabolic rate (FMR) (Haufler and Servello 1994).

A large and growing body of literature reports on FMR as determined using doubly labeled water (Nagy 1987; Nagy and Obst 1989; Speakman and Racey 1987). In birds, for example, FMR estimates (in kJ/day) include 118 in the 22-g tree swallow (*Tachycineta bicolor*), 343 in the 220-g Eurasian kestrel (*Falco tinnunculus*), 997 in the 1089-g little penguin (*Eudyptula minor*), and 2401 in the 3706-g gray-headed albatross (*Diomedea chrysostoma*) (Nagy and Obst 1989).

Table 7.2 presents mean values for the rate of energy expenditure associated with the activities given for the shrike in table 7.1. From such data, one can determine how energetic costs vary with observed variations in behavior in the field, and further, these energy costs can be related to survival, reproductive output, and body condition. For example, Mock (1991) used daily allocation of time and energy of western bluebirds (*Sialia mexicana*) to examine the trade-off between parental survival and survival of their young. She concluded that the species regulates overall daily energy expenditures through the differential use of thermal environments and activity budgets. Speakman and Racey (1987); see also Entwistle et al. 1995) used doubly labeled water to study the energetics of the brown long-eared bat (*Plecotus auritus*). Other methods of determining energy costs include estimating mass loss after activity periods (e.g., long flights), oxygen consumption, and heart-rate telemetry (see Goldstein 1990 for review).

Table 7.2. Cost of activity in loggerhead shrikes

	Number of shrikes	Number of observations	kJ/hr	Multiple of H_b
Basal metabolism (H_b)	9	27	1.79 ± 0.20	1.0
Alert perching	6	68	3.51 ± 0.60	1.98
Preening	3	4	3.87 ± 0.71	2.18
Eating	5	21	3.87 ± 0.63	2.19
Hopping	2	5	4.05 ± 0.67	2.28
Flying	—	—	23.7	13.2

Source: Weathers et al. 1984: table 3; reproduced by permission of the American Ornithologists' Union

Foraging and Diet

Optimal Foraging Theory

Foraging behavior research has advanced our understanding of animals and animal management in many important ways: in assessing habitat and designing reserves, in livestock husbandry and grazing management, in better understanding the limitations of resources to herbivores and matching of their dietary preferences with resource availability, in understanding and predicting foragers' impact on their resources and environment, and in controlling pests (Ash et al. 1996). This knowledge has provided the basis for advances in many aspects of wildlife management. For example, we now know that herbivore diet is closely linked to habitat and patch use; habitats and patches are objects that can be manipulated.

Animals are generally adapted to exploit specific types of foraging substrates most efficiently. A substrate is the specific location and surface at which the animal directs its foraging. The animal must compare the risks associated with foraging in a particular manner with the risks associated with other methods and other locations. The choice of foraging method and location is based not only on the number of prey present but also on the quality of that prey. That is, do numerous low-quality items that are easy to obtain have a net energy benefit over scarce but high-quality items that are difficult to obtain? Schoener (1969) hypothesized that animals can achieve this net balance by two extreme strategies: energy maximization or time minimization. Energy maximizers try to obtain the greatest amount of energy possible within a given period of time. Time minimizers seek to minimize the time required to obtain a given amount of food. Time minimizers thus have more time available for other activities, such as grooming and parental care (see also Morse 1980:53–54). As Stephens and Krebs (1986) recognized, however, maximization of net energy is not necessarily a desirable goal. The confusion, they observed, concerns equating the net rate of energy consumption with the ratio of benefit to cost (or foraging efficiency). They noted that maximizing efficiency ignores the time required to harvest resources, and it fails to distinguish tiny gains made at a small cost and larger gains made at a larger cost. For example, a gain of 0.01 calories at a cost of 0.001 calories gives the same benefit:cost ratio as a gain of 10 calories costing 1. The 10-calorie alternative, however, yields 1000 times the net profit of the 0.01 alternative (Stephens and Krebs 1986:8–9).

These and related considerations of costs and benefits form the basis of a large area of scientific investigation collectively known as optimal foraging theory. The term *optimal foraging* is a poor choice of terms, however, because it implies that there is some optimal strategy to follow; foraging theory per se makes no such claim (Stephens 1990). Although some have questioned the usefulness of optimal foraging theory (e.g., Pierce and Ollason 1987), it is our contention that all studies must be firmly based in theory, and that optimal foraging theory provides a guide for development of a research strategy. In addition, comparing actual field data with an "optimal" model allows determination of how divergent the optimal and realized strategies are.

Stephens and Krebs's (1986) book provides an excellent coverage of foraging theory. Here we will examine these theories and their associated models only in the ways that they relate to our descriptions of how animals might perceive their environment. Ideas about perception lead, in turn, to how we should observe and record animals as they exploit food resources. Thus, we will briefly explore how models of foraging behavior can help us design our studies.

In considering how animals determine whether or not to forage in an area, we have two rather distinct alternative views: Is the animal selecting from among various items of prey distributed throughout a generally suitable area, or is the animal distinguishing between various patches of prey? Foraging theorists thus distinguish between prey-choice models and patch-choice models. An animal can forage "optimally" within either of these two basic constructs. These models have clear implications for the design of foraging studies. That is, if prey items are distributed in a patchy manner, then our assessment of prey abundance or availability must be of proper spatial and temporal scale to recognize these patches. An overall average of prey abundance over a large area would fail to identify their patchy nature. Likewise, studies conducted at too small a scale could fail to identify any patches or could identify patches that the animal might not recognize.

Both prey-choice and patch-choice models assume that a foraging animal sequentially searches for prey, encounters the prey, and then decides whether or not to try to consume the prey. However, the form of the decision taken by the animal upon encountering a prey item differs between these two models. In the prey-choice model the animal must decide whether it should take the item or continue searching. One can then state various rules on which the animal should base its decision (e.g., prey size). In contrast, in the patch-choice model the animal decides how long it should forage in the patch encountered. Both models consider how an animal can best make these decisions with the goal of maximizing the long-term average rate of energy intake.

These models thus develop a general theoretic constraint upon which we can base our foraging studies. Clearly, the rate and method of searching, the frequency of encountering a prey item, and the type of attack used, all tell us a great deal about the ecology of a species. Further, variations in search, encounter, and prey consumption may lend insight into the current physiological condition of the animal. Thus, measuring aspects of foraging rate, the frequency and types of encounter, and related aspects of foraging are important in understanding animal distribution, survival, and reproductive success in occupied locations.

Foraging theory predicts that, when the animal considers prey items, the decision to take a specific item will depend not only on the item's abundance but also on the abundance of other food items. Food items can be ranked by their ratio of food value to handling time (Morse 1980:54). Handling time is the amount of time necessary to pursue, capture, and consume the prey (Stephens and Krebs 1986:14). Thus, foraging theory suggests that we should be concerned not only with how an animal goes about foraging but also with the types and relative ranking of abundance and quality of prey items encountered by the animal. Sampling the abundance of only the items consumed by the animal tells us little about the reasons for its decision.

Foraging theory has been applied as a framework for the design of studies of herbivore ecology. As outlined by Wilmshurst et al. (1995), optimal foraging theory assumes that foraging decisions by herbivores should be strongly influenced by physiological and environmental constraints on the rates of nutrient intake. Two such constraints commonly invoked for vertebrate grazers are: the effect of plant density on the short-term rate of food intake (the availability constraint), and the effect of digestive capacity on the long-term rate of energy assimilation (the processing constraint). The short-term rate of food intake should be positively related to plant size, bite size, and plant density. Using this model as a guide, we would thus design our studies to measure plant size and density and to relate these environmental variables to bite size and actual food intake.

Working with elk, Wilmshurst et al. (1995) hypothesized that, at low biomass, the processing rate of forage should be high but the short-term rate of intake low, whereas at high forage biomass the processing rate should be low but the short-term rate of intake high. The foraging maturation hypothesis states that the net rate of energy intake should be maximized accordingly on patches of intermediate plant biomass. Wilmshurst and colleagues concluded that the preference for grass patches of intermediate biomass and fiber content could help explain patterns of animal aggregations and seasonal migration.

Schmitz (1992) tested whether white-tailed deer select their diet in accordance with a foraging model that predicts that, to maximize fitness, deer have to balance a trade-off between maximizing growth and offspring production on the one hand and minimizing risk of starvation on the other. He found that deer appear to show plasticity in their diet in response to temporal and spatial changes in perceived risks and gains. Deer appear to balance gains in fitness due to reproduction with losses in fitness due to energetic shortfalls. When starvation risk is eliminated, deer tend to select diet breadths that simply maximize their mean rate of energy intake.

Foraging theory directly relates to wildlife management. As developed by Nudds (1980) for ungulates, the type of foraging model most closely followed by an animal (energy maximizing, equal food value, nutrient optimizing, unequal food value) can guide land management decisions concerning the amounts and types of food to emphasize. Nudds concluded that deer, as well as other temperate-latitude ungulates, are primarily habitat specialists but become food generalists in winter. The foraging behavior of deer in winter adheres most closely to the predictions of the energy maximizing models; it seems energetically less costly to remain in sheltered areas and fast than to forage in exposed areas. Translating these conclusions to a management scenario, Nudds suggested that manipulating winter habitats of deer by increasing only the abundance of "preferred" food is not warranted. Management would be more beneficially directed toward the physical structure of the winter habitat. Although some of Nudd's suggestions have been criticized (Jenkins 1982; but see Nudds 1982), his 1980 paper is important in that it directly links theory with management. Likewise, Kotler et al. (1994) showed how the study of patch use in Nubian ibex (*Capra ibex*) can be used by managers to modify habitat to reduce predation risk, thus allowing the animals to use available food more efficiently. As noted by Jenkins (1982), foraging theory, combined with good empirical work on food preference, may lead to valuable new insights about problems in wildlife management.

Foraging models used within a management context should address the spatial and temporal scales on which management can be controlled. These management scales are usually larger than the scales of the models; management scales are often determined by socioeconomic factors not under the control of the researcher. Although management may not specifically address the details of grazing behavior, for example, an understanding of the details of the foraging process is essential to provide a contextual framework for decision making (Ash et al. 1996).

Diet

A dichotomy exists in the literature regarding the emphasis placed on quantification of animal diets. Wildlife biologists and economic entomologists have expended much effort to determine the actual food items consumed by animals. Korschgen (1980) observed that, in the late 1800s and early 1900s, studies of food habits examined the economic importance of bird feeding habits, concentrating on the plunder of agricultural crops, poultry, and livestock. The greatest activity in food-habit studies took place in the 1930s and 1940s, emphasizing waterfowl and upland game birds. Regarding ungulate food habits, papers dealing with diets dominated the literature prior to 1950. The proportion of research reporting on food availability, food digestibility, and food requirements has grown steadily since that time.

In contrast, scientists studying the ecological relationships of animals seldom attempt to quantify the occurrence of specific prey items in the diet. Rather, ecological studies have concentrated on indirect measures of food use, such as foraging location (Hutto 1990; Rosenberg and Cooper 1990). Although morphological differences between species undoubtedly reflect some degree of evolutionary response to resources, they may not necessarily be good predictors of species' diets, especially under local environmental conditions (Rosenberg and Cooper 1990).

Although studies of food habits abound, most are single-species studies from single locations that were conducted over a short period of time. Thus, little generalization is possible. Further, as noted by Rosenberg and Cooper (1990), one of the reasons that studies of avian diets have been neglected by modern ornithologists is that researchers fear the detail, tedium, and technological expertise thought to be necessary for such studies. Regardless of the reasons, there is a shortage of literature on the diets of most species of animals in the world that is useful in describing and especially predicting their patterns of habitat use.

Sampling Techniques

Methods used to study diets of vertebrates can be divided into three basic categories: those involving collection of individual animals, those involving capture or other temporary disturbance of individual animals, and those requiring little or no disturbance of individuals (Rosenberg and Cooper 1990).

Several reviews of dietary assessments are readily available. Although many of these studies directly concern specific groups of animals (such as seabirds), many of the methods also apply to other groups of animals. Rosenberg and Cooper (1990) provided a thorough review of methods used to sample bird diets. Ratti et al. (1982:765–913) reprinted papers on the food habits and feeding ecology of waterfowl; they included a bibliography of other important references on diet. Each new edition of the Wildlife Society's *Wildlife Management Techniques Manual* includes reviews of methods for birds and mammals (see Haufler and Servello 1994; Litvaitis et al. 1994). Riney (1982:124–37) summarized studies of mammalian food habits, and many authors in Cooperrider et al. (1986) covered diet studies in wildlife.

The most frequently used method of sampling diets is direct examination of stomach contents. The primary advantage of such sampling is that an adequate number of stomachs is usually relatively easy to obtain by collecting animals through trapping or shooting. With shooting, an individual animal can be collected after the researcher has observed its specific foraging behavior; one can then attempt to relate the specific food items in the stomach to those sampled from the foraging substrate and to the animal's behavior used to gather the food. For game animals, researchers often take stomach samples from hunter check stations. Another advantage of gut sampling is that the entire contents of the stomach can be obtained. Kill sampling, however, has numerous disadvantages: The animal obviously cannot be resampled at a later date, preventing quantification of temporal (and possibly spatial) changes in food habits. The researcher has a potentially substantial impact on the population under study, negating studies of other aspects of the population's ecology (i.e., abundance, reproductive performance, behavior). Finally, the researcher is often subjected to severe criticism from certain segments of the public when the killing of animals is included in research.

Nondestructive methods of sampling food habits are available for wildlife. Live-caught animals can be forced to regurgitate using a variety of chemical emetics. Although some mortality can occur from the use of emetics, methods are available that will minimize losses (see Rosenberg and Cooper 1990 for review). For many animals the most easily obtained samples of diet come from their feces, collected either from the environment or during live trapping. In live-trapping studies, droppings can be obtained year-round from animals of any age or any reproductive state, and repeated sampling from known individuals is possible. Ralph et al. (1985) described a technique for collecting and analyzing bird droppings. This and related techniques have been used successfully in many studies of birds (Davies 1976, 1977a,b; Waugh 1979; Waugh and Hails 1983; Tatner 1983; Ormerod 1985) and small mammals (Meserve 1976; Dickman and Huang 1988). Many studies of ungulates have detailed methods of fecal collection and analysis (Riney 1982:129–31; Haufler and Servello 1994).

Nondestructive methods of determining food habits also involve observation of animal foraging behavior and analysis of food removal rates. Direct observation of food eaten by animals is possible with some species in some vegetation types. Many studies have been designed to quantify the amount of plant material removed by foraging ungulates. Researchers assess the height, weight, and condition of plants over periods of time and then relate the results to the type and amount of food consumed (Dasmann 1949; Severson and May 1967; Willms et al. 1980). Large ungulates can sometimes be observed grazing or browsing: "bite counts," calculated as the number of bites per plant species, are recorded (Willms et al. 1980; Thill 1985). Studies of food removal and bite counts can be combined to develop a picture of food habits. Diurnal birds of prey, such as eagles, vultures, and large hawks, can be observed when they forage in open areas.

Unfortunately, nondestructive methods of assessing food habits do not present a panacea for the researcher. In most cases, foraging animals cannot be observed closely without adversely influencing their behavior, thus negating direct observation of food habits. Because trapping most species is difficult, food in the stomach is often highly digested, one does not know where or when the animal was feeding, and there is little control over which animals are captured.

As with most aspects of wildlife research, a combination of carefully selected methods is usually necessary. A useful strategy is first to determine if the species under study shows any foraging behaviors that depart markedly from other closely related species. For example, studies of foraging behavior in the Sierra Nevada showed that many small insectivorous birds (e.g., chickadees, *Parus* spp.) significantly increased their use of bark foraging during winter; a closer examination of food availability and food habits was warranted (Morrison et al. 1989). Preliminary sampling (using direct or indirect methods) of stomach samples also indicates the sample sizes necessary and the time required to analyze those samples, as well as the likely level of resolution possible after a full study is conducted. There is no excuse for simply collecting large numbers of stomachs that will either sit on the laboratory shelf or yield no useful information for the management of a species.

Differential digestion rates of food items impose a large potential bias in any study of gut contents. Different kinds of foods consumed at about the same time are often digested at different rates. Furthermore, steps must be taken to prevent excessive postmortem digestion of food. For example, small-bodied insects may be gone from the gizzard within 5 minutes, whereas hard seeds may persist for several days (Swanson and Bartonek 1970). Several authors have developed correction factors for the differential rates of digestion shown by animals (Mook and Marshall 1965). Differential digestion of food is not confined to the intertaxonomic level: Rosenberg and Cooper (1990) discussed data that showed that second and third instar moth larvae were digested in less than half the time it took to digest fourth and fifth instars. As noted throughout this book, the goals of a specific study will determine the level of identification required to reach useful conclusions.

Level of Identification

The topic of the proper taxonomic identification of prey items has received little attention in the literature. The taxonomic level selected for diet analysis can have substantial impacts on the ecological interpretation of results. This problem is analogous to the selection of variables for use in habitat models; that is, how finely should we divide categories? If identification to the species level were a simple matter, then this issue would be of minor concern. Many food items are difficult to identify even when in excellent condition; the mastication and digestion of food further complicate the task. The level at which foods are identified is likely to affect similarity measures and conclusions drawn from them.

Greene and Jaksic (1983) studied the influence of prey identification level on measures of niche breadth and niche overlap in raptors, carnivores, and snakes. They calculated niche metrics for the finest prey identification levels (usually specific or generic) reported in diet studies and then recalculated the metrics after combining the prey lists to the ordinal level. They found that niche breadth was consistently larger at the finer prey identification levels than for the ordinal level of classification for all vertebrate groups examined. Calculations using the ordinal levels underestimated niche breadth at higher-resolution levels by 17–242 percent and underestimated niche-breadth scores for single species even more extremely. Food-niche overlaps based on the ordinal level overestimated higher-resolution overlap. Overestimates ranged to infinity when two species did not coincide in the use of any prey species but appeared to do so because of an ordinal level of identification. Greene and Jaksic clearly showed that using the ordinal level of prey classification can result in serious misinterpretations of some of the potentially most important food-niche and community parameters in assemblages of many animals.

Cooper et al. (1990) offered several guidelines regarding the level of taxonomic identification to choose in a study. Taxonomic levels that contain enough observations to make analysis meaningful should be identified. There are several practical considerations here. First, variables (prey categories) with high numbers of zero counts will not be normally distributed and usually cannot be transformed to normality. Thus, multivariate statistical procedures lose validity (see chapter 6). Second, one should decide if dividing a particular order into finer levels will result in any benefit. That is, if ecological and behavioral characteristics of two groups within an order do not differ substantially, then it is unlikely that subdivision will provide much additional information. Food items do not necessarily need to be identified using Linnaean nomenclature. Phenotypically distinguishable taxa (such as grasshopper A, B, and so on) can substitute for Linnaean identification (see also Greene and Jaksic 1983; Wolda 1990). The level of food identification chosen for a study should be made on the basis of the goals of the study, not *a posteriori* on the basis of funds, time, or difficulty in identifying the items. Simple preplanning prevents later disappointments and, more important, potential waste of animal life when kill sampling is involved.

Analyses

Statistical analyses of behavioral data should be an integral part of the initial study design. A large and varied number of methods of statistical analysis have been used with behavioral data, largely on the basis of the goals of the researcher and the form of the data. Clearly, sufficient and appropriate planning should guide the analyses used.

We will not review the many statistical techniques that can be applied to behavioral data. Readers can consult virtually any general statistical text for direction in analyzing behavioral data. Especially useful techniques are those of Siegel (1956), Conover (1980), Snedecor and Cochran (1980), Zar (1984), and Sokal and Rohlf (1995). Texts dealing more specifically with quantitative methods in ethology include Hazlett (1977), Colgan (1978), Siegel and Castellan (1988), Weinberg and Goldberg (1990), Sommer and Sommer (1991), and Lehner (1996). Martin and Bateson (1993: chapter 9) presented a concise but thorough review of fundamental univariate techniques for the study of animal behavior. Our chapter 6 deals with multivariate methods. Although not dealing specifically with statistical methods, Kamil (1988) discussed the application of experimental methods from the perspective of a behavioralist. Although concentrating on applications to foraging studies, many papers in Morrison et al. (1990) can be applied to analysis of any type of behavioral data. Roa (1992) and Manly (1993) discussed interesting analyses for use in experiments of food preference.

Behavioral data are usually recorded as categorical variables, such as various activity types (e.g., walking, running). Continuous data are often later classified into categories for analysis, such as speed of movement. When data are so classified, the result is a contingency table, the cells of which contain frequencies of the various category combinations (e.g., activity type by sex). The null hypothesis of homogeneity of categories is then tested using contingency analysis, such as the chi-square or *G* log–likelihood statistic. When three or more categories are compared, multidimensional contingency table analysis is used (see Colgan and Smith 1978).

Researchers usually record behaviors of animals in some sequential fashion, for example, walk-pause-walk-bite-swallow-walk. Analysis of such data using contingency tables and chi-square and related analyses may not be strictly valid, however, because sequential observations are likely not to be independent and thus violate the important assumption of independence of most statistical tests (Raphael 1990). The often sequential nature of data collection can, however, be a benefit in the elucidation of behavioral data. Examining the sequence of behaviors of an individual provides potentially much more information on how the animal exploits its environment than an overall lumping or averaging of behaviors does.

A sequence in which the behavioral pattern always occurs in the same order is termed deterministic. Classical behavioralists refer to such sequences as fixed action patterns. Vertebrates seldom if ever repeat behaviors in the same order, of course, but they exhibit some amount of variability that may be predictable. These sequences are considered "stochastic" (or probabilistic, to identify the statistical probability that can be assigned to each behavior). Sequences that show no temporal pattern (i.e., the component behavior patterns are sequentially independent) are considered random sequences. In a random sequence, one behavior or set of behaviors can be followed by any other behavior with equal probability. The conditional probability that one behavior follows another (i.e., the probability that behavior B follows behavior A, given that A has occurred), is called the transitional probability and is denoted as *P* (B/A) (Martin and Bateson 1993:152–54).

There are several methods for analyzing sequential data, such as time-series analyses. Of particular interest to us are analyses involving Markov chains. Markov analysis is a method for distinguishing whether a sequence is random or whether it contains some degree of temporal order. A first-order Markov process is one in which the probability of occurrence for the next event depends only on the immediately preceding event. If the probability depends on the two preceding events, then the process is considered second-order. Higher-order processes are involved when additional events are considered. One analyzes sequences by comparing the observed frequency of each transition with the frequency of transitions that would be expected if the sequences were random (Martin and Bateson 1993:152). A simple example of the Markov analysis was given by Martin and Bateson (1993:153) and is repeated here in figure 7.3. Raphael (1990) presented a review and a detailed example of the application of Markov analysis to foraging data; see also Colwell 1973; Riley 1986; Diggle 1990; and Gottman and Roy 1990 for analyses of sequential data.

Analyzing the sequential nature of data can identify changes in the behavior of individuals that might be obscured at least initially by the examination of only overall averages. Further, such analyses can identify which aspect of an individual's behavior is being impacted by a change (natural or human induced) in its environment. One can also relate each step in a sequence of behaviors to measures of the environment encountered during each step. For example, the behavioral sequence of a foraging bird (e.g., hop-hop-glean-probe-hop-hop) can be related to the foraging substrate being encountered (e.g., ground-ground-leaf-bark-

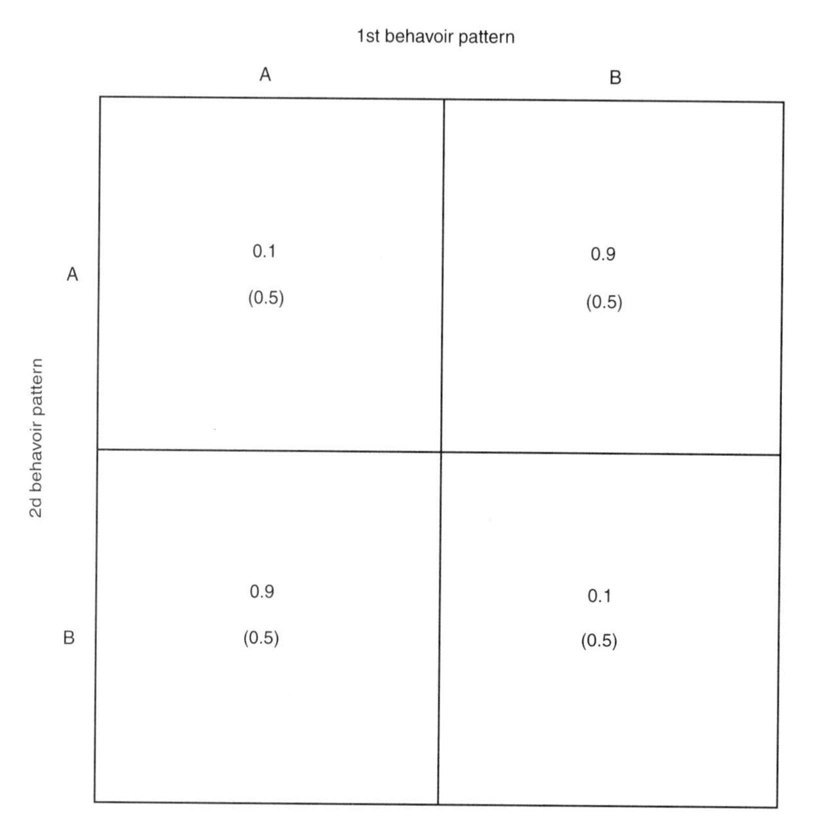

Figure 7.3. A highly simplified transition matrix, analyzing the sequence shown above it, which comprises only two different behavior patterns (A and B). The matrix shows the empirical transition probabilities for the four different types of transition (A|A, B|A, A|B, B|B). For example, the lower left cell shows that the conditional probability of B, given the occurrence of A (B|A), is 0.9 (9 out of 10 transitions after A has occurred). For comparison, the transition probabilities under a random model (0.5 for each type in this example) are shown in parentheses. The matrix confirms that A and B tend to alternate (the probabilities of B|A and A|B are high), while repeats are rare (the probabilities of B|B and A|A are low). (From P. Martin and P. Bateson, *Measuring Behaviours*, 2d ed. [Cambridge: Cambridge University Press © 1993], fig. 9.5; reprinted with the permission of Cambridge University Press)

ground-ground). Behavioral sequences might lend insight into the response of an animal to treatment effects that might not be evident using averages. Yet, such changes in behavior might explain changes in activity times and perhaps even changes in fecundity. Here again, the goal of the study should determine the detail required.

Indices

Scientists have developed a type of methodology to quantify the use of resources in relation to their availability. Widely known as preference or electivity indices, these measures seek to compare frequency of use of a resource with the availability of that resource in an animal's environment by representing the data as a single index value. Many of these indices have received a good deal of attention in the wildlife literature, especially with regard to analysis of food habits of ungulates; they have also been applied to vegetation and other measures of the environment (Morrison 1982). Because of this attention and the potential application of indices to a wide variety of situations, we will discuss some of the important considerations in their use.

The general approach to using electivity indices is to establish a ratio between frequency of an item used by an animal and the abundance of that item available for use. For example, Ivlev's electivity index (Ivlev 1961) compares relative availability of an item in the environment (p) with an animal's relative use of that item (r): $E_i = (r_i - p_i)/(r_i + p_i)$. Other indices have similar form, as reviewed by Lechowicz (1982). If r and p are equal for all items, then the animal is choosing the items randomly. If r and p differ, then one usually concludes that the animal is either avoiding (a negative index value) or selecting (positive value) an item.

The most straightforward indices simply consist of the estimated percentage of use of an item divided by the total number of all items available for use. Values usually range from −1 to 0 for avoidance and 0 to infinity for preference. Such indices have been termed the forage ratio (Jacobs 1974) and have been widely used (e.g., Heady and Van Dyne 1965; Chamrad and Box 1968; Petrides 1975; Hobbs and Bowden 1982). The major drawback to these simple indices, however, is their intrinsic asymmetry, that is, their unbounded positive values. A log transformation is used for the forage ratio (Jacobs 1974; Strauss 1979; Lechowicz 1982). Unfortunately, the forage ratio also changes with the relative abundance (p) of items in the environment. Thus, this index cannot be used if one wishes to examine the relationship between the relative abundance of items and the animal's preference for those items (Jacobs 1974; Lechowicz 1982).

Readers need only remember our earlier discussions of the great difficulties in accurately quantifying use and availability of resources to identify the overriding factor influencing index values. Clearly, proper determination of the variables to compare in such analyses will largely determine the results. An example modified from Johnson (1980) will illustrate this point. Suppose an investigator collects an animal and finds that its gut contains food items A, B, and C in the percentages shown in the top panel of table 7.3 under the first "Usage" column. A sample of the animal's feeding area reveals that the items were present in the percentages shown under the first "Availability" column. Many investigators would conclude that item A was avoided because its use was less than its availability, while items B and C were preferred. But suppose other investigators do not believe that item A is a valid food item; perhaps it was ingested only accidentally while the animal consumed other foods. They would then consider the data in the bottom panel of table 7.3, obtained by deleting item A from the analysis. Now, although item C is still deemed preferred, the assessment of item B has changed from preferred to avoided. Thus, we see that conclusions drawn from such analyses will depend markedly upon the array of items thought by the investigator to be available to the animal. Naturally, this caution applies whether or not indices are used. The level of prey identification used (as discussed earlier) will also markedly impact conclusions. Here, it is critical that the researcher use preliminary studies to identify valid items.

Table 7.3. Results of comparing usage and availability data when a commonly available but seldom-used item (A) is included and excluded from consideration

Item	Usage (%)	Availability (%)	Conclusion	Rank: Usage	Rank: Availability	Rank: Difference
			Item A Included			
A	2	60	Avoided	3	1	+2
B	43	30	Preferred	2	2	0
C	55	10	Preferred	1	3	−2
			Item A Excluded			
B	44	75	Avoided	2	1	+1
C	56	25	Preferred	1	2	−1

Source: Johnson 1980: table 1; reproduced by permission of the Ecological Society of America

Table 7.4. Various indices of electivity or feeding preference based on the proportions of food in the diet (r_i) and in the environment (p_i)

Algorithm	Comment	Reference
1. $E_i = (r_i - p_i)/(r_i + p_i)$	Ivlev's electivity index	Ivlev 1961
2. $E'_i = r_i/p_i$	Ivlev's forage ratio	Ivlev 1961
3. $D_i = \dfrac{r_i - p_i}{(r_i + p_i) - 2r_ip_i}$		Jacobs 1974
4. $Q_i = \dfrac{r_i(1 - p_i)}{p_i(1 - r_i)}$	Use of $\log_{10}Q$ recommended	Jacobs 1974
5. $L_i = r_i - p_i$		Strauss 1979
6. $\alpha_i = W_i = \dfrac{r_i/p_i}{\Sigma_i r_i/p_i}$	Chesson's *alpha,* Vanderploeg and Scavia's selectivity coefficient	Chesson 1978; Vanderploeg and Scavia 1979a
7. $E^*_i = \dfrac{W_i - (1/n)}{W_i + (1/n)}$	n = number of kinds of food items	Vanderploeg and Scavia 1979b

Source: Lechowicz 1982: table 1

The work of Ivlev (1961), Jacobs (1974), Chesson (1978, 1983), Strauss (1979), Vanderploeg and Scavia (1979a,b), and Johnson (1980) described the development of the more widely used electivity indices. Lechowicz (1982) compared the characteristics of seven of the electivity indices proposed by these researchers; most of these indices are permutations of Ivlev's original index (see table 7.4). Lechowicz found that Ivlev's, Strauss's, and Jacob's indices could not potentially obtain the full range of index values for all values of r and p. The index values for intermediate values of r and p depend on the relative abundance of other items in the environment or of those used by the animal. A critical problem with most of the indices is that direct comparisons between indices derived from samples differing in relative abundance are inappropriate (but see Chesson [1983] for exceptions).

To avoid the problems associated with inclusion or exclusion of specific items in the calculation and evaluation of indices, Johnson (1980) developed a procedure based on ranks. He proposed using the difference between the rank of use and the rank of availability. If we use the earlier example with item A included (table 7.3), the differences in rank of use and availability are +2, 0, and −2 for items A, B, and C, respectively. Excluding item A from the analysis results in values of +1 and −1 for B and C, respectively. Although the index values themselves change, the difference between B and C remains +2. The loss of information regarding the absolute difference between items realized when using ranks is likely of little consequence. Statistical methods based on ranks are nearly as efficient as methods based on the original data; this is especially true if the assumptions necessary to treat the original data are not met (e.g., assumption of normality). Johnson also provided methods for determining the statistical significance among components of the data. The method of compositional analysis developed earlier (chapter 5) can also be used to analyze diet and activity data.

Literature Cited

Altmann, J. 1974. Observational study of behavior: Sampling methods. *Behaviour* 49:227–67.

Ash, A., M. Coughenour, J. Fryxell, W. Getz, J. Hearne, N. Owen-Smith, D. Ward, and E. A. Laca. 1996. Second international foraging behavior workshop. *Bulletin of the Ecological Society of America* 77:36–38.

Bachman, G. C. 1993. The effect of body condition on the trade-off between vigilance and foraging in Belding's ground squirrels. *Animal Behaviour* 46:233–44.

Baulu, J., and D. E. Redmond, Jr. 1978. Some sampling considerations in the quantification of money behavior under field and captive conditions. *Primates* 19:391–400.

Berdoy, M., and S. E. Evans. 1990. An automatic recording system for identifying individual small animals. *Animal Behaviour* 39:998–1000.

Bernstein, I. S. 1991. An empirical comparison of focal and ad libitum scoring with commentary on instantaneous scans, all occurrence and one-zero techniques. *Animal Behaviour* 42:721–28.

Block, W. M. 1990. Geographic variation in foraging ecologies of breeding and nonbreeding birds in oak woodlands. *Studies in Avian Biology* 13:264–69.

Brennan, L. A., and M. L. Morrison. 1990. Influence of sample size on interpretation of foraging patterns by chestnut-backed chickadees. *Studies in Avian Biology* 13:187–92.

Caraco, T., S. Martindale, and T. S. Whittham. 1980. An empirical demonstration of risk-sensitive foraging preferences. *Animal Behaviour* 28:820–30.

Carranza, J. 1995. Female attraction by males versus sites in territorial rutting red deer. *Animal Behaviour* 50:445–53.

Chamrad, A. D., and T. W. Box. 1968. Food habits of white-tailed deer in south Texas. *Journal of Range Management* 21:158–64.

Chesson, J. 1978. Measuring preference in selective predation. *Ecology* 59:211–15.

Chesson, J. 1983. The estimation and analysis of preference and its relationship to foraging models. *Ecology* 64:1297–1304.

Colgan, P. W., ed. 1978. *Quantitative ethology.* New York: John Wiley and Sons.

Colgan, P. W., and J. T. Smith. 1978. Multidimensional contingency table analysis. In *Quantitative ethology,* ed. P. W. Colgan, 145–74. New York: John Wiley and Sons.

Colwell, R. K. 1973. Competition and coexistence in a simple tropical community. *American Naturalist* 107:737–60.

Conover, W. J. 1980. *Practical nonparametric statistics.* 2d ed. New York: John Wiley and Sons.

Cooper, R. J., P. J. Martinat, and R. C. Whitmore. 1990. Dietary similarity among insectivorous birds: Influence of taxonomic versus ecological categorization of prey. *Studies in Avian Biology* 13:104–9.

Cooperrider, A. Y., R. J. Boyd, and H. R. Stuart, eds. 1986. *Inventory and monitoring of wildlife habitat.* USDI Bureau of Land Management Service Center, Denver.

Dasilva, G. L. 1992. The western black-and-white colobus as a low-energy strategist: Activity budgets, energy expenditure and energy intake. *Journal of Animal Ecology* 61:79–91.

Dasmann, W. P. 1949. Deer-livestock forage studies in the interstate winter deer range in California. *Journal of Range Management* 2:206–12.

Davies, N. B. 1976. Food, flicking, and territorial behavior of the pied wagtail (*Motacilla alba yarelli* Gould). *Journal of Animal Ecology* 45:235–52.

Davies, N. B. 1977a. Prey selection and social behavior in wagtails (Aves: Motacillidae). *Journal of Animal Ecology* 46:37–57.

Davies, N. B. 1977b. Prey selection and the search strategy of the spotted flycatcher (*Muscicapa striata*): A field study on optimal foraging. *Animal Behaviour* 25:1016–33.

Defler, T. R. 1995. The time budget of a group of wild woolly monkeys (*Lagothrix lagotricha*). *International Journal of Primatology* 16:107–20.

de Vries, H., W. J. Netto, and P. L. H. Hanegraaf. 1993. Matman: A program for the analysis of sociometric matrices and behavioural transition matrices. *Behaviour* 125:157–75.

Dewsbury, D. A. 1992. On the problems studied in ethology, comparative psychology, and animal behavior. *Ethology* 92:89–107.

Dickman, C. R., and C. Huang. 1988. The reliability of fecal analysis as a method for determining the diet of insectivorous mammals. *Journal of Mammalogy* 69:108–13.

Diggle, P. J. 1990. *Time series: A biostatistical introduction.* Oxford: Oxford University Press.

Entwistle, A. C., J. R. Speakman, and P. A. Racey. 1995. Effect of using the doubly labelled water technique on long-term recapture in the brown long-eared bat (*Plecotus auritus*). *Canadian Journal of Zoology* 72:783–85.

Finch, D. M. 1984. Parental expenditure of time and energy in the Abert's towhee (*Pipilo aberti*). *Auk* 101:473–86.

Ford, H. A., L. Bridges, and S. Noske. 1990. Interobserver differences in recording foraging behavior of fuscous honeyeaters. *Studies in Avian Biology* 13:199–201.

Goldstein, D. L. 1990. Energetics of activity and free living in birds. *Studies in Avian Biology* 13:423–26.

Gottman, J. M., and A. K. Roy. 1990. *Sequential analysis: A guide for behavioral researchers.* Cambridge: Cambridge University Press.

Greene, H. W., and F. M. Jaksic. 1983. Food-niche relationships among sympatric predators: Effects of level of prey identification. *Oikos* 40:151–54.

Hall, L. S., P. R. Krausman, and M. L. Morrison. 1997. The habitat concept and a plea for standard terminology. *Wildlife Society Bulletin* 25:173–182.

Hamilton, W. D. 1971. Geometry for the selfish herd. *Journal of Theoretical Biology* 31:293–311.

Harcourt, A. H., and K. J. Stewart. 1984. Gorillas' time feeding: Aspects of methodology, body size, competition and diet. *African Journal of Ecology* 22:207–15.

Haufler, J. B., and F. A. Servello. 1994. Techniques for wildlife nutritional analysis. In *Research and management techniques for wildlife and habitats,* ed. T. A. Bookhout, 307–23. 5th ed. Bethesda, Md.: The Wildlife Society.

Hazlett, B. A., ed. 1977. *Quantitative methods in the studies of animal behavior.* New York: Academic Press.

Heady, H. F., and G. M. Van Dyne. 1965. Botanical composition of sheep and cattle diets on a mature animal range. *Hilgardia* 36:465–92.

Hejl, S. J. Verner, and G. W. Bell. 1990. Sequential versus initial observations in studies of avian foraging. *Studies in Avian Biology* 13:166–73.

Hobbs, N. T. 1989. Linking energy balance to survival in mule deer: Development and test of a simulation model. *Wildlife Monograph* 101:1–39.

Hobbs, N. T., and D. C. Bowden. 1982. Confidence intervals on food preference indices. *Journal of Wildlife Management* 46:505–7.

Hurlbert, S. H. 1984. Pseudoreplication and the design of ecological field experiments. *Ecology* 54:187–211.

Hutto, R. L. 1990. On measuring the availability of food resources. *Studies in Avian Biology* 13:20–28.

Immelmann, K., and C. Beer. 1989. *A dictionary of ethology.* Cambridge, Mass.: Harvard University Press.

Ivlev, V. S. 1961. *Experimental ecology of the feeding of fishes.* New Haven: Yale University Press.

Jacobs, J. 1974. Quantitative measurement of food selection. *Oecologia* 14:413–17.

Jenkins, S. H. 1982. Management implications of optimal foraging theory: A critique. *Journal of Wildlife Management* 46:255–57.

Johnson, D. H. 1980. The comparison of usage and availability measurements for evaluating resource preference. *Ecology* 61:65–71.

Kamil, A. C. 1988. Experimental design in ornithology. In *Current ornithology,* Vol. 5, ed. R. F. Johnston, 312–46. New York: Plenum Press.

Kamil, A. C., and T. D. Sargent, eds. 1981. *Foraging behavior: Ecological, ethological, and psychological approaches.* New York: Garland STPM Press.

Kamil, A. C., J. R. Krebs, and H. R. Pulliam, eds. 1987. *Foraging behavior.* New York: Plenum Press.

Kenward, R. 1987. *Wildlife radio tagging.* New York: Academic Press.

Kildaw, S. D. 1995. The effect of group size manipulation on the foraging behavior of black-tailed prairie dogs. *Behavioral Ecology* 6:353–58.

Korschgen, L. J. 1980. Procedures for food-habitat analyses. In *Wildlife management techniques manual,* ed. S. D. Schemnitz. 4th ed. Washington, D.C.: The Wildlife Society.

Kotler, B. P., J. E. Gross, and W. A. Mitchell. 1994. Applying patch use to assess aspects of foraging behavior in Nubian ibex. *Journal of Wildlife Management* 58:299–307.

Lechowicz, M. J. 1982. The sampling characteristics of electivity indices. *Oecologia* 52:22–30.

Leger, D. W., and I. A. Didrichsons. 1994. An assessment of data pooling and some alternatives. *Animal Behaviour* 48:823–32.

Lehner, P. N. 1979. *Handbook of ethological methods.* New York: Garland STPM Press.

Lehner, P. N. 1996. *Handbook of ethological methods.* 2d ed. Cambridge: Cambridge University Press.

L'Heureux, N., M. Lucherini, M. Festa-Bianchet, and J. T. Jorgenson. 1995. Density-dependent mother-yearling association in bighorn sheep. *Animal Behaviour* 49:901–10.

Litvaitis, J. A., K. Titus, and E. M. Anderson. 1994. Measuring vertebrate use of terrestrial habitats and foods. In *Research and management techniques for wildlife and habitats,* ed. T. A. Bookhout, 254–74. 5th ed. Bethesda, Md.: The Wildlife Society.

Lomnicki, A. 1988. *Population ecology of individuals.* Princeton: Princeton University Press.

Losito, M. P., R. E. Mirarchi, and G. A. Baldassarre. 1989. New techniques for time-activity studies of avian flocks in view-restricted habitats. *Journal of Field Ornithology* 60:388–96.

Loughry, W. J. 1993. Determinants of time allocation by adult and yearling black-tailed prairie dogs. *Behaviour* 124:23–43.

Machlis, L., P. W. D. Dodd, and J. C. Fentress. 1985. The pooling fallacy: Problems arising when individuals contribute more than one observation to the data set. *Zeitschrift fuer Tierpsychologie* 68:201–214.

McNay, R. S., J. A. Morgan, and F. L. Bunnell. 1994. Characterizing independence of observations in movements of Columbian black-tailed deer. *Journal of Wildlife Management* 58:422–29.

Manly, B. F. J. 1993. Comments on design and analysis of multiple-choice feeding-preference experiments. *Oecologia* 93:149–52.

Martin, P., and P. Bateson. 1993. *Measuring behaviour.* 2d ed. Cambridge: Cambridge University Press.

Masataka, N., and B. Thierry. 1993. Vocal communication of Tonkean macques in confined environments. *Primates* 34:169–80.

Mech, L. D. 1983. *Handbook of animal radio-tracking.* Minneapolis: University of Minnesota Press.

Meserve, P. L. 1976. Food relationships of a rodent fauna in a California coastal sage community. *Journal of Mammalogy* 57:300–319.

Mock, P. J. 1991. Daily allocation of time and energy of western bluebirds feeding nestlings. *Condor* 93:598–611.

Moen, A. N. 1973. *Wildlife ecology.* San Francisco: W. H. Freeman.

Mook, L. J., and H. W. Marshall. 1965. Digestion of spruce budworm larvae and pupae in the olive-backed thrush, *Hylocichla ustulata swainsoni* (Tschudi). *Canadian Entomologist* 97:1144–49.

Morrison, M. L. 1982. The structure of western warbler assemblages: Ecomorphological analysis of the black-throated gray and hermit warblers. *Auk* 99:503–13.

Morrison, M. L. 1988. On sample sizes and reliable information. *Condor* 90:275–78.

Morrison, M. L., C. J. Ralph, J. Verner, and J. R. Jehl, Jr., eds. 1990. *Avian foraging: Theory, methodology, and applications.* Studies in Avian Biology no. 13.

Morrison, M. L., D. L. Dahlsten, S. M. Tait, R. C. Heald, K. A. Milne, and D. L. Rowney. 1989. *Bird foraging on incense-cedar and incense-cedar scale during winter in California.* USDA Forest Service Research Paper PSW-195.

Morse, D. H. 1980. *Behavioral mechanisms in ecology.* Cambridge, Mass.: Harvard University Press.

Nagy, K. A. 1987. Field metabolic rate and food requirement scaling in mammals and birds. *Ecological Monographs* 57:111–28.

Nagy, K. A., and B. S. Obst. 1989. Body size effects on field energy requirements of birds: What determines their field metabolic rates? *International Ornithological Congress* 20:793–99.

Newton, P. 1992. Feeding and ranging patterns of forest hanuman langurs (*Presbytis entellus*). *International Journal of Primatology* 13:245–85.

Nudds, T. D. 1980. Foraging "preference": Theoretical considerations of diet selection by deer. *Journal of Wildlife Management* 44:735–40.

Nudds, T. D. 1982. Theoretical considerations of diet selection by deer: A reply. *Journal of Wildlife Management* 46:257–58.

Ormerod, S. J. 1985. The diet of dippers *Cinclus cinclus* and their nestlings in the catchment of the River Wye, Mid-Wales: A preliminary study of faecal analysis. *Ibis* 127:316–31.

Paterson, J. D., P. Kubicek, and S. Tillekeratne. 1994. Computer data recording and DATAC6, a BASIC program for continuous and interval sampling studies. *International Journal of Primatology* 15:303–15.

Petrides, G. A. 1975. Principal foods versus preferred foods and their relation to stocking rate and range condition. *Biological Conservation* 7:161–69.

Pierce, G. J., and J. G. Ollason. 1987. Eight reasons why optimal foraging theory is a complete waste of time. *Oikos* 49:111–18.

Poysa, H. 1991. Measuring time budgets with instantaneous sampling: A cautionary note. *Animal Behaviour* 42:317–18.

Pulliam, H. R. 1973. On the advantages of flocking. *Journal of Theoretical Biology* 38:419–22.

Quera, V. 1990. A generalized technique to estimate frequency and duration in time sampling. *Behavioral Assessment* 12:409–24.

Ralph, C. P., S. E. Nagata, and C. J. Ralph. 1985. Analysis of droppings to describe diets of small birds. *Journal of Field Ornithology* 56:165–74.

Raphael, M. G. 1988. A portable computer-compatible system for collecting bird count data. *Journal of Field Ornithology* 59:280–85.

Raphael, M. G. 1990. Use of Markov chains in analysis of foraging behavior. *Studies in Avian Biology* 13:288–94.

Ratti, J. T., L. D. Flake, and W. A. Wentz. 1982. *Waterfowl ecology and management: Selected readings.* Bethesda, Md.: The Wildlife Society.

Remsen, J. V., and S. K. Robinson. 1990. A classification scheme for foraging behavior of birds in terrestrial habitats. *Studies in Avian Biology* 13:144–60.

Riley, C. M. 1986. Foraging behavior and sexual dimorphism in emerald toucanets (*Aulacorhynchus prasinus*) in Costa Rica. M. S. thesis, University of Arkansas, Fayetteville.

Riney, T. 1982. *Study and management of large mammals.* New York: John Wiley and Sons.

Roa, R. 1992. Design and analysis of multiple-choice feeding-preference experiments. *Oecologia* 89:509–15.

Roberts, G., and P. R. Evans. 1993. Responses of foraging sanderlings to human approaches. *Behaviour* 126:29–43.

Roberts, W. A., and S. Mitchell. 1994. Can a pigeon simultaneously process temporal and numerical information? *Journal of*

Experimental Psychology: Animal Behavior Processes 20:66–78.
Rosenberg, K. V., and R. J. Cooper. 1990. Approaches to avian diet analysis. *Studies in Avian Biology* 13:80–90.
Rosenthal, R. 1976. *Experimenter effects in behavioral research.* New York: Irvington.
Ruggiero, L. F., R. S. Holthausen, B. G. Marcot, K. B. Aubry, J. W. Thomas, and E. C. Meslow. 1988. Ecological dependency: The concept and its implications for research and management. *Transactions of the North American Wildlife and Natural Resources Conference* 53:115–26.
Sakai, H. F., and B. R. Noon. 1990. Variation in the foraging behaviors of two flycatchers: Associations with stage of breeding cycle. *Studies in Avian Biology* 13:237–44.
Samuel, M. D., and M. R. Fuller. 1994. Wildlife radiotelemetry. In *Research and management techniques for wildlife and habitats,* ed. T. A. Bookhout, 370–418. 5th ed. Bethesda, Md.: The Wildlife Society.
Schleidt, W. M., G. Yakalis, M. Donnelly, and J. McGarry. 1984. A proposal for a standard ethogram, exemplified by an ethogram of the bluebreasted quail (*Coturnix chinensis*). *Zeitschrift fur Tierpsychologie* 64:193–220.
Schmitz, O. J. 1992. Optimal diet selection by white-tailed deer: Balancing reproduction with starvation risk. *Evolutionary Ecology* 6:125–41.
Schoener, T. W. 1969. Optimal size and specialization in constant and fluctuating environments: An energy time approach. *Brookhaven Symposium in Biology* 22:103–14.
Severson, K. E., and M. May. 1967. Food preferences of antelope and domestic sheep in Wyoming's Red Desert. *Journal of Range Management* 20:21–25.
Shettleworth, S. J. 1993. Varieties of learning and memory in animals. *Journal of Experimental Psychology: Animal Behavior Processes* 19:5–14.
Siegel, S. 1956. *Nonparametric statistics for the behavioral sciences.* New York: McGraw-Hill.
Siegel, S., and N. J. Castellan. 1988. *Nonparametric statistics for the behavioral sciences.* 2d ed. New York: McGraw-Hill.
Slater, P. J. B. 1978. Data collection. In *Quantitative ethology,* ed. P. W. Colgan, 7–24. New York: John Wiley and Sons.
Snedecor, G. W., and W. G. Cochran. 1980. *Statistical methods.* 7th ed. Ames: Iowa State University Press.
Sokal, R. R., and F. J. Rohlf. 1995. *Biometry.* 3d ed. San Francisco: W. H. Freeman.
Sommer, B., and R. Sommer. 1991. *A practical guide to behavioral research: Tools and techniques.* 3d ed. New York: Oxford University Press.
Speakman, J. R., and P. A. Racey. 1987. The energetics of pregnancy and lactation in the brown long-eared bat, *Plecotus auritus.* In *Recent advances in the study of bats,* ed. M. B. Fenton, P. A. Racey, and J. M. V. Rayner, Cambridge: 367–95. Cambridge University Press.
Stamps, J. A., and V. V. Krishnan. 1994. Territory acquisition in lizards: I. First encounters. *Animal Behaviour* 47:1375–85.
Stephens, D. W. 1990. Foraging theory: Up, down, and sideways. *Studies in Avian Biology* 13:444–54.
Stephens, D. W., and J. R. Krebs. 1986. *Foraging theory.* Princeton: Princeton University Press.
Strauss, R. E. 1979. Reliability estimates for Ivlev's electivity index, the forage ratio, and a proposed linear index of food selection. *Transactions of the American Fisheries Society* 108:344–52.
Swanson, G. A., and J. C. Bartonek. 1970. Bias associated with food analysis in gizzards of blue-winged teal. *Journal of Wildlife Management* 34:739–46.
Swihart, R. K., and N. A. Slade. 1985. Testing for independence of observations in animal movements. *Ecology* 66:1176–84.
Tatner, P. 1983. The diet of urban magpies, *Pica pica. Ibis* 125:90–107.
Thill, R. A. 1985. Cattle and deer compatibility on southern forest range. In *Proceedings of a conference on multispecies grazing,* ed. F. H. Baker and R. K. Jones, 159–77. Morrilton, Ark.: Winrosk International.
Vanderploeg, H. A., and D. Scavia. 1979a. Two electivity indices for feeding with special reference to zooplankton grazing. *Journal of Fisheries Research Board of Canada* 36:362–65.
Vanderploeg, H. A., and D. Scavia. 1979b. Calculation and use of selectivity coefficients of feeding: Zooplankton grazing. *Ecological Modelling* 7:135–49.

Waugh, D. R. 1979. The diet of sand martins in the breeding season. *Bird Study* 26: 123–28.

Waugh, D. R., and C. J. Hails. 1983. Foraging ecology of a tropical aerial feeding bird guild. *Ibis* 125:200–217.

Weathers, W. W., W. A. Buttemer, A. M. Hayworth, and K. A. Nagy. 1984. An evaluation of time-budget estimates of daily energy expenditures in birds. *Auk* 101:459–72.

Webb, J. K., and R. Shine. 1993. Prey-size selection, gape limitation and predator vulnerability in Australian blindsnakes (Typhlopidae). *Animal Behaviour* 45:1117–26.

Weckerly, F. W. 1994. Selective feeding by black-tailed deer: Forage quality or abundance? *Journal of Mammalogy* 75:905–13.

Weinberg, S. L., and K. P. Goldberg. 1990. *Statistics for the behavioral sciences.* Cambridge: Cambridge University Press.

Weisman, R., S. Shackleton, L. Ratcliffe, D. Weary, and P. Boag. 1994. Sexual preferences of female zebra finches: Imprinting on beak colour. *Behaviour* 128:15–24.

White, G. C., and R. A. Garrott. 1990. *Analysis of wildlife radio-tracking data.* San Diego: Academic Press.

Whitehead, H. 1995. Investigating structure and temporal scale in social organization. *Behavioral Ecology* 6:199–208.

Wikelski, M., and F. Trillmich. 1994. Foraging strategies of the Galapagos marine iguana (*Amblyrhynchus cristatus*): Adapting behavioral rules to ontogenetic size change. *Behaviour* 128:255–79.

Williams, P. L. 1990. Use of radiotracking to study foraging in small terrestrial birds. *Studies in Avian Biology* 13:181–86.

Willms, W., A. McLean, R. Tucker, and R. Ritchey. 1980. Deer and cattle diets on summer range in British Columbia. *Journal of Range Management* 33:55–59.

Wilmshurst, J. F., J. M. Fryxell, and R. J. Hudson. 1995. Forage quality and patch choice by wapiti (*Cervus elaphus*). *Behavioural Ecology* 6:209–17.

Wolda, H. 1990. Food availability for an insectivore and how to measure it. *Studies in Avian Biology* 13:38–43.

Zar, J. H. 1984. *Biostatistical analysis.* 2d ed. Englewood Cliffs, N.J.: Prentice-Hall.

8 Of Habitat Patches and Landscapes: Habitat Heterogeneity and Responses of Wildlife

Introduction

Landscapes are the great environmental integrators. It is within landscapes where individuals interact as populations, where species mix in communities, and where communities overlap with each other and with their abiotic contexts in ecosystems. Landscapes integrate all these factors across spatial scales ranging from as large as home ranges of large vertebrate carnivores to as small as interstices of soil particles in which symbiotic fungi infuse tree rootlets. It is within landscapes where dynamics and disturbance regimes interplay with the ever-changing biota across scales ranging from sweeping catastrophic fires and volcanic eruptions to soil pits and sunflecks created from single tree-falls. It is within landscapes where we are often challenged, as wildlife researchers or managers, to truly test our understanding of the complexity of wildlife in their native environments.

It is within landscapes where our habitat management designs appear in stark clarity. Research and management designed to understand and alter mechanistically the parts of ecosystems in dissociation from the whole are put to the harsh field test in the full ecological context that is the landscape.

In general, landscape ecology is the scientific study of species, communities, and ecosystems across geographic areas typically defined by hydrological boundaries. As a distinct discipline, landscape ecology is a relatively recent phenomenon, although landscape-level studies have been around for decades. The field of landscape design and architecture, once a discipline focused on design of entirely human-altered environments, has begun to converge with landscape ecology. This convergence has been propelled by two related disciplines. One discipline is that of environmental psychology, which is perhaps one facet of human ecology and which determines the patterns and even evolutionary basis for human preference for various natural or altered environments. The other discipline is that of landscape design of environments as applied to seminatural conditions such as managed forests.

Working on the scale of landscapes—and we will propose some rigor in defining such amorphous terms as *scale* and *landscape*—poses special problems for both researchers and managers. Researchers must contend with studying subjects that usually have no replicates, no controls, and all the environmental noise that complicates data analysis and confounds interpretation. Managers must contend with the plethora of ecological conditions and disturbance events across wide scales of space and time. Landscapes are entirely open systems in the way that rivers, subsurface water, air, and organisms interchange with adjacent regions. Distributions of many wildlife species do not respect the cartographer's or manager's boundaries.

So why do we study complex community and ecosystem processes within landscapes and not just continue to focus research and management on the pieces? The devil is in the details, so to speak, and in the whole system. Wildlife often responds to broad patterns as well as individual patches of resources and environments. Changes in broad patterns of resource patches can insidiously disrupt population functions, which would not become evident by examining merely local expressions of habitat conditions and demographic vital rates of specific species. Populations and habitats alike can "unravel" from their overall tapestry within landscapes and within the communities in which they are found. Sometimes such disruptions occur with lag times that belie the true cause unless we cast our examination to broader scales. And humans greatly affect various aspects of environmental processes when we disrupt or channel flows of substances, nutrients, energy, fluids, and even wildlife, by changing vegetation conditions, topographies, and the soil itself.

Some ecological processes are more "emergent" across landscape-scale areas, and are not evident or cannot be easily understood with smaller-area studies. Such landscape processes include: geological events of formation and wearing down of landforms; hydrological events of surface and subsurface flows; pedological events of soil formation, change, and erosion; and biological events of population dynamics and species exchanges, invasions, speciation, and extinction. How humans occupy landscapes and use resources can have drastic effects on each of these processes.

In this and the following chapter, we explore the relation of wildlife habitat research and management to landscapes. We begin with some definitions. We then discuss heterogeneity of resources and environments within landscapes, including patterns, scales, fragmentation, and measures. Next, we explore responses of wildlife to such heterogeneity, including wildlife use of patchy environments, effects of size and patterns of resource patches on community structure, edge effects, response of populations to resource fragmentation, and some challenging questions about fragmentation management. We then discuss the ecology and dynamics of islands and isolated resource patches in oceanic and continental settings, including interpreting species-area relations. We compare and contrast continental environments and resource patches with oceanic islands. Next, we discuss dynamics of environments in landscapes, including dynamics of resource patches and types of disturbance regimes, followed by a discussion of managing for species-specific habitat corridors and connectivity and the value of remnant patches of natural environments. Finally, we end with a discussion of monitoring environments and providing for wildlife in landscapes.

Definitions and Kinds of Landscapes

Definition and Classification of Landscapes

A *landscape* is ". . . a part of the space on the earth's surface, consisting of a complex of systems, formed by the activity of rock, water, air, plants, animals and [people] and that by its physiognomy forms a recognizable entity" (Zonneveld 1979). By this general definition, landscapes do not necessarily have a specific size, and might be considered on various scales depending on the area over which unique constellations of ecological processes operate to form "recognizable entities."

Although it may not be generally recognized, much work in the traditional field of vegetation synecology provides the basis for this definition. In particular, the Zurich-Montpellier School of Phytosociology in European circles trained plant synecologists to view the land as broadly delineated "entities" composed of repeating patterns of dominant species, topographies, and landforms. The Zurich-Montpellier School taught that such patterns can be identified on the ground or discovered by analyzing field survey data, and represented in maps through the "entitation" process, which seeks common, repeating patterns of similar biotic and abiotic conditions. Eventually, such an approach has become incorporated, consciously or by convergent thinking, into the definitions and classifications of landscapes.

The entitation procedure, whether for vegetation communities or landscape types, nonetheless allows for the individual response of component species (*sensu* Gleason 1926; Grime 1977; also see chapter 3) and elements of ecological processes. For example, a number of species might converge in their use of elevation, aspect, soil condition, and location and appear as a vegetation community entity, even though each species has unique response characteristics to precipitation, temperature, and nutrient regimes. In the same way, landscape entities might contain a common set of elements, including plant and animal communities, ecological processes, and disturbance regimes, which coincide in a particular geographic setting, but each element may extend beyond the area in individual and unique ways. However, species of a community entity often respond to the presence of other species, so the individual response hypothesis can be taken only so far. Similarly, some landscape factors mediate or exacerbate others. One example in grasslands is how weather conditions, coupled with agricultural development, may alter native fire regimes. This may encourage exotic, pioneer plants to invade and dominate a landscape. Dominance by exotics can then alter the course of future fire regimes. Thus, individual components and forces of landscapes do not act in isolation but can be mutually determining.

Although it is not explicit in the definition, the landscape concept may be applied to both terrestrial and aquatic environments including marine systems. We can see that a coastal, estuarine, lacustrine (lake), or even oceanic or marine environment may contain "recognizable entities" or landscapes if we just look at the right elements. For example, cliff-nesting seabirds may be described as inhabiting coastal landscapes consisting of recognizable entities defined by climatic and physiographic components. Pelagic environments used by albatrosses and whales can be described as marine landscapes (or "waterscapes") defined by current cells, air circulation patterns, upwelling patterns, benthic topography, and weather systems, even if we cannot readily see these elements without measuring devices or help from satellite imagery. Much of this chapter, however, will focus on landscapes in terrestrial situations.

Understanding the patterns and reasons for the distribution of species and landscape elements is the challenge of landscape ecology. Thus, a functional definition of *landscape ecology* is the study of the response of organisms, species, ecological communities, and ecosystem processes to each other in spatially heterogeneous and temporally changeable environments across a broad area typically delineated by local or regional hydrological boundaries.

Forman and Godron (1986) broadly classified terrestrial landscapes into six general types on the basis of the degree of human disturbance: natural, managed, cultivated, suburban, urban, and megalopolis. Natural landscapes are perhaps a special condition in that they provide sources or refugia of species sensitive to human disturbance and afford benchmarks from which to measure biodiversity, community structure, population dynamics, and ecosystem processes changed by human intervention. Managed and cultivated landscapes, including managed forests, livestock grazing lands, and agricultural lands, collectively constitute what is now more frequently called seminatural landscapes. Suburban and urban landscapes offer the challenge of providing parks, greenbelts, tree cover, and other elements studied in the discipline of urban wildlife management (Adams 1994). And the megalopolis landscape offers perhaps the greatest challenge to habitat managers as geographic areas in which much of native biota is lost and in which mostly exotic species thrive.

In the twenty-first century, city and county planners may be challenged to redefine the often uncontrolled sprawl of urban and megalopolis landscapes to better incorporate what we have learned are critical environmental elements found in suburban and seminatural landscapes (e.g., Robbins 1991; Beatley 1994). According to the World Resources Institute (WRI 1996), the number of people who live in urban areas is expected to double to more than 5 billion globally between 1990 and 2025. One commonly used measure of urban growth is the "magacity," defined as a city with a population exceeding 8 million. In 1950, just two megacities existed: New York, with a population of 12.3 million, and London, with 8.7 million. By 1990, there were 21 megacities, 16 of them in the developing world. In 2015, there will be 33 megacities, 27 in the developing world (WRI 1996). Over the next several decades, human population growth will be greatest in urban and megalopolis landscapes, particularly in developing nations, although continued metastatic settlement patterns such as in the Amazon Basin also will pose serious threats to conserving the increasingly insular and dwindling natural landscapes of the world. Perhaps one place to look for part of the answer to habitat management in urban landscapes is the provision of city greenbelts.

On Scale

One of the major advantages of the discipline of landscape ecology is the integration of ecological study across various scales of space and time. However, the term *scale* is often used ambiguously in landscape ecology studies (and most others) and begs clarity of definition. In table 8.1 we list six dimensions of scale along

Table 8.1. Various aspects of scale and examples at three levels of magnitude

Aspect of scale	Broad-scale magnitude	Midscale magnitude	Fine-scale magnitude
Geographic extent	Entire major drainage basins, or entire ecoregions ("large-scale" study)	Subbasins or local watersheds, or more local physiographic provinces; large groups of vegetation patches	Areas smaller than subbasins or local watershed; small groups of individual vegetation patches or substrates
Map scale	"Small-scale" maps, e.g., ≥1:1,000,000	"Medium-scale" maps, e.g., 1:100,000	"Large-scale" maps, e.g., 1:24,000
Spatial resolution	Typically coarse-grained, such as for characterization of vegetation and environmental conditions at 1-km^2 pixel size	Environmental patches within subbasins or local watersheds, e.g., ≥10 ha	Fine-grained, e.g., <10-ha patch sizes
Time period	Paleoecological past and evolutionary future	Historic past and approximately one century into the future	Very recent management past and current conditions projected only a few years into the future
Administrative hierarchy	International treaties, such as on biodiversity and plant and animal commerce; also national laws and land management or resource regulations	Individual agency or tribal-specific policies and legal guidelines; allusions to state- or provincial-level agency policies, industry or corporation policies, and private landholder resource management goals	Local management unit-level operational policies, down to project-level operations
Level of biological organization	General abundance of vegetation communities, cover types, and structural stages; mapped locations of ecoregions or ecosystems; with inference to ecological communities and species assemblages	Distribution and abundance of individual species or species groups	Species; gene pools or demes; subspecies or varieties; morphs or ecotypes

Note: The terms *scale* and *large scale* are often used ambiguously in landscape ecology studies, but rigor in definition can bring clarity. The levels of magnitude presented here are intended as general guidelines.

with suggested guidelines for three levels of magnitude of each dimension.

Geographic extent is one dimension of scale. Typically, the term *large scale* is used in landscape studies to connote a large geographic extent. This term is unfortunate, because it is contrary to the cartographically correct use of this term for *map scale,* where "large-scale maps" refer to larger values of map scale ratios and thus to maps that generally cover a small geographic extent. Thus, we suggest replacing the ambiguous term *large scale* with *large geographic extent* or just *large extent.*

Another dimension of scale is *spatial resolution,* which can range from images with coarse-grained pixels or large vector polygons to those with very fine-grained resource patches or point locations of conditions. Whereas geographic extent and map scale are roughly (inversely) correlated—that is, large-scale maps tend to cover smaller geographic extents, depending on the physical size of the map—spatial resolution can be a more independent dimension. That is, a small-scale map (e.g., 1:2,000,000 scale) covering a major drainage basin (e.g., the St. Lawrence Seaway) might be represented by coarse-grained spatial resolution (e.g., 1-km^2 pixels), such as that which is available from many remote sensing satellite images, or by fine-grained spatial resolution, such as vector polygons ≥ 10 ha. Thus, it is important to denote the spatial resolution as well as the geographic extent of a particular map image or a study.

Time period is another dimension of scale that is seldom made explicit. But time is vital to interpreting geographic data and landscape simulations, particularly for disturbance events and response by organisms and populations. Ideally, disturbances should be empirically depicted temporally by their duration and frequency for the best interpretation of recurrent patterns and potential influence from human activities. Also, the time period over which data were gathered for geographic analysis (or for a particular map) in landscape studies is important to report, because some ecosystem processes, including disturbance regimes, may occur beyond the time period studied.

Another dimension of scale, more pertinent to management use, not often explicitly addressed is *administrative hierarchy.* This refers to the breadth of political, social, cultural, or even economic mandates and policies of governments, which may play important roles in some studies or management plans or in interpreting observations. Particularly for management, it could be useful to explicate which levels of organization hierarchy apply. For example, at the broad-scale magnitude, studies of marine "landscapes" might pertain to international fishing treaties beyond the coastal sovereign fishing zones of individual nations and may also involve national policies for protection of marine mammals, fish stocks, or coral reefs. It could be important to be clear which "scales" of legal mandates and resource management policies pertain.

Finally, *level of biological organization* refers to the biological dimension of scale and whether a study or plan pertains to ecosystems, communities, assemblages, species, or more finely defined entities such as gene pools or ecotypes. This dimension could also refer to classification levels of vegetation communities or ecosystems, such as plant associations, vegetation types, and ecoregions. Note that a landscape study or management plan might pertain to a fine-scale magnitude of biological organization such as an inventory of ecotypes, but across a broad geographic extent such as a drainage basin. In this way, the various dimensions of scale may be applied at different magnitudes for a given purpose.

In conclusion, we suggest that landscape studies (and management plans) clearly identify the magnitudes of geographic extent, map scale, spatial resolution, time period, organizational hierarchy (if appropriate), and levels of biological organization addressed and evaluated. In this way, much confusion over terms and methods can be avoided.

Scales of Disciplines and Management Issues

Another useful way to consider scale is to think of a "zoning map" showing time and geographic extent pertinent to scientific disciplines, as shown in figure 8.1. Mapped out in this figure are "zones" of various scientific disciplines useful in landscape ecology studies. For example, in the upper left corner, one would study biogeography of animals (zoogeography) and plants (phytogeography) roughly over a time period of the recent past to the distant geological past, and over a spatial extent of roughly 10^6 to 10^8 ha (certainly, this can vary depending on the study). Study of ecological biogeography might extend from the recent past to the near future and perhaps to small spatial extents, say, down to 10^5 ha. And researchers of climate change are interested in projecting global or regional climatic and biome trends from the present to perhaps centuries or a millennium in the future over roughly the same spatial extent.

In the next tier down, studies of paleoecology and evolutionary ecology typically pertain to ≥10,000 years in the past, whereas late prehistoric paleoecology (e.g., of Holocene events) and early historic ecology may peer back only a century to 10 millennia, over a spatial extent of perhaps a 100 to 1 million ha, depending on the specific study. The analysis of population viability, including stability of metapopulations (see chapter 3), might pertain to hindcasting a few decades to a century into the past and projecting upwards of a millennium into the future, although most viability studies project to only decades or a century or so; and the spatial extent is slightly less than that of paleoecology studies, again depending on the range of the species and environments studied.

Studies of population ecology and species life history typically are concerned with the recent past and the very near future, perhaps on the order of just a few decades and on more localized geographic extents. Studies of ecological succession usually are site-specific but can address conditions a century or so into the past and a century or so into the future. Finally, soil genesis typically occurs or is studied mostly on a site-specific basis but about a millennium or more into the past, whereas prospects for soil renewal may not begin for centuries into the future. Certainly, other disciplines also can be added to the zoning map.

This type of space-time zoning map may be useful for identifying realms of duration and geographic extent over which managers should address issues of planning effects. This is shown in figure 8.2, where ecosystem management issues are plotted in correspondence to the disciplines shown in figure 8.1. Beginning in the upper left corner, at distant-past time periods and areas of large geographic extent, the manager might need to understand geographic origins of species, including centers of origin, centers and routes of species spread, and the role of refugia. More than for academic interest, such understanding can greatly help identify locations of evolutionary significance deserving conservation consideration. An example can be found in the complex island archipelago of southeastern Alaska and northern British Columbia, where unique subspecies and species

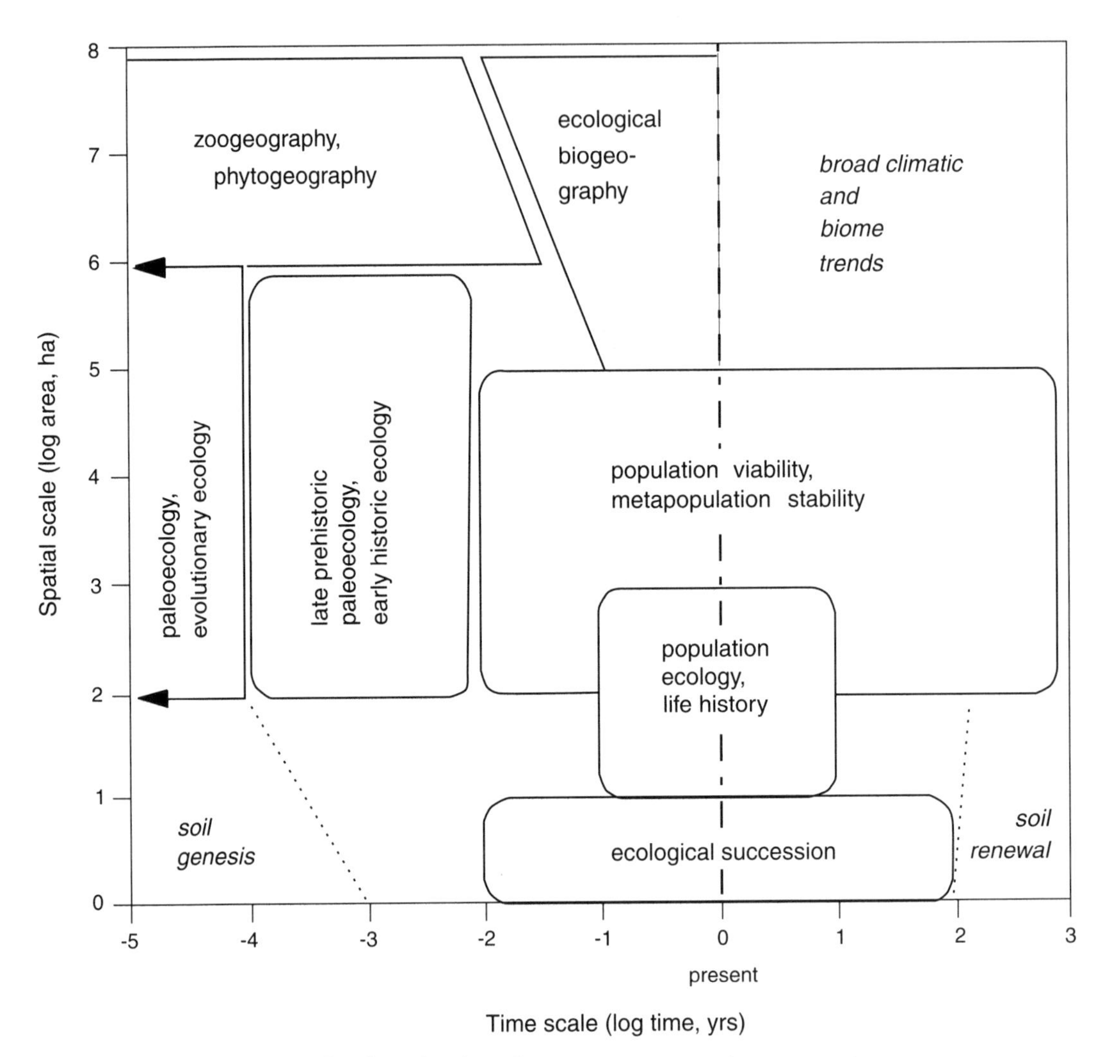

Figure 8.1. A zoning map of ecological scale and species assessment, plotting spatial area against time (note log axes). This figure illustrates the spatial and temporal extents of various disciplines that may be useful in wildlife habitat management and land use planning.

have evolved and paleoecological conditions have brought large carnivores into sympatry. If ecosystem managers are concerned with representing the range of current broad-scale conditions, then they will look to the recent past and near future across wide geographic areas. And still across a wide geographic extent, the manager might peer into the near future (a century to a millennium) to project climatic or biome effects when considering potential influences of climate change, acid precipitation, and ozone depletion.

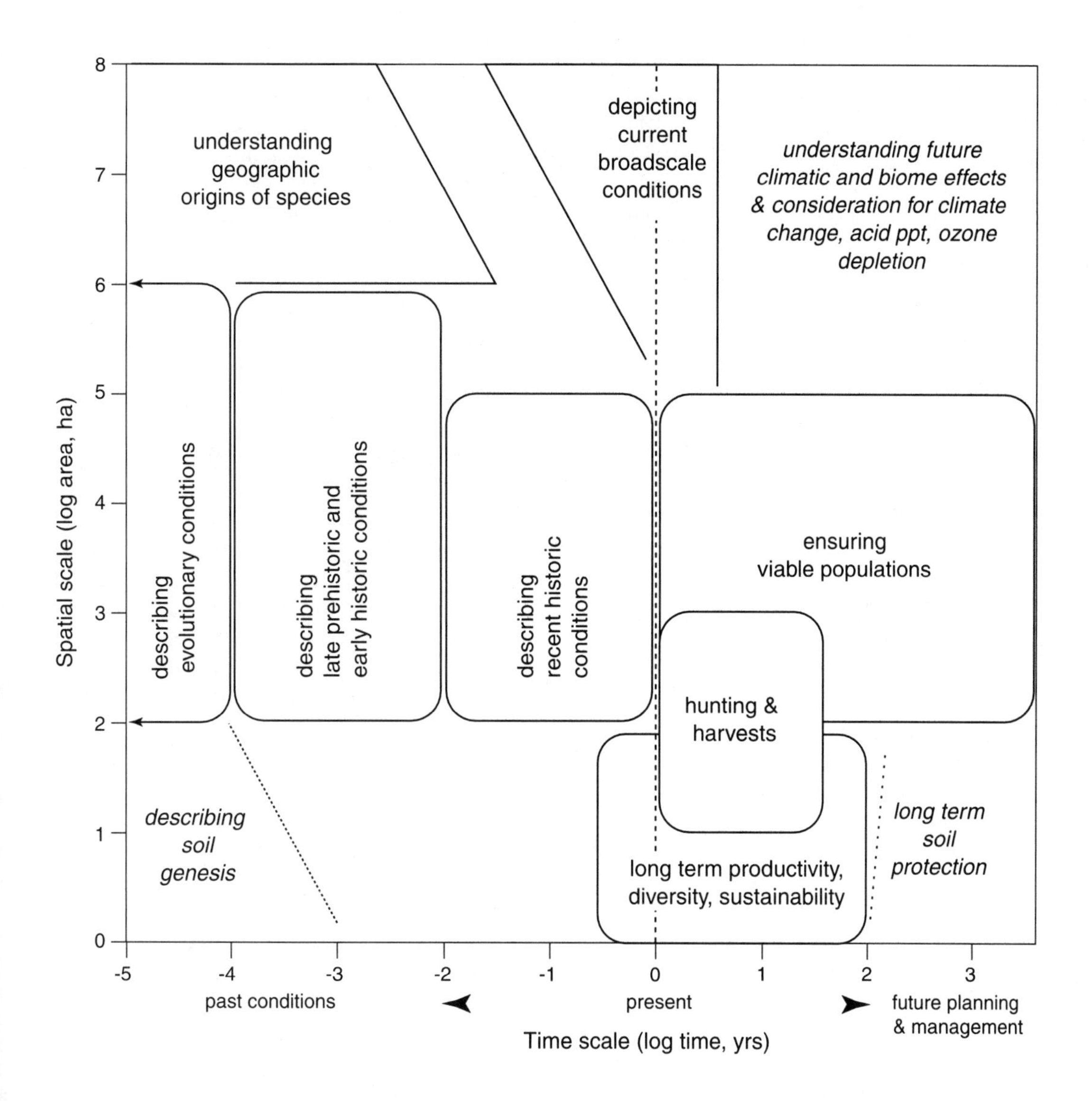

Figure 8.2. A zoning map of ecological scale and ecosystem management issues, plotting spatial area against time (note log axes). Compare with figure 8.1; this figure suggests some of the issues that managers may wish to address in wildlife habitat plans.

Over a lesser geographic extent, the manager might need to describe evolutionary, late prehistoric, early prehistoric, or recent historic conditions. This would be useful for determining the range of natural conditions in an area and whether historic conditions truly represent environments in which species persisted over the long term or even evolved (see chapter 2). The goal of ensuring viable populations would prompt the manager to look into the future (on the order of a century to a millennium), if projections allow, and over a geographic extent of perhaps up to half a million hectares or so, depending on the environment and species of interest. Concern for near-term levels of the harvest of game animals should be nested within broader concerns for maintaining long-term harvestability and viability of the target species, but these concerns pertain only to the near future and over small geographic areas. Management of long-term productivity, diversity, and sustainability of ecosystems can draw from all these scales and issues. Description of soil genesis is a site-specific issue that looks back a millennium or more in the past, and description of long-term soil protection should look centuries to millennia into the future.

Last, figure 8.3 illustrates the temporal and spatial extents of typical land management plans and the degree to which plans might study the past and project future cumulative effects. Most land resource management plans are designed to operate for perhaps a decade. Future plan revisions would come later as part of the subsequent planning cycle. But even if the formal duration of a plan is short, one should peer back in time and project into the future to understand historical conditions better and predict future, long-term effects, lest we continue to chip away at our resource base by what may be termed the long-term tyranny of short-term actions. Comparing figure 8.3 with figures 8.1 and 8.2 suggests some management issues that ought to be addressed at each planning level and the scientific disciplines that should be brought to bear in informing managers about those issues.

As an example, in the United States, a plan developed for an entire national forest or grassland (the second solid box down from the top in fig. 8.3) may be in formal effect for a decade or so and pertain to an area on the order of 10^4 to 10^5 ha. Issues that the plan would proximately deal with (fig. 8.2) might include ensuring viable populations and to some extent representing the full range of ecological or environmental conditions. However, in constructing the plan, managers should peer into the past and project cumulative effects into the future, perhaps a millennium or so each way. In so doing, for the past they need to describe late prehistoric, early historic, and recent historic conditions by use of paleoecology and early historic ecology (see chapter 4 for a brief discussion of retrospective studies). To ensure future viable populations they might employ population viability analysis (see chapters 3 and 10).

Further, such an approach by itself would not constitute an adequate ecosystem management plan, just as issues addressed by local project plans alone do not address all ecosystem management needs. It is our view that "ecosystem management" is nothing less than a full set of policies or plans tiered across all geographic extents (the vertical column of solid boxes in fig. 8.3), which account for past and current conditions and future cumulative effects of human activities. The full set of policies and plans must embody consistent implementation standards, management goals, evaluation criteria, and inventory and monitoring activities across all geographic extents and administrative hierarchies from national policies to local projects. In this way, all issues and conditions, including past conditions and future

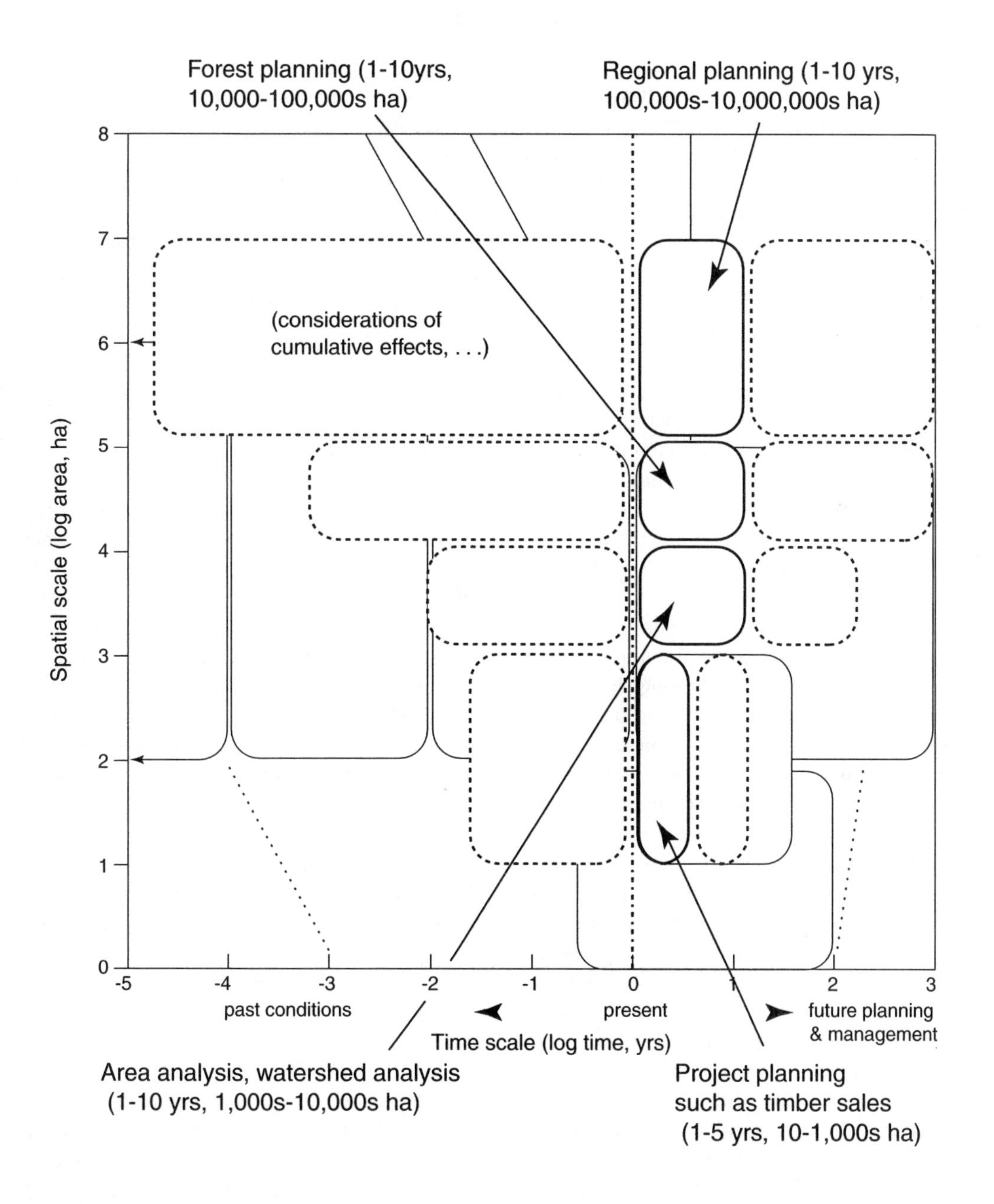

Figure 8.3. The spatial and temporal extent of four levels of habitat or land management planning: regional, administrative unit (e.g., national forest), area or watershed, and project. Solid boxes denote the typical spatial and temporal extent explicitly addressed at each planning level. Dotted boxes denote the extent of past or future conditions that should be taken into account for each planning level as consideration of historical and future cumulative effects. Compare with figures 8.1 and 8.2 for overlap with ecological considerations and management issues.

effects, can be consistently treated and monitored across the full breadth of management issues and cumulative effects of past, current, and future activities. (We develop the concepts of ecosystem management further in chapter 11.)

This discourse on planning on multiple scales illustrates that human influence as well as natural disturbance regimes can alter the types, spatial patterns, and development of wildlife habitats. For this reason, it is vital to understand the influence of habitat heterogeneity on wildlife, explored in the next section.

Patterns of Habitat Heterogeneity

The heterogeneity of resource patches in landscapes has been discussed by different authors in various ways (table 8.2). Among the aspects of resource patch heterogeneity are patch type richness and diversity, patch dynamics, patch connectivity, patch isolation, fragmentation of environments, vegetation corridors, edge effects, edge contrast, and edge permeability.

We define *habitat heterogeneity* as "the degree of discontinuity in environmental conditions across a landscape for a particular species." Environmental conditions can include vegetation composition and structure, as well as more dynamic flows of energy, nutrients, resources, and fluids (water and air). Discontinuities can occur as *ecotones,* relatively sharp breaks in environmental conditions, or as *ecoclines,* broader gradations in conditions. Discontinuities can occur naturally, as with changes in soil type or presence of water bodies, or artificially, as with plowed grasslands or burned forests.

Fragmentation refers to the degree of heterogeneity of habitats, usually vegetation patches, across a landscape, particularly related to isolation and size of resource patches. Since it refers to habitat, fragmentation is necessarily a species-specific condition. In the literature, the term *fragmentation* has typically and often implicitly referred to cutting of old or native forests, although the term should also apply to heterogeneity of any kind of habitat, as caused by human or other means. Also, the literature often refers to landscape fragmentation; this term is strictly incorrect, because it is environments or resources (habitats for specific species) that become fragmented within landscapes, not entire landscapes per se.

Various kinds or degrees of species-specific habitat heterogeneity can be described. In the extreme case, resource or vegetation patches can be isolated as islands surrounded by vastly different conditions. The response of species and communities to island situations has been the subject of much ecological study and is discussed in detail below.

Partial isolation of habitats and environments may pose greater challenges in research and management than islands do. Partial isolation may have a gradient effect on the viability of metapopulations (see chapter 3), such as by incrementally lowering a population's crude density (numbers of animals of a population per unit area of unsuitable and suitable environments in a landscape) and by lowering the effective (expected interbreeding) size of populations, even though the absolute number of animals in an area may remain the same.

Another kind of heterogeneity is temporal fragmentation, sometimes called *ecological continuity.* This refers to the degree to which a particular environment, such as an old forest, occupies a specific area through time. Even if an old forest ecosystem is allowed to regrow after having been interrupted, such as with widespread forest conversion or cutting, much of the original species closely associated with

Table 8.2. Types of environmental heterogeneity in landscapes

Type	Description	Example	Source
Patch type richness and diversity	Number and relative area of habitat types for a species within a landscape	Number of patches of different types affect species richness, diversity, and numeric dominance in a community	Kitching and Beaver 1990
Patch dynamics	The incursion and melding of patches over time as a function of disturbance events and successional growth of vegetation	Distribution of vegetation patches over time as affected by stand-replacing fires and subsequent regrowth	Wu and Loucks 1995
Patch connectivity	Degree of adjacency of patches with similar conditions in a landscape	Connectivity of fencerow edge habitat in Ohio farmlands differentially affecting wildlife with different long-range dispersal capabilities	Demers et al. 1995
Patch isolation	The distance from one type of patch to the next nearest patch of the same type	Isolated patches less often colonized by species that do not disperse easily through unsuitable environments, such as Bachman's sparrow (*Aimophila aestivalis*) in South Carolina pine woodlands	Dunning et al. 1995
Fragmentation	The breaking up of contiguous environmental patches into smaller and more disjunct or isolated patches	Fragmentation of grasslands differentially affecting bird species with varying sensitivity to habitat area, in Illinois; differential effects of forest fragmentation on population viability of arboreal marsupials in Australia	Herkert 1994; Lindenmayer and Lacy 1995
Corridors	Linear arrays of environments in a landscape	Movement corridors in undisturbed riparian woodland for cougars (*Felis concolor*) to support populations in Santa Ana Mountain range of southern California; riparian woodland and shelterbelt corridors in North Dakota supporting populations of migratory birds; strips of rain forest supporting populations of insects in Australia	Beier 1993; Haas 1995; Hill 1995

Table 8.2 Continued

Type	Description	Example	Source
Edge effect	Incursion of microclimate and vegetation into a patch, typically forested, from a disturbed edge or opening	Clearcuts causing reduced tree stocking density, increased growth and reproduction of dominant trees, higher tree mortality, and incursion of warmer, drier microclimates into adjacent old-growth forests in the Washington Cascade Mountains	Chen et al. 1992, 1995
Edge contrast	The degree of difference in vegetation structure between two adjacent patches	Great contrast in daily average air and soil temperatures, wind velocity, short-wave radiation, and air and soil moisture differ significantly between clearcuts and old-growth forests in the Washington Cascade Mountains	Chen et al. 1993
Edge permeability	The degree to which wildlife will travel through edges and among patches	Dynamics of modeled populations affected by types of habitat patch edges in relation to dispersal and movement capability	Stamps et al. 1987

Note: Most examples cite simulations of populations or combinations of field observations with simulations. Much empirical work remains to be done on most types of heterogeneity listed here.

such environments may nonetheless be lost. Thus, regional and site histories are important to interpreting community composition and species occurrence. Some research on the potential problems of ecological continuity, particularly of old forests, has been done in Europe, although the concept is still relatively new in land management in North America. European researchers have discovered that vascular plants have differential adaptations to the degree of ecological continuity in their taiga habitats of Scandinavia (Delin 1992). Some cryptogams (lichens and bryophytes, including mosses) are adversely affected by temporal disruption of old forest conditions, and thus can serve as indicators of ecological continuity of such environments (Tibell 1992; Selva 1994). Much work remains to be done in determining sensitivity of wildlife species to ecological continuity both within resource patches and across landscapes.

Heterogeneity and fragmentation can also refer to subtle discontinuities in environmental conditions other than gross vegetation structure. One example is the horizontal separation of vegetation within a stand, such as among canopies of large trees. This kind of fragmentation has sometimes been called within-stand patchiness, or alpha-diversity of vegetation structure (Kitching and Beaver 1990). This kind of fragmentation in forests would probably adversely affect arboreal-dwelling species requiring contiguous canopy structures, such as some primates and rodents that rely on complex, three-dimensional runways through forest canopies. Species that are likely to require such conditions include liontailed macaques (*Macaca silenus*), a highly endangered primate of wet evergreen *shola* rain forests in southern India; red tree voles (*Arborimus longicaudus*) of western North American conifer forests; and Indian giant squirrels (*Ratufa indica*) of deciduous and moist evergreen forests of peninsular India. Other species that require dense forest canopy conditions but are perhaps less vulnerable to forest gaps and canopy separation may include Nilgiri langurs (*Trachypithecus johni*) and common giant flying squirrels (*Petaurista petaurista*) of south India and northern flying squirrels (*Glaucomys sabrinus*) of North America. (Doubtless, other species can be added to these lists.) The effect and the number of species that would be included on these lists depend on the degree of facultative or obligate use by the species (see discussion in next section).

Another poorly studied and subtle aspect of fragmentation is the vertical separation of vegetation layers such as forest canopies and understories. The correlation of the degree of heterogeneity of vertical forest stand structure with bird species diversity is well known (chapter 3; Anderson et al. 1979; Pearson 1971). This degree of heterogeneity may also influence the use of stands and landscapes by forest-dwelling raptors that fly and forage below the canopy, such as broad-winged hawks (*Buteo platypterus*), northern goshawks (*Accipiter gentilis*), and some forest owls. Vegetation layer diversity and separation can be greatly affected by vegetation management, such as silvicultural thinning of forests and burning of woodlands and grasslands.

Huston (1994) posited that physical heterogeneity of the environment can interact with species competition and the mobility and size of organisms to determine species diversity (fig. 8.4). In this model, spatial heterogeneity is less effective in preventing competitive exclusion among mobile organisms than among sessile ones. Thus, species diversity is highest in environments with high physical heterogeneity and with species of low mobility or small body size and low competition intensity.

Response of Wildlife to Habitat Heterogeneity

Habitat heterogeneity can be described by the geometry of patch patterns for specific species. In oceanic and continental settings alike, individual environmental patches can be described as islands, peninsulas, isthmuses, or as part of the general landscape matrix. Patterns of multiple patches can be described by use of many indices (table 8.2; O'Neill et al. 1988; Turner and Gardner 1990; also see chapter 10).

However, it seems that our ability to depict and measure mathematically the various aspects of heterogeneity patterns, including fragmentation, has outstripped our empirical understanding of wildlife response. Our understanding derives from a mixture of presumption from observational studies, mathematical modeling, and, far less commonly, statistical inference from rigorously applied experimental studies in the field.

Perhaps one root of studying heterogeneity of environments can be found in studies of foraging behavior. Studies of optimal foraging theory traditionally have assessed how organisms select resource patches for foraging and have related return in prey taken to time and energy expended in its location (e.g., Krebs et al. 1974; also see chapter 7). Foraging energetics can determine species presence in an area and overall community structure. For instance,

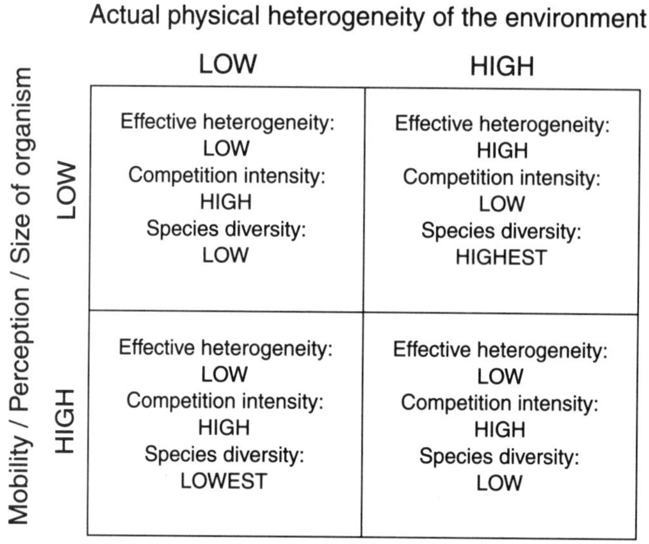

Actual physical heterogeneity of the environment

Mobility / Perception / Size of organism	LOW	HIGH
LOW	Effective heterogeneity: LOW Competition intensity: HIGH Species diversity: LOW	Effective heterogeneity: HIGH Competition intensity: LOW Species diversity: HIGHEST
HIGH	Effective heterogeneity: LOW Competition intensity: HIGH Species diversity: LOWEST	Effective heterogeneity: LOW Competition intensity: HIGH Species diversity: LOW

Figure 8.4. The effects of species mobility, physical heterogeneity of the environment, and competition on species diversity. Effective heterogeneity is the combination of the first two elements. Mobility can also be depicted as the ability of organisms to perceive their environment and might be crudely correlated with body size. (Reproduced from M. A. Huston, *Biological Diversity: The Coexistence of Species on Changing Landscapes* [Cambridge: Press Syndicate of Cambridge University © 1994], 95, fig. 4.4, with the permission of Cambridge University Press)

Maurer (1990) evaluated the utility of optimal foraging theory for determining the structure of insectivorous bird communities.

A rich literature exists on the effects of resource heterogeneity on population dynamics and community structure. The literature has amply demonstrated that habitat area and distance from colonizing species pools affect the number of species occurring in specific habitats and the likelihood of occupancy by individual species using the habitat. Much of the literature suggests that species' attributes—mainly, dispersal ability, demographics, body size and correlated home range size, and degree of habitat specialization—greatly influence how habitat distance and area affect individual fitness and population demography.

One aspect of this that has garnered much study is the influence of isolation on population viability and community structure. The effects of isolation depend upon the degree of habitat specialization and the use of intervening environments by organisms. Often, assumptions are made about species' obligate use of particular environments, such as forest interiors, edges, and old-growth forests, with scant empirical evidence. The real world is often more complicated, as we discuss next.

What Is an Obligate?

Applying principles of island biogeography to patches of environments within landscapes often begs the assumption that some species can be identified as obligate users, that is, species

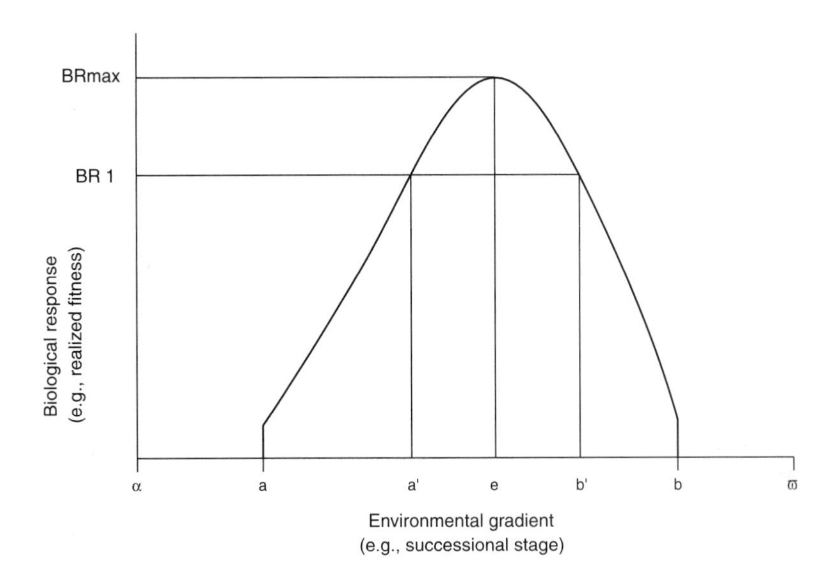

Figure 8.5. Generalized depiction of a species' biological response to an environmental gradient. See text for description of points.

that highly specialize in the use of a particular environment or structure. Many assessments of habitat fragmentation, for example, have assumed that wildlife species can be identified as obligate users of forest interiors, forest edges, old-growth forests, or other environments or landscape conditions. While in many practical ways this is true for some species, and simple assumptions are most parsimonious, the question is more complicated than may first appear and bears closer inspection.

The response pattern of individual organisms to environmental continua, such as temperature or vegetation structure along a sere, is typically some bell-shaped curve (fig. 8.5). Assuming that in this figure "response" is something like realized fitness (or some index thereto), the high point of the curve (point e) marks the best environmental condition for maximum fitness (BRmax), and the two trailing edges of the curve suggest increasingly unsuitable conditions. The maximum potential range of the environmental gradient in the ecosystem also can be depicted (points α and ω), even if not used by the organism. Maximum ranges of tolerance by the organism are marked by truncation of the curve at each side (points a and b, but there may be circumstances where a = α and b = ω). In some cases, the truncation may be more abrupt or the curve may be asymmetric, but the basic pattern still holds.

Traditional ecological principles that have used this fundamental precept include Shelford's law of tolerance and Leibig's law of the minimum, as well as measures of Hutchinsonian niche breadth, overlap, and community structure. This curve probably differs in specific shape for different species in the same community and for individuals of the same species in

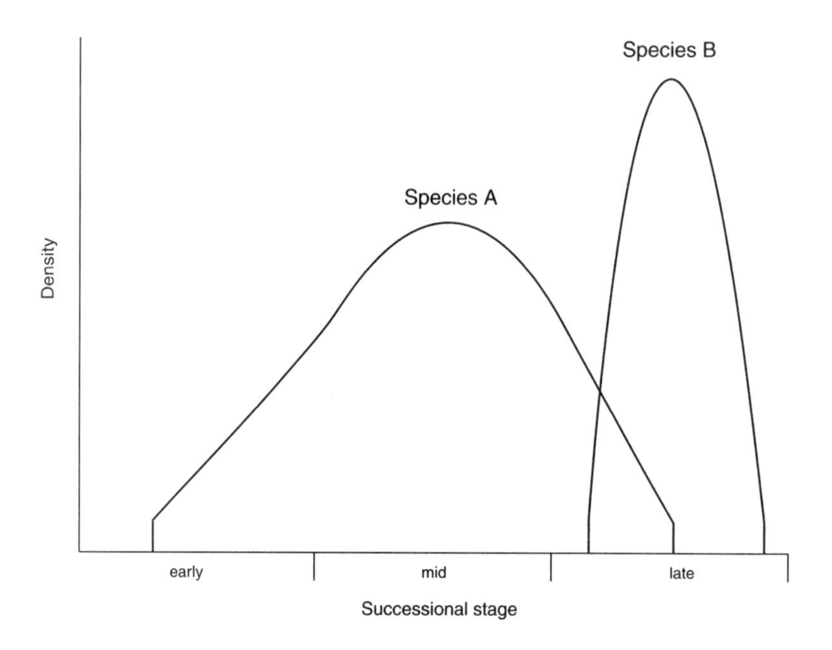

Figure 8.6. Generalized depiction of two species' biological responses to successional forest stages. Species A is a generalist, and Species B is a specialist.

different ecosystems. Moreover, the curve presented here represents only a single organism; biological response of all members of a population would be represented by tolerance intervals spanning each side of the curve.

So what is an obligate? The most common view of an obligate species is one defined by a species' narrow (stenotopic) overall use of some resource. That is, the species selects environments with only a specific narrow range (a–b). In one sense, every species is an obligate of some range of conditions along some environmental gradient(s). The term, though, needs quantitative rigor to be applied consistently. Most intended uses of the term refer to a range of tolerance *far smaller* than the available range of the environmental gradient. *Far smaller* can be defined as "a specific proportion of the overall range," such as 10 percent or less (that is, $[a–b]/[\alpha–\omega] \leq 0.10$; or, with discretely defined environmental conditions, the species uses $\leq$10 percent of available or potential conditions).

Two ranges of tolerance are depicted in figure 8.6. Species A has a broad tolerance for an environmental gradient (is eurytopic), in this case let us say successional forest stages, whereas Species B has a narrower tolerance (is stenotopic) for only one stage. In this example, Species B is an obligate user of late successional stages, whereas Species A uses several stages and is thus only a facultative user of the late stages. Species B is shown with a taller curve under the precept that at least some habitat specialists might have a greater realized fitness (density being an index to fitness, although this could be misleading) than that of generalists in the same environment, although this is not necessarily the rule. Species B could be naturally rare and always have densities lower than Species A.

The problem is that not many species may demonstrate this simple pattern of absolute adherence to a single environmental condition, such as successional stages, for the range of environmental conditions typically considered in landscape ecology studies. That is, many stenotopic species will also make at least facultative or opportunistic use of other conditions. An example is a forest interior species or an old-growth forest species that also uses edges or earlier successional stages if such edges or stages contain some specific elements of older forests (although in their study of birds in northeastern U.S. forests, DeGraaf and Chadwick [1987] strictly defined stand condition obligates and forest type obligates as species using only one condition or type). Such edge or early successional conditions may be less than optimal (biological response < BRmax) for species that reach their greatest abundance in forest interiors or old-growth forests. But edge or early successional environments can still serve to provide the species with some degree of connectivity, across landscapes, between nodes of more optimal habitats. In real-world management, such "exceptions" may be the more useful and interesting cases.

For instance, in a study of forest birds, one of us (Marcot 1985) found that mature forest species such as the brown creeper (*Certhia americana*) and hermit warbler (*Dendroica occidentalis*)—sometimes thought of as mature forest obligates—also occurred, but in lower numbers and with greater variability, in mid-successional forest stages where a few (≥7/ha) scattered old Douglas-fir (*Pseudotsuga menziesii*) trees or snags remained in the overstory (fig. 8.7). In the strict interpretation, neither of these species is an absolute obligate user of mature forest, although their greater mean density and lower variability in mature forest strongly suggest that this environment provides their optimal conditions and that the species should not be called forest habitat generalists. Note too that the pattern of mean density in each of these species roughly matches the general pattern of Species B in figure 8.6.

This example also demonstrates the interplay between three complementary metrics of biological response illustrated in the data shown in figure 8.7: (1) Mean density reaches its greatest value in the best environment (mature forest stage), although it could be nonzero in other, suboptimal stages. (2) The coefficient of variation (CV) of density among replicate study plots is lowest in the best environment. (3) The percentage of occurrence (PO) of each species among replicate study plots is greatest in the best environment. The data on the brown creeper (fig. 8.7a) show these traits well. The data on the hermit warbler (fig. 8.7b) also show these patterns, although, if one were to inspect just the PO patterns alone, one would conclude that pole and medium tree stages are equally suitable, whereas the data on mean density and CV suggest an incrementally better environment in the medium tree stage (greater mean density and lower variation in density than in the pole tree stage).

Presence of these two species in early successional stages of this study, although in lower numbers, may represent some occasional or intermittent use for dispersal, nesting, or foraging, but only more intensive autecological studies on reproductive response would reveal the specific kind of use. Nonetheless, such early successional stages are used (especially if they contain old-forest components), which could

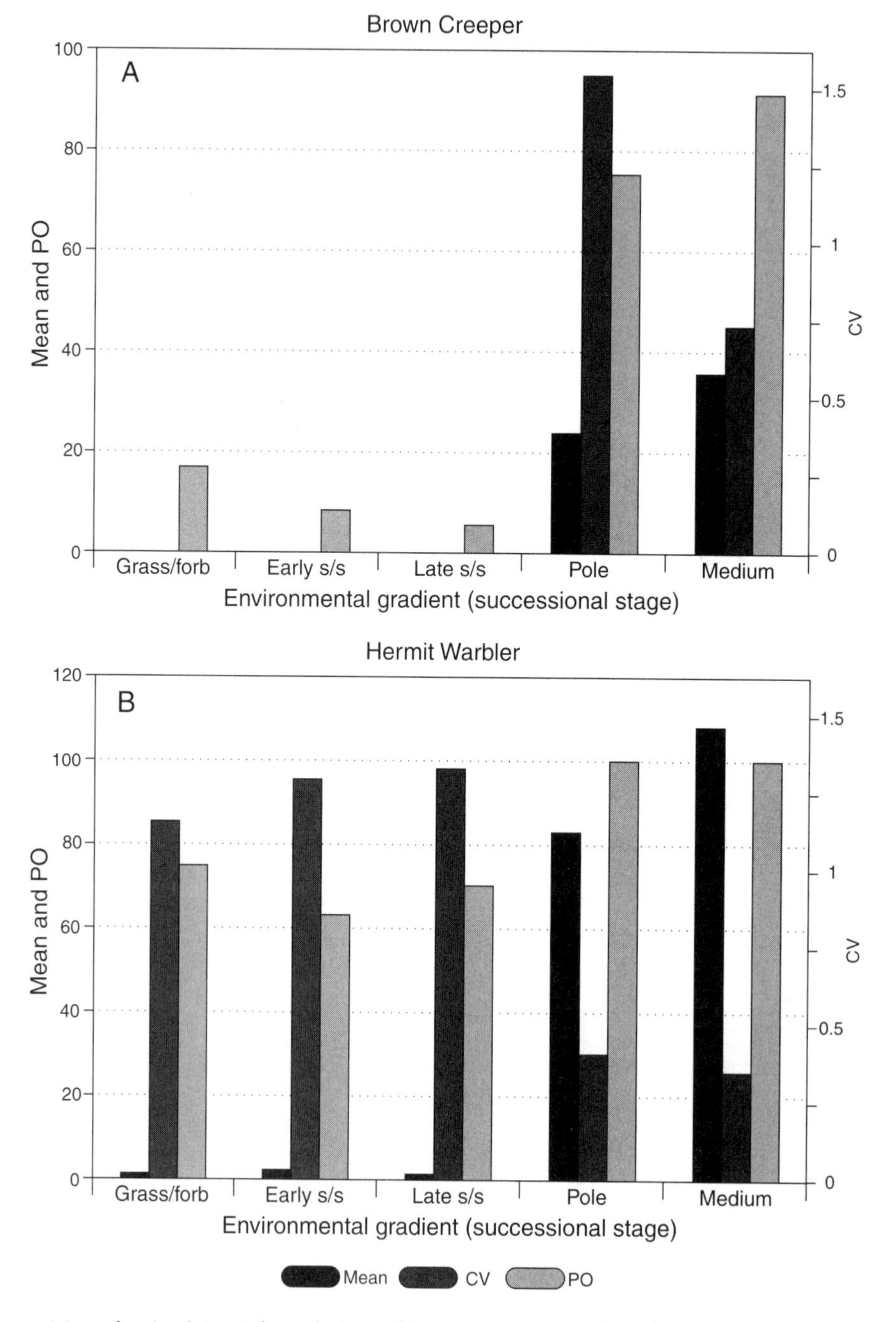

Figure 8.7. Mean density (*n*/40.5-ha index), coefficient of variation (CV) of density among replicate study plots, and percentage of occurrence (PO, percentage of replicate study plots occupied) of (A) brown creepers (*Certhia americana*) and (B) hermit warblers (*Dendroica occidentalis*) among five successional stages of Douglas-fir (*Pseudotsuga menziesii*) forest in northwestern California. Stages are: grass/forb, early shrub/sapling, late shrub/sapling, pole, and medium tree. (From Marcot et al., in press)

contribute to at least some degree of habitat connectivity for these "mature forest" species (and others like them) across a landscape. Investigations of density, variation in density, and percentage of occupancy in early successional stages with and without old-forest components can help reveal which elements are important and can help in developing silvicultural guidelines for young forest management (Marcot 1985).

This example also serves to illustrate that degree of specificity of use defines the level of obligate relationship. A more fluid definition of obligate use might allow for some threshold biological response to be identified (point BR1 in fig. 8.5), for which a specific range of an environmental gradient can be correlated (points a′ and b′). In this case, the threshold could be identified as some percentage of maximum possible response (e.g., BR1 = 80 percent of BRmax), and *far smaller* (as mentioned above) would now be defined on the basis of this response range (that is, $[a'-b']/[\alpha-\omega] \leq 0.10$, or the species uses ≤10 percent of available or potential discrete conditions). In this way, species can be identified as having varying degrees of obligate relation to specific environmental conditions.

A simple example is shown in figure 8.8. Plotted here is the cumulative number of terrestrial vertebrate species that use old forest structural stages of the interior Columbia River basin, in the northwestern United States (based on Marcot et al., in press), as a function of the

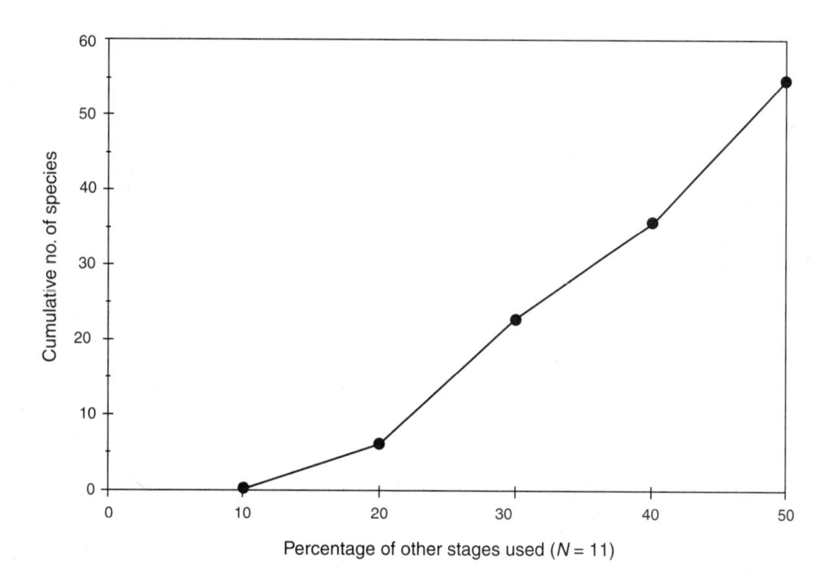

Figure 8.8. Cumulative number of terrestrial vertebrate species (amphibians, reptiles, birds, and mammals) that use old forest structural stages as a function of the percentage of all other vegetation structural stages used, in the interior Columbia River basin. For example, 36 species use old forest structural stages and about 40 percent of all other structural stages (i.e., four other stages). (Based on data in Marcot et al., in press)

percentage of all other vegetation structural stages used (up to 50 percent of them). The figure shows that no species occurs solely in old forest structural stages. Six species occur in old forest structural stages, but each uses about 20 percent of all other stages as well (i.e., two other stages). An additional 17 species use old forest stages and about 30 percent of all other stages too. Thus, there is no species that specializes only in old forest stages, and there is an increase in the number of species with increasing generalization in other vegetation stages. The cutoff point for "obligate" old forest species used by Marcot et al. (in press) was the 20 percent level, but this was a subjectively derived threshold for comparing the degree of specialization of species among other vegetation structural stages and vegetation types.

In summary, use of environmental conditions—be they vegetation types, structural or successional stages, substrates, or other conditions—can be viewed as a gradient of specialization to generalization. In identifying obligate users of some condition, the range of tolerance of individual organisms, the mean and variation in use among organisms, and the use of complementary indices such as mean density, coefficient of variation of density, and percentage of occurrence, can all provide useful information (Ruggiero et al. 1988). Recognizing at least facultative use of some environmental conditions can be quite helpful in devising management guidelines for suboptimal, "connectivity habitat" to better link optimal habitats or environmental conditions across a landscape.

Edge Effects and Changing Perspectives

The ecological value of ecotones and ecoclines, as with other aspects of environmental heterogeneity, depends on the management objective. For example, an objective may be to maintain native species and communities in naturally continuous forest environments. Then, introduction of forest openings, as with patch cutting, runs the risk of invasion by exotic plants (Fraver 1994), invasion by nest parasites such as brown-headed cowbirds (*Molothrus ater*), or increases in avian predators (Andren and Angelstam 1988; Angelstam 1986; Paton 1994; Rudnicky and Hunter 1993a; Yahner et al. 1989). However, if the objective is to provide some early-seral vegetation to enhance big game, then some proportion of the landscape in openings may prove beneficial (Krefting 1962; Kremsater and Bunnell 1990; Lyon and Jensen 1980). It is instructive to trace the history of how researchers and managers have addressed changing objectives for habitat edges and fragmentation.

Going back at least to Leopold's *Game Management* (Leopold 1933), we see that ecotones and edges between vegetation types have long been recognized as areas of particularly high wildlife concentration. Some early studies on edges include Petrides' (1942) work on the relation of hedgerows to wintering wildlife in New York, Johnston's (1947) work on breeding birds of forest edges in Illinois, and Sammalisto's (1957) study of birds along woodland–open peatland edges in south Finland. Prior to the 1980s, edges and woodland openings were viewed in the literature as largely beneficial to wildlife, with little discussion of potential adverse effects. As early as 1938, Lay posed the question, How valuable are woodland clearings to wildlife? and concluded that they enhance game populations. Kelker (1964:180) reaffirmed Leopold's "accepted ideas" on wildlife management, including the "laws of interspersion and dispersion of habitat features."

In the 1970s, much excellent work was done in determining the use of forest-opening edges for game, principally deer and elk (Thomas et al. 1979; Ffolliott et al. 1977), and in providing silvicultural and forestry management guidelines to enhance such conditions (Black and Thomas 1978; Edgerton and Thomas 1978; Hall and Thomas 1979). However, also during that decade, some researchers were beginning to determine the influence that logging and land use can have on fragmenting environments and adversely influencing wildlife populations and communities. For example, Whitcomb et al. (1977) studied long-term turnover of species in forest fragments created by logging; Robbins (1979) discovered adverse effects of forest fragmentation on songbird populations; and Simberloff and Abele (1976) posed how refuges could be designed on the basis of the use of island biogeography theory to offset adverse effects of fragmenting environments.

By the 1980s, for several reasons serious challenges were being made to the assumption that edges and openings are "by law" beneficial to wildlife (Harris 1984). One, in wildlife biology and management there was a growing emphasis on nongame species, some of which are not necessarily benefited by edges, openings, and fragmentation of environments. Two, studies suggested that habitat patch size and isolation, as affected by fragmentation, can greatly influence the presence of some species and the composition of wildlife communities (e.g., Freemark and Merriam 1986; Lehmkuhl 1990). In one sense, this was anticipated by Leopold's (1933) law of habitat interspersion and dispersion, but now the emphasis was on negative correlations with wildlife populations. Three, the acceleration in exploitation of native environments, including ancient forests in western North America and rain forests in the tropics, prompted great concern for the survival of area-sensitive habitat specialists. Also, some researchers were determining that habitat heterogeneity affects some wildlife species in prairie and grassland environments (Wiens 1974) as well as in forests. Far from the previous view of its benefits to game animals, fragmentation was being heralded as the great bearer of species extinction (Wilcox and Murphy 1985).

Concern also turned to the proportion of landscapes that were being altered by human activities, particularly for urbanization, agriculture, grazing, and timber harvest. The small woodland openings previously touted by Lay and the rich habitat edges described by Johnston and Sammalisto were now being seen as dominating some native grassland, woodland, and forest landscapes in boreal, temperate, and tropical regions alike (e.g., Franklin and Forman 1987). Research attention in some areas turned to the need for conserving remnant patches of native vegetation for area-sensitive and rare species and declining native communities (e.g., Howe et al. 1981; Saunders et al. 1987; Mills 1995).

In light of such research, Reese and Ratti (1988) and Guthery and Bingham (1992) raised the need to reevaluate the traditional concept of edge. Soon the conceptual pendulum swung swiftly the other way, and many researchers and managers came to equate fragmentation with loss of native environments, mostly old forests, due to human exploitation. Of course, measures and studies of fragmentation and heterogeneity should equally apply to other environments and conditions, including prairies (Burger et al. 1994) and clearcuts and other disturbed areas (Marcot 1985; Rudnicky and Hunter 1993b).

In the 1990s, a great deal of research has focused on the mechanisms of environmental fragmentation and species extinction of a wide variety of plant, invertebrate, and vertebrate taxonomic groups. Research has also focused on the effects of land management activities on fragmentation and species response. Examples include studies of effects of forest fragmentation on the pileated woodpecker (*Dryocopus pileatus*) (Aubry and Raley 1990), capercaillie grouse (*Tetrao urogallus*) (Wegge et al. 1992), and northern spotted owl (*Strix occidentalis caurina*) (McKelvey et al. 1993). Management has generally followed suit and addressed means of providing for species-specific habitat corridors and matrix connections and protecting remnant native environments. Recently, wildlife habitat management has embraced a revival of systems ecology in the guise of "ecosystem management," with Aldo Leopold apotheosized as the founder of these precepts (Knight 1996). Gone, however, are the references to the simpler laws of habitat edge, interspersion, and dispersion.

Perhaps this brief review illustrates that conservation issues should be placed in their historical context to understand best the evolution of research foci and current management concerns. In historical perspective, it is useful to note that, as recently as the 1970s, Jack Ward Thomas and his colleagues were battling the adage that good timber management is necessarily good wildlife management (Thomas 1979). Integration of silviculture and wildlife management has come far (e.g., Mannan et al. 1994) and can go even further. And it is likely that the early researchers of woodland openings and habitat edges did not discount the evils of excess, but that conditions then did not warrant their being highlighted. Perhaps this account, then, is also a statement of the speed with which native environments have been altered in recent decades, as well as a caveat to future researchers and managers to continue to question what seem to be immutable principles of conservation research and management. In reality, such principles often reside as much in historical and cultural contexts as they do in scientific ones.

Wildlife and Environments along Ecotones and Ecoclines

Edges—ecotones and ecoclines—are where different communities commingle, so that plant and animal species richness along edges is often greater than within homogeneous resource patches. This may be caused by the physical distribution and spatial overlap of species assemblages along edges as much as it is caused by ecological orientations of some organisms to edge environments per se. For example, species richness is often disproportionately high in riparian areas. Riparian environments likely do offer resources, including cover, water, and food, not found in as high abundance per unit area in upland situations. However, some proportion of the species richness in riparian environments may simply be due to riparian environments being linear features that segment a landscape and act as controlling nodes, or "cut points," across the quilt of resource patches in a landscape; thus, many species must travel through riparian environments as they traverse a landscape, which serves to inflate the species richness of riparian environments disproportionate to their area.

There are few known cases of wildlife species that are absolute obligate users of, and specially adapted to, edge environments, although a large number of species will take advantage of edge conditions. Many cases of so-called edge species may simply be species using different vegetation communities. Thus, their occurrence near edges may be merely a result of needing proximity to different environments within their home range areas. In such cases species may be termed facultative edge users. Examples include the Townsend's solitaires

(*Myadestes townsendi*) that establish winter territories along pinyon–juniper–ponderosa pine edges in the southwestern United States (Salomonson and Balda 1977) and hawks (*Buteo* spp.) coexisting along prairie-parkland ecotones (Schmutz et al. 1980).

Other research on edge environments has revealed complex microclimatic gradients and vegetation response. Microclimate is an important factor determining foraging site selection by some species, including wintering mountain chickadees (*Parus gambeli*), which selected sites with higher air temperatures and lower wind speeds than generally available in forests of south-central Wyoming (Wachob 1996). Chen et al. (1993) found that in clearcuts and along clearcut-forest edges, daily average air and soil temperature, wind velocity, and short-wave radiation were consistently higher, and soil and air moisture were lower, than within old-growth forest interiors. Proximate weather patterns affected these differences. Microclimate was more variable temporally along the edge than within clearcuts or forest interiors. In a subsequent study, Chen et al. (1995) determined that microclimate influences extended from 30 to >240 m into the forest, the shallowest influence from gradients in short-wave radiation and the deepest from humidity. In oak-chestnut forests of the eastern United States, Matlack (1993) found significant edge effects of light, temperature, litter moisture, vapor pressure deficit, humidity, and shrub cover up to 50 m from the edge. He also reported that aspect influences some microclimatic variables and that a large proportion of forest in small to medium-sized fragments is climatically altered by edge.

Vegetation response along edges was researched by Chen et al. (1992), who determined that clearcut logging causes reduced tree stocking, increased growth of dominant tree species, and higher tree mortality along edges of old-growth conifer forests of western North America. They found that the "depth-of-edge influence" zone, defined as the point along the clearcut-forest gradient at which a given variable returned to a condition representing two-thirds of the interior forest environment, ranged 16–137 m from the edge, depending on the variable. Canopy cover had shallow depth of edge; tree stem density (stems per ha >6-cm diameter) and tree diameter at breast height had medium depth of edge; and total basal area had high depth of edge. They noted that there is no "interior" forest for a patch <10 ha if depth-of-edge influence is 137 m and that edge effects were influenced by topographic position.

Matlack (1993) and Fraver (1994) studied forest edge effects in eastern U.S. forests. On the basis of the cover of native and exotic plant species, Fraver found that, along edges between agricultural lands and mixed hardwood forests of North Carolina, edge effects in forests occurred up to 50 m on south-facing edges and only 10–30 m on north-facing edges. Along Australian wheat belt edges, Hester and Hobbs (1992) found a greater density and cover of nonnative plant species, decreasing "rapidly" with increasing distance from the edge into native vegetation reserves.

Yet the specific effects of microclimatic and vegetation gradients along edges on wildlife populations and reproductive success have been poorly studied. Despite this, in many management circles and in research modeling, much inference has been made that edges harm wildlife associated with forests. This is clearly an area needing further rigorous exploration of true associations and causal mechanisms.

Edges have also been called predator traps by some researchers finding evidence of increased avian predation in such situations. Some researchers have reported increased incidence of avian nest parasitism and predation along edge environments and openings (Wilcove 1985). In forested landscapes of central Vermont, Coker and Capen (1995) reported that brown-headed cowbirds occurred in 46 percent of openings surveyed and that cowbird presence was related to opening area, distance to closest chronic disturbance opening, and number of livestock areas within 7 km of the opening. Their work suggested, though, that small (about 4 ha), remote forest openings are unlikely to attract cowbirds in the landscapes studied. This last conclusion contrasts with findings by Hahn and Hatfield (1995), who reported that nest parasitism rates by brown-headed cowbirds were significantly higher in forest interiors than old-field communities. They also reported differences in host selection compared with that found in other studies and concluded that cowbirds vary regionally in host and habitat use.

Dynamics of Colonization: Founders and Rescuers

One of the basic factors that determine the occupancy rate of habitats is the dynamics of local colonization and extinction. Much of the literature in the field of island biogeography has addressed these dynamics in both island and continental situations. Colonization of islands, and, to some degree, of continental resource patches and landscapes, may occur more frequently with the following types of species: those that are more vagile, that is, capable of longer-distance movements such as home range movement, natal dispersal, irruptive movement, and seasonal migration (chapter 3); those that produce more "disseminules," or dispersers, per breeding season or unit time (but see Shoener and Shoener 1983, below); and those that are able to withstand unsuitable conditions between islands or resource patches. Colonization by a species also occurs more frequently when the island or resource patch is: larger or contains more area of suitable environmental conditions (Cole 1981); closer to species source pools; more heterogeneous, supporting a variety of microhabitats; and, in terrestrial situations, at least somewhat connected to other resource patches by habitat corridors (Dunning et al. 1995) or habitat remnants or components scattered throughout the intervening matrix lands.

Many researchers view specific colonization events largely as stochastic incidents dependent on the chances of organisms finding environments with sufficient resources for survival and reproduction. However, Smith and Peacock (1990) proposed that colonization of resource patches by vertebrates in terrestrial systems can be much more deterministic and predictable, and is often strongly influenced by the presence of conspecifics, including potential mates, in "recipient" patches.

One aspect of colonization dynamics of organisms in patchy or insular environments pertains to how local populations get started and their subsequent genetic diversity and viability. Colonizers providing the seeds of "starter" populations are called *founders* or founder populations. Founder populations can begin in newly suitable environments and thus serve to extend the distribution of a species, or they can act to recolonize previously occupied sites left vacant by temporary dips in site suitability or by random local extinction of the species. The contribution of founders to overall metapopulation stability depends on the environmental stability of the site for maintaining the species once it appears; the proportion of valuable disseminules, or dispersers, of the source population that is tapped; and the value of the site as a stepping stone for other patch occupancy, even if the site itself provides suboptimal conditions and is occupied only intermittently. Also important to founder dynamics are the number of founder organisms, the rate of subsequent interbreeding with adjacent populations, and the rate of immigration contribution from source populations, which are factors that affect the genetic diversity of the local population. If the number of founders is low and subsequent breeding exchange or immigration is low, then the population will undergo a genetic bottleneck (see chapter 2). In more extreme cases, it may then suffer deleterious inbreeding effects. This is called the *founder effect.*

However, there are conditions where founders may serve as "nature's experiments" by successfully occupying suitable environments and then evolving unique ecotypes, morphs, subspecies, or, in some island situations, entire suites of new species. Cases of adaptive radiation of founders into new species complexes have been documented for many situations, including Hawaiian honeycreepers (subfamily Drepanididae of family Fringillidae) and the most famous case of Darwin's finches. Darwin's finches consist of 13 species of 4 genera (Fringillidae) that have evolved in isolation on the Galapagos Islands of Ecuador. But even richer are the Hawaiian honeycreepers, which currently consist of 23 species among 12 genera (9 of these species are likely moribund or possibly already extinct) and another 4 genera and 6 species known to have become extinct. Each of these species suites on Hawaii and the Galapagos likely derived from but a single ancestral finch-like founder whose descendants later dispersed among the islands and evolved a wide array of amazing bill shapes and foraging modes.

The colonization abilities of such nonmigratory species across oceanic environments represent extreme cases of what has been called sweepstakes events, or chance random traverses of widely unsuitable environments to distant habitats. Another interesting case of *sweepstakes colonization* and subsequent *adaptive radiation* is found in the little-studied New Zealand lizard fauna, as discussed in box 8.1.

New Zealand is an ideal natural laboratory for studying sweepstakes colonization, founder dynamics, adaptive radiation, ecology of sibling species, and reproductive isolating mechanisms among herps, birds, and bats. New Zealand split from the ancient continent of Gondwana about 80 million years ago, spawning the Tasman Sea, and today represents some of the oldest exposed rock on earth. Much of its flora and fauna is unique and of ancient ancestry. Its colonization history includes the ancient ancestors of the short-tailed bat (*Mystacina tuberculata*), rock wren (*Xenicus gilviventris*), rifleman (*Acanthisitta chloris*), kokako (*Callaeas cinerea*), and saddleback (*Philesturnus carunculatus*), each of which perhaps derived from Australian immigrants finding their way across the early Tasman Sea. Later, New Zealand rafted farther north, and this along with climatic warming set the stage for colonization by plants and animals from warmer latitudes. For a brief time it was colonized and occupied by tropical biota including coral reefs, but then it returned to more temperate and cooler conditions and biota (Bishop 1992). Colonization of New Zealand by birds has occurred many times over its geological history and continues today. Many founders likely fell extinct, but those that rooted produced some remarkable descendants. On New Zealand alone, the kakapo (a parrot), takahe (*Porphyrio mantelli,* a rail), kiwis (*Apteryx* spp.), and the late moas, all represent the evolution, in at least four distinct families, of both gigantism and flightlessness, traits sometimes occurring in birds in isolated environments with few predators.

Box 8.1 An example of adaptive radiation of lizards of New Zealand

New Zealand has approximately 37 species of native lizards, 11 of which are rare, vulnerable, or recommended for Red Data Book listing (Towns 1985). All 37 species are endemic and include 16 species of geckos (among three genera—*Naultinus, Heteropholis,* and *Hoplodactylus*) and 21 species of skinks (among two genera—*Cyclodina* and *Leiolopisma*). (Taxonomy of all species is not firm, and color variants of some species may yet prove to be new species.) This rather remarkable endemic species diversity likely derived from far fewer ancestral founders, perhaps one gecko and two skinks, although it is not known from where; geckos and skinks are cosmopolitan and occur widely throughout the Australian and South Pacific regions.

Some evidence suggests that the geckos occurred on New Zealand before it split from the ancient supercontinent Gondwana. Regardless, many of the native lizard taxa of New Zealand currently constitute *sibling species* complexes (also called *cryptic species*)—closely related taxa only recently diverged into separate species that are still nearly identical in appearance. For example, in some of the New Zealand geckos, mouth and tongue color is required for species identification (Towns 1985). These geckos display the tongue and open mouth as aggressive responses, and thereby color might serve as species identification signals to congeners and conspecifics. Many native lizard species have quite limited distributions within New Zealand, but in some parts of the country some sibling species overlap in range. Such overlap suggests that interspecific identification signals and perhaps other reproductive isolating behaviors have evolved sufficiently to permit sympatry without interbreeding and *genetic swamping* (loss of species' genetic identity). Sibling species that occur in sympatry also may use different resources and select different microenvironments, but these dynamics need study for this particular set of lizard species.

Even in recent times, many natural colonizers to New Zealand have been recorded (intentional introductions by humans aside) including the welcome swallow (*Hirundo neoxena,* sometimes treated as a race of the Pacific swallow, *H. tahitica*) and silvereye (*Zosterops lateralis*). These are two interesting examples of different colonization routes from the same source, Australia, along the "Roaring Forties" westerly winds that have brought so many other dispersers to the New Zealand islands over the millennia. Mapping first-appearance records (Falla et al. 1991) suggests that the welcome swallow likely colonized New Zealand perhaps three separate times over a period of less than four decades (fig. 8.9): once in 1920 in the far north, once in 1943 on Aukland Island in the far south, and once on the extreme northwest corner of South Island in 1955 (the 1955 colonizers may have come from the first two founders, but their location on the exposed windward corner of South Island, which also attracts many other vagrant dispersers, suggests an independent, long-distance colonization event). Before 1958 the swallow was considered a rare vagrant, but eventually it spread throughout the two main islands at a rate averaging 50 km/yr. By 1975 they appeared 800 km farther east on Chatham Islands (probably a minor sweepstakes dispersal event from one or both of the two main New Zealand islands).

In contrast, the silvereye probably first successfully colonized New Zealand, roughly in 1856, on the northern half of South Island and the southwestern corner of North Island and then expanded outward from there (fig. 8.10). On South Island, they spread a distance of 275 km in 4 years from mid-Canterbury to the Otago Peninsula by 1860, thus at an overland rate of 69 km/yr, and on North Island they spread 640 km in 11 years for a rate of 58 km/yr, slightly faster than the welcome swallow. Silvereyes are from eastern and southeastern Australia and Tasmania, where only a portion of their populations migrate. At least several dispersal events occurred before they success-

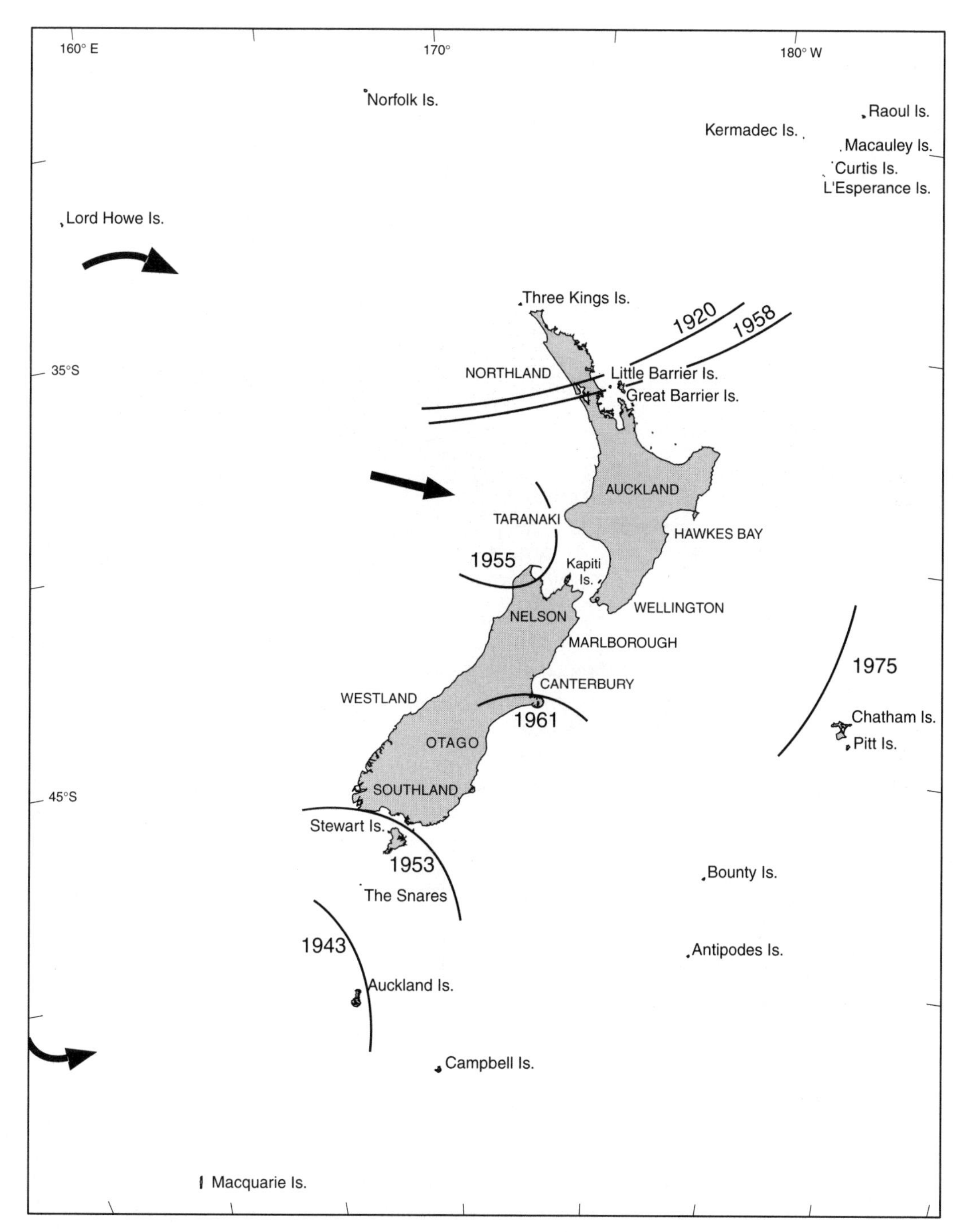

Figure 8.9. Colonization dynamics of the welcome swallow (*Hirundo neoxena*) from Australia to New Zealand. See text for description. (Based on data in Falla et al. 1991)

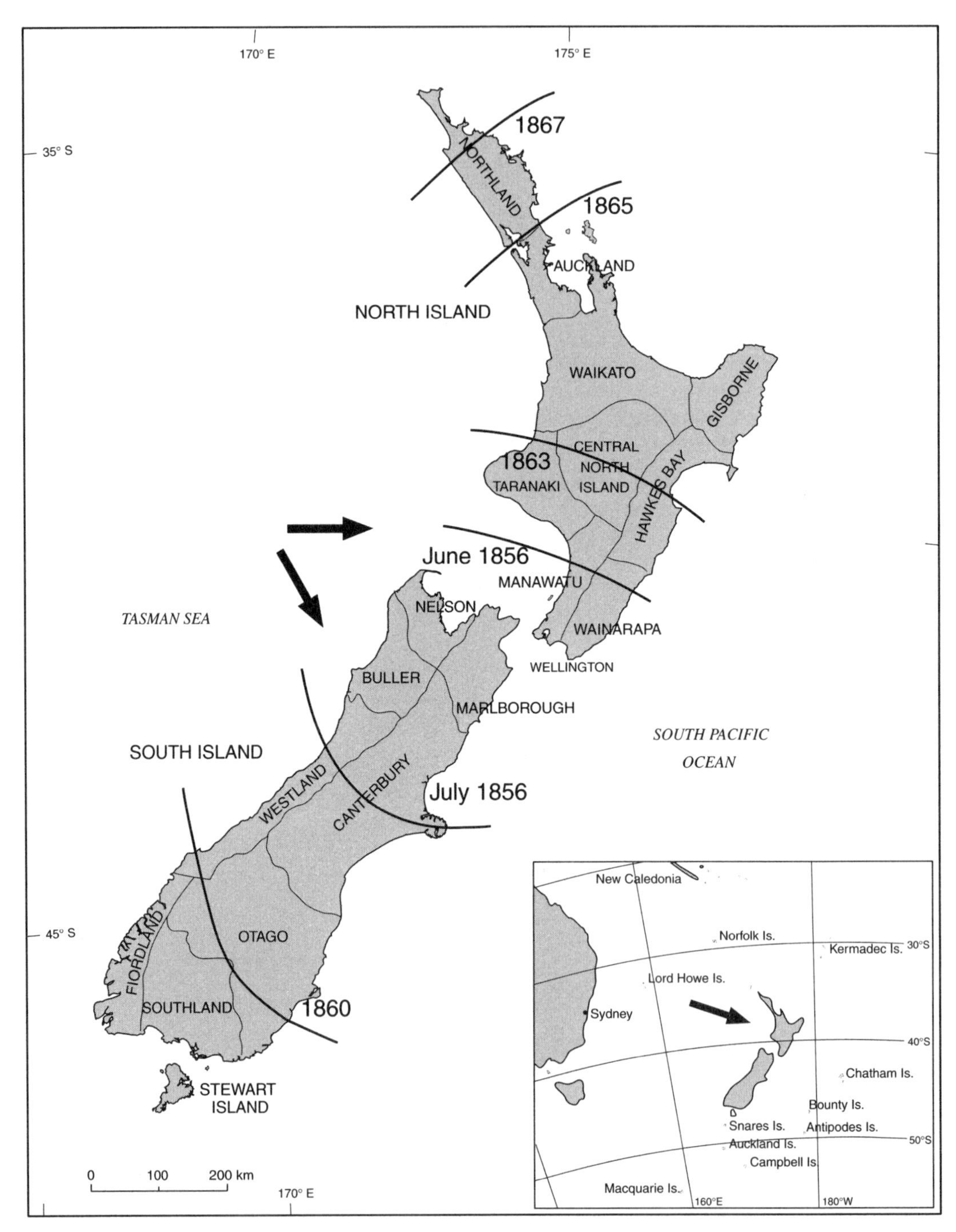

Figure 8.10. Colonization dynamics of the silvereye (*Zosterops lateralis*) from Australia to New Zealand. See text for description. (Based on data in Falla et al. 1991)

fully settled; before 1856 there were scattered records of silvereyes in Otago and Southland, the earliest being in 1832 from Milford Sound near the southwestern end of South Island.

More recently, however, massive compromises to the native New Zealand wildlife have resulted from human-caused habitat changes and introductions of exotic species, including possums, hedgehogs, rats, and cats, among others, in much the same story as that of Hawaii and many other islands and archipelagoes of the world. In Hawaii, it is estimated that 5 new plant and 18 new exotic arthropod species are introduced annually, many of which have become serious pests (Maxfield 1996).

Some exotic species become introduced for purposes of biological control, soil stabilization, and horticulture, or as pets, but can wreak ecological havoc on native biota. In Hawaii, the rosy snail (*Euglandina rosea*) was introduced from Florida to control populations of another exotic, the African giant snail (*Achatina fulica*), but the rosy snail's diet also includes Hawaii's endemic tree snails. Likewise, the endangered Hawaiian coot (*Fulica americana alai*) may be outcompeted for food plants, and the endangered Hawaiian stilt (*Himantopus mexicanus knudseni*), for food insects, by the tilapia (*Sarotherodon mossambicus*), an African fish introduced to control growth of algae and weeds in reservoirs and irrigation ditches (Maxfield 1996). Such problems are widespread among islands of the world, making the less disturbed archipelagoes, such as the Galapagos and portions of coastal British Columbia and southeast Alaska, important for scientific study, although these locations too are feeling more human pressures over time. And those flightless birds of evolutionary uniqueness have too often proved to be maladaptive to this onslaught of human pressure and attendant exotic species, often falling extinct, as with moas, or becoming moribund, as with the takahe and kiwi.

Welcome swallows and silvereyes are only two of a relatively recent set of colonizers to New Zealand, mostly from Australia. What has induced this recent flurry of successful avian colonization? Perhaps several conditions contribute (F. Schmechel, pers. comm.). Understanding their dynamics sets the stage for better understanding of the dynamics of colonization among continental environments and the potential effects of introducing exotic species. First, recent human land uses in New Zealand have provided new areas of disturbed environments favored by some bird species, such as pastures and suburban areas, favored by various introduced finches, and grasslands, favored by self-invading spur-winged plovers (*Lobibyx novaehollandiae*) and other shorebirds.

Second, introduced (mostly mammalian) predators have decimated many of the native bird species that may have outcompeted the colonizers. Removal of such key competitors have been implicated in changes in community composition of other ecosystems (Brown 1988; Cody 1974; Colwell 1973). An example from New Zealand is the pied stilt (*Himantopus leucocephalus*), a recently successful self-invader from Australia. The endemic, native black stilts (*H. novaezelandiae*) of New Zealand, which likely derived from an ancient stilt colonizer (maybe a very early successful invasion of the pied stilt?), can outcompete pied stilts. But black stilts are highly endangered from introduced mammalian predation and are essentially missing from much of their previous distribution in New Zealand, giving pied stilts a foothold.

Third, habitats and source populations of the invaders themselves may be increasing. In this example, the source populations of bird species that are associated with disturbed environments in eastern Australia and Tasmania may be increasing, thus sending out a greater number of dispersers and potential colonists per unit of time. When these three conditions are combined with time and luck, a greater number of intermittent invaders are likely to be successful. Parallels with these three conditions can be found in continental situations to help explain the spread of starlings, egrets, finches, mynas, house mice, Polynesian rats, and many other self-introduced or artificially introduced wildlife species.

Islands often contain interesting endemic biota and are fascinating systems to study. The native biota of Hawaii, for example, consists of many endemics; nearly all insects and terrestrial mollusks have evolved on site and are endemic, with high endemism rates also of flowering plants, birds, ferns and allies, and other species groups (fig. 8.11). Distance to species sources affects colonization and also endemism. The rate of successful colonization is inversely related to distance to species pools. But farther distance promotes genetic isolation from subsequent outbreeding or genetic swamping, so the more isolated the environment the greater the proportion of endemics, as found on Hawaii. In some cases it is a race between two

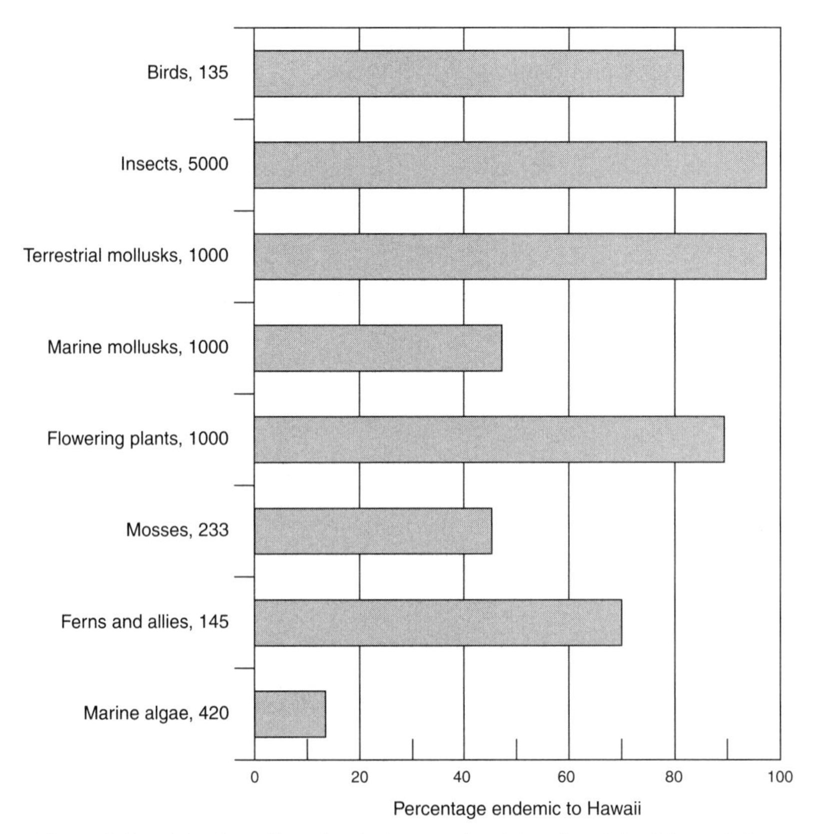

Figure 8.11. Number of species (values to the right of each bar) in and the percentage that are endemic to Hawaii (Based on data in Sohmer and Gustafson 1987)

opposing forces: on the one hand, genetic differentiation and speciation through adaptive radiation and, on the other hand, local extinction from inbreeding depression, genetic drift, demographic stochasticity, and environmental catastrophes such as major storms.

The mean colonization rate of species on the Galapagos Islands is on the order of once per 3000 years, whereas Hawaii receives a colonizer once per 70,000 years (some estimates put it at 100,000 years), assuming an initial colonizer biota of about 1000 species in each archipelago (Loope 1989) (fig. 8.12). So, all else being equal, the Galapagos should have 23 times the biodiversity of Hawaii. Both island systems are subtropical or tropical, and both contain tall volcanic mountains. But all things are never equal in ecological systems, and Hawaii is by far the more diverse system for several reasons. Hawaii, at 70 million years, is far older than the more youthful, 3 million-year-old Galapagos and thus has had more time to receive colonizers, develop a wider array of environments, and allow for speciation through adaptive radiation. Hawaii is also twice as large as the Galapagos and thus can support a greater variety of environments and species, and per unit of distance is a bigger target for arriving colonists. Also, Hawaii is more diverse in climate than the Galapagos, which are dominated largely by arid scrub habitats and are located equatorially near cold upwelling zones. To continue the comparison, New Zealand in turn dwarfs both the Hawaiian and Galapagos systems in species richness and area, although it is intermediate in distance to species pools (fig. 8.13).

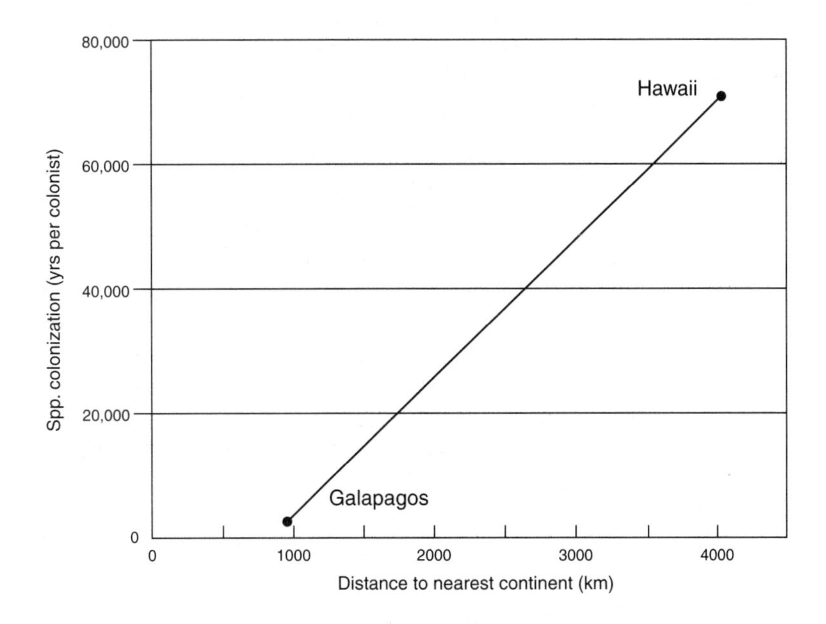

Figure 8.12. Colonization rate of species on two archipelagoes as affected by distance to nearest continental species source pool (Based on data in Loope 1989:3–4)

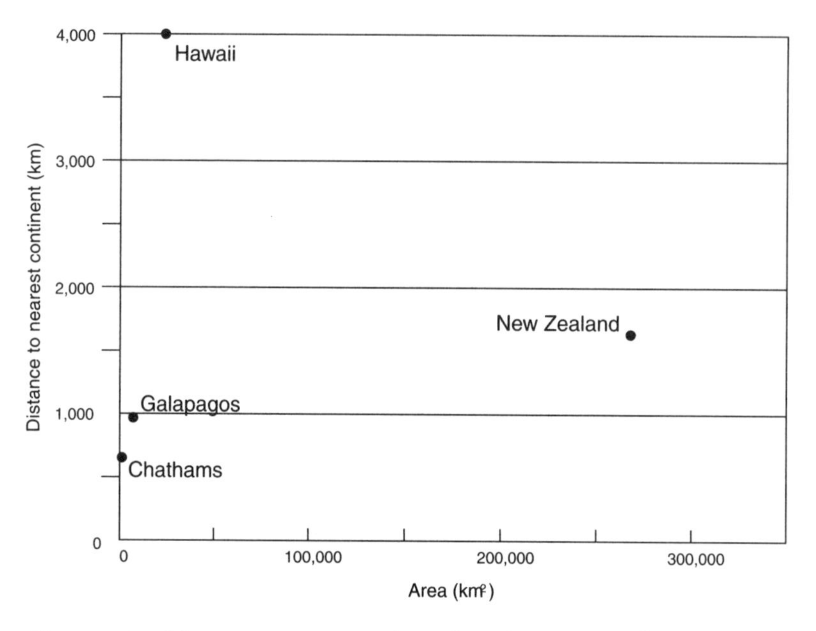

Figure 8.13. Distance to nearest continental species source pool as a function of total archipelago area for four island archipelagoes

Not considering developmental history and environmental heterogeneity, one would expect a higher resident biodiversity and higher colonization rate on larger islands and on islands that are closer to species pools. Plotting distance to the nearest continent as a species pool against the total island area shows that, of the four island groups discussed in this section, New Zealand meets these criteria best (fig. 8.13), followed by the Galapagos and Chathams, and then Hawaii. The ancient age of Hawaii accounts for its biodiversity; through time it has developed a wide variety of macro- and microenvironments and a high degree of adaptive radiation of founders.

Since colonization and persistence of vegetation and wildlife are at least somewhat allied, it is instructive to know something about dispersal and colonization dynamics of plants as well. Mechanisms of relatively rare dispersal events of plants onto islands tend to vary significantly from those of rare plants in continental settings. For example, data on the dispersal mode of plant immigrants to the Hawaiian Islands suggest that 75 percent are dispersed by birds, 23 percent by oceanic drift, and 1 percent by air flotation (fig. 8.14). Plant propagules that are dispersed by birds travel by being eaten and carried internally or by being mechanically attached, embedded in mud on feet, or embedded in feathers. In contrast, rare vascular plants occurring in a fully continental setting exhibit a fuller array of dispersal means, including gravity, wind, growth or reproduction, and travel by vertebrates (mostly birds and mammals) as the major dispersal means, with water and insects as less common means (fig. 8.15).

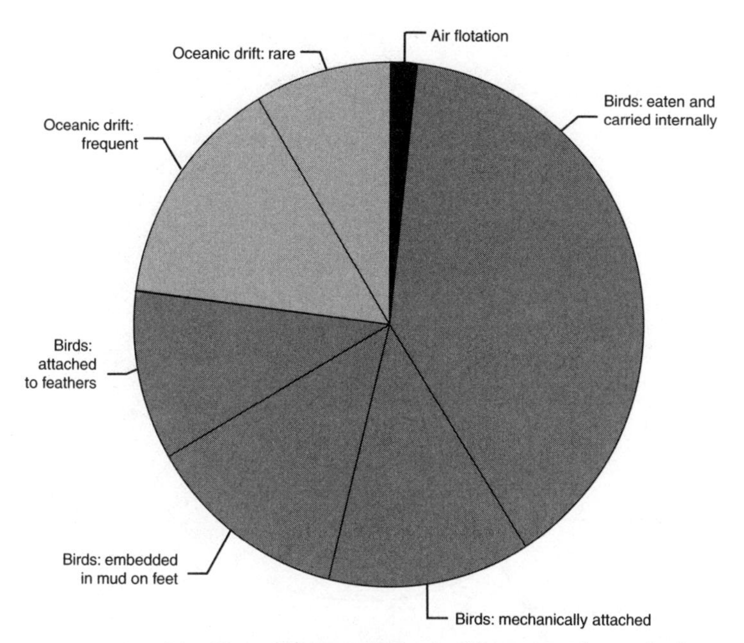

Figure 8.14. Modes of long-distance dispersal of flowering plant colonists assumed to have given rise to the presently known native flora of the Hawaiian Islands (not including plants carried by humans) (Based on data in Carlquist 1974)

Sweepstakes (and more frequent) dispersal of wildlife can take place in continental as well as oceanic biomes. In some cases, management depends on such intermittent events to afford outbreeding of local populations and to avoid genetic or demographic problems of small, isolated populations. One example is the hoped-for chance exchanges of breeding individuals between semi-isolated populations of California spotted owls (*Strix occidentalis occidentalis*) in California (Anderson and Mahato 1995). In another example, local colonies of red-cockaded woodpeckers (*Picoides borealis*) in the southeast United States occur largely in forests of longleaf pine (*Pinus palustris*) on federal lands (mostly national forests). The woodpecker colonies and the forest habitats they use often occur as small patches scattered across landscapes containing environments that are mostly unsuitable because of human activities such as large-scale reduction or conversion of native forests. It is assumed in some cases that, through random and intermittent dispersal events, the woodpeckers can interchange breeders and genes between some colonies and populations (Haig et al. 1993; Maguire et al. 1995). In cases of more extreme isolation, biologists have artificially augmented local populations by translocating birds. In other examples, dispersal of organisms among isolated, montane environments may occur only infrequently; successful dispersal may be affected by stochastic events of weather and the chance location of suitable

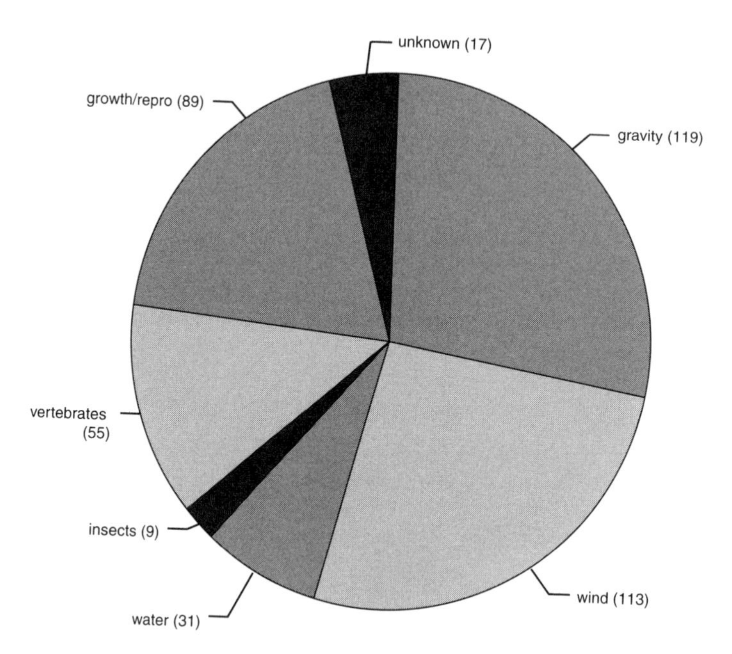

Figure 8.15. Modes of dispersal of rare vascular plants in the interior Columbia River basin east of the Cascades crest, northwestern United States (Based on data in Marcot et al., in press)

environments (Johnson 1975; Lomolino et al. 1989).

The founder principle is most important for captive breeding situations and zoo populations. A thorough discussion of tracking pedigrees, ensuring correct outbreeding levels, and other related topics is beyond the scope of this chapter. The reader is directed to other, more specialized literature on this topic (e.g., Backus et al. 1995; Grier and Barclay 1988; Senner 1980; Tudge 1992).

Colonizers also play a critical role in the "rescue" of small local populations potentially imperiled by lack of genetic diversity and too few breeders. Two kinds of rescue can be defined: *Genetic rescue* refers to supplementing a local population with new genes, which helps avoid problems of the fixation of deleterious alleles and an increase in genetic homozygosity caused by genetic drift, inbreeding, or the founder effect (also see discussion in chapter 3). *Demographic rescue* involves supplementing the local effective population with new individuals so that the population does not become extinct through random variations in individual breeding success or number of offspring. As a general rule of thumb, the number of immigrants needed for demographic rescue is approximately an order of magnitude greater than the number needed for genetic rescue, but this depends on generation time, lifespan and age-specific reproductive value, degree of environmental change and stress, and other factors.

In this chapter we have provided a foundation for considering heterogeneity of environments and resources. We continue the discussion in the next chapter, where we address in greater depth the effects on wildlife species and communities, extinction dynamics of species, disturbance dynamics of landscapes, and guidelines for monitoring patches and landscapes.

Literature Cited

Adams, L. W. 1994. *Urban wildlife habitats: A landscape perspective.* Minneapolis: University of Minnesota Press.

Anderson, M. C., and D. Mahato. 1995. Demographic models and reserve designs for the California spotted owl. *Ecological Applications* 5(3):639–47.

Anderson, S. H., H. H. Shugart, and T. M. Smith. 1979. Vertical and temporal habitat utilization within a breeding bird community. In *The role of insectivorous birds in forest ecosystems,* ed. J. G. Dickson, R. N. Conner, R. R. Fleet, and J. C. Kroll, 203–16.

Andren, H. P., and P. Angelstam. 1988. Elevated predation rates as an edge effect in habitat islands: Experimental evidence. *Ecology* 69:544–47.

Angelstam, P. 1986. Predation on ground-nesting birds' nests in relation to predator densities and habitat edge. *Oikos* 47:365–73.

Aubry, K. B., and C. M. Raley. 1990. Landscape-level responses of pileated woodpeckers to forest management and fragmentation. *Northwest Environmental Journal* 6:432–33.

Backus, V. L., E. H. Bryant, C. R. Hughes, and L. M. Meffert. 1995. Effect of migration or inbreeding followed by selection on low-founder-number populations: Implications for captive breeding programs. *Conservation Biology* 9(5):1216–24.

Beatley, T. 1994. *Habitat conservation planning: Endangered species and urban growth.* Austin: University of Texas Press.

Beier, P. 1993. Determining minimum habitat areas and habitat corridors for cougars. *Conservation Biology* 7(1):94–108.

Bishop, N. 1992. *Natural history of New Zealand.* Aukland, New Zealand: Hodder and Stoughton.

Black, H., and J. W. Thomas. 1978. Forest and range wildlife habitat management: ecological principles and management systems. In *Proceedings workshop on nongame bird habitat management in the coniferous forests of the western United States,* ed. R. M. DeGraff, 47–55. USDA Forest Service General Technical Report PNW-64. Portland, Oreg.

Brown, J. S. 1988. Patch use as an indicator of habitat preference, predation risk, and competition. *Behavioral Ecology and Sociobiology* 22(1):37–47.

Burger, L. D., L. W. Burger, Jr., and J. Faaborg. 1994. Effects of prairie fragmentation on predation on artificial nests. *Journal of Wildlife Management* 58(2):249–54.

Carlquist, S. 1974. *Island biology.* New York: Columbia University Press.

Chen, J., J. F. Franklin, and T. A. Spies. 1992. Vegetation responses to edge environments in old-growth Douglas-fir forests. *Ecological Applications* 2(4):387–96.

Chen, J., J. F. Franklin, and T. A. Spies. 1993. Contrasting microclimates among clearcut, edge, and interior of old-growth Douglas-fir forest. *Agricultural and Forest Meteorology* 63:219–37.

Chen, J., J. F. Franklin, and T. A. Spies. 1995. Growing-season microclimatic gradients from clearcut edges into old-growth Douglas-fir forests. *Ecological Applications* 5(1):74–86.

Cody, M. L. 1974. *Competition and the structure of bird communities.* Princeton, N.J.: Princeton University Press.

Coker, D. R., and D. E. Capen. 1995. Landscape-level habitat use by brown-headed cowbirds in Vermont. *Journal of Wildlife Management* 59(4):631–37.

Cole, B. J. 1981. Colonizing abilities, island size, and the number of species on archipelagoes. *American Naturalist* 117:629–38.

Colwell, R. K. 1973. Competition and coexistence in a simple tropical community. *American Naturalist* 107:737–60.

DeGraaf, R. M., and N. L. Chadwick. 1987. Forest type, timber size class, and New England breeding birds. *Journal of Wildlife Management* 51(1):212–17.

Delin, A. 1992. Kärlväxter i taigan i Hälsingland—deras anpassningar till kontinuitet eller störning [Vascular plants of the taiga—adaptations to continuity or to disturbance]. *Svensk Botanisk Tidskrift* 86:147–76.

Demers, M. N., J. W. Simpson, R. E. J. Boerner, A. Silva, L. Berns, and F. Artigas. 1995. Fencerows, edges, and implications of changing connectivity illustrated by two contiguous Ohio landscapes. *Conservation Biology* 9(5):1159–68.

Dunning, J. B., Jr., R. Borgella, Jr., K. Clements, and G. K. Meffe. 1995. Patch isolation, corridor effects, and colonization by a resident sparrow in a managed pine woodland. *Conservation Biology* 9(3):542–50.

Edgerton, P. J., and J. W. Thomas. 1978. Silvicultural options and habitat values in coniferous forests. *Proceedings workshop on nongame bird habitat management in the coniferous forests of the western United States,* ed. R. M. DeGraff, 56–65. USDA Forest Service General Technical Report PNW-64. Portland, Oreg.

Falla, R. A., R. B. Sibson, and E. G. Turbott. 1991. *Collins guide to the birds of New Zealand and outlying islands.* Auckland, New Zealand: Harper Collins Publishers.

Ffolliott, P. F., R. E. Thill, W. P. Clary, and F. R. Larson. 1977. Animal use of ponderosa pine forest openings. *Journal of Wildlife Management* 41(4):782–84.

Forman, R. T. T. 1983. Corridors in a landscape: Their ecological structure and function. *Ekologia* 2(4):375–87.

Forman, R. T. T., and M. Godron. 1986. *Landscape ecology.* New York: John Wiley and Sons.

Franklin, J. F., and R. T. T. Forman. 1987. Creating landscape patterns by forest cutting: Ecological consequences and principles. *Landscape Ecology* 1:5–18.

Fraver, S. 1994. Vegetation responses along edge-to-interior gradients in the mixed hardwood forests of the Roanoke River basin, North Carolina. *Conservation Biology* 8(3):822–32.

Freemark, K. E., and H. G. Merriam. 1986. Importance of area and habitat heterogeneity to bird assemblages in temperate forest fragments. *Biological Conservation* 36:115–41.

Gleason, H. A. 1926. The individualistic concept of the plant association. *Bulletin of the Torrey Botanical Club* 53:7–26.

Grier, J. W., and J. H. Barclay. 1988. Dynamics of founder populations established by reintroduction. In *Peregrine falcon populations,* ed. T. J. Cade, J. H. Enderson, C. G. Thelander, and C. M. White, 698–700. Boise, Idaho: Peregrine Fund.

Grime, J. P. 1977. Evidence for the existence of three primary strategies in plants and its relevance to ecological and evolutionary theory. *American Naturalist* 111:1169–94.

Guthery, F. S., and R. L. Bingham. 1992. On Leopold's principle of edge. *Wildlife Society Bulletin* 20(3):340–44.

Haas, C. A. 1995. Dispersal and use of corridors by birds in wooded patches on an agricultural landscape. *Conservation Biology* 9(4):845–54.

Hahn, D. C., and J. S. Hatfield. 1995. Parsitism at the landscape scale: Cowbirds prefer forests. *Conservation Biology* 9(6):1415–24.

Haig, S. M., J. R. Belthoff, and D. H. Allen. 1993. Population viability analysis for a small population of red-cockaded woodpeckers and an evaluation of enhancement strategies. *Conservation Biology* 7(2):289–301.

Hall, F. C., and J. W. Thomas. 1979. Silvicultural options. In *Wildlife habitats in managed forests: the Blue Mountains of Oregon and Washington,* ed. J. W. Thomas, 128–47. USDA Forest Service Agricultural Handbook No. 553.

Harris, L. D. 1984. *The fragmented forest.* Chicago: University of Chicago Press.

Herkert, J. R. 1994. The effects of habitat fragmentation on midwestern grassland bird communities. *Ecological Applications* 4(3):461–71.

Hester, A. J., and R. J. Hobbs. 1992. Influence of fire and soil nutrients on native and non-native annuals at remnant vegetation edges in the Western Australian Wheatbelt. *Journal of Vegetation Science* 3(1):101–8.

Hill, C. J. 1995. Linear strips of rain forest vegetation as potential dispersal corridors for rain forest insects. *Conservation Biology* 9(6):1559–66.

Howe, R. W., T. D. Howe, and H. A. Ford. 1981. Bird distributions on small rainforest remnants in New South Wales. *Australian Wildlife Research* 8:637–51.

Huston, M. A. 1994. *Biological diversity: The coexistence of species on changing landscapes.* Cambridge: Press Syndicate of Cambridge University.

Johnson, N. K. 1975. Controls of number of bird species on montane islands in the Great Basin. *Evolution* 29:545–67.

Johnston, V. R. 1947. Breeding birds of the forest edge in Illinois. *Condor* 49:45–53.

Kelker, G. H. 1964. Appraisal of ideas advanced by Aldo Leopold thirty years ago. *Journal of Wildlife Management* 28(1):180–85.

Kitching, R. L., and R. A. Beaver. 1990. Patchiness and community structure. In *Living in a patchy environment,* B. Shorrocks and I. R. Swingland, 147–76. Oxford: Oxford University Press.

Knight, R. L. 1996. Aldo Leopold, the land ethic, and ecosystem management. *Journal of Wildlife Management* 60(3):471–74.

Krebs, J. R., J. C. Ryan, and E. L. Charnov. 1974. Hunting by expectation or optimal foraging? A study of patch use by chickadees. *Animal Behavior* 22:953–64.

Krefting, L. W. 1962. Use of silvicultural techniques for improving deer habitat in the Lake States. *Journal of Forestry* 16:40–42.

Kremsater, L. L., and F. L. Bunnell. 1990. Creating black-tailed deer winter range in second-growth forests. *Northwest Environmental Journal* 6:387–88.

Lay, D. W. 1938. How valuable are woodland clearings to wildlife? *Wilson Bulletin* 50:254–56.

Lehmkuhl, J. F. 1990. The effects of forest fragmentation on vertebrate communities in western Oregon and Washington. *Northwest Environmental Journal* 6:433–34.

Leopold, A. 1933. *Game management.* New York: Scribners.

Lindenmayer, D. B., and R. C. Lacy. 1995. Metapopulation viability of arboreal marsupials in fragmented old-growth forests: Comparison among species. *Ecological Applications* 5(1):183–99.

Lomolino, M. V., J. H. Brown, and R. Davis. 1989. Island biogeography of montane forest mammals in the American Southwest. *Ecology* 70:180–94.

Loope, L. L. 1989. Island ecosystems. In *Conservation biology in Hawai'i,* ed. C. P. Stone and D. B. Stone, 3–6. Honolulu: University of Hawaii.

Lyon, L. J., and C. E. Jensen. 1980. Management implications of elk and deer use of clear-cuts in Montana. *Journal of Wildlife Management* 44:352–62.

McKelvey, K., B. R. Noon, and R. H. Lamberson. 1993. Conservation planning for species occupying fragmented landscapes: The case of the northern spotted owl. In *Biotic interactions and global change,* ed. P. M. Kareiva, J. G. Kingsolver, and R. B. Huey, 424–52. Sunderland, Mass.: Sinauer Associates, Inc.

Maguire, L. A., G. F. Wilhere, and Q. Dong. 1995. Population viability analysis for red-cockaded woodpeckers in the Georgia piedmont. *Journal of Wildlife Management* 59(3):533–42.

Mannan, R. W., R. N. Conner, B. G. Marcot, and J. M. Peek. 1994. Managing forestlands for wildlife. In *Research and management techniques for wildlife and habitats,* ed. T. A. Bookhout, 689–721. Washington, D.C.: Wildlife Society.

Marcot, B. G. 1985. Habitat relationships of birds and young-growth Douglas-fir in northwestern California. Ph.D. dissertation. Oregon State University, Corvallis Oreg.

Marcot, B. G., L. K. Croft, J. F. Lehmkuhl, R. H. Naney, C. G. Niwa, W. R. Owen, and R. E. Sandquist. 1998. *Macroecology, paleoecology, and ecological integrity of terrestrial species and communities of the interior Columbia River basin and portions of the Klamath and*

Great basins. USDA Forest Service General Technical Report PNW-GTR-410.

Matlack, G. R. 1993. Microenvironment variation within and among forest edge sites in the eastern United States. *Biological Conservation* 66:185–94.

Maurer, B. A. 1990. Extensions of optimal foraging theory for insectivorous birds: implications for community structure. *Studies in Avian Biology* 13:455–61.

Maxfield, B. A. 1996. The Hawaiian Islands, 20 years later. *Endangered Species Bulletin* 21(4):18–21.

Mills, L. S. 1995. Edge effects and isolation: Red-backed voles on forest remnants. *Conservation Biology* 9(2):395–403.

O'Neill, R. V., J. R. Krummel, R. H. Gardner, G. Sugihara, B. Jackson, D. L. DeAngelis, B. T. Milne, M. G. Turner, B. Zygmunt, S. W. Christensen, V. H. Dale, and R. L. Graham. 1988. Indices of landscape pattern. *Landscape Ecology* 1(3):153–62.

Paton, P. W. C. 1994. The effect of edge on avian nest success: How strong is the evidence? *Conservation Biology* 8(1):17–26.

Pearson, D. L. 1971. Vertical stratification of birds in a tropical dry forest. *Condor* 73:46–55.

Petrides, G. A. 1942. Relation of hedgerows in winter to wildlife in central New York. *Journal of Wildlife Management* 6(4):261–80.

Reese, K. P., and J. T. Ratti. 1988. Edge effect: A concept under scrutiny. *North American Wildlife Natural Resources Conference* 53:127–36.

Robbins, C. S. 1979. Effect of forest fragmentation on bird populations. In *Proceedings workshop on management of north central and northeastern forests for nongame birds,* ed. R. M. DeGraff and K. E. Evans, 198–212. USDA Forest Service General Technical Report NC-51.

Robbins, C. S. 1991. Managing suburban forest fragments for birds. In *Challenges in the conservation of biological resources: A practitioner's guide,* ed. D. J. Decker, M. E. Krasny, G. R. Goff, C. R. Smith, and D. W. Gross, 253–64. Boulder, Colo.: Westview Press.

Rudnicky, T. C., and M. L. Hunter, Jr. 1993a. Avian nest predation in clearcuts, forests, and edges in a forest-dominated landscape. *Journal of Wildlife Management* 57(2):358–64.

Rudnicky, T. C., and M. L. Hunter, Jr. 1993b. Reversing the fragmentation perspective: Effects of clearcut size on bird species richness in Maine. *Ecological Applications* 3(2):357–66.

Ruggiero, L. F., R. S. Holthausen, B. G. Marcot, K. B. Aubry, J. W. Thomas, and E. C. Meslow. 1988. Ecological dependency: The concept and its implications for research and management. *North American Wildlife Natural Resources Conference* 53:115–26.

Salomonson, M. G., and R. P. Balda. 1977. Winter territoriality of Townsend's solitaires (*Myadestes townsendi*) in a pinon-juniper-ponderosa pine ecotone. *Condor* 79:148–61.

Sammalisto, L. 1957. The effect of the woodland–open peatland edge on some peatland birds in south Finland. *Ornis Fennica* 34:81–89.

Saunders, D. A., G. W. Arnold, A. A. Burbidge, and J. M. Hopkins. 1987. *Nature conservation: The role of remnants of native vegetation.* Chipping Norton, New South Wales, Australia: Surrey Beatty and Sons Limited.

Schmutz, J. K., S. M. Schmutz, and D. A. Boag. 1980. Coexistence of three species of hawks (*Buteo* spp.) In the prairie-parkland ecotone. *Canadian Journal of Zoology* 58:1075–89.

Schoener, T. W., and A. Schoener. 1983. The time to extinction of a colonizing propagule of lizards increases with island area. *Nature* 302:332–34.

Selva, S. B. 1994. Lichen diversity and stand continuity in the northern hardwoods and spruce-fir of northern New England and western New Brunswick. *Bryologist* 97(4):424–29.

Senner, J. W. 1980. Inbreeding depression and the survival of zoo populations. In *Conservation biology: An evolutionary-ecological perspective,* ed. M. E. Soule and B. A. Wilcox, 209–24. Sunderland, Mass.: Sinauer Associates.

Simberloff, D. S., and L. G. Abele. 1976. Refuge design and island biogeography theory: Effects of fragmentation. *American Naturalist* 120:40–41.

Smith, A. T., and M. M. Peacock. 1990. Conspecific attraction and the determination of metapopulation colonization rates. *Conservation Biology* 4:320–23.

Sohmer, S. H., and R. Gustafson. 1987. *Plants and flowers of Hawai'i.* Honolulu: University of Hawaii Press.

Stamps, J. A., M. Buechner, and V. V. Krishnan. 1987. The effects of edge permeability and habitat geometry on emigration from patches of habitat. *American Naturalist* 129:533–52.

Thomas, J. W., ed. 1979. *Wildlife habitats in managed forests: The Blue Mountains of Oregon and Washington.* USDA Forest Service Agricultural Handbook No. 553.

Thomas, J. W., C. Maser, and J. E. Rodiek. 1979. Edges. In *Wildlife habitats in managed forests: The Blue Mountains of Oregon and Washington,* ed. J. W. Thomas, 48–59. USDA Forest Service Agricultural Handbook No. 553.

Tibell, L. 1992. Crustose lichens as indicators of forest continuity in boreal coniferous forests. *Norwegian Journal of Botany* 12(4):427–50.

Towns, D. R. 1985. *A field guide to the lizards of New Zealand.* New Zealand Wildlife Service Occasional Publication No. 7. Wellington, New Zealand. 28 pp.

Tudge, C. 1992. *Last animals at the zoo: How mass extinction can be stopped.* Covelo, Calif: Island Press.

Turner, M. G., and R. H. Gardner. 1990. *Quantitative methods in landscape ecology.* New York: Springer-Verlag.

Wachob, D. G. 1996. The effect of thermal microclimate on foraging site selection by wintering mountain chickadees. *Condor* 98:114–22.

Wegge, P., J. Rolstad, and I. Gjerde. 1992. Effects of boreal forest fragmentation on capercaillie grouse: empirical evidence and management implications. In *Wildlife 2001: Populations,* D. R. McCullough and R. H. Barrett, 738–49. London: Elsevier Applied Science.

Whitcomb, B. L., R. F. Whitcomb, and D. Bystrak. 1977. Long-term turnover and effects of selective logging on the avifauna of forest fragments. *American Birds* 31:17–23.

Wiens, J. A. 1974. Habitat heterogeneity and avian community structure in North American grasslands. *American Midlands Naturalist* 91:195–213.

Wilcove, D. S. 1985. Nest predation in forest tracts and the decline of migratory songbirds. *Ecology* 66:1212–14.

Wilcox, B. A., and D. D. Murphy. 1985. Conservation strategy: the effects of fragmentation on extinction. *American Naturalist* 125:879–87.

World Resources Institute. (WRI). 1996. *A guide to the global environment.* Washington, D.C.: World Resources Institute. (On World Wide Web at http://www.wri.org/wri/wr-96-97/96tocful.html.)

Wu, J., and O. L. Loucks. 1995. From balance of nature to hierarchical patch dynamics: A paradigm shift in ecology. *Quarterly Review of Biology* 70:439–66.

Yahner, R. H. 1988. Changes in wildlife communities near edges. *Conservation Biology* 2:333–39.

Yahner, R. H., T. E. Morrell, and J. S. Rachael. 1989. Effects of edge contrast on depredation of artificial avian nests. *Journal of Wildlife Management* 53:1135–38.

Zonneveld, I. S. 1979. *Land evaluation and land(scape) science.* Enschede, The Netherlands: International Training Center.

9 Of Habitat Patches and Landscapes: Habitat Isolation, Dynamics, and Monitoring

Introduction

In this chapter we continue the story of habitat patches and landscapes begun in the previous chapter. Here we focus on the response of organisms, species, populations, and communities to conditions of patchy environments and landscape dynamics.

Extinctions, Isolation, and Fragmentation

Extinction Dynamics

Frequency of local extinctions is affected by the reverse conditions of those listed in chapter 8 (in "Dynamics of Colonization") for colonization, as well as time since isolation of the island or resource patch. In one study, Schoener and Schoener (1983) reported that time to extinction of colonizing anole lizards (*Anolis sagrei*) in Bahamian islands increased monotonically with island area and that rapid colonization occurred above a certain island size, but that number of initial propagules (lizard colonizers) had no effect. In general, the interplay between colonization and extinction dynamics results in specific levels of species richness and diversity in both island and continental situations.

Burkey (1995) found that extinction rates of lizards, birds, and mammals on a set of small islands were higher than on a single large island, and they were higher in more fragmented systems. Burkey concluded that some degree of habitat connectivity would help reduce extinction rates in isolated habitat remnants.

The Effects of Isolation

Important effects of isolation include the depression of species diversity in isolated areas (fig. 9.1). With small populations in insular (island or continental) situations, isolation can increase the likelihood of adverse genetic effects such as fixation of deleterious alleles, increasing homozygosity, and overall decline in allelic diversity of the gene pool, as caused by genetic drift. Inbreeding depression—including depressed fertility and fecundity, increased natal mortality, and decreasing age of reproductive senescence—is one manifestation of small effective population size (see chapter 3). Small colonizer populations are subject to founder effects, which set the stage for loss of genetic and phenotypic diversity with subsequent isolation from outbreeding.

At the community level, recent isolation of previously rich environments or those better connected to species source pools can result in a decline in species richness over time. When isolation of environments causes the loss of wildlife species, it is termed *faunal relaxation* (figs. 9.1, 9.2). This dynamic has

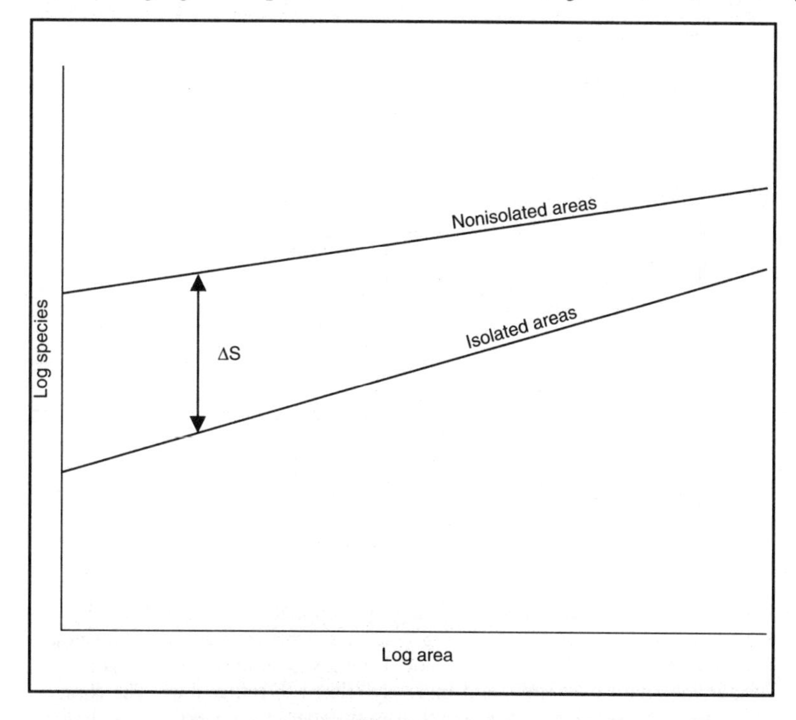

Figure 9.1. Species-area relations of isolated and nonisolated areas. Species loss, ΔS, represents relaxation effects occurring when land bridges become submerged and isolate islands or when continental reserves change in status from nonisolated to isolated. (From W. J. Boecklen and D. Simberloff, "Area-based Extinction Models in Conservation," in *Dynamics of Extinction*, ed. D. K. Elliott [New York: John Wiley and Sons © 1986], 255, fig. 2; reprinted by permission of John Wiley and Sons, Inc.)

been documented for oceanic islands isolated from mainlands by submergence of land bridges (Newmark 1987), as well as for patches of rain forest isolated by slash-and-burn or clearcut timber harvesting in the tropics. Eventually, the fauna reaches an equilibrium species richness where local extinction and emigration of species equal immigration and colonization (fig. 9.3).

A classic example of faunal relaxation is the well-studied fauna of Barro Colorado Island (BCI) in Panama, where the creation of the Panama Canal isolated a large patch of native rain forest (Karr 1990). Karr reported that local extinction of birds on BCI after isolation involved species with lower survival rates in the adjacent mainland forest; these species disappeared earlier from the island than species with

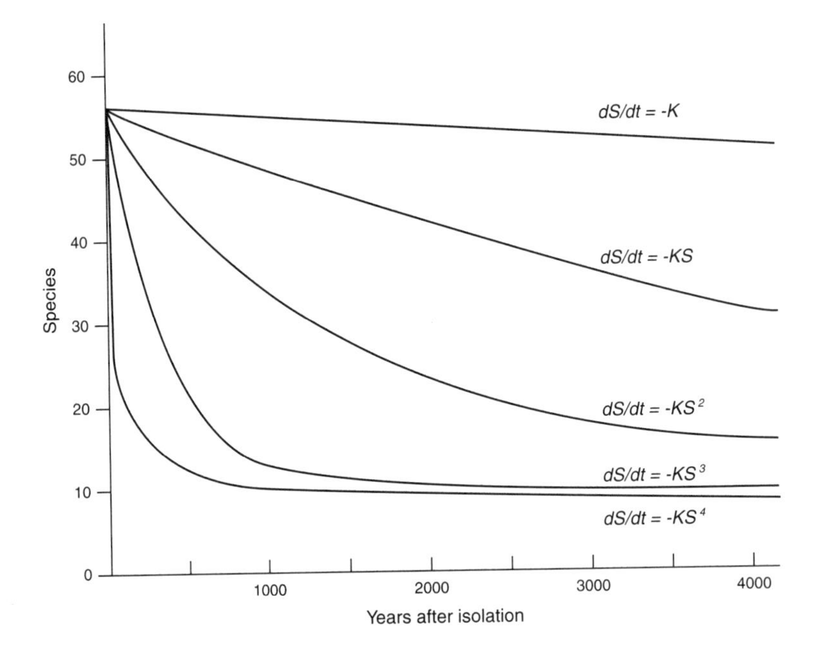

Figure 9.2. Faunal collapse of the Nairobi National Park as predicted by five faunal collapse species models. S = species richness (number of species, ordinate), t = time, K = extinction coefficient. K is assumed to be taxon specific, invariant over time, and inversely related to refuge area. The choice of the model and value of K is as yet arbitrary, because no extinctions have been observed to occur. The disparate results among models at longer simulation times strongly suggest the need for taxon-specific field studies of demography, genetics, and population dynamics of individual species. (From W. J. Boecklen and D. Simberloff, "Area-based Extinction Models in Conservation," in *Dynamics of Extinction*, ed. D. K. Elliott [New York: John Wiley and Sons © 1986], 258, fig. 3; reprinted by permission of John Wiley and Sons, Inc.)

higher survival rates did. The culprits contributing to extinction on BCI were species with reduced reproductive success caused by high nest predation and/or altered landscape dynamics, combined with naturally low adult survival rates. Karr did not find a correlation of extinction with population size (but see chapter 3 for evidence of such correlation in native Hawaiian birds).

Isolation of reserves and parks has been a concern for some biologists who suspect that relaxation effects are causing declines in native wildlife in protected areas (Newmark 1986). In some cases, legal and ecological boundaries of protected areas, such as national parks, do not necessarily coincide (Newmark 1995), although it is imperative in such analyses to determine the actual status of the environment and populations in lands adjacent to such protected areas. Newmark (1995) analyzed extinction of mammal populations in national parks in western North America and concluded that extinction rates have exceeded colonization rates and are higher in smaller park units. He also

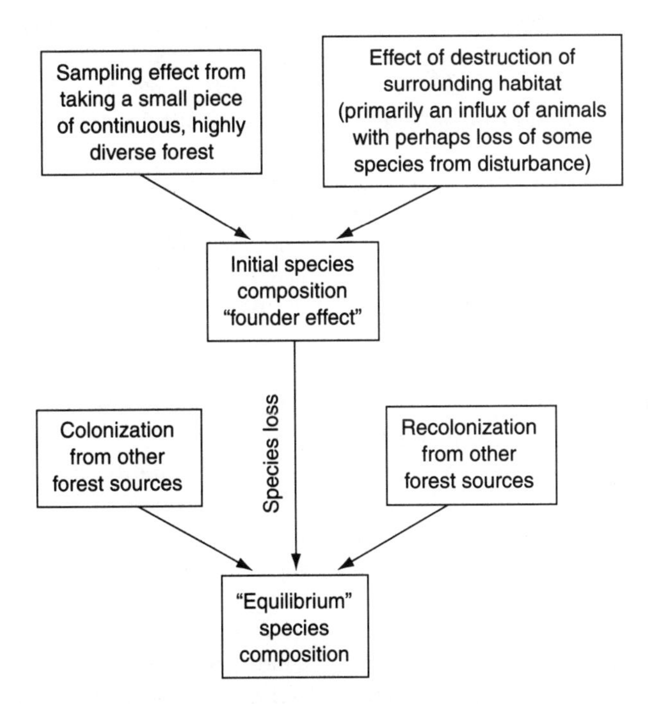

Figure 9.3. Basic elements of the consequences of habitat isolation and subsequent faunal relaxation (species loss) to new equilibrium levels of diversity. Additional factors (not shown here) of habitat isolation affecting equilibrium levels include changes in habitat patch microclimate, habitat patch edge effects, and effects on foraging efficiency and energetics of organisms. (From T. E. Lovejoy, J. M. Rankin, R. O. Bierregard, Jr., K. S. Brown, Jr., L. H. Emmons, and M. E. Van der Voort, "Ecosystem Decay of Amazon Forest Remnants," in *Extinctions,* ed. M. E. Nitecki [Chicago: University of Chicago Press © 1984], 299, fig. 1; reproduced by permission of the University of Chicago Press)

reported that the major factors affecting higher rates of extinction of mammals were, in decreasing order of importance, small initial population size, occurrence within the order Lagomorpha, and shorter age of maturity (which is positively related to shorter population generation time).

Relaxation effects can be displayed with the use of incidence functions, which depict the presence or absence of species in resource patches of varying size. When projected over time, incidence functions also portray changes in species richness within patches and metapopulation dynamics of resource patch occupancy. Whittam and Siegel-Causey (1981) used incidence functions to describe species richness of Alaskan seabird colonies. Hanski et al. (1996) developed incidence functions to depict metapopulation dynamics of an endangered butterfly, the Glanville fritillary (*Melitaea cinxia*), in meadow resource patches on Åland Island, Finland. Many other examples can be found in the literature.

Related to faunal relaxation is the delayed loss of competitive, rare species in islands and environment isolates. The loss can take place over a period of time following isolation or alteration of environments, so that extinction eventually "catches up" with the conditions. Thus, Loehle and Li (1996) termed the phenomenon *extinction debt* and concluded that the effect is real and should be accounted for in conservation and reserve design.

Over millennia, extreme isolation and protection from new predators and competitors can result in adaptive radiation of new species complexes in developing environments and ecological vacuums, as discussed in chapter 8. This can occur in montane and other isolated terrestrial continental environments as well as on oceanic islands. Additional effects of isolation from competitors or predators, also mentioned in chapter 8, include evolution of gigantism, flightlessness, and in some cases polymorphism or, more extreme, adaptive radiation of new species complexes.

Isolation of environmental conditions in continental settings also contributes to the development of *relictual faunas*, or ancient species persisting because of suitable environments. By definition, relicts persist in refugia. Relicts can be of any taxonomic group; several that occur on New Zealand were discussed in the previous chapter. Relicts can also occur in continental settings. Welsh (1990) reported that the Del Norte salamander (*Plethodon elongatus*), Olympic torrent salamander (*Rhyacotriton olympicus*), and tailed frog (*Ascaphus truei*) are paleoecological relicts that have long been associated with ancient primeval conifer forests of the U.S. Pacific Northwest. Marcot et al. (in press) compared the extant fauna of the interior western United States with Tertiary fossil fauna and reported the persistence of relicts of 7 Tertiary genera (represented by 32 extant species) and 20 Tertiary families (represented by 55 extant genera). Unlike the relict salamanders reported by Welsh, the relict genera and families of the interior West occupy a wide range of environmental conditions including native grasslands, shrublands, and forests.

It may be tempting to use relict faunas as management or ecological indicators, but some caveats need to be addressed. Relicts tend to occur in odd and disjunct locations so do not necessarily represent zonal or climatic climax conditions. Insofar as relicts are holdovers from earlier environments, their distribution does not necessarily reflect the suitability of current conditions. Thus, one should be wary of defining habitat and landscape requirements on the basis of current habitat use patterns of relicts without knowledge of the paleoecological history of the population and the site. Nonetheless, both floral and faunal relicts are often of scientific interest and deserve special conservation consideration (Millar and Libby 1991).

Is Isolation and Fragmentation Always Bad?

Isolation of environments and populations is not inherently evil. There are circumstances where isolation is an advantage for conservation. One advantage is avoiding spread of disease (Hess 1994), parasites, and pathogens. Another advantage is the establishment of several founder populations in sites with different disturbance dynamics and thus different likelihoods of success. In this case, overall persistence of the metapopulation is enhanced if there is little correlation of potentially disastrous environmental disturbances among the populations centers, and if populations are large enough to avoid genetic problems or if there is occasional gene exchange (outbreeding) between populations. Yet another advantage of isolation is the maintenance of relict faunas naturally isolated by changes in climate, vegetation, or landform.

In many cases, naturally developed island faunas are best left isolated from exotic species, especially introduced predators and competitors. Species at the edges of their range might occupy some isolated and very different environments through long-distance colonization events or because of relict distributions in refugia. Peripheral environments may prove important for long-term dispersion of the species and for evolution of unique morphs, subspecies, new species lineages, or entire species complexes. Often, it is important to describe carefully the spatial and temporal scale of isolation, as well as the causes, to determine appropriate management actions to maintain the condition or to ameliorate undesirable isolation problems caused by human activities.

Likewise, fragmentation of environments in landscapes is not necessarily always undesirable. On some scale, nearly all environments and species-specific habitats are fragmented, that is, noncontiguous, if only because they occur on different continents or land masses. Environments or resources are often naturally fragmented through time, as with seasonality of fruits in tropical forests, or through space, as with the occurrence of obligate symbionts such as some pollinators. In some circumstances, natural spatial and temporal fragmentation of environments or resources can lead to the evolution of new morphs or life forms. As with isolation, it is important to determine the causes and effects of fragmentation of environments to help direct appropriate management action.

What should be done about isolation or fragmentation of habitats for specific species that is clearly caused by human activities? The simple answer is to provide habitat linkages, including corridors or dispersed environments or resources, and to block habitats in the future. But these are blanket solutions that do not necessarily always meet multiple conservation objectives or fit the capability of the land or ownership patterns. We provide some guidelines below for considering conservation activity in cases of fragmented or isolated habitats.

Habitat Isolates

In this section we explore in greater depth the conditions and results of the isolation of environments, including kinds of habitat isolates and species-area relations. We compare resource patches with oceanic islands for applying principles of island biogeography and the results of environmental fragmentation and isolation to continental landscape conditions.

Kinds and Effects of Habitat Isolates

A habitat isolate is a set of isolated environmental conditions that are used by specific species. The term *isolate* may refer to oceanic islands, continental habitat islands (such as patches of montane forest), and individual patches of resources or environments occurring in a continental landscape setting. Habitat isolates also can include a variety of fascinating *azonal* conditions (i.e., environmental conditions atypical of a particular region). Some interesting examples include:

- *polinias,* or open water areas used by bearded seals (*Erignathus barbatus*) and walruses (*Odobenus rosmarus*) of the high Arctic;
- *gator holes,* or water-holding depressions created by American alligators (*Alligator mississippiensis*) and used by a variety of wildlife in the Everglades and Big Cypress Swamp in southern Florida;
- *kipukas,* or remnant native forests persisting on mounds in a landscape of relatively recent lava flow on shield volcanoes of Hawaii above about 1000 m elevation, in which persist Hawaiian honeycreepers and other native birds such as the 'oma'o, or Hawaiian thrush (*Myadestes obscurus*);
- *tapuis,* or ancient, colossal, erosional-remnant limestone pillars rising from Venezuelan rain forests, upon which have evolved unique plants, insects, and amphibians;
- *blue holes* (the Caribbean) and *cenotes* (Central America), or steep water-filled sinkholes occurring in karst landscapes, which are used for watering by many wildlife species and nesting by some including the turquoise-browed motmot (*Eumomota superciliosa*) in Yucatan, Mexico; and
- *nunataks,* or tundra islands surrounded by coalescent glaciers, which support relict plants derived from interglacial periods such as in the Juneau ice field of southeast Alaska.

It is relatively easy to identify most of these kinds of azonal habitat isolates and, where necessary, to establish species associations and conservation guidelines for them. Thus, the rest of this discussion will focus on oceanic islands, continental habitat islands, and resource patches.

Few studies have empirically demonstrated specific demographic effects of resource patch isolation or the improvement of habitat connectivity on dynamics of small populations. In one study of Bachman's sparrow (*Aimophila aestivalis*) populations in landscapes with linear patches of pine woodlands in South Carolina, Dunning et al. (1995) found that isolated woodland patches were less likely to be colonized than unisolated patches were. They also reported that woodland corridor configurations aided successful colonization of newly created woodland patches. They concluded that occupancy of woodland habitats at a regional scale can be aided by the design of habitat corridors, particularly for species that do not disperse well through inhospitable environments. We discuss habitat corridors further below (in "Utility of Habitat Corridors and Related Connectors").

In southwestern Oregon, Mills (1995) found that populations of California red-backed voles (*Clethrionomys californicus*) were virtually confined to isolated patches of forest remnants, seldom using the intervening clearcuts. Also, he found that trapping rates of voles were six times higher in remnant interiors than on the edge, likely corresponding to a greater density of the vole's preferred food, hypogeous (underground fruiting) sporocarps of mycorrhizal fungi. Thus, in this landscape, forest habitat isolates offered the only refuge for this species, which also showed evidence of being a forest-interior specialist. The long-term viability of such isolated vole colonies, however, is unknown.

Island faunas can be particularly vulnerable to isolation effects. Overall, many of the 40 or more bird species native to Hawaii, including honeycreepers and other species groups known only from the fossil record, became extinct after the islands were originally settled by Polynesians. Then, after Europeans later settled the islands two centuries ago, another 22 native species or subspecies of all taxonomic groups became extinct. These extinctions have been attributed to hunting by humans; introduction of disease, rats, cats, mongooses, grazing livestock, exotic birds, and game mammals; felling of native forests; urbanization; and spread of exotic vegetation.

Species-Area Relations

Community composition within isolated environments is dictated in part by individual species' relation to their habitat area. Species-area relations have formed one of the hearts of classic island biogeography, and a great deal has been published on this topic in many geographic areas and for many taxonomic groups.

At its simplest, the number of species, S, is affected by resource patch area A according to $S = CA^z$, where C = a scaling constant that varies by taxon and location, and z = the rate at which the number of species increases with increasing area. When plotted on a log-log relationship, where $\log S = \log C + z \log A$, the function appears as a straight line, and z becomes the slope of the line. From a number of studies of species richness on oceanic archipelagoes, z varies from approximately 0.24 for breeding land and freshwater birds in the West Indies and land vertebrates on islands of Lake Michigan to 0.49 for breeding land and freshwater birds on islands of the Gulf of Guinea (MacArthur and Willson 1967). In consequence of the z factor, as a general rule of thumb, twice the number of species seem to require 10 times the area (Darlington 1957; see also Harris 1984).

Howe et al. (1981) reported that area size of small rain forest remnants in New South Wales, Australia, was the best single predictor of bird species richness. Their species-area relationship suggested a z value of 0.50. Patch isolation, disturbance by livestock, and distance to water, however, all tended to reduce the number of resident species found. Lomolino (1984) reported that species-area and species-isolation relations of terrestrial mammals on 19 archipelagoes were consistent with basic predictions from the equilibrium theory of island biogeography (where species richness can be predicted given island or habitat isolate area, distance from species colonization pool, and rates of immigration, emigration, and local extinction). Also, Lomolino noted that the species-area relationship weakened (the z value declined) with more isolated islands and with less-vagile species.

Many cautions have been leveled at interpreting species-area relations and incidence functions of individual species. Perhaps the foremost caution is that of sampling effect, or the chance inclusion of a given species in a large area simply due to more area "sampled." A parsimonious approach to interpreting species-area and incidence functions would demand first assuming no causal relation or underlying mechanism, even if correlations are statistically significant. However, much anecdotal and some experimental evidence suggests strong ecological causes of such functions, although in some cases they are poorly understood.

Another caution is to remember that most studies provide only a snapshot of systems in transition. Species richness and occurrence of individual species are often in flux, particularly in continental habitat isolates. The snapshot does not reveal the trend of richness levels or occupancy rates, which may be influenced by seasonal changes in habitat use, or longer-term dynamics such as relaxation or tension (the increase in richness in newly suitable environments).

Environmental isolates in continental landscapes differ from oceanic islands in several functions: (1) Movements of organisms between habitat patches within landscapes may be more complex than between islands, because intervening environments in landscapes may provide varying degrees of required resources, and thus may serve as stepping-stone or sink areas. (2) Colonization and extinction processes may be more complex in landscapes than on islands. (3) Smaller environmental patches in landscapes may cease to function as habitat for a species if patch size and area of resources are small in relation to key life-history requirements. Also, as we have seen, edge effects can render small patches unusable for some interior-dwelling species, and the context of the patch in the landscape matrix can greatly affect its use. (4) Species-specific dispersal habitat in landscapes may be suboptimal but usable, whereas ocean stretches seldom provide useful dispersal habitat for terrestrial organisms. All four of these conditions can contribute to greater species richness and less pronounced faunal relaxation effects in environmental patches occurring in a continental setting than on oceanic islands having the same area.

An Example of Species-Area Relations in U.S. National Park Service Units

With these differences in mind, we can explore and interpret species-area relations of terrestrial vertebrates in administrative units of the U.S. National Park Service (NPS). Some 256 NPS units from all regions of the United States were included in this analysis. The units range in area from 0.01 to 33,716 km^2 (fig. 9.4) with a mean size of 1111 km^2 and a median size of 26 km^2. The size distribution is highly leptokurtic (kurtosis = 36.8), which means that the area of most units is clustered closer to the mean value than they would be in a normal distribution. The size distribution is also positively skewed (skewness = 5.7), which together with the kurtosis trend means that there are few very large units; rather, most are relatively moderate to small in size. The very large units occur mostly in Alaska, with a few of moderately large size in the Rocky Mountains and in the general western region (fig. 9.4). As we will see, it is important to understand the underlying regional bias in size distributions.

We can first inspect general patterns of the species richness of amphibians, reptiles, birds, and mammals within NPS units without regard to unit size, as shown in figure 9.5. Frequency-abundance histograms (the labeled diagonal cells of fig. 9.5) suggest that amphibians and reptiles are seminormally distributed (i.e., their abundance histograms suggest a truncated normal distribution), whereas those of birds and mammals are more fully normally distributed. The seminormal pattern of herps reflects that many NPS units have few or no herp species, some have a moderate number of species, and a few units have many species. This pattern may result from an overall greater herp species richness in the southeastern United States, which has many small NPS units (as well as at least one large unit, Everglades National Park). It might also result from the widespread absence of some herp species in the larger NPS units and in northern regions of the country (e.g., in Alaska, where NPS units tend to be large and herps absent). However, it also can be

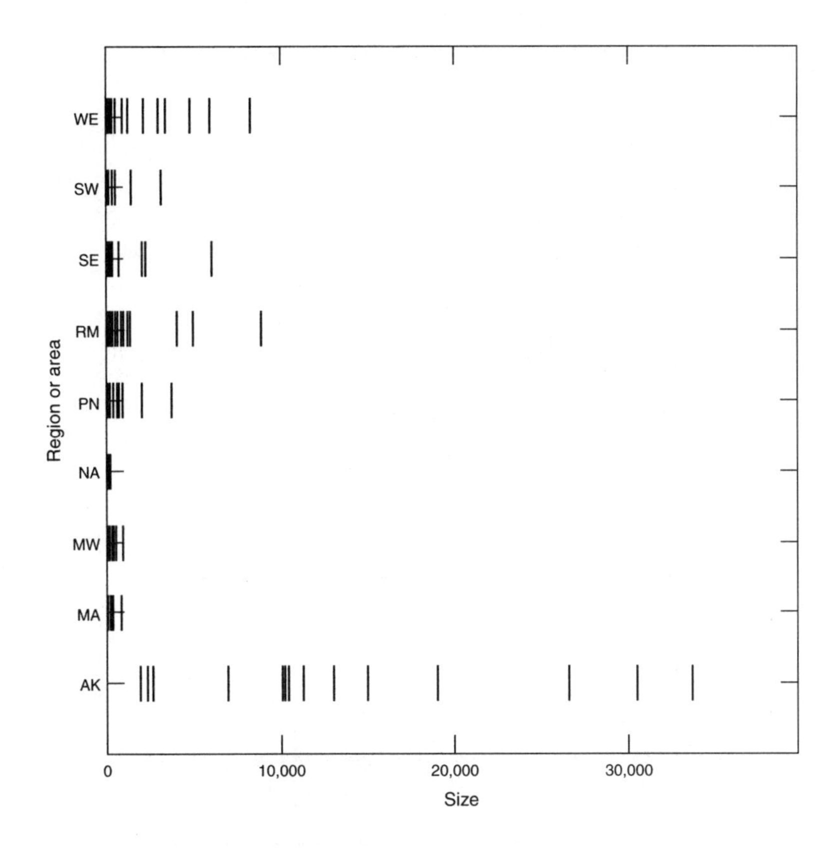

Figure 9.4. Size (km²) of 256 administrative units of the U.S. National Park Service, by geographic area and region. Regions: WE = West, SW = Southwest, SE = Southeast. Areas: RM = Rocky Mountains, PN = Pacific Northwest, NA = North Atlantic, MW = Midwest, MA = Mid-Atlantic, AK = Alaska. (Data from USDI National Park Service, Washington, D.C.)

caused by the small body size, small home range size, and thus small area requirements of some herps relative to the size of NPS units that contain them. Probably a combination of these effects contributes to the noted pattern. On the other hand, the frequency-abundance histogram of birds (labeled cell in fig. 9.5) suggests a more normal pattern, where the highest frequency is of intermediate bird species richness. These bird assemblages likely occur in NPS units of intermediate area, whereas the (largest size) NPS units occurring farthest north in Alaska hold the fewest bird species (see discussion on species-latitude correlations in chapter 2). The frequency-abundance histogram of mammals is nearly flat except for a modal frequency for low richness values and fewer units providing for many species (again, which may be a latitudinal effect).

The normality of the frequency-abundance histograms can be tested by use of probability plots. (Knowing the statistical distribution of

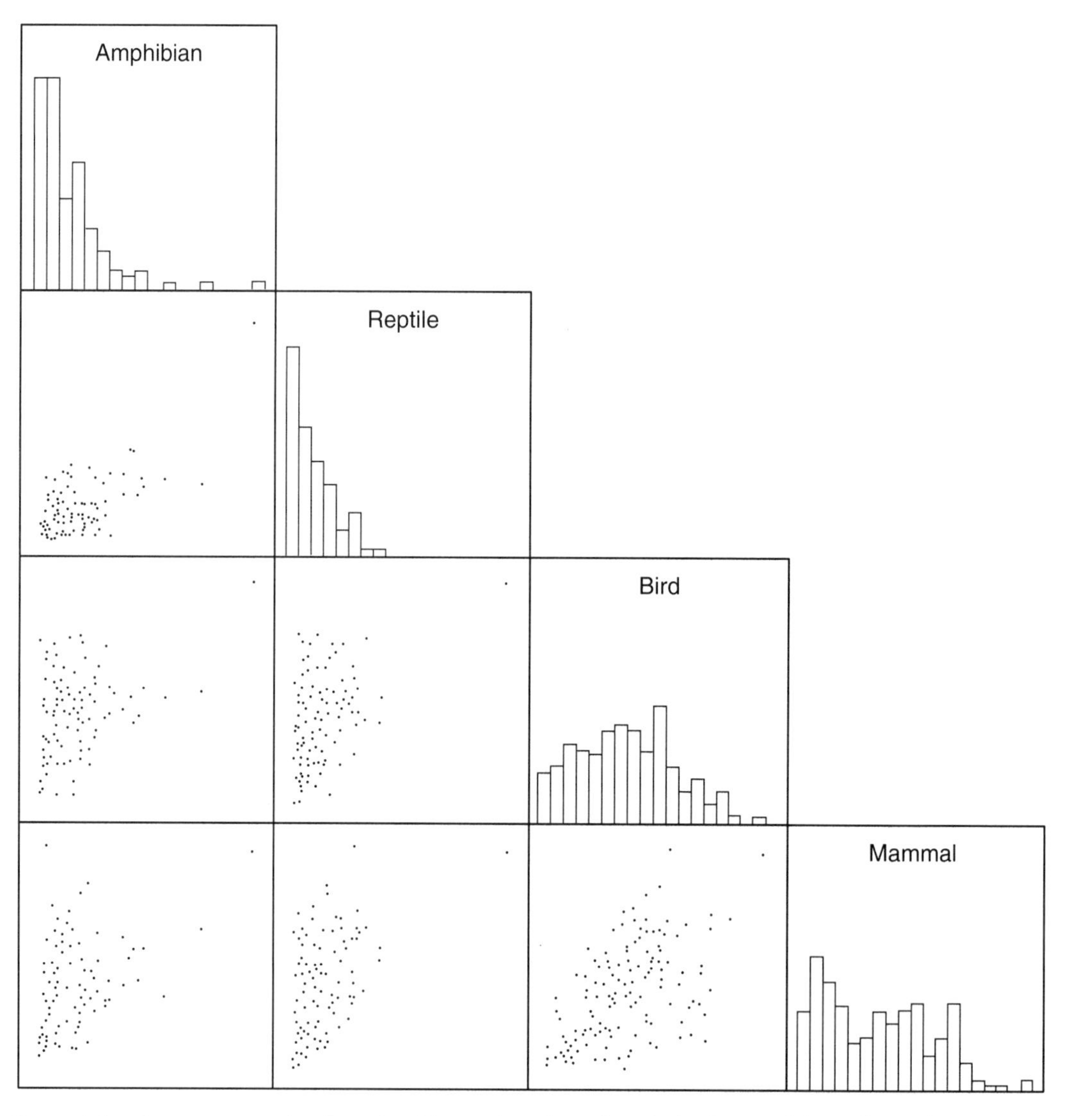

Figure 9.5. Scatterplot matrix of species richness (number of species) of terrestrial vertebrates in 256 administrative units of the U.S. National Park Service, by taxonomic class. The labeled diagonal cells show histograms of frequencies as a function of species richness (increasing along the abscissa). Note that amphibians and reptiles are seminormally distributed, whereas birds and mammals are more fully normally distributed. The unlabeled cells show data scatters of each taxa. (Data from USDI National Park Service, Washington, D.C.)

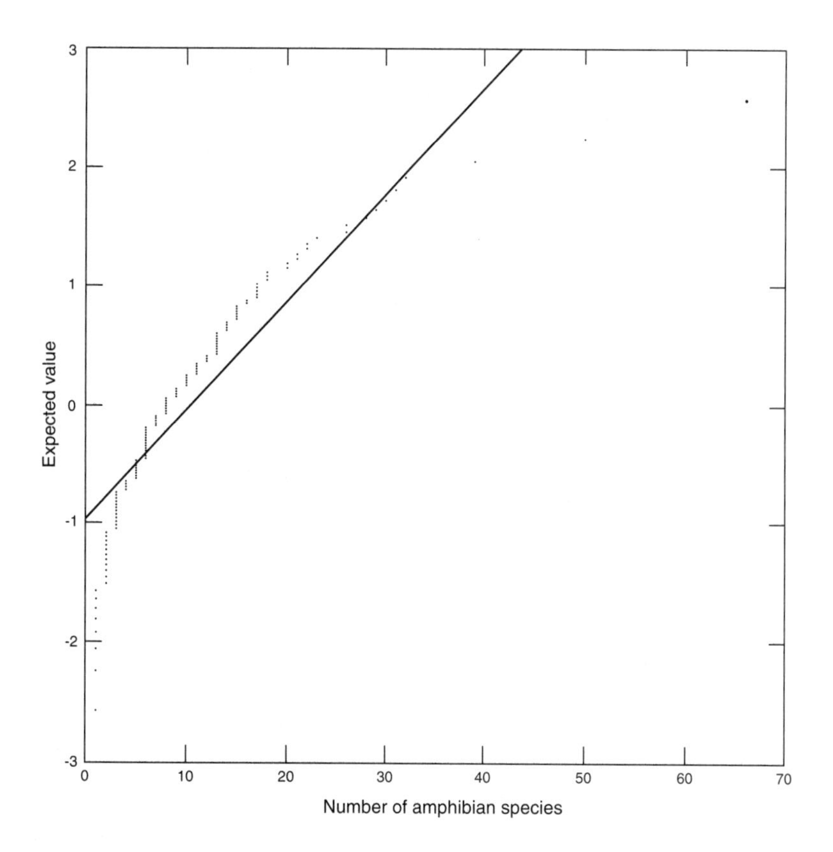

Figure 9.6. Normal-probability plot of amphibian species richness in administrative units of the U.S. National Park Service, to test normality of the distribution of the richness values. The lack of correspondence of the data points to the best-fit linear regression line suggests that the richness values are nonnormally distributed. (Data from USDI National Park Service, Washington, D.C.)

richness values could be important for other analyses that assume normally distributed data, and also for interpreting the primary richness patterns themselves.) For example, a normal-probability plot of the amphibian data (fig. 9.6) shows a highly nonlinear pattern; therefore, the data on amphibian species richness among NPS units are not normally distributed. Instead, the amphibian richness data better fit an exponential-probability plot (fig. 9.7). This suggests that the amphibian richness data come from a truncated normal distribution (as mentioned above) or, what can be computationally equivalent, from an exponential distribution of the form $f(y) = 1 - e^{(-y/s)}$, where s = the slope of the line on the exponential-probability plot. (Also, with log-normal species abundance distributions, as

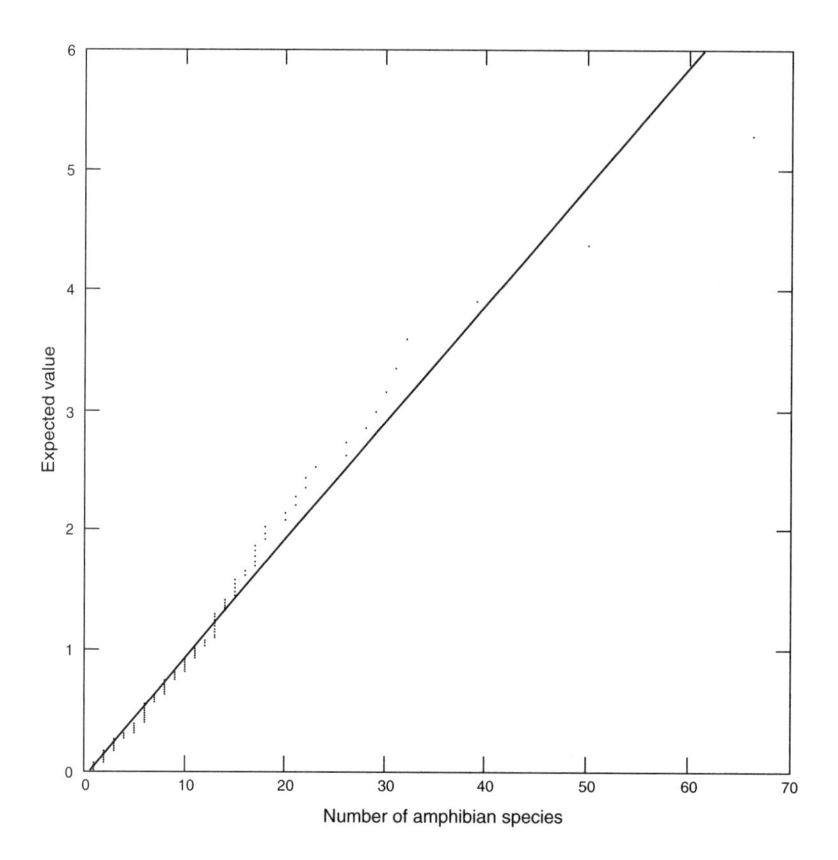

Figure 9.7. Exponential-probability plot of amphibian species richness in administrative units of the U.S. National Park Service, to test normality of the distribution of the richness values. The close correspondence of the data points to the best-fit linear regression line suggests that the richness values are seminormally distributed. (Data from USDI National Park Service, Washington, D.C.)

with the herp examples here, an exponential-probability plot can be approximated by a gamma-probability plot whose shape parameter = 0.5. A gamma distribution is a transformed chi-square with real degrees of freedom.) A similar pattern is present for reptile richness, but bird and mammal richness are better represented with normal-probability plots, suggesting normally distributed richness values. A box plot (fig. 9.8) of the amphibian richness values also suggests a skewed and nonnormal distribution, because outside values and outliers occur only on the higher richness side of the axis, and the median value lies to the left of center within the 1- and especially the 1.5-interquartile ranges. This is yet another, nonparametric, way to test for normality in the distribution of richness values.

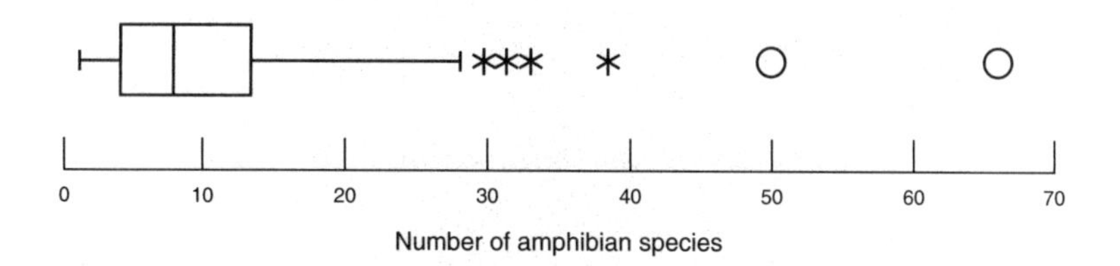

Figure 9.8. Box plot of amphibian species richness in administrative units of the U.S. National Park Service. The vertical line within the box is the median; the left and right edges of the box represent the 1-interquartile range; the arms represent the 1.5-interquartile range; the stars represent outside values; and the open circles represent outliers. (Data from USDI National Park Service, Washington, D.C.)

Next, as shown in the unlabeled cells of figure 9.5, we find that species richness within NPS units is significantly correlated ($P < 0.005$, Pearson correlation coefficient) between all taxa. This is not surprising given the range in size of NPS units, which span the minimal area requirements of the smallest herp or small mammal to the largest wide-ranging raptor or mammalian carnivore.

The analysis so far sets the stage now for interpreting species-area relations, as shown in figure 9.9. Species richness of each of the four taxonomic orders is significantly correlated with NPS unit area (table 9.1). The slopes of the regression lines represent the species-area *z* values, which range from about 0.10 for amphibians and birds to 0.16 for mammals, with reptiles intermediate. These slopes are substantially flatter (the *z* values are lower) than those from many oceanic island studies (cited above), suggesting that a moderating effect of species richness is occurring in the NPS units. However, these *z* values fit well those estimated for wildlife in reserves in other landscape settings. For example, the *z* value was 0.176 for breeding and wintering birds in reserves of the Royal Society for the Protection of Birds in Great Britain (Rafe et al. 1985). But many factors influence species occurrence and richness in reserves. Usher (1985) suggested using species-area relations only as a first approximation for wildlife reserve design and selection and thereafter studying dynamics of community composition and richness.

The four factors distinguishing environmental isolates as listed in the final paragraph of the previous section probably all play a role here. Likely to be of particular importance is the species-specific effect of habitat contributed by adjacent lands. That is, the NPS units may not act as true habitat isolates for each species; often units share boundaries with adjacent environments suitable for and containing populations of many of the same wildlife species. This boundary effect can confound simple interpretation of the effects of NPS management policies on wildlife (as is true of the policy effects of other land management agencies or owners), but must be considered when devising broad-area conservation plans that account for cumulative effects across ownership boundaries.

Boundary effects, species richness–latitude correlations, and differences in NPS unit size by geographic area and type of unit, all likely

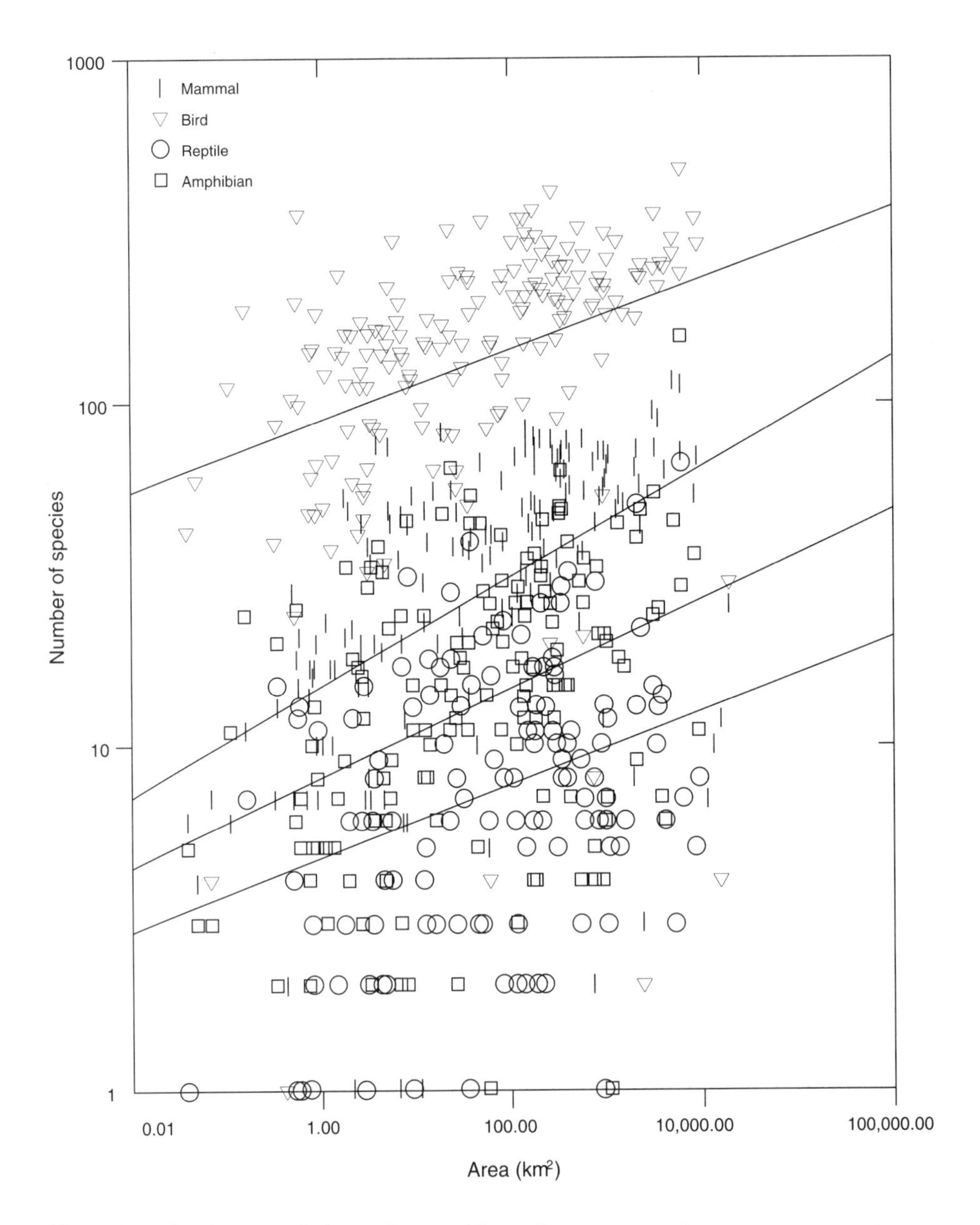

Figure 9.9. Species-area relations of terrestrial vertebrates in 256 administrative units of the U.S. National Park Service, by taxonomic class. Note that axes are $\log_{10}$ transformed. Diagonal lines are best-fit linear regressions and, from top to bottom, represent birds, mammals, reptiles, and amphibians. (Data from USDI National Park Service, Washington, D.C.)

Table 9.1. Species-area correlations and linear regressions of terrestrial vertebrate taxa in U.S. National Park Service units

Taxon	Regression constant	Regression x-coefficient (species-area z factor)	Degrees of freedom[a]	r	P
Amphibians	0.667	0.109	131	0.311	<0.0005**
Reptiles	0.899	0.131	150	0.384	<0.0005**
Birds	1.941	0.105	176	0.320	<0.0005**
Mammals	1.168	0.160	173	0.494	<0.0005**

Note: Correlations are based on log transformations of area and richness values.
[a]Degrees of freedom are based on pair-by-pair deletion of missing data
**Highly significant regression

contribute to the regression slopes and data scatter shown in figure 9.9. We can explore some of these relations by using amphibians as an example. The occurrence of amphibians is richer in NPS units in the Southeast than in the Southwest or West, and there is a higher median number of species per NPS unit in the North Atlantic area than in other areas (fig. 9.10). The Rocky Mountain area consistently has the fewest amphibians per NPS unit, except for the Alaska area, which has none. This matches general patterns of amphibians being richest in the Southeast and usually absent in the far North. The z slope coefficient for amphibians is significantly greater in national parks ($z = 0.227$) than in national monuments ($z = 0.111$) (both regressions and the differences between the two slopes are statistically significant at $P \leq 0.05$). This is because national parks tend to be larger or more variable in size, and thus collectively contain a greater diversity of habitats for amphibians than national monuments do. Perhaps by virtue of their size or location, national parks also tend to adjoin other landownerships (e.g., national forests) that additionally provide suitable environments for amphibians, more often than do national monuments, which tend to be more isolated from such environments, but this hypothesis needs testing.

Border Effects

Border effects of the Great Smoky Mountains National Park in the Appalachian Mountains of the eastern United States were studied by Ambrose and Bratton (1990). They reported that, from 1940 to 1978, the density of forest patches and cleared patches increased, the size of forest patches decreased, and there was little change in the percentage of forest cover. In contrast, inside the park borders, the density of forest patches and cleared patches decreased, and the size of forest patches and the percentage of forest cover increased. Thus, during this time period, landscape heterogeneity and connectivity of habitat for some forest-dwelling species changed significantly. Furthermore, different sections of the park border varied in these changes, depending on local variations in land use history.

Many of the other NPS units included in the above analysis are probably undergoing similar, variable changes in border effects. The greater the contrast in environments along the border, the more that parks (or other natural areas) may begin to act as habitat isolates and as

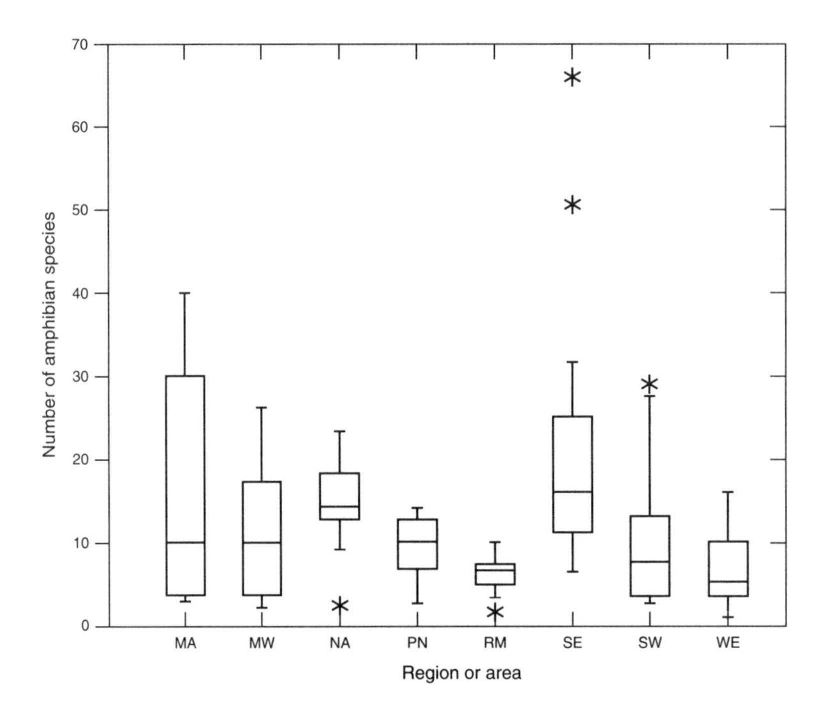

Figure 9.10. Amphibian species richness in administrative units of the U.S. National Park Service, by geographic area (see fig. 9.4 for names of areas and fig. 9.8 for a description of box plot symbols) (Data from USDI National Park Service, Washington, D.C.)

Freeman's (1986) "genetic islands" for some species. Diligent protection of parks and natural areas from undue or unplanned external threats can help protect wildlife resources (Goggins 1987) within parks and connectivity of populations and habitats across park borders. Also, in some cases, habitat isolates for some species may require considerable management of conditions within them if they are to remain useful in contributing to biodiversity. A hands-off approach will not always be the best effective management method in the short- or long-term.

Species-area relations, of course, are only one type of tool of potential utility for land managers, who should also strive to understand species-specific dynamics of border effects, as well as colonization, extinction, dispersal, migration, demographics, interspecific relations, and other factors discussed in this chapter and elsewhere in this volume. No one said that understanding and conserving wildlife in continental landscapes with patchy environments is easy.

The Conservation Value of Remnant Patches of Natural Environments

Our discussion of potential adverse effects of isolation and fragmentation of environments should not be taken to mean that remnant patches of natural environments have little or no conservation value. Quite the contrary, remnant patches may be all that is left in a landscape, watershed, or geographic region from which to rebuild a more natural biota or ecosystem (Saunders et al. 1987). Thus, remnants can have a disproportionately high conservation value per unit area compared with that of large natural areas, depending on the landscape context and management needs.

Often, remnant old-growth forests may be the final local bastion for species closely associated with such environments, including species of fungi, lichens, bryophytes, vascular plants, and invertebrates, as well as vertebrates. Protection of small, isolated old-forest remnants may be worthwhile; in some situations, such conservation measures may be highly efficient, in that they would entail only a small land area but would have great benefits. Although small old-growth forest remnants would not support all life history requirements of mid-sized and large species such as some mustelids and other wide-ranging carnivores, they can nonetheless provide valuable species source pools for propagules and inocula of many plants and small animals vital to the function of native ecosystems and soil productivity (Amaranthus et al. 1994). And forest fragments can provide at least some resources used by native vertebrates (Howe et al. 1981).

In some countries, small remnants of native forests are maintained for their value as environmental benchmarks, for scenic interest, or as sites of rich floral or faunal diversity. In India, old forest remnants in landscapes of managed teak (*Tectona grandis*) plantations are termed preservation plots and serve, in part, as restoration benchmarks. In many cases, remnants of natural environments can serve as connectors between larger patches, as we discuss next.

Remnant patches of native environments can render vital service to conserving sources of associated plants and animals and providing stepping-stone connectivity of a habitat for a species throughout a landscape. Beyond this, native remnants can serve a purely utilitarian and anthropocentric role such as providing sources of valuable plants, pharmaceuticals, and foods (Schelhas 1995). Remnants can also provide valuable learning experiences for restoration management. We applaud any efforts to consider the conservation value of remnant native environments in the full context of these and other values.

Utility of Habitat Corridors and Related Connectors

Use of habitat corridors for linking populations of a species was discussed in chapter 3. A number of hallmark or summary papers on habitat corridors are available (e.g., Forman 1983; Harris 1988; MacClintock et al. 1977).

Corridors can include linear strips of environment that link larger habitat blocks for particular species. One type of a linear habitat corridor is a riparian forest buffer. Riparian forest buffers have long been part of many silvicultural management prescriptions for providing movement corridors for wildlife species in managed forest landscapes (e.g., Chapel et al. 1992), and they have been proposed as a means for maintaining regional biodiversity (Naiman et al. 1993; Harris 1984), although such roles have been little studied.

Darveau et al. (1995) reported that in boreal balsam fir (*Abies balsamea*) stands in Quebec Province, Canada, bird densities increased 30–70 percent in riparian forest strips the year following timber harvest and decreased thereafter to approximately pretreatment levels, representing an initial enticement of disturbance-tolerant birds and a subsequent faunal relaxation. Forest-dwelling bird species were less abundant than ubiquitous species. Four forest birds—the golden-crowned kinglet (*Regulus satrapa*), Swainson's thrush (*Catharus ustulatus*), blackpoll warbler (*Dendroica striata*), and black-throated green warbler (*D. virens*)—were virtually absent in 20-m-wide strips but present in 60-m-wide strips. This incidence function suggests an effective riparian forest buffer size for forest bird conservation, but data are needed on longer-term demographic response.

Some biologists have posed that riparian buffer strips can serve to attract predators, perhaps at undesirable density and diversity. Vander Haegen and DeGraaf (1996) found higher predation rates on both ground and shrub bird nests in riparian buffer strips created by commercial clear-cutting than in intact forests. Predation rates were similar in main stem and tributary buffer strips. The predators consisted of six mostly forest-dwelling species that used the buffers to forage and perhaps to travel. The authors recommended buffer strips ≥150 m wide along riparian zones to reduce edge-related nest predation, especially in landscapes where buffers constitute a significant portion of the existing remnant forest. However, the effect of higher predation on bird species fitness in the clear-cut-created buffers is unknown and needs empirical study. Evidence of higher predation, as with competition, does not necessarily mean that populations of target species are no longer viable.

Corridors can also include specific habitat components more or less linearly arranged across a landscape, such as perch poles in the desert. In the Mojave Desert of California, Knight and Kawashima (1993) found higher densities of common ravens (*Corvus corax*) along highway and powerline transects than in control areas with no highways or powerlines within 3.2 km, and raven nests were more abundant along powerlines. Ravens may have been attracted to highways for road-kill carrion. Red-tailed hawks (*Buteo jamaicensis*) and their nests were more abundant along powerline transects than along highway or control transects. The authors recommended that land managers evaluate possible effects on vertebrate populations and species interactions when assessing future linear right-of-way projects.

Another form of corridor is transmission-line cuts, which can open forest or woodland canopies, provide lush grass, forb, or shrub cover, and much linear edge across a landscape that serves to intersect other resource patches. Effects of transmission-line corridors on wildlife were studied by Anderson (1979), Chasko and Gates (1982), Kroodsma (1982), and many others. Chasko and Gates found that, in a Maryland oak-hickory forest, the corridor was dominated by mixed-habitat bird species rather than grassland birds. The authors defined mixed-habitat bird species as those that use two or more vegetation conditions, such as grasslands and shrubs. They also found that the few isolated shrub patches occurring in the grassy corridor provided "habitat islands," where nest density and fledging success were high. Predators apparently were not able to exploit patchily distributed shrub nests in the corridor; therefore, the authors recommended managing for increased vegetation heterogeneity within transmission-line corridors to increase nest density and success of mixed-habitat bird species.

Corridors and habitat connections have been posed for large mammals as well. Silvicultural prescriptions designed to provide deer habitat corridors have been popular for some time (e.g., Wallmo 1969). Beier's (1993) simulation study of cougar habitat merged considerations of minimum habitat area and corridor use. He concluded that habitat areas as small as 2200 km^2 could support cougars with low extinction risk if demographic rescue rates along habitat corridors were on the order of one to four immigrants per decade.

Studying arboreal marsupials in southeastern Australia, Lindenmayer and Nix (1993) found that designing wildlife habitat corridors on the basis of suitable habitat, species home range, and predictions of minimum corridor width alone was insufficient. They posed that additional design criteria should include site context and connectivity and the social structure, diet, and foraging patterns of desired species. In particular, they found that the key variable affecting arboreal marsupials' use of linear strips of vegetation was number of trees with hollows as potential nest sites.

Habitat isolates, corridors, and connectors do not, however, exist in static form. In most ecosystems, they are subject to systematic and planned changes and to stochastic disturbance events, which often render their long-term conservation problematic and particularly challenging. In the next section, we discuss such dynamics of habitats in landscapes.

Dynamics of Habitats in Landscapes

The study of dynamics of habitats in landscapes covers many topics that range widely across scales of space and time (Delcourt et al. 1983), including soil dynamics, fire ecology, vegetation succession, meteorology, climatology, and paleoclimatology. To best understand effects on wildlife, disturbances should be depicted according to their frequency, intensity, duration, location, and geographic extent. In this way, disturbances to habitats can be incorporated into spatially explicit models of population demography to project effects on species distribution and viability (Conroy et al. 1995; Dunning et al. 1995; Holt et al. 1995; Turner et al. 1995).

Four major classes of disturbance can be identified according to their intensity and geographic extent (fig. 9.11). Type I disturbances

<table>
<tr><td colspan="2"></td><td colspan="2">Geographic area affected</td></tr>
<tr><td colspan="2"></td><td>Widespread
(>1000 ha)</td><td>Local
(1–1000 ha)</td></tr>
<tr><td rowspan="2">Degree of disturbance</td><td>high</td><td>Type I
Major environmental catastrophe
(volcanoes, major fires, hurricanes)</td><td>Type II
Local environmental disturbance
(wind, ice storms, insects, disease)</td></tr>
<tr><td>low</td><td>Type III
Chronic or systematic change over wide area
(predators, competition, forestry)</td><td>Type IV
Minor environmental change
(local fires, developments)</td></tr>
</table>

Figure 9.11. Four types of disturbance shown by degree, or intensity, and geographic area affected

are major environmental catastrophes that are relatively short-term and intense, affecting large areas. They include volcanoes, major fires, floods, and hurricanes (or typhoons in the Pacific Ocean). Hurricanes can have major effects on wildlife habitat and populations, and in some cases habitat refugia can serve as important protection zones for populations.

Hurricanes can affect species and resources differentially (Boucher et al. 1990). For example, Pierson et al. (1996) found that two severe cyclonic storms on Samoa caused a more severe population decline in the more common and widely distributed species of flying fox (*Pteropus tonganus*) than in the endemic species (*P. samoensis*). They attributed this difference to greater susceptibility of *P. tonganus* to hunting mortality in villages, because the species foraged there on flowers and fruits after the storms. The greater susceptibility was also because of the greater proportion of foliage in the diet of *P. samoensis*, whereas *P. tonganus* is more highly frugivorous, and overall, poststorm density of flowers and fruits was depressed. The endemic species was protected from storm effects in rain forest refugia, particularly in areas of high topographic relief such as volcanic cones and steep valleys protected from wind damage, whereas the more widespread species occurred in the less protected village environments, which were more severely damaged.

One rather well-studied storm was Hurricane Hugo, which can serve here as an example of major effects of Type I storm disturbances on wildlife and habitat. Hugo hit South Carolina, in the southeastern United States, on 21 September 1989 with sustained winds of 217 km/hr, gusts of 282 km/hr, and a storm surge of ≥5.8 m. Damage to forests by Hugo was greater than that of Hurricane Camille, the eruption of Mount St. Helens, and the recent Yellowstone fires combined. Hugo damaged forests on over 17,800 km^2, with damage greatest in coastal Francis Marion National Forest, where about 75 percent of commercially marketable pine trees were felled by the storm (Ehinger 1991). Many forests of South Carolina hit by Hugo had held some of the densest colonies of the endangered red-cockaded woodpecker found on the U.S. east coast. Prior to Hugo, Francis Marion supported about 25 percent of all southeastern red-cockaded woodpeckers, including one of the world's largest populations of approximately 500 breeding pairs (West 1989). In Francis Marion, most of the trees damaged were mature pines favored by the woodpeckers; Hugo reduced the woodpecker populations there by 63 percent and destroyed 87 percent of known active cavity trees and 50–60 percent of foraging trees (West 1989). Recovery of forest conditions for the woodpeckers could take 75 years (Hamrick 1991), and demographic consequences to woodpecker populations will likely be long-lasting (Hooper et al. 1990).

Hugo also passed directly over the 11,300-ha Caribbean National Forest on Puerto Rico and heavily damaged 8900 ha, including breakage and blowdown of many trees and almost complete defoliation of the remaining trees (West 1989). Initial censuses showed that only 23 of the original colony of 46 wild Puerto Rican parrots (*Amazona vittata*), a highly endangered species, remained. However, the captive flock of 53 parrots was unhurt, which is a prime example of the value of captive breeding of at least some highly endangered species subject to periodic, major disturbance events in the wild. After the storm, many predators were observed in areas of heavy treefall; together with destruction of nest structures caused by Hugo, predation can seriously depress some populations. However, the parrot population has since recovered to population levels approaching that of pre-Hugo, thanks to the surviving and actively reproducing birds, the availability of improved natural nest cavities, and the effectiveness of the enhanced nest management program. Since Hugo, some Puerto Rican parrots have nested at lower elevations and used cavities in tree species never reported since parrot observations began in 1968 (Vilella and Garcia 1995). Thus, this major catastrophic event changed the abundance of tree cavities and structure of vegetation such that the availability of suitable sites for this endangered species was altered. This resulted in changes in geographic areas used by the birds, which in turn helped maintain the population.

Effects of Hurricane Hugo on other forest bird species in Puerto Rico were also studied by Wunderle (1995), who reported that canopy dwellers shifted to foraging in forest understories and openings, and bird assemblages became far less distinct by microhabitat conditions than they had been before the hurricane. Such habitat displacement of birds also included movement of frugivorous birds into preexisting gaps and forest invasion by forest-edge or shrubby second-growth species. Wunderle speculated that it may take many years for resources and structures in forest understory and gaps, and associated unique birds assemblages as observed prior to the storm, to become distinct again.

Other examples of Type I disturbances that have been studied in North America include the 1980 eruption of Mount St. Helens (Wissmar et al. 1982), the 1988 Yellowstone National Park fire (Mishall et al. 1989; Pearson et al. 1995; Wakimoto 1990), and the 1993 Mississippi River floods (Custer et al. 1996; Miller and Nudds 1996). Each kind of Type I disturbance can bring very different changes to environments and species composition, and deserves individual attention in habitat planning that accounts for such infrequent events (chapter 11). Some Type I disturbance events occur in multiple cycles, such as fire regimes or floods with various repeat frequencies. In one example, river discharge events can occur in multiple, overlapping frequencies (figs. 9.12, 9.13).

Type II disturbances include locally intense environmental changes from such events as wind storms, ice storms, and local outbreaks of defoliating insects. Forest canopy gaps, sometimes called microserules, undergo local succession of plant species and are important contributors to overall forest stand structure and composition. For example, Lawton and Putz (1988) reported that wind-formed canopy gaps of tropical montane elfin forests of Monteverde, Costa Rica, tended to be small, occur frequently, and annually affect small proportions (≤1 percent) of the overall forest cover, and that they promoted growth of shade-intolerant plant species, tree saplings on nurse logs and in mineral soil disturbed by uprooting trees, and plants from soil seed banks more than from seeds dispersed into the gap. The researchers also found that many saplings in gaps originated from epiphytes in the crowns of the trees that fell. In temperate forests of northern Minnesota, Webb (1989) found that tree damage from thunderstorm winds (25–35 m/sec) related to tree size, species, species wood strength, and incidence of species-specific fungal pathogens. He found many within-stand differences in wind-caused mortality and subsequent plant development of the canopy gap due to differences in shade tolerance, initial forest structure, gap size, and windfirmness of shade-tolerant understory plants.

Gap size also influences nitrogen cycling dynamics, as reported by Parsons et al. (1994), in forests of lodgepole pine (*Pinus contorta*) in southwestern Wyoming. In these forests,

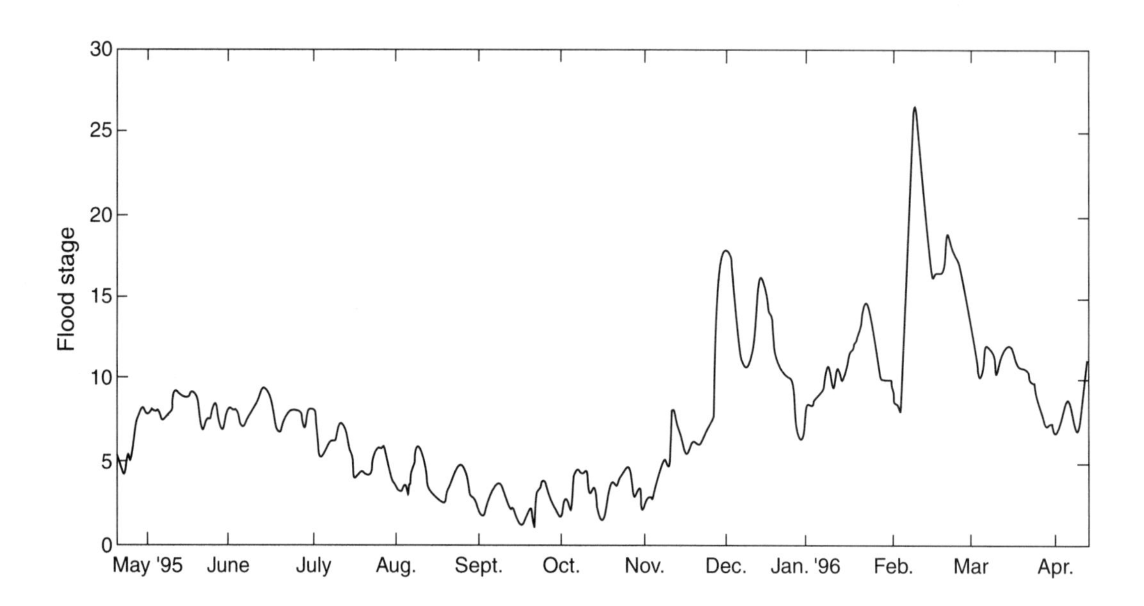

Figure 9.12. An example of how disturbance regimes can overlap in frequency and intensity. This is a plot of flood stages of the Columbia River at Portland, Oregon, from May 1995 to April 1996 (flood stage is an index combining flow and height above mean river surface level). Overlapping fluctuations in river flood stages, in order of decreasing frequency and increasing intensity, are evident in this chart as follows: semimonthly peaks spanning approximately 3 flood stages; annual peaks spanning approximately 10–15 flood stages; and major 50–100-year flood events spanning >10 flood stages (peaks in February) superimposed on the annual flood stage cycle. This hydrograph also represents the key features of hydrological regimes, including magnitude, frequency, duration, timing, and rate of change (Richter et al. 1996). Compare with figure 9.13. (Data from University of Oregon, Eugene, Oregon)

removal of 15–30 tree clusters represented a threshold above which significant losses of available nitrogen to the groundwater occurred.

Outbreaks of defoliating insects such as the spruce budworm (*Choristoneura fumiferana*, Lepidoptera: Tortricidae) are another Type II disturbance and can cause local to extensive changes in forest conditions. Complex species interactions can determine control of such insect pests. For example, in spruce-fir forests in northern New Hampshire and western Maine, Crawford and Jennings (1989) reported that the entire bird community showed significant functional responses (increased foraging) to increasing budworm density, whereas only two species, the Canada warbler (*Wilsonia canadensis*) and golden-crowned kinglet (*Regulus satrapa*), also showed numeric responses (increased reproduction). The researchers concluded that insectivorous birds are capable of dampening the amplitude of budworm infestations when habitats are suitable for supporting adequate bird populations. This corroborates other similar findings such as those

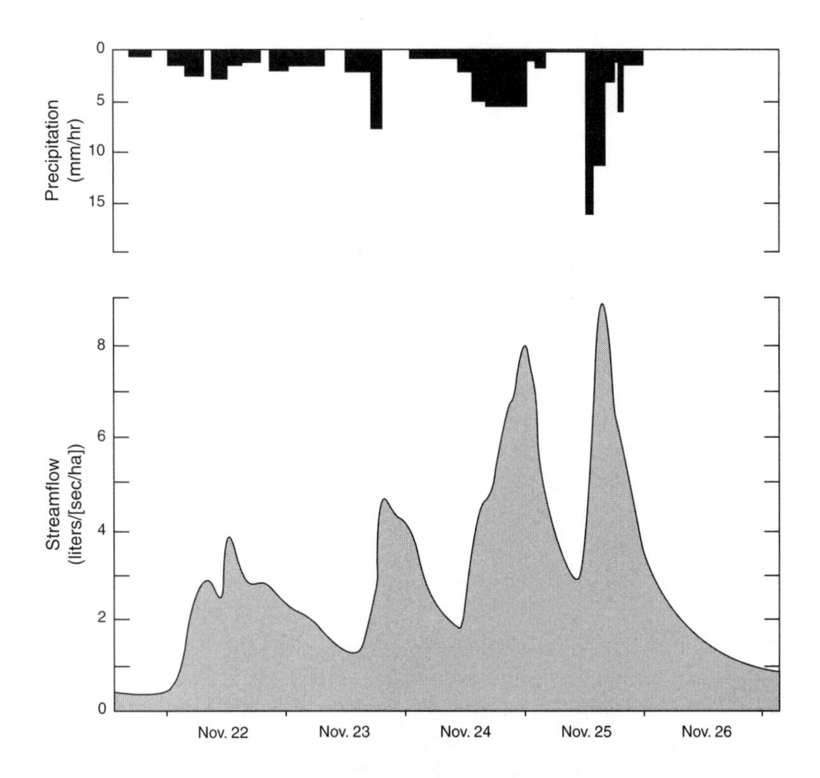

Figure 9.13. An example of streamflow dynamics over the course of five days, or at a much finer time resolution than the hydrograph presented in figure 9.12. Such localized and short-duration fluctuations may nonetheless affect hourly or daily use of stream resources by aquatic species such as stream salamanders, frogs, dippers, ducks, and mustelids. Data originally from H. J. Andrews Experimental Forest, Oregon, 1977. Note the rapid response of streamflow to precipitation onset and cessation. (From Swanston 1991:149, fig. 5.5)

suggested for woodpecker foraging upon bark beetles in northern Colorado (Koplin 1969).

Like pandemic levels of native pathogens (e.g., root rot) and defoliators (e.g., budworm), relatively intense fires in grasslands, shrublands, and forests often tend to leave spotty footprints rather than entirely denuding large areas of vegetation (Johnson 1992; Agee 1993). In this way, as with smaller canopy gaps created by windfall of trees, fires often produce complex patterns of dynamics within and among vegetation patches. As vegetation patches change in time, so do responses by animals in their use of patches for breeding, foraging, refuge from predators, resting, and dispersal. Wildlife species evolve in concert with native disturbance regimes and take optimal advantage of resources distributed through space and time in shifting patches. Thus, the further that human activities alter native patch disturbance dynamics, the greater may be the discontinuity with the evolved habitat selection behaviors of some species. In some cases, even if a suitable environment is present but is greatly altered in patch distribution pattern and temporal occurrence, an associated wildlife species may be excluded. This aspect of coevolution with native disturbance regimes, native resource patch dynamics, habitat selection behaviors of wildlife, and effects of changes in native disturbance regimes on wildlife viability and fitness is little studied and deserves research attention.

Type III disturbances are chronic changes over wide areas, including slow alteration of native landscapes for human habitation, ecological succession, and long-term climate change. Wildlife relations to Type III disturbances include changes in species abundance along seres, as studied by Anderson et al. (1980) for herbivorous mammals in montane environments and by Bock et al. (1978) for postfire seral changes in bird community structure in the Sierra Nevada. Climate changes have the potential to affect distribution of vegetation and wildlife massively over broad areas (chapter 2). Even climate fluctuations lasting on the order of a few years can greatly affect productivity of some wildlife populations, sometimes far away. One example is the effect of El Niño warm water cycles in the eastern Pacific Ocean, which interrupted the 1982–1983 breeding chronology and reproductive attempts of seabird communities on Christmas Island in the central Pacific Ocean (Schreiber and Schreiber 1984). The entire seabird community failed reproductively and temporarily disappeared from this equatorial atoll as a result of the distant upwelling effect on food resources. Effects of that particular El Niño may have also extended to nonmarine species far from the Pacific.

Type IV disturbances include minor environmental changes. These are low-intensity and local effects of such events as spot fires, low-density rural developments along edges of natural landscapes, and gap dynamics of vegetation canopies (Acevedo et al. 1995). In particular, gaps can be caused by natural plant death or mortality agents such as insect defoliators, plant pathogens, and plant diseases, as well as by fire and weather conditions such as ice or wind storms. Individual treefalls in forests' open canopies change local microclimates at the forest floor and afford sun-tolerant plants a foothold. Treefalls uproot soil masses and redistribute litter, duff, and upper soil layers along with their associated microfauna and microflora. Across larger areas, other Type IV disturbances can include the erosion of soils and substrates over short time spans and the erosion of landforms over long time spans.

Across watersheds and landscapes, disturbances affecting vegetation patch structure and composition tend to alter such ecosystem processes as surface water discharge, nutrient runoff, organic matter input to soils, net productivity, and microclimate. In turn, these changes can influence species composition associated with soil, ground surface, plant canopy, and other substrates. Changes in vegetation patches can alter energy balances of individual organisms, such as by changing food content or values in foraging patches, thereby tipping the balance of foraging efficiency and affecting successful reproduction and fitness of organisms. Such may be the case with northern spotted owls nesting in forest landscapes heavily fragmented from clearcut logging (Meyer et al. 1992).

Managing native vegetation conditions in landscapes that are subject to relatively frequent, intensive Type II and IV disturbances, such as stand-replacing fires, and in landscapes where native vegetation occurs only in small, isolated patches can be a great challenge. In such cases, there may not be a sufficient area or number of patches of native vegetation to provide for resilience in the face of intense disturbances, so high cost and much effort must be expended to prevent changes. In some circumstances, such as the fire-prone inland of the U.S. West, disturbances will inevitably occur regardless of, or in this case because of, efforts to thwart them (Everett et al. 1994).

Often, land use history, such as decades of fire suppression with a resulting buildup of fuels, high-grade selection logging of large old trees, and mining activities resulting in changes in soil and riparian systems, accounts largely for current or impending dynamics of vegetation. Rates of return to predisturbance conditions vary by vegetation type. For example, following forest clearcutting in Hubbard Brook Experimental Forest in New Hampshire, quick regrowth of dense stands of pin cherry (*Prunus pensylvanica*), a short-lived, early successional tree, was found to reduce loss of nutrients from the ecosystem and to help return much of the nutrient standing crop (N, K, Mg, Na, but not Ca) by only six to eight years postcutting (Marks and Bormann 1972).

What management lessons can be drawn from this brief review of major categories of disturbance and dynamics of resource patches in landscapes? First, in all but the simplest ecosystems, dynamics of vegetation and environmental factors consist of a complex medley of changes occurring on multiple "schedules." Landscapes must be assessed individually to determine site histories and likely vegetation responses to any disturbance regimes induced or altered by management activities. For example, fire behavior can be influenced by historical factors such as recent fire occurrence in the area, by proximate factors such as weather and topography, by vegetation factors such as canopy gap openings (which in turn are influenced by windfirmness of the vegetation, orientation to prevailing winds, and other factors), and by many other factors. In turn, fire, like other disturbance regimes, influences the likelihood of other changes such as succession, and can greatly alter suitability of the environment for wildlife.

Second, wildlife can play a major role in affecting disturbance regimes and how habitats respond to disturbances, such as through predation or transportation (phoresis) of disturbance agents (e.g., budworms and pathogens) and herbivory influence on vegetation otherwise susceptible to disturbances. Such ecological functions of species often are not, but should be, considered in management plans that alter or introduce disturbances.

Third, management may wish to study more fully how activities change native disturbance regimes and how wildlife may respond behaviorally (functionally) and demographically (numerically) to such changes.

And fourth, the specific future conditions of most ecosystems incurring disturbances operating at multiple scales of space, intensity, and time are not very predictable. Rather, what can be better predicted are short- and long-term disturbance regimes operating at various spatial scales. In this sense, management can then craft a set of desired future dynamics, perhaps to reconstruct or mimic native disturbance regimes in which some species may have evolved optimal habitat selection behaviors. This last point is discussed further in chapter 11.

Monitoring Habitats and Providing for Wildlife in Landscapes

For decades, if not centuries, European forest and woodland managers have been managing some aspects of wildlife habitat on the landscape scale. Successful placement of salt licks and watering holes for ungulates, nest boxes for waterfowl, and nesting platforms for raptors and many other such activities implicitly account for landscape patterns of vegetation and habitat selection behaviors of animals. What is relatively new in habitat management, however, is the focus on ecosystem processes and multiple species and the need for explicitly monitoring environments and wildlife response on landscape scales. Monitoring landscape effects on wildlife populations can include tracking trends in populations and habitat occupancy and determining habitat selection behaviors and numeric and functional responses of organisms.

Understanding how populations and organisms occur in resource patches and use fragmented landscapes can help managers set realistic expectations for wildlife presence and population density. Since not all resource patches may be occupied by a species simultaneously, those patches that are small, that contain less than the full complement of resources for the species, and that are located peripherally could be monitored across seasons and years to determine true trends in occupancy patterns. Such monitoring should occur before changing management direction in a way that would severely reduce habitat quality, particularly for at-risk species. At the very least, preactivity monitoring sets the baseline from which changes can be gauged (see chapter 7).

In general, to help ensure viable populations, management guidelines should provide for the full range of habitat conditions and habitat quality needed to maintain a well-distributed population. Certainly, not all species are equally well-distributed, nor is a given species equally abundant across its distributional range. Thus, results of monitoring and implementing management guidelines should be tempered with knowledge of the natural abundance patterns of species. Providing a full array of resources and environments to help ensure persistence of a population should entail providing: seasonal and range movement habitats; habitats for migration, dispersal, and resting (including waterfowl loafing); and habitats to maintain genetic diversity by ensuring a large effective population and connected subpopulations, where natural conditions provide. In addition, providing for peripheral habitat, suboptimal or secondary habitats, sinks, corridors, key links, and other environments critical to maintaining metapopulations (see table 3.3) may be in order. Such management guidelines are best developed from local empirical knowledge of the life history and ecology of the species in the field. No model or generalized set of guidelines can replace basic field zoology and autecological studies.

Monitoring would best serve its purpose when keyed to "early warning signals" of impending changes in populations. However, many long-lived wildlife species or species with high site fidelity may not respond quickly to initial changes in environmental and habitat conditions (see also the discussion of faunal relaxation and extinction debt in "The Effects of Isolation," above). Population responses, functional or numeric, might show lag effects. By the time the population abundance has responded, it may be too late to alter the course of environmental change. This can be solved by selecting population parameters that respond quickly to habitat changes. Such parameters might include changes in breeding attempts, foraging behaviors, selection of foraging substrates, juvenile sex ratios and population age structure, or the proportion of nonreproductive and nonterritorial floaters in the population. Also, functional responses of populations might occur more quickly than numeric responses. Thus, foraging activities, nest-site switching, or congregations of breeding units (e.g., pairs, colonies, herds, or leks) may serve as better early warning signals than breeding rates and overall population density would. Ultimately, the purpose is to index and ensure long-term high fitness of individuals, that is, the reproductive vitality of offspring.

Other early warning signals might include dysfunctions in social or individual behaviors, such as increases in "divorce" rates of otherwise tightly pair-bound species such as cranes and some owls, or increases in the rate at which highly territorial organisms such as raptors abandon seemingly suitable areas. Increases in the nonbreeding segment of the population (including floaters or helpers) and declines in individual health (e.g., percentage of body fat) may also signal decline in environmental suitability.

The trick in successful monitoring is not only to detect adverse changes in time to correct them but also to determine the cause. This is deceptively difficult and one reason that passive adaptive management (learning from management activities not designed as experiments) often fails to serve as a good monitoring or learning tool. For instance, causes of the starvation of woodland caribou (*Rangifer tarandus*) during winter can be proximate, stemming from a low density of lichens fallen from tree canopies that provide food, or ultimate, stemming from a widespread decline in air quality caused by industrial and automobile emissions that adversely affect the growth and reproduction of lichens and increase their mortality.

Monitoring for proximate or ultimate causes means beginning with an understanding of how the system works, representing that understanding in a conceptual and diagramatic model (see chapter 10), and then crafting a statistically correct sampling design to determine trends of the parameters of interest and the factors that most likely influence such trends. Statistical considerations can be integrated into an active adaptive management approach (learning from management activities that are designed as experiments). Only by teasing out causal factors can we determine which aspects of management guidelines to support or amend.

Literature Cited

Acevedo, M. F., D. L. Urban, and M. Ablan. 1995. Transition and gap models of forest dynamics. *Ecological Applications* 5(4):1040–55.

Agee, J. K. 1993. *Fire ecology of Pacific Northwest forests.* Washington, D.C.: Island Press.

Amaranthus, M., J. M. Trappe, L. Bednar, and D. Arthur. 1994. Hypogeous fungal production in mature Douglas-fir forest fragments and surrounding plantations and its relation to coarse woody debris and animal mycophagy. *Canadian Journal of Forest Research* 24(11):2157–65.

Ambrose, J. P., and S. P. Bratton. 1990. Trends in landscape heterogeneity along the borders of Great Smoky Mountains National Park. *Conservation Biology* 4:135–43.

Anderson, D. C., J. A. MacMahon, and M. L. Wolfe. 1980. Herbivorous mammals along a montane sere: Community structure and energetics. *Journal of Mammalogy* 61:500–519.

Anderson, S. H. 1979. Changes in forest bird species composition caused by transmission-line corridor cuts. *American Birds* 33:3–6.

Beier, P. 1993. Determining minimum habitat areas and habitat corridors for cougars. *Conservation Biology* 7(1):94–108.

Bock, C. E., M. Raphael, and J. H. Bock. 1978. Changing avian community structure during early post-fire succession in the Sierra Nevada. *Wilson Bulletin* 90:119–23.

Boecklen, W. J., and D. Simberloff. 1986. Area-based extinction models in conservation. In *Dynamics of extinction*, ed. D. K. Elliott, 247–76. New York: Wiley and Sons.

Boucher, D. H., J. H. Vandermeer, K. Yih, and N. Zamora. 1990. Contrasting hurricane damage in tropical rain forest and pine forest. *Ecology* 71:2022–24.

Burkey, T. V. 1995. Extinction rates in archipelagoes: Implications for populations in fragmented habitats. *Conservation Biology* 9(3):527–41.

Chapel, M., A. Carlson, D. Craig, T. Flaherty, C. Marshall, M. Reynolds, D. Pratt, L. Pyshora, S. Tanguay, and W. Thompson. 1992. *Recommendations for managing late-seral-stage forest and riparian habitats on the Tahoe National Forest.* USDA Forest Service, Pacific Southwest Region, Tahoe National Forest. 31 pp. + appendix.

Chasko, G. G., and J. E. Gates. 1982. Avian habitat suitability along a transmission-line corridor in an oak-hickory forest region. *Wildlife Monographs* 82:1–41.

Conroy, M. J., Y. Cohen, F. C. James, Y. G. Matsinos, and B. A. Maurer. 1995. Parameter estimation, reliability, and model improvement for spatially explicit models of animal populations. *Ecological Applications* 5(1):17–19.

Crawford, H.S., and D. T. Jennings. 1989. Predation by birds on spruce budworm *Choristoneura fumiferana:* Functional, numerical, and total responses. *Ecology* 70(1):152–63.

Custer, T. W., R. K. Hines, and C. M. Custer. 1996. Nest initiation and clutch size of great blue herons on the Mississippi River in relation to the 1993 flood. *Condor* 98(2):181–88.

Darlington, P. J. 1957. *Zoogeography: The geographic distribution of animals.* New York: John Wiley and Sons.

Darveau, M., P. Beauchesne, L. Belanger, J. Huot, and P. Larue. 1995. Riparian forest strips as habitat for breeding birds in boreal forest. *Journal of Wildlife Management* 59(1):67–78.

Delcourt, H. R., T. A. Delcourt, and T. Webb. 1983. Dynamic plant ecology: The spectrum of vegetation change in space and time. *Quantative Science Review* 1:153–75.

Dunning, J. B., Jr., D. J. Stewart, B. J. Danielson, B. R. Noon, T. L. Root, R. H. Lamberson, and E. E. Stevens. 1995. Spatially explicit population models: Current forms and future uses. *Ecological Applications* 5(1):3–11.

Ehinger, L. H. 1991. Hurricane Hugo damage. *Journal of Arboriculture* 17(3):82–83.

Everett, R., P. Hessburg, J. Lehmkuhl, M. Jensen, and P. Bourgeron. 1994. Old forests in dynamic landscapes: Dry-site forests of eastern Oregon and Washington. *Journal of Forestry* (Jan.):22–25.

Forman, R. T. T. 1983. Corridors in a landscape: Their ecological structure and function. *Ekologia* 2(4):375–87.

Freeman, J. 1986. The parks as genetic islands. *National Parks* 60:12–17.

Goggins, G. C. 1987. Protecting the wildlife resources of national parks from external threats. *Land and Water Law Review* 22:1–27.

Hamrick, D. 1991. Assisting homeless woodpeckers. *Birds International* 3(1):18–27.

Hanski, I., A. Moilanen, T. Pakkala, and M. Kuussaari. 1996. The quantitative incidence function model and persistence of an endangered butterfly metapopulation. *Conservation Biology* 10(2):578–90.

Harris, L. D. 1984. *The fragmented forest.* Chicago: University of Chicago Press.

Harris, L. D. 1988. Landscape linkages: The dispersal corridor approach to wildlife conservation. *North American Wildlife Natural Resources Conference* 53:595–607.

Hess, G. R. 1994. Conservation corridors and contagious disease: A cautionary note. *Conservation Biology* 8(1):256–62.

Holt, R. D., S. W. Pacala, T. W. Smith, and J. Liu. 1995. Linking contemporary vegetation models with spatially explicit animal population models. *Ecological Applications* 5(1):20–27.

Hooper, R. G., J. C. Watson, and R. E. F. Escano. 1990. Hurricane Hugo's initial effects on red-cockaded woodpeckers in the Francis Marion National Forest. *North American Wildlife Natural Resources Conference* 55:220–24.

Howe, R. W., T. D. Howe, and H. A. Ford. 1981. Bird distributions on small rainforest remnants in New South Wales. *Australian Wildlife Research* 8:637–51.

Johnson, E. A. 1992. *Fire and vegetation dynamics, studies from the North American boreal forest.* Cambridge: Cambridge University Press.

Karr, J. R. 1990. Avian survival rates and the extinction process on Barro Colorado Island, Panama. *Conservation Biology* 4:391–97.

Knight, R. L., and J. Y. Kawashima. 1993. Responses of raven and red-tailed hawk populations to linear right-of-ways. *Journal of Wildlife Management* 57(2):266–71.

Koplin, J. R. 1969. The numerical response of woodpeckers to insect prey in a subalpine forest in Colorado. *Condor* 71(4):436–38.

Kroodsma, R. L. 1982. Edge effect on breeding forest birds along a power-line corridor. *Journal of Applied Ecol.* 19:361–70.

Lawton, R. O., and F. E. Putz. 1988. Natural disturbance and gap-phase regeneration in a wind-exposed tropical cloud forest. *Ecology* 69(3):764–77.

Lindenmayer, D. B., and H. A. Nix. 1993. Ecological principles for the design of wildlife corridors. *Conservation Biology* 7(3):627–30.

Loehle, C., and B. Li. 1996. Habitat destruction and the extinction debt revisited. *Ecological Applications* 6(3):784–89.

Lomolino, M. V. 1984. Mammalian island biogeography: Effects of area, isolation and vagility. *Oecologia* 61:376–82.

Lovejoy, T. E., J. M. Rankin, R. O. Bierregaard, Jr., K. S. Brown, Jr., L. H. Emmons, and M. E. Van der Voort. 1984. Ecosystem decay of Amazon forest remnants. In *Extinctions*, ed. M. H. Nitecki, 295–325. Chicago: University of Chicago Press.

MacArthur, R. H., and M. Willson. 1967. *The theory of island biogeography.* Princeton, N.J.: Princeton University Press.

MacClintock, L., R. F. Whitcomb, and B. L. Whitcomb. 1977. Island biogeography and "habitat islands" of eastern forest. II. Evidence for the value of corridors and minimization of isolation in preservation of biotic diversity. *American Birds* 31:6–12.

Marcot, B. G., L. K. Croft, J. F. Lehmkuhl, R. H. Naney, C. G. Niwa, W. R. Owen, and R. E. Sandquist. 1998. *Macroecology, paleoecology, and ecological integrity of terrestrial species and communities of the interior Columbia River basin and portions of the Klamath and Great basins.* USDA Forest Service General Technical Report PNW-GTR-410.

Marks, P. L., and F. H. Bormann. 1972. Revegetation following forest cutting: Mechanisms for return to steady-state nutrient cycling. *Science* 176:914–15.

Meyer, J. S., L. L. Irwin, and M. S. Boyce. 1992. *Influence of habitat fragmentation on spotted owl site location, site occupancy, and reproductive status in western Oregon.* Progress report, 31 January 1992, to USDA Forest Service, Pacific Northwest Regional Office, Portland, Oregon.

Millar, C. I., and W. J. Libby. 1991. Strategies for conserving clinal, ecotypic, and disjunct population diversity in widespread species. In *Genetics and conservation of rare plants,* ed. D. A. Falk and K. E. Holsinger, 149–70. New York and Oxford: Oxford University Press.

Miller, M. W., and T. D. Nudds. 1996. Prairie landscape change and flooding in the Mississippi River valley. *Conservation Biology* 10(3):847–53.

Mills, L. S. 1995. Edge effects and isolation: Red-backed voles on forest remnants. *Conservation Biology* 9(2):395–403.

Mishall, G. W., J. T. Brock, and J. D. Varley. 1989. Wildfires and Yellowstone's stream ecosystems. *BioScience* 39(10):707–15.

Naiman, R. J., H. Decamps, and M. Pollock. 1993. The role of riparian corridors in maintaining regional biodiversity. *Ecological Applications* 3(2):209–12.

Newmark, W. D. 1986. Mammalian richness, colonization, and extinction in western North American national parks. Ph.D. dissertation, University of Michigan, Ann Arbor.

Newmark, W. D. 1987. A land-bridge island perspective on mammalian extinctions in western North American parks. *Nature* 325:430–32.

Newmark, W. D. 1995. Extinction of mammal populations in western North American national parks. *Conservation Biology* 9(3):512–26.

Parsons, W. F. J., D. H. Knight, and S. L. Miller. 1994. Root gap dynamics in lodgepole pine forest: Nitrogen transformations in gaps of different size. *Ecological Applications* 4(2):354–62.

Pearson, S. M., M. G. Turner, L. L. Wallace, and W. H. Romme. 1995. Winter habitat use by large ungulates following fire in northern Yellowstone National Park. *Ecological Applications* 5(3):744–55.

Pierson, E. D., T. Elmqvist, W. E. Rainey, and P. A. Cox. 1996. Effects of tropical cyclonic storms on flying fox populations on the South Pacific islands of Samoa. *Conservation Biology,* 10(2):438–51.

Rafe, R. W., M. B. Usher, and R. G. Jefferson. 1985. Birds on reserves: The influence of area and habitat on species richness. *Journal of Applied Ecology* 22:327–35.

Richter, B. D., J. V. Baumgartner, J. Powell, and D. P. Braun. 1996. A method for assessing hydrologic alteration within ecosystems. *Conservation Biology* 10(4):1163–74.

Saunders, D. A., G. W. Arnold, A. A. Burbidge, and J. M. Hopkins. 1987. *Nature conservation: The role of remnants of native vegetation.* Chipping Norton, New South Wales, Australia: Surrey Beatty and Sons Limited.

Schelhas, J. 1995. Conserving the biological and human benefits of forest remnants in the tropical landscape: Research needs and policy recommendations. In *Integrating people and wildlife for a sustainable future,* ed. J. A. Bissonette and P. R. Krausman, 53–56. Bethesda, Md.: The Wildlife Society.

Schoener, T. W., and A. Schoener. 1983. The time to extinction of a colonizing propagule of lizards increases with island area. *Nature* 302:332–34.

Schreiber, R. W., and E. A. Schreiber. 1984. Central Pacific seabirds and the El Niño southern oscillation: 1982 to 1983 perspectives. *Science* 225:713–16.

Swanston, D. N. 1991. Natural processes. In *Influences of forest and rangeland management on salmonid fishes and their habitats.* ed. W. R. Meehan, 139–79. American Fisheries Society Special Publication 19. Bethesda, Md. 751 pp.

Turner, M. B., G. J. Arthaud, R. T. Engstrom, S. J. Hejl, J. Liu, S. Loeb, and K. McKelvey. 1995. The usefulness of spatially explicit population models in land management. *Ecological Applications* 5(1):12–16.

Usher, M. B. 1985. Implications of species-area relationships for wildlife conservation. *Journal of Environmental Management* 21:181–91.

Vander Haegen, W. M., and R. M. DeGraaf. 1996. Predation on artificial nests in forested riparian buffer strips. *Journal of Wildlife Management* 60(3):542–50.

Vilella, F. J., and E. R. Garcia. 1995. Post-hurricane management of the Puerto Rican parrot. In *Integrating people and wildlife for a sustainable future;* ed. J. A. Bissonette and P. R. Krausman, 618–21. Bethesda, Md.: The Wildlife Society.

Wakimoto, R. H. 1990. The Yellowstone fires of 1988: Natural process and natural policy. *Northwest Science* 64:239–42.

Wallmo, O. C. 1969. *Response of deer to alternate-strip clearcutting of lodgepole pine and spruce-fir timber in Colorado.* USDA Forest Service Resource Note RM-141. Rocky Mountain Forest Range Experiment Station, Fort Collins, Colo. 4 pp.

Webb, S. L. 1989. Contrasting windstorm consequences in two forests, Itasca State Park, Minnesota. *Ecology* 70:1167–80.

Welsh, H. H. 1990. Relictual amphibians and old-growth forest. *Conservation Biology* 4:309–19.

West, A. J. 1989. Concerning damages to federal, state, and private forest resources and the USDA Forest Service participation in the relief efforts following Hurricane Hugo. Statement of Allan J. West, Deputy Chief for State and Private Forestry, Forest Service, United States Department of Agriculture, before the Subcommittee on Forests, Family Farms and Energy Committee on Agriculture, United States House of Representatives. Presented 6 November 1989, Moncks Corner, S.C. Manuscript with USDA Forest Service, Southeast Regional Office, Atlanta, Ga.

Whittam, S., and D. Siegel-Causey. 1981. Species interactions and community structure in Alaskan seabird colonies. *Ecology* 62:1515–24.

Wissmar, R. C., A. H. Devol, J. T. Staley, and J. R. Sedell. 1982. Biological responses of lakes in the Mount St. Helens blast zone. *Science* 216:178–81.

Wunderle, J. M., Jr. 1995. Responses of bird populations in a Puerto Rican forest to Hurricane Hugo: The first 18 months. *Condor* 97:879–96.

10 Modeling Wildlife-Habitat Relationships

Introduction

In this chapter we explore the basis and use of models of wildlife-habitat relationships. First, we discuss the basic utility and objectives for modeling wildlife-habitat relationships. Then, we offer definitions and a classification of wildlife-habitat models. We discuss how scientific uncertainty affects wildlife management and how model development and application should be used in light of uncertainties. We then review current model forms used in predicting wildlife-habitat relationships and describe how vegetation and wildlife models may be linked. Knowledge-based and decision-aiding models are then introduced. Finally, we discuss the often overlooked topic of model validation.

Use and Types of Models

Objectives for Modeling

The main objectives for developing models of wildlife-habitat relationships are (1) to *formalize,* or describe, our current understanding about a species or an ecological system, (2) to *understand* which environmental factors affect the distribution and abundance of a species, (3) to *predict* further distribution and abundance of a species, (4) to *identify* weaknesses in our understanding, and (5) to *generate* hypotheses about the species or system of interest.

Not all these goals are equally attainable. For example, many observational field studies may result in statistical descriptions of patterns of wildlife-habitat relationships. Such observational descriptions are pertinent to specific locations, environmental conditions, and time periods. They should not be assumed to provide much power to predict conditions beyond those contexts, but such studies are often used for this purpose. At best, they can meet goal 5, to generate hypotheses. Typically, though, most interest in modeling wildlife-habitat relationships does deal with prediction. In this book, predictive modeling refers to estimating the presence, distribution, or abundance of a wildlife species or group of species, given information on actual or possible environmental and habitat conditions.

Types of Predictions

There are two main types of prediction that may be made from models. One is *hindcasting,* which identifies key environmental variables, typically those of vegetation structure, that account for observed variation in species variables such as abundance. Hindcasting is used to explain patterns observed in species occurrence and abundance and is pertinent, strictly speaking, only to the time and place at which the original data were gathered.

The other class of prediction is *forecasting,* which is an explicit attempt to predict species conditions, given environmental conditions at a time or place not represented by the field data used to generate the model in the first place. Many workers use results of hindcasting, such as those obtained with the use of correlation, regression, or multivariate statistics, to predict further species conditions. However, without proper design of the initial investigation and without validation studies, predictions from hindcasting may be quite unreliable, because environmental, demographic, and ecological conditions may vary significantly among locations or over time. At best, using hindcasting models for prediction in new situations entails the assumption that factors not accounted for in the prediction model are insignificant or are unchanged.

What is the best means of predicting species responses to environmental conditions? Proper forecasting techniques account for autocorrelation of a variable over some time series or over spatial (such as environmental) gradients. Forecasting should consider causes of the distribution and abundance of a species, rather than simply correlations. The degree to which observational data should be used to tell us something about habitat selection and resource requirements of a species, for example, should be examined very carefully before such extrapolations are made.

What Is a Model?

A model is any formal representation of some part of the real world. Hall and Day (1977) suggested that a model may be conceptual, diagramatic, mathematical, or computational. These forms may also be viewed as stages in a logical model-building process.

Developing a *conceptual model* may entail synthesizing current scientific understanding, field observations, and professional judgment of a particular species or habitat, and proposing a few hypotheses to explain the species' distribution and abundance. Even (especially) at the conceptual stage, an explicit statement of assumptions and simplifications is necessary for the model to be true or useful. The *diagramatic stage* takes a conceptual model one step further by explicitly showing interrelationships between various environmental parameters and species behaviors. The *mathematical stage* quantifies these relationships by applying coefficients of change and formulae of correlation or causality. Finally, the *computational stage* aids in exploring or solving the mathematical relationships by analyzing the formulae on computers.

The conceptual and diagramatic stages of modeling are often the most difficult and the most revealing stages of building ecological theories and enhancing understanding. They must derive from a well-shaped statement of modeling goals and objectives.

Types of Predictive Wildlife-Habitat Relationship Models

Various types of predictive wildlife-habitat models are shown in figure 10.1. Predictive models can be largely empirically based, as from field studies, or theoretically based, as from mathematical representations. Empirical models can be purely descriptive, as derived from case studies (and thus provide predictions only for specific locations), or statistical, as based on a sampling protocol (see chapter 4). Statistical empirical models can be used more generally than case studies can and can be of varying complexity depending on the number of habitat variables considered.

Theoretical models can be qualitative, as with diagramatic models (an example appears in fig. 10.2), or quantitative, as with mathematical models of population trends. Quantitative theoretical

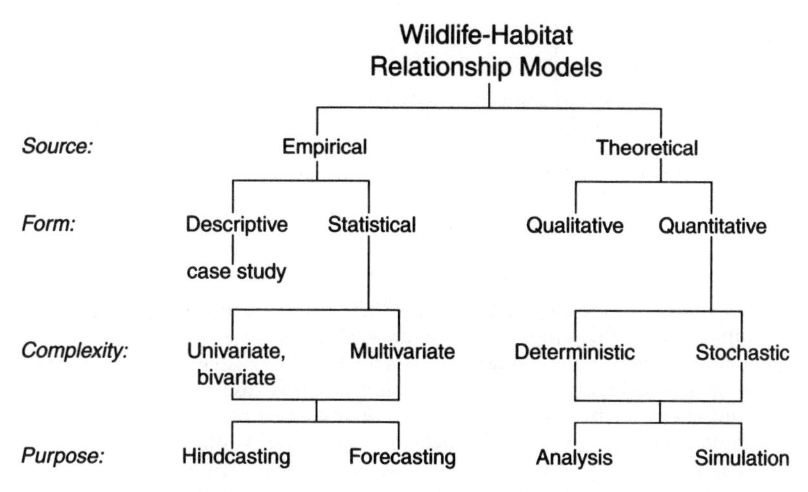

Figure 10.1. Classification of predictive wildlife-habitat models according to source of data and utility of the model

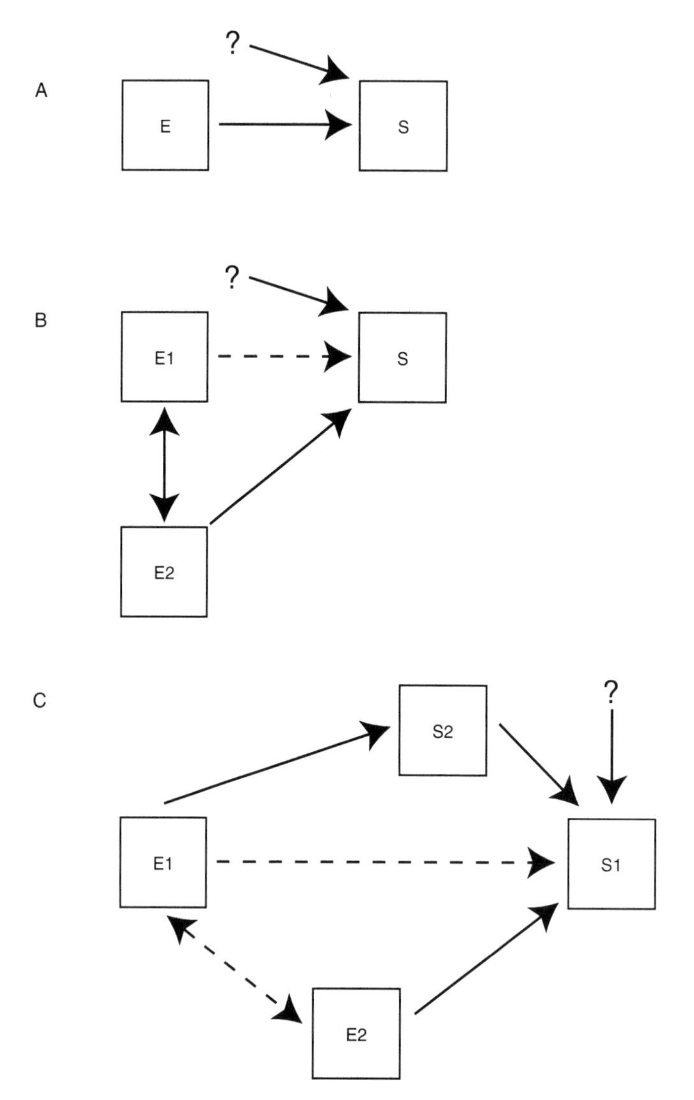

Figure 10.2. Causes and correlates in wildlife-habitat relationships—three examples: (A) In this simplest case, some wildlife response, S, such as population presence or abundance, is assumed to be explained and caused by some environmental factor, E. (B) In a more complex case, we may be measuring one environmental factor, E1, when the real cause is another environmental factor, E2. (C) Getting closer to the real world, a second species, S2, may be part of the cause. In all parts of this figure, the unexplained variation (?) is due to measurement error, experimental error, or effects of other environmental or species factors; solid arrows denote causal relations, and dotted arrows denote correlational relations.

models can be deterministic, as with analytic solutions, or stochastic, which take into account variations in habitat or wildlife variables. Many stochastic, quantitative theoretical models are used in simulations of wildlife response to changes in habitat conditions.

Empirical models overall may provide more specific understanding and predictability of wildlife response to habitat conditions than theoretical models do, depending on how variables are selected for field study (see chapters 5 and 6). Thus, empirical predictive models may be more accurate but less robust and less general than theoretical predictive models.

So which type of model is best for predicting wildlife-habitat relationships? The easy answer is: all of them, taken together. The various types of modeling approaches are complementary, not competitive. In any predictive venture, especially those with high stakes—high opportunity cost of resources or high likelihood of population declines or extirpations—the best approach is to develop, run, and test several models of various types. Models are like politicians: support them, use them, but don't unquestioningly trust them. Compare them. Find out if several models of the same environment or species tell you the same story, and if they don't, find out why; are you learning more about the structure of the models—their state variables and relations—than the real-world biological entity of interest?

One example of the need to use multiple models was provided by Mills et al. (1996), who tested viability predictions of a population of grizzly bears (*Ursus arctos horribilis*) using four different simulation models on the same demographics data set. They found that "striking differences" occurred in estimates of extinction rates and expected population size among the models and that inclusion and type of density-dependence relations accounted for some of the major differences between model outputs. Each model accounted for and mathematically represented a different set of population parameters. There was no clear indication of which model was "correct." Instead, the most *appropriate* model (or, better, suites of models) may be different for different species or management questions. Thus, such modeling tools should be viewed as providing heuristic or learning value, or at best a relative ranking of alternative management scenarios, rather than explicit answers on exact population extinction probabilities. The lesson here is: Know thy models and thy species.

A Classification of Wildlife-Habitat Relationship Models

Table 10.1 represents a hierarchic classification of wildlife-habitat relationship models. The major model classes that most managers and researchers may deal with are models of vegetation structure, species response to vegetation structure, population dynamics and viability, landscapes, and ecosystems. Additional classes of models of use in management include models of indicator species, models for monitoring species and habitats, and knowledge-based and decision-aiding models. The bases and examples of each of these kinds of models are discussed below.

Selecting Models

A number of models of habitat-wildlife relationships are available to wildlife biologists. Later in this chapter we review models listed by their function (table 10.1). Not included in this list are a number of models taken from strictly theoretical or highly mathematical literature.

If the purpose of modeling is to assist management, then one might consider a multiscale approach to model development and selection. A first step may be to identify clearly the management scales (table 8.1) in terms of

Table 10.1. Classification of wildlife-habitat relationship models

Models of Research and Management Interest

- Models of vegetation structure
  - Stand growth models
  - Succession models
- Models of species response to vegetation structure
  - Single-species models
    - Life history models
    - Habitat preference models
    - Optimal foraging models
    - Correlation models
    - Multivariate statistical models
    - Habitat suitability index (HSI) models
    - Habitat capability (HC) models
    - Habitat evaluation procedures (HEP)
    - Bayesian and pattern recognition (PATREC) models
  - Multiple-species models
    - Coarse-filter and fine-filter models
    - Species-habitat matrices
    - Guild and life form models
    - Community structure models
    - Community and ecosystem simulation models
    - Gap analysis models
    - Biodiversity models
    - Hierarchy models
- Models of population dynamics and viability
- Models of landscapes
  - Vegetation disturbance models
  - Fragmentation models
  - Cumulative effect assessments
  - Models of insular biogeography
- Models of ecosystems
  - Models of nutrient cycling and energy flow
  - Modeling historic range of natural conditions

Models of Management Interest

- Models of indicator species
- Models for monitoring species and habitats
- Knowledge-based and decision-aiding models
  - Decision support models
  - Expert systems

geographic extent, map scale, spatial resolution, time period, administrative hierarchy, and levels of biological organization at which the management activity or plan is directed. Then, identify the key areas of scientific inquiry (fig. 8.1) and management issues (fig. 8.2) that may pertain to a particular management activity or plan, along with the expected duration of the activity or plan and the need for assessing prehistoric or historic conditions and cumulative effects (fig. 8.3). From this, turn to the classification of models by their function (table 10.1) and select the kinds of models most pertinent to the questions, issues, and scales. Of course, other factors will guide model selection, such as what information and which models are available, and the need to integrate with other management activities or plans.

Of Correlates and Causes

When we build a model, including a statistical evaluation (hindcast) based on an empirical observational study, it is not always evident which factors are correlates and which are true causes. In many cases, however, it is important to identify true causes in order to know appropriate ways to amend management activities to meet objectives better. But teasing apart the "causal web" of wildlife-habitat relationships may be trickier than what first meets the eye.

Consider three progressively complex situations as depicted in figure 10.2. In the first instance (fig. 10.2A), some species response, S, such as population presence or abundance, is assumed to be directly caused and explained by some environmental variable, E. There also may be some degree of unexplained variation in species response (shown as ? in fig. 10.2). The unexplained variation (?) is due to measurement error, experimental error, or the effect of other environmental factors not included in the study. It can be quantified such as by calculating residuals in a regression analysis or by partitioning error terms in an analysis of variance.

However, as depicted in figure 10.2B, in the real world we may be measuring one environmental variable, E1, when the real causal factor is another, unmeasured environmental variable, E2. In this case, the two environmental variables are themselves correlated, and there may or may not be a causal relation between them. E1 and E2 may be vastly different kinds of environmental factors and may operate at different spatial or temporal scales as well. We think we have explained the biological response of the species S by the observed correlation with E1, but we may be greatly mistaken in presuming that management of E1 will necessarily affect the species in the way that we wish. Further, the unexplained variation (?) in S then is due to the less-than-perfect correlation between E1 and E2 as well as to measurement and experimental error and to factors beyond E1 and E2 not addressed in the study.

As an example, we may find that mean fecundity rates (S) in a population of Townsend's voles (*Microtus townsendii*) to be negatively correlated with food abundance during the previous season (E1), and we may thereby infer that high food abundance leads to high population density, which in turn suppresses mean fecundity levels as a population regulation mechanism. However, upon closer inspection, it may turn out that the real culprit (E2) causing lower mean fecundity of voles is parasitism by botflies (*Cuterebra grisea*) (e.g., Boonstra et al. 1980). It may turn out that both food resources and botfly incidence are affected by weather, so that environmental factors E1 and E2 are correlated but there is no direct causal link between them. If our study were to focus only on food resources and vole fecundity, we would conclude that food abundance is the cause and that managing for higher vole densities could be afforded by managing for more consistent food resource levels. This would be in error. Also, the unexplained variation in vole fecundity would be caused by the less-than-perfect correlation between food resources and botfly incidence, as well as by additional factors beyond food or botflies not addressed in the study.

However, the real world is often even more complicated than this. As shown in figure 10.2C, another species, S2—potentially a competitor, predator, or symbiont—may also play a role in affecting the species of interest, S1. (We included the parasite as an environmental factor in the last example.) Following our example above, mean vole fecundity may be influenced by some (hypothetical) predator that selectively removes high-fecundity individuals from the vole population. S2 itself may share some environmental factor (E1) that correlates with (but does not cause) S1. (S2 also may be influenced beyond E1 by other environmental factors not shown in fig. 10.2C.)

It is important to recognize such *causal webs* of organisms and environmental factors. We should at least challenge ourselves to draw a causal diagram (the diagramatic phase of modeling as outlined above) so that we can hypothesize which are causal factors and which are merely correlative. Differentiating between causes and correlates is critical for guiding costly habitat management activities to respond to complex environmental issues such as changes in air quality or regional climate, and for establishing an appropriate monitoring scheme including identifying and tracking key indicators.

Path regression analysis is a statistical technique that can aid in quantifying the relative contribution of causal factors. Path regression is used to determine the partial correlation coefficients of individual factors that can influence a species population or, in some uses, a management objective. An example of a path regression analysis is shown in figure 10.3.

Dealing with Uncertainty and Unknowns in Modeling

Biological models do not predict species distribution and abundance without error. Rather, modeling wildlife-habitat relationships, like managing species habitats, typically entails dealing with the following kinds of obstacles:

- imprecise data
- uncertain inferences
- limiting and fallacious assumptions
- unforeseen environmental, administrative, and social circumstances
- risks of failure

Imperfections are often present in habitat analyses and management decisions, but they are especially critical when there is risk of reducing a wildlife population or eliminating a species. Uncertainty may be encountered when analyzing biological data, when making inferences about species responses to environmental conditions, and when selecting and instituting a management plan.

Types of uncertainty may be classified as scientific uncertainty and decision-making uncertainty. Just as analyzing species and habitats entails a different process from that used to make decisions on resource management, so are the kinds and implications of uncertainties from the analysis process distinct from those in a decision-making process. Results of a technical study, such as a risk analysis of population viability, may be part (but only part) of the information used by a decision maker in developing a habitat management plan.

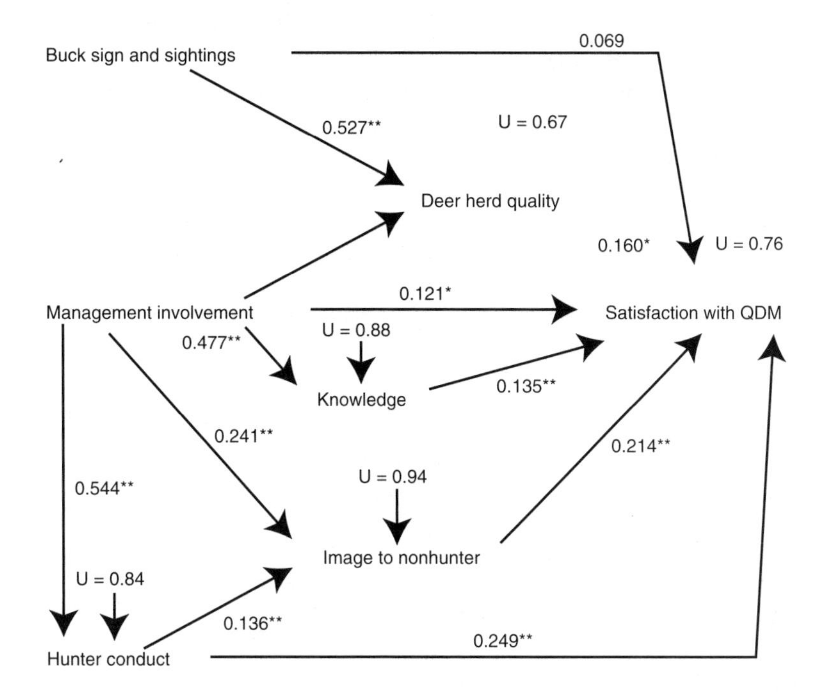

Figure 10.3. An example of a path regression analysis that partitions the various factors that account for variation in public satisfaction with quality deer management (QDM). The values are partial correlations (i.e., the correlation of each factor once the contribution of all other factors is accounted for). U = unexplained variation, calculated as $(1 - R^2)^{1/2}$; $*P < 0.05$, $**P < 0.0001$. (Reproduced from Woods et al. 1996 [fig.1], by permission of the Wildlife Society)

Types of Uncertainty

Scientific uncertainty in habitat modeling refers to the nature of the data and the ways in which information on species and habitats is represented and applied. Scientific uncertainty essentially means that our predictions of how species respond to environmental conditions are not perfect. Uncertainty may occur because (1) the system itself is naturally variable and very complex, and thus difficult to predict; (2) the process of estimating values of parameters in the habitat model entails a degree of error; (3) models used to generate predictions are in some sense invalid; or (4) the scientific question being asked is ambiguous or incorrect.

Variability of Natural Systems—Noise in the Message

Many aspects of natural systems vary over time. Predicting attributes of the system—the "message" we are trying to interpret—may often involve observing and modeling traits that are influenced from outside factors (fig. 10.2), that is, "noise" inherent in the message. Such noise introduces variation in measurements and uncertainty in estimating and predicting attributes of the system. In statistical models of habitat relationships, noise is typically depicted as unexplained variation in the occurrence or abundance of a species. One kind of unexplained statistical variation is the value of "residuals" in linear regression models. Sometimes this kind of noise in the system can be a useful source of information itself, as described in box 10.1.

Uncertainty of Empirical Information—Errors of Estimation

Values of environmental parameters are typically estimated from a sample set of observations. A parameter, for example, may be the average number of tree stems per hectare or the variance in litter sizes of black-footed ferrets (*Mustela nigripes*) to the extent that these can be attributed to individual and environmental variation. When a parameter is estimated from

Box 10.1. The use of statistical noise: a two-stage approach to determining how vegetation patches and patch patterns across the landscape individually contribute to species abundance by using estimation errors

Elsewhere in this text, we have asserted that evaluations of species-habitat relationships should strive not to mix different spatial scales in the same statistical analysis. This is particularly true in multivariate or multifactor analysis, in which mixing will confound or mask important relationships. Some authors have presented frameworks for modeling vegetation on various scales of geographic extent (e.g., Bourgeron et al. 1994). But there are few guidelines available for assessing wildlife response across various spatial scales (but see Holling 1992), particularly to determine the relative contribution of within-vegetation patch conditions and among-patch landscape patterns.

One approach may involve a two-stage statistical analysis. In the first stage, one uses a multivariate correlation analysis, such as multiple linear regression or logistic regression. Predictor variables are within-patch attributes (e.g., vegetation structure, flora, microclimate, or presence of substrates), and the response variable pertains to site-specific effects on wildlife (e.g., presence or abundance within study plots). The result of this first stage in the analysis is the identification of the within-patch attributes that most account for variation in the wildlife response variable. The values of the residuals for each sample (study plot) are saved in a file for each species being analyzed. They'll come in handy in the next step.

In the second stage, one uses the same multivariate correlation analysis, but this time the predictor variables are among-patch landscape attributes (e.g., juxtaposition of patch types, distance to nearest patch of the same type, or topographic or physiographic conditions), and the wildlife response variables are the residual values, of each species at each site, as saved from the first stage of the analysis. Remember, the residuals represent the variation in wildlife response that is *not* accounted for by within-patch conditions (technically, *residuals* refers to a corresponding sum of squares term that involves error plus any interaction effects that may not be zero [Milliken and Johnson 1992]). Thus, this analysis represents asking the question, What landscape-level attributes account for that portion of wildlife response that cannot be explained at the within-patch scale, and what proportion of the unexplained variation is so accounted? The result of the second analysis stage is identification of landscape attributes that significantly correlate with, and thus statistically explain, the variation in the residuals for each species.

Here is an example. In a study conducted by one of us (Marcot 1985) in young-growth Douglas-fir forests of northwestern California, the relative abundance of birds in 55 study plots was determined by using the variable-radius circular plot (VCP) census technique. Also centered at each census plot were vegetation plots to sample within-patch conditions of vegetation structure. Each census plot was located in a broader vegetation patch or stand for which various landscape attributes of topography and patch proximity were measured. The analysis was conducted for different seasons and by partitioning out two general young-growth conditions: early grass/forb and shrub stages were assessed separately

(box continued on following page)

a sample set of observations, uncertainty or errors in estimation may occur, from a statistical viewpoint. The estimation may be: *biased* if each of the values of the observations is consistently less or greater than actual (unknown) values; *inaccurate* if the estimated value of the parameter of interest (such as a mean or a variance) is substantially different from the true value; or *imprecise* if values of individual observations vary widely. Each of these errors in estimating the value of a parameter constitutes a different kind of scientific or statistical uncertainty.

Such errors of estimation can arise from a number of sampling problems. These may include inadequate sample size, observations taken from disparate times or places, and samples taken nonrandomly or nonsystematically, depending on the assumptions of the estimator being used. Errors of estimating the

Box 10.1 Continued

from young forested stages, because results would otherwise represent the obvious contributions of presence or absence of tree canopies.

Results of the first stage suggested specific multiple linear regression models of individual species with within-patch vegetation attributes. As one example, during the breeding season, song sparrows (*Melospiza melodia*) occurred in the early grass/forb and shrub stages. Results of multiple linear regression suggested that five within-patch attributes significantly affected song sparrow abundance: percentage of ground cover in green vegetation <10 cm tall, down wood mass, and number of plant species ≥ 2 m tall (positive coefficients in the linear regression model); and litter depth and foliage volume of deciduous and evergreen plants 0.1–2 m tall (negative coefficients).

Next, the residual values from the regression were regressed against the topography and landscape attributes of each study site. Results suggested that, across the landscape, proximity to permanent water and thus to riparian vegetation was the one key factor (out of those analyzed) that additionally accounted for song sparrow abundance, once within-patch attributes were accounted for.

Similar analyses were conducted for all other bird species for which data were available. Surprisingly, some 71 percent of all bird species showed no additional associations with among-patch conditions once within-patch conditions were accounted for. Species showing additional among-patch effects included winter wrens (*Troglodytes troglodytes*), for which in forested stages during the breeding season some 62 percent of the variation in its residuals (measured by adjusted R^2) was accounted for by slope position and slope angle. Also, distance to the nearest habitat patch of a similar vegetation structural stage accounted for some of the unexplained variation of bird abundances in the shrub stages (but not in the forested stages) for the western wood-pewee (*Contopus sordidulus*), warbling vireo (*Vireo gilvus*), and western tanager (*Piranga ludoviciana*) during the breeding season, and the chestnut-backed chickadee (*Parus rufescens*), cedar waxwing (*Bombycilla cedrorum*), and evening grosbeak (*Coccothraustes vespertinus*) during the winter season. (The lack of correlation in the forested stages for these species may be due to the fairly well-connected nature of the forest stages within the general landscape studied, compared with the shrub stages, which occurred as more isolated patches.) From the analysis, predictive models were built that suggested bird species abundances resulting from both within-patch vegetation structure and among-patch landscape conditions.

This two-stage-analysis approach can help determine the appropriate landscape context for habitat conservation or restoration activities. There is no reason why this approach could not be extended beyond two stages, but many wildlife response variables become far less meaningful on inappropriate scales. For example, variation in site-specific abundance should not be explained by regional climate conditions, but rather by vegetation structure or local topographic conditions, which more immediately affect local microclimates. As with any study or modeling approach, it is vital to match appropriate scales for both predictor and response variables.

value of parameters may also arise from applying the wrong kind of estimator, such as in applying a formula for calculating variance. If correct use of the formula assumes that observations were made independently and randomly—when they were actually made over a time series or systematically, such as at even intervals over a transect—then an error of applying the wrong kind of estimator has been made.

Uncertainty of Model Structure—Model Validation

Model validity refers to a broad spectrum of performance standards and criteria. Examples are model credibility, realism, generality, precision, breadth, and depth (Marcot et al. 1983). The various criteria refer to such attributes of models as the number of parameters in a model and their interactions, the context within which a model was developed or should be used, and the underlying and simplifying assumptions of the model structure (see table 10.2 for definitions of model validation criteria). A parameter that is estimated precisely, accurately, and with out bias may still be used inappropriately, as in a model that is applied to the wrong environment, location, season, or species.

Appropriateness of the Problem—Asking the Right Question

The context in which a theory is applied or a model is used may introduce yet another source of uncertainty. Even given that a model has been validated—that is, has been shown to be a useful tool and to generate acceptable predictions according to particular criteria—it may still be applied to the wrong problem. In some cases, this may be unavoidable if no other models are available.

For example, a life table model assuming that the sexes occur in equal ratios and that adults breed each year may generate acceptable predictions for use with Dall sheep (*Ovis dalli*), but may generate grossly inaccurate predictions when used for species with variable or quite different social breeding organizations, such as the pronghorn (*Antilocapra americana*). This would call into question the reliability of the model when used with certain species or under certain circumstances.

Further, the hypothesis or problem being addressed by using a particular model may be ambiguous or even unanswerable. For example, a model of species-habitat relationships that describes vegetation types may not provide a particularly useful foundation for answering questions about landscape dynamics necessary for maintaining viable populations of the species.

Accounting for Error in Modeling

One of the major problems in using models of wildlife-habitat relationships is that of propagation of error. Error can arise from model structure, missing data, mismatched scales of geographic extent and spatial resolution, and other systematic sources, as well as from measurement error and the stochastic nature of biological systems. How do all such errors compound in a particular model? The problem of error propagation has been poorly addressed in the statistical and modeling literature and needs much work. One approach to depicting the compounding of error is to partition the variance associated with model output into additive factors, each representing the major sources of error. This is analogous to methods used in analysis of variance, in which mean square errors are partitioned into sampling error and experimental error. This approach may entail an analytic formulation for summing variance and covariance terms (box 10.2). Other approaches may invite use of model sensitivity analysis. We further discuss errors in modeling below under "Validating Wildlife-Habitat Models."

Table 10.2. Criteria useful for validating wildlife-habitat relationship models

Criterion	Explanation
Precision	A model's capability to replicate particular system parameters
Generality	A model's capability to represent a broad range of similar systems
Realism	Accounting for relevant variables and relations
Precision	The number of significant figures in a prediction or simulation
Accuracy	The degree to which a simulation reflects reality
Robustness	Conclusions that are not particularly sensitive to model structure
Validity	A model's capability to produce all empirically correct predictions
Usefulness	The existence of at least some empirically correct model predictions
Reliability	The fraction of model predictions that are empirically correct
Adequacy	The fraction of pertinent empirical observations that can be simulated
Resolution	The number of parameters of a system that the model attempts to mimic
Wholeness	The number of biological processes and interactions reflected in the model
Heurism	The degree to which the model usefully furthers empirical and theoretical investigations
Adaptability	The possibilities for future development and application
Availability	The existence of other, simpler, validated models that perform the same function
Appeal	Matching our intuition and stimulating thought, and practicability
Breadth	Proportional to the number and kinds of variables chosen to describe each (habitat) component
Depth	Proportional to the number and kinds of variables chosen to describe each (habitat) component
Face validity	A model's credibility
Sensitivity	The match of model variables and parameters with real-world counterparts, their variation causing outputs that match historical data; also, the dependence of model output on specific variations of variables
Hypothesis validity	The realism with which subsystem models interact
Technical and operational validity	The identification and importance of all divergence in model assumptions from perceived reality, as well as the identification and importance of the validity of the data
Dynamic validity	The analysis of provisions for application to be modified in light of new circumstances

Source: Marcot et al. 1983; reproduced by permission of the Wildlife Management Institute

Decision Making Involving Uncertainty and What to Do about It

Probably all management decisions dealing with wildlife habitat are made under some uncertainty of current conditions or future effects. Decision-making uncertainty can arise from imprecise data, uncertain inferences, limiting and fallacious assumptions, and unforeseen environmental, administrative, and social circumstances. Each of these factors can contribute to risks of failure in not meeting desired management goals.

How might the manager or decision maker proceed with such uncertainties? A host of decision-analysis techniques are available that aid in assessing the value of perfect information, the value of sample information, the credibility of information, and quantitative measures of the state of knowledge (Rubinstein 1975). We review some decision-aiding models below. Using these approaches to identify areas and degrees of decision-making uncertainty may be useful also for establishing management activities as adaptive management experiments. In a sense, uncertainty is an opportunity for testing management hypotheses about outcomes of actions, as long as basic tenets of adaptive management are not violated (see chapter 11).

Balancing Theory with Empiricism

We have dedicated several chapters in this book to reviewing study design and measures of habitat and wildlife behavior. Empirical field studies, whether observational descriptions or experimental tests, can be used to develop models of wildlife-habitat relationships. Theory,

Box 10.2. The sticky problem of propagation of error in modeling wildlife-habitat relationships: How does error compound in a wildlife-habitat model and why should we worry about it?

Every variable and function in a wildlife-habitat model can have several kinds of associated error, including measurement error, experimental error, and random error. Although it may be feasible to estimate such error for each variable or simple relation in a univariate sense, it is the compounding of error among variables and in complex functions that may seriously affect final model output. The result of such *error propagation* may be model output that is significantly biased, imprecise, or inaccurate. Thus, it may be critical to understand how error terms compound.

The biostatistical literature has poorly addressed the estimation of error propagation because it is such a wicked analytic problem. The classic approach to the problem is to dissect the variance of some variable, y, into its Taylor series expansion terms of component measured quantities, $F(x_1, x_2, x_3, \ldots)$ (Kotz et al. 1982:549):

$$var(y) = \sum \left(\frac{\partial F}{\partial x_i}\right)^2 var(x_i) + 2\sum_{ij} \left(\frac{\partial F}{\partial X_j}\right)\left(\frac{\partial F}{\partial x_j}\right) cov(x_i x_j)$$

The wicked part of this problem is not in estimating the first variance term but in estimating the covariance terms, $cov(x_i x_j)$, which may be extremely difficult to impossible to measure from empirical data. This is especially true with real-world studies of landscapes, ecosystems, and populations in conditions that are poorly replicable or unreplicable and for which control conditions are not feasible.

However, it is the covariance between ecological variables that may often be a major source of variation and error in model output. Even in the simple case of two interacting variables, x and y, the appropriate variance estimator involves the wicked covariance term: for the term xy, variance is calculated as $y^2 var(x) + x^2\ var(y) + 2xy\ cov(x,y)$ (Kotz et al. 1982:549) Empirically, covariance between all such key variables in a wildlife-habitat model may be impossible to estimate empirically and calculate analytically.

Propagation of error also means that initial errors may have a fatal effect on the final results; that is, small changes in initial data may produce large changes in final results. Such problems are called *ill-conditioned* and may include population models that exhibit chaotic or initially unpredictable behavior (Hassell et al. 1991; Morris 1990). Additional kinds of error, particularly those in computer models and those that can propagate across functions, include rounding errors and truncation errors; other possible analytic approaches to dissecting propagated errors include forward and backward analysis (see Fröberg (1969:3–9).

What is a modeler to do? One tractable approach to evaluating error propagation is to conduct sensitivity analyses of the model, whereby changes in outputs are plotted as a function of incremental changes in input variables (Overton 1971). This can help identify the domains over which models exhibit chaotic behavior, such as when populations exhibit irregular cycles or suddenly crash or expand with only minor changes in the input variables. Sensitivity analysis can be one phase of model validation (this is discussed further in the text).

however, often plays an important role in model development. Theoretical models may tend to be more robust and general, whereas empirical models may be more locally accurate and precise. Each complements the other. Ultimately, empirical models can be used to induce more general theoretical ones, and theoretical models can help guide the specific development of empirical models.

Using Models to Generate Research Hypotheses

The modeling process should be a means by which we challenge ourselves to articulate explicitly what we think we know about some system. To this end, model output can be used to generate research hypotheses. In an adaptive management context, management activities can be crafted to test the more important assumptions or to provide information on the key unknowns. When management activities are applied on the ground, these assumptions and unknowns may be termed management hypotheses. To test management hypotheses scientifically in order to evaluate their correctness, we should craft management activities to follow the guidelines for correct study design, including evaluation of baseline conditions, provision of controls, adequate study and treatment duration, and adequate replication of controls and treatments. Then effects of the management activities can be analyzed to validate the original assumptions and provide new information to revise or reaffirm the management hypotheses and guidelines. In turn, the models used to suggest the original management activities would be updated and new activities suggested, if warranted by the findings. Thus, the ideal adaptive management process is cyclic, not linear, and entails strict adherence to correct experimental design.

As models become more complex, and as management objectives broaden to include landscapes and ecosystems, use of models for generating research hypotheses should become more salient in the decision-making process. The real-world constraints of uncertain future research budgets, changing management goals, and balancing the need for short-term publications and long-term studies must be addressed in this use of models.

In the next section we review specific kinds of wildlife-habitat relationship models.

A Review of Models of Habitat Relationships

Models of Vegetation Structure

Stand Growth Models

A number of models have been developed that display and predict composition and structure of vegetation stands. These include silvicultural *stand growth and yield models* (e.g., Avery and Burkhart 1983; Clutter et al. 1983), some of which are: CLIMACS (Dale and Hemstrom 1984), Douglas-fir Simulator (DF-SIM) (Curtis et al. 1981), FORCYTE (Kimmins 1987), FOREST (Ek and Monserud 1974), FREP (Leary 1979), Prognosis (Wykoff et al. 1982), Stand Projection System (SPS) (Arney 1985), STEMS (Belcher et al. 1982), and WOODPLAN (Williamson 1983) (see Mannan et al. 1994 for an additional review of forest models). Growth and yield information is generally available for a variety of forest types on commercial forest land.

Ramm and Miner (1986) reviewed 14 growth and yield programs in the north-central region of the United States. Meldahl (1986) compared and critiqued alternative modeling methodologies for growth and yield prediction models, including yield tables, multiple regression models, diameter distribution models, differential and difference equation models, and individual tree models. He concluded that a manager must first describe what inputs are available and what kinds of information are really needed in growth and yield projection systems before selecting the correct model.

Many stand growth and yield models of forests assume even-age silvicultural management of a single-species stand. Stand growth is typically depicted in such models as beginning at final harvest, such as clear-cutting. However, data are available for other methods of forest regeneration, such as shelterwood cutting. The user may specify such parameters as stand origin (such as artificial planting or natural seeding), seedling spacing, fertilization, presence and degree of precommercial thinning, number and intensity of commercial thinnings, sanitation entries, other intermediate treatments, and timing of final harvest. Such models typically describe the expected structure of a forest stand in terms of stem density per acre, stem volume (board feet or cubic feet), stem basal area per acre, quadratic mean diameter at breast height (dbh), and tree height. A few stand growth models, such as SPS, provide stem dbh distributions as well as quadratic mean dbh.

More recently, forest ecosystem models have been developed that explicitly incorporate nutrient cycling and net primary production functions. Running (1994) tested hydrological, carbon, and nitrogen cycle processes in such a model, the FOREST-BGC ecosystem model, by using data from the Oregon Transect Ecological Research project (OTTER). Running found high correlation between model predictions and empirical estimates of aboveground net primary production and average leaf nitrogen concentration but much lower correlations for predawn leaf water potential and equilibrium leaf area index. Successful simulation of these last two variables requires accurate data for soil water-holding capacity. Running concluded that defining the water-holding capacity of the rooting zone and the maximum surface conductance for photosynthesis and transpiration rates proved to be critical system variables that defy routine field measurement. The many additional processes simulated in FOREST-BGC had no repeated field data sets for validations. This example illustrates that as vegetation models become more complex, as in explicitly representing many ecosystem processes, their parameterization and validation become increasingly problematic.

Unfortunately, stand growth and yield models are often used outside the geographic scope, stand age, and stand conditions from which the empirical growth and yield data were gathered. Thus, such models may produce unreliable, and possibly optimistic, predictions of stand conditions.

Some stand growth and yield models provide estimates of suppression mortality, which may be useful for predicting the density and size of future snags or down wood in the stand. For example, Neitro et al. (1985) used DF-SIM to model the occurrence of snags in even-age stands of Douglas-fir (*Pseudotsuga menziesii*) in western Washington and Oregon. Several of the other models listed above also predict the rate of suppression mortality and may be used similarly.

Dennis et al. (1985) reviewed problems associated with mathematically modeling growth and yield of renewable resources. In particular, they point out that variation in yield estimates, aside from sampling variations inherent in inventories, is usually large and uncontrollable. Techniques to help plan for use levels, such as with linear programing or forest stand growth models, cannot account for such variation. And even if they could, we seldom have data on the frequency and effects of factors such as storms, insect epidemics, and fires, which affect resource amounts and growth rates. Such random events can greatly influence optimal solutions for best rate of use of resources. Also, spatial variation is seldom accounted for in growth and yield models, such as specific tree stem spacing as it affects mortality functions.

Succession Models

Another class of vegetation growth and change models tracks changes in acreage of *successional or structural vegetation stages.* Examples are the DYNamically Analytic Silviculture Technique (DYNAST) (Boyce 1980) and FORPLAN, which are used in the USDA Forest Service for habitat and general resource planning (Kirkman et al. 1984). These models operate on current acreage of various forest types and their growth stages and the rates of succession for each type. The output displays the acreage of each forest type and growth stage over time.

Often, such models are used to calculate habitat capability for a variety of wildlife species, as discussed below (also see fig. 10.4). One shortcoming of such models is that they typically lack explicit sensitivity to spatial patterns of forest stages; the same array of acreage of various forest developmental stages will produce the same habitat capability estimates regardless of habitat patch sizes or arrangements. This may be of little consequence to wildlife species that use early- to midsuccessional stages if such patches are widely distributed, but may produce great errors when predicting responses of species requiring scarce or declining habitats, such as specific edge or forest interior conditions in some locations.

Another kind of vegetation structural model is that of forest gap models (Acevedo et al. 1995). These are models of forest growth and structural transition due to canopy gap openings and tree mortality events. They can be based on Markovian (probabilistic) chains of successive vegetation structures (Horn 1975). Forest canopy gaps are important in that they affect understory response, stand structural profiles, and dominant and subdominant vegetation (Halle et al. 1978), as well as nutrient cycling (Parsons et al. 1994) and wildlife response.

Models of Species Response to Vegetation Structure

Modeling species response to vegetation structure ultimately entails linking models of vegetation growth, structure, and succession with predictions of occurrence and abundance of wildlife species. A variety of model forms serve this purpose. Such models may be categorized generally as single-species and multiple-species models.

Single-Species Models

Life history models. Ecological models depicting and predicting life history characteristics of species have been proposed for understanding how behavioral and phenotypic traits evolve (Barclay and Gregory 1981; Holm 1988; Schlosser 1990). A variant of life history models relates body size of organisms to resource needs, with implications for conservation (Gaston and Blackburn 1996a), species rarity (Gaston and Blackburn 1996b), and population density (Juanes 1986). Such models help us understand the ecological roles of species and species groups and the relations between body size, food needs, and resource levels.

Holling (1992) studied the relation between bird and mammal body size and home range size, and concluded that birds and mammals of all trophic levels use foraging resources in the

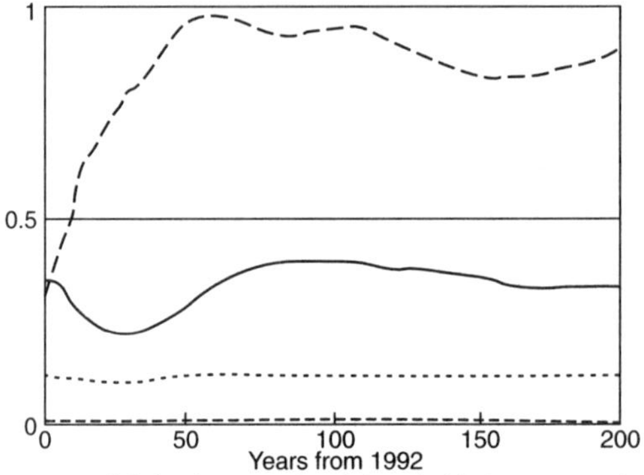

Relative amounts of timber harvested and three habitats as craggy landscape is turned into a forest reserve (no timber harvest)

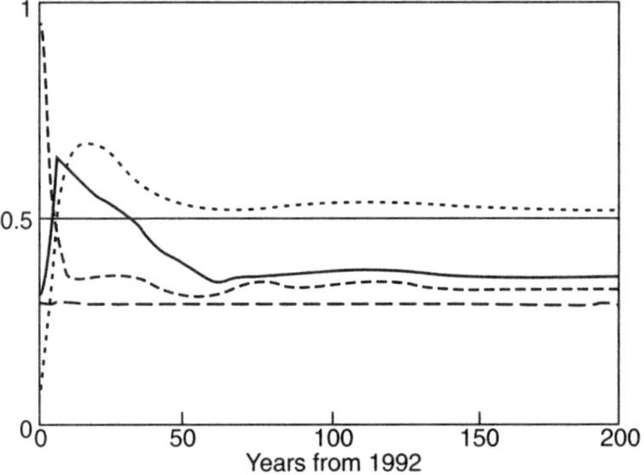

Relative amounts of timber harvested and three habitats as traditional forestry is imposed on craggy landscape

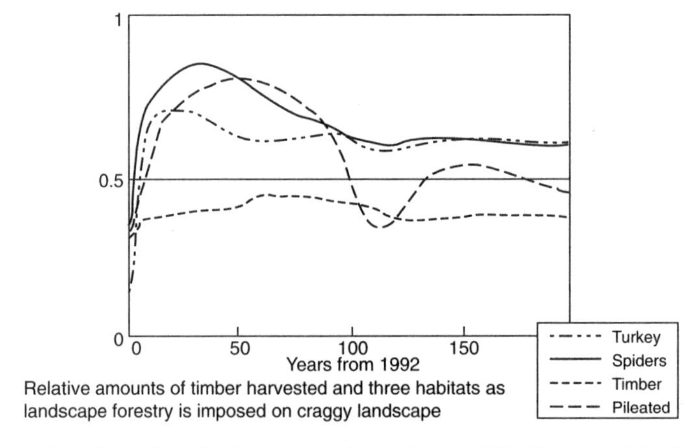

Relative amounts of timber harvested and three habitats as landscape forestry is imposed on craggy landscape

Figure 10.4. Examples of running the forest growth simulator DYNAST to model relative amounts of habitat for three wildlife species and timber harvest levels, under three different forest harvest scenarios in the southern Appalachian Mountains in western North Carolina. (From Boyce and McNab 1994: figs. 5, 6, 7; reprinted from the *Journal of Forestry* [vol. 92, no.1] published by the Society of American Foresters, 5400 Grosvenor Lane, Bethesda, Md. 20814-2198; not for further reproduction)

same manner by measuring the spatial grain of habitat patches with a resolution defined as a function of their size. He also concluded that landscapes form a well-defined hierarchy of discrete clumps of resource and substrate sizes, proximities, and textures, which in turn induces a parallel hierarchy in animals. His work can provide a basis for evaluating wildlife response to multiple-scale changes in environmental conditions.

Habitat preference models. The study of animals' selection and preference for food and habitats is essentially a study of the adaptive advantages of particular behaviors. For example, Fujita (1900) assessed the habitat preference of birds for various vegetation structures in the Russian Far East, and Wilson et al. (1990) determined color preferences of fruit-eating birds. Porter and Church (1987) assessed the effects of environmental pattern on analysis of habitat preference. Rosenzweig (1987) reviewed how habitat selection contributes to biological diversity.

The concepts of ecological dependency and preference were critically evaluated by Carey (1984) and Ruggiero et al. (1988). Dependency and preference can only be inferred from statistics on habitat selection, that is, the differential use of specific resources or environments as compared with their general availability.

Optimal foraging models. Related to life history and habitat preference models are models of foraging behavior (Pyke 1984). For example, Belovsky (1987) related foraging requirements to body size to help explain the life history traits of home range area. Maurer (1990) related optimal foraging theory to explain the structure of insectivorous bird communities.

Other models of animal behavior can be useful for understanding habitat relations. For example, Caraco (1979) reviewed the ecological response of animal group size to environmental conditions. Such a model may be useful for interpreting the herding behavior of ungulates and how environmental conditions might affect herd size, composition, and distribution.

Correlation models. Correlation models display the degree to which species parameters are explained by environmental parameters. Correlations in such models are typically based on empirical data and are best used for hindcasting rather than for forecasting species' responses unless validated. Too often, however, unvalidated correlation models are taken as predictive models. Garsd (1984) reviewed various pitfalls of depicting and using correlation models, including the common mistake of interpreting spurious correlation as a causal relation.

Multivariate statistical models. Modeling species-habitat relationships with multivariate statistics is a common practice (e.g., papers in Capen 1981). In general, multivariate models help identify significant combinations of environmental parameters that account for observed variation in the distribution and abundance of wildlife species (see chapter 6).

One shortcoming of a multivariate approach is that results can be difficult to interpret. Mathematically, many multivariate statistical techniques combine several environmental parameters into one collapsed function, which is then correlated with species distribution and abundance. Biologically, however, it is not always clear what the collapsed functions or the correlations mean, particularly if axes are rotated and if data are standardized and transformed to increase the degree of correlation. Nevertheless, multivariate models are indispensable for exploring patterns in (i.e., for hindcasting) large empirical data sets and for understanding relations between wildlife species and environmental variables.

Habitat suitability index (HSI) models. One of the most popular approaches to modeling environmental conditions is the use of habitat suitability index models. HSI models are used extensively by the USDI Fish and Wildlife Service (Schamberger et al. 1982) and other federal resource management agencies. These models typically denote habitat suitability of a species as the geometric mean of *n* environmental variables deemed to most affect species presence, distribution, or abundance. The general model form of the HSI is:

$$HSI = (V_1 \times V_2 \times \ldots V_n)^{1/n}$$

where the *V*'s represent *n* key environmental variables. Each variable and the resulting HSI values are scaled from 0 to 1. The overall HSI value assumedly represents the final response of the species to the combination of the values of the environmental parameters.

For example, the three environmental variables denoted in an HSI model for the yellow warbler (*Dendroica petechia*) are the percentage of deciduous shrub crown cover, the average height of deciduous shrub canopy, and the percentage of shrub canopy composed of hydrophytic shrubs (Schroeder 1983). The resulting suitability index in the yellow warbler model represents relative habitat values for reproduction. HSI models have been constructed for a wide variety of species in the United States.

HSI models are useful for representing in a simple and understandable form the major environmental factors thought to influence most the occurrence and abundance of a wildlife species. However, HSI models are best viewed as hypotheses of species-habitat relationships rather than as causal functions (Schamberger et al. 1982). Their value lies in documenting a repeatable assessment procedure and providing an index to particular environmental characteristics that can be compared between alternative management plans. They do not provide information on population size, trend, or behavioral response by individuals to shifts in resource conditions, and seldom include interaction or error terms.

Habitat capability (HC) and habitat effectiveness (HE) models. Closely allied to habitat suitability index models are habitat capability and habitat effectiveness models. These models essentially perform the same function as HSI models but may vary slightly in structure. HC models typically provide an estimate of the total area within which resources for a particular species can be found, or rank a given area for the relative capability of supporting a species, given a few key environmental factors. HE models rank resources in an area according to the degree to which maximum use or carrying capacity can be met.

An HE model was constructed for assessing habitat effectiveness for the winter range of the Rocky Mountain elk (*Cervus elaphus nelsoni*) in the Blue Mountains of eastern Oregon and Washington (Wisdom et al. 1986). This model calculates an elk habitat effectiveness index as the geometric mean of four environmental variables, including distance from cover-forage edge, miles of road open to motorized traffic per square mile of habitat, habitat condition and successional stage, and type of management treatment. The model was evaluated by Holthausen et al. (1994) by use of expert opinion.

With HSI, HC, and HE models, it is difficult to interpret if the resulting index value is intended to represent environmental conditions or population response. Also, the sensitivity of the resulting habitat index values to any one environmental variable is diminished as more variables are added to the model. This behavior is a function of the mathematics of a geometric mean model and may not accurately reflect actual habitat use or population response. Finally, as with HSI models, HC and HE models should be used to represent relative environmental conditions and as a means of generating hypotheses about species-habitat relationships, rather than as evidence of causal relations or as reliable predictions of species response. Bender et al. (1996) provided a procedure for evaluating confidence intervals for HSI models; this approach may be extended to HC and HE models as well.

Habitat evaluation procedures (HEPs). Habitat evaluation procedures have been used extensively by the USDI Fish and Wildlife Service to assess environmental conditions at the species level (Flood et al. 1977; U.S. Fish and Wildlife Service 1980). The procedure is based on habitat units (HUs), which are defined as the product of habitat quality (on a 0 to 1 index, as from a habitat suitability index) and habitat quantity. HEP models may require much field data on specific environmental attributes, such as forage quality or quantity. However, the procedure provides a structured way to document a repeatable assessment of environmental conditions. HEPs are often used to evaluate impacts of and mitigations for proposed projects on environmental conditions for species of special interest. Roberts and O'Neil (1985) provided a procedure for selecting species for HEP assessments. Rewa and Michael (1984) provided a way of evaluating environmental quality for ecological guilds by using an HEP approach.

Wakeley and O'Neil (1988) presented methods to increase efficiency in applying HEPs. Their suggestions included: delineating cover types by using remote imagery and combining types; choosing wildlife species to model for which there is available inventory information; choosing model forms that make best use of available inventory data and that focus on the most important life history components; designing field sampling for environmental conditions to be cost effective and tailored to the range of modeled conditions; and using computers to aid in collecting and analyzing field inventory data and conducting model analysis.

Bayesian and pattern recognition (PATREC) models. Pattern recognition models are useful for predicting effects of changes in environmental conditions on wildlife in the form of a risk analysis. PATREC is based on the use of Bayes's theorem (Williams et al. 1977) and calculates the probability of a species' response, such as population density, given the probabilities of the presence of various environmental conditions. The general form of the model is: $P(S|E) = P(E|S)\ P(S)/P(E)$, where $P(S|E)$ = the posterior probability of species density (S), given that specific environmental conditions (E) are present; $P(E|S)$ = the prior probability of the presence of environmental conditions, given a specific species density; $P(S)$ = the unconditional probability of the species having a specific density; and $P(E)$ = the unconditional probability of specific conditions in the environment. PATREC models have been used in forest planning by integrating them with vegetation response models (Kirkman et al. 1984). An example of a PATREC model is the evaluation of bighorn sheep (*Ovis canadensis*) habitat by Holl (1982).

There are a few key problems with PATREC and Bayesian approaches. Values of prior probabilities, $P(E|S)$, are often difficult to determine empirically and are usually represented as best professional judgment, which may suffer from bias. Also, posterior probabilities, $P(S|E)$, are often quite sensitive to values of the prior probabilities. Thus, if priors are not known with much certainty, posteriors may be wrought with a large and unknown degree of error or bias.

However, a Bayesian approach can be useful for modifying sample wildlife surveys by analyzing interim results. This so-called empirical or sequential Bayesian approach (Marks and Woodruff 1979; Morris 1983) was illustrated by Johnson (1985, 1989) for use with waterfowl samples. In his study, the design of the U.S. Fish and Wildlife Service sampling of waterfowl was modified by analyzing results of surveys to date, so that estimators of waterfowl density are more accurate. The increase in accuracy was gauged by a reduction of an average 30 percent in the mean square errors of one density estimate (a Bayesian ratio estimator). However, the bias of the density estimates also increased. In addition, the precision of estimates was greater or less than that of the usual estimators, but the authors deemed the decrease in precision insignificant. Gazey and Staley (1986) used a sequential Bayesian analysis to estimate population size from a mark-recapture study, and Haas et al. (1994) applied a temporal Bayes network to model aspen stand growth.

In other uses of the Bayesian approach, Haas developed a Bayes network model of the decision-making process of USDA Forest Service district rangers. He concluded that the actual, or "normative," decision-making procedure differs from the optimal or theoretical procedure in that rangers (1) use forest management computer models to aid in forming their hypotheses of future forest states under different management options rather than to provide definitive estimates of management risk, and (2) consider many additional factors in their decision-making process other than explicit future outcomes.

Multiple-Species Models

Recent advances in multiple-species modeling and evaluations have begun to bridge the gap between single-species models and general models of vegetation or landscape conditions. Many issues of biodiversity management have benefited from newer approaches, including use of species-habitat matrices, guild and life form models, and community structure models. In general, multiple-species models have an advantage over single-species models in simultaneously assessing potentially conflicting species requirements.

Coarse-filter and fine-filter models. One approach to accounting for multiple species simultaneously is the so-called coarse filter approach (Hunter 1991; Haufler et al. 1996). This assumes that supplying general environmental conditions at coarse-grained resolutions will provide for the host of associated wildlife species on multiple spatial scales. An example is supplying old-growth forest conditions by specifying general conditions of forest stand age and assuming that the needs of all species closely associated with old-growth forest ecosystems would be met.

The shortcoming of this approach is that it has seldom been explicitly tested, although it is often touted as the way to provide for wildlife communities without investment of species-specific analyses and knowledge. It seems reasonable to expect that many species would "fall between the cracks," especially species that are locally endemic, that occur in atypical and scarce environments or substrates, or that require resources not explicitly supplied by coarse-scale descriptions of environments.

Because of this shortcoming, the coarse-filter approach should be used as an initial step in an analysis process. It provides a broad-extent, coarse-grained, initial phase of assessment or definition of management activities in what should be a multiscale approach. To this end, Rastetter et al. (1992) suggested ways to use fine-scale knowledge to develop coarser-scale models of ecosystem attributes.

Species-habitat matrices. One simple form of representing relationships between wildlife species and their environments is species-habitat matrices. These are tables listing vegetation types and environmental conditions with which wildlife species are associated. Often the data are qualitative and derived from a combination of field studies and professional judgment. Examples from the Wildlife Habitat Relationships Program of USDA Forest Service include species-habitat matrices for amphibians, reptiles, birds, and mammals in California (Verner and Boss 1980). Such matrices or information bases are useful for predicting sets of wildlife species associated with specific environmental conditions.

Validation of a wildlife-habitat relationship model by Raphael and Marcot (1986) suggested that such information bases are probably best used to predict the occurrence of species in general vegetation types and environmental conditions across broad regions rather than on the individual stand scale. A shortcoming of such models is that they generally do not quantify population response. Thus, such models cannot be used to gauge population density or to quantify population trend.

Guild and life form models. Guild and life form models denote the response of a set of species with similar characteristics to changes in environmental conditions (e.g., DeGraaf et al. 1985). Such models simplify the assessment of many species by referring to fewer sets of species. Guilds and life forms may be defined *a priori* as sets of species with common attributes, as with the models of Short (1983). They may also be defined by multivariate analysis of empirical data on species abundance and distribution, as with the assessment of Hubbard Brook bird guilds by Holmes et al. (1979).

The guild approach may prove useful when environmental conditions and target species are well defined (e.g., Knopf et al. 1988). However, individual species of a guild may vary disparately in response to environmental conditions while the guild as a whole shows little or no variation (Hariston 1981; Block et al. 1987). The conclusion is that, although grouping species by guild might be useful for depicting groups of species with similar functions or trophic relations, the guild approach is not useful for specifically predicting responses of individual species to environmental conditions and changes. In this case, it may be more appropriate to model individually each species within the guild and combine results.

Community structure models. Community structure models describe wildlife species distribution, abundance, and diversity as a function of the structure of the environment. Multivariate statistics are commonly used to assess these relationships (e.g., Scott et al. 1987), such as in the assessment of diversity of wildlife in late successional forests. A shortcoming of this approach, like the guild approach, is that individual species' responses may vary considerably while the composite measures of community structure remain more or less constant. Also, habitat management objectives seldom are adequately expressed in terms of wildlife species diversity indices. Species richness mapping can aid in identifying key areas for conservation consideration (Conroy and Noon 1996; Williams et al. 1996), although results must be interpreted very carefully (see "Gap analysis models," below).

Community and ecosystem simulation models. Community and ecosystem simulation models are used to evaluate stochastic population and system responses to variable environmental conditions, effects of catastrophic changes in vegetation and environments on populations, and other aspects of species, community, and ecosystem function. In the 1960s, ecological systems analysis was a popular approach (e.g., see Watt 1966). A number of reviews (e.g., Biesterfeldt 1984) or introductory treatises (e.g., Eberhardt 1977; Law and Kelton 1982) on simulation modeling and systems ecology are available.

Among the many ecosystem models described in the literature are those constructed for grasslands (Bledsoe and Jamieson 1996) and wetlands (Poiani and Johnson 1993). Forest simulation models have been developed by Smith et al. (1981) for nongame bird management. Hall et al. (1977) developed a "circuit language" for depicting energy flow in ecosystems.

Gap analysis models. The U.S. Fish and Wildlife Service has instituted a Gap analysis procedure for identifying areas of high richness (number of wildlife species) and mapping habitats for species on the basis of coarse-grained vegetation attributes such as type and structure (Scott et al. 1993). Results of the Gap program have been used to identify vegetation and wildlife habitat management needs, such as in Idaho (Caicco et al. 1995). The Gap program is providing state-specific databases and species range maps based on mapped vegetation conditions. The information on species occurrence is typically compiled to produce isocline maps showing species richness and "hot spots" of high richness centers. These centers are then overlaid on maps of natural areas of various designations to determine which centers do not coincide with "protected" environments, in order to promote changes in land allocations and ownerships to provide such protection.

Since the Gap program and similar assessments (such as in Costa Rica [Powell et al. 1995–1996] and Mexico [Bojorquez-Tapia et al. 1995]) have garnered much interest and momentum, it may be useful to examine some of the underlying assumptions. First, species richness by itself does not reveal much about species composition; low richness areas could have rare, endemic, or spatially constrained species not found in species-rich areas. Second, species-rich areas might occur in ecoclines or ecotones where communities of different species intermix, rather than occurring in the centers of vegetation communities more representative of such types. Ecoclines or ecotones might include prairie-forest interfaces or remnant patches or corridors of native environments. A corollary to this second point is that ecoclines of high richness might not include centers of species ranges where population densities are highest (see chapter 3 on "bull's-eye" range distribution patterns); what is depicted may be sink or marginal habitats rather than source or primary habitats. And third, species-rich areas might occur in areas of high human disturbance with many common, weedy, pioneer or invader species, rather than in areas of low human disturbance, as is commonly assumed.

An obvious solution to some of these problems is to represent species composition explicitly in the Gap analyses and maps. That is, know thy community composition; know where rare species occur; know where endemics occur; and use richness mapping only as one step or phase in a conservation priority-setting program in forming initial or tentative working hypotheses. To its credit, much of the recent use of the Gap program information has addressed at least some of these concerns by partitioning out rare species, endemics, and other species categories.

Biodiversity models. Models of species richness, Gap-type species-habitat relation models and maps, and information on community diversity have been used to assess and map biodiversity (e.g., Miller 1994). However, in most cases, biodiversity models merely constitute species richness models. There is a need to further develop models of other aspects of biodiversity, particularly for mapping and modeling land use effects on demes and population genetics (e.g., Richards and Leberg 1996), ecological communities, ecological functions of species (chapter 11), and ecosystem processes (chapters 8 and 9).

Hierarchy models. Ecological systems also have been represented as a series of hierarchies of elements and functions (Allen and Hoekstra 1984; Allen et al. 1984; Kolasa 1989). The hierarchy approach has been used to develop classifications of aquatic environments (Maxwell et al. 1995) and for land management planning of U.S. national forests (Church and Murray 1994). The advantage of a hierarchic model is ease of representing environments and species on various spatial scales and levels of biological organization.

Models of Population Dynamics and Viability

Models of population dynamics and viability were discussed in chapter 3. Such models aid in predicting potential risks of extinction of populations. Burgman et al. (1988) reviewed the use of some extinction models for conserving species. Roozen (1987) and Strebel (1985) explored how variability in population parameters and environmental conditions affect extinction likelihoods. Quinn and Hastings (1987) explored how isolated patches of habitat enhance extinction probabilities. Taylor (1990) reviewed concepts of how metapopulation dynamics, dispersal, and predation affect population viability. Conner (1988) argued for conservation of populations at high levels to ensure long-term viability, rather than at minimally adequate levels.

A growing area of model development deals with spatially explicit or geographically referenced population models, such as the RAMAS/GIS model, as reviewed by Kingston (1995). In population models, individual demography is linked to location-specific environmental conditions, including quality and extent of habitat (Kareiva and Wennergren 1995; Pulliam and Dunning 1995). Spatially explicit models predict breeding individuals' occupancy of habitat patches in heterogeneous landscapes and various population and metapopulation trends. For example, Akçakaya et al. (1995) assessed the effect of spatial patterns of habitats on the viability of helmeted honeyeater populations. Dunning et al. (1995), Holt et al. (1995), and Turner et al. (1995) reviewed several spatially explicit population models in development. Their general conclusion was that such models can fundamentally aid basic ecological knowledge of landscape phenomena and the application of landscape ecology to conservation and management, and that the time is ripe for a better merging of vegetation and animal spatially explicit models. Wahlberg et al. (1996) reviewed the utility of spatially explicit population models for predicting occurrence of endangered species in fragmented landscapes. One caveat worth remembering, though, when using spatially explicit population models is that many factors usually not explicitly included in the model—such as density-dependent demographic effects, competitors, predators, and effects of harvest—often have strong influences on population dynamics.

Models of Landscapes

Included in the category of landscape models are those that display spacing and patterns of habitat patches explicitly on the *landscape scale* (also see the immediately preceding review of spatially explicit population models). Models of landscapes include vegetation disturbance models, fragmentation models, and cumulative effects assessments.

Vegetation Disturbance Models

Vegetation disturbance models simulate the extent and distribution of vegetation stages across a landscape given the frequency and intensity of disturbances such as fire and timber harvesting. Such models (e.g., Shugart 1984; Shugart and Seagle 1985; Pickett and White 1985) can be used to plan habitat management (Smith et al. 1981; Karr and Freemark 1985).

Many kinds of models have been advanced for predicting changes in vegetation conditions. These include models of vegetation cover in forest stands (Moeur 1985), succession (see review above in "Succession Models"), and disturbance and patch dynamics (Pickett and White 1985). Ellison and Bedford (1995) used a spatial simulation model to determine the response of a Wisconsin wetland plant community to anthropogenic (human-induced) disturbance.

Fragmentation Models

Models displaying species response to fragmentation and isolation of habitat patches include incidence functions, which denote the probability of a species occurring in a particular patch given its size. The greater the degree of fragmentation, the higher the likelihood of the extinction of species associated with forest interiors (see chapter 3).

Fragmentation models address a variety of factors that may influence species persistence in a landscape: patch size (Askins et al. 1987), isolation of patches (Mills 1995), competition and structure of environments (Dueser and Porter 1986), as well as the interaction of environmental fragmentation with parasites (Dobson 1988) and predators (Martin 1988) in the ways that these affect population regulation. The use of insular biogeography theory for predicting effects of forest fragmentation on population persistence and quality of forest environments has served as the basis for some management guidelines (Franklin and Forman 1987). The computer model FRAGSTATS (McGarigal and Marks 1995) provides perhaps the widest array available of indices of habitat connectivity, diversity, and fragmentation. Other authors have applied graph theory to modeling vegetation configuration and fragmentation (Marcot and Chinn 1982; Cantwell and Forman 1993).

Cumulative Effects Assessments

Cumulative effects models are used to assess effects on environments (Grant and Swanson 1991) and distribution and abundance of wildlife species across a landscape. The term *cumulative effects model* is a generic one. It may be applied to any assessment of the effects of management activities or natural disturbances on wildlife species across a geographic area or over time. Salwasser and Samson (1985) discussed development of cumulative effects models, noting that the major steps are stating management goals and standards, representing major environmental factors, projecting changes in environments, and estimating wildlife effects.

Weaver et al. (1985) presented a cumulative effects simulation model for assessing grizzly bear habitat in the Yellowstone ecosystem. Their model consists of three submodels used to assess suitability of environments, bear displacement from human activities, and bear mortality under one or more planning scenarios. Each submodel in turn is composed of variables deemed important to the presence of grizzly bears. For example, the habitat submodel accounts for food and thermal cover, habitat diversity, seasonal equity, and denning suitability. The cumulative effects of all environmental conditions, displacement, and mortality influences on the presence of grizzly bears are combined for a particular location and set of management conditions to produce habitat effectiveness values and a bear mortality risk index. The grizzly bear cumulative effects model is sizable and computation is intensive, so it requires at least a large minicomputer to run.

Models of Insular Biogeography

Models of insular biogeography, born of theory or empirical data, depict the effects of habitat fragmentation and isolation on species presence and diversity, or population extinction (Rosenzweig and Clark 1994), on islands or in landscapes. For example, Blake and Karr (1987) studied the effects of the isolation of woodlots on breeding birds.

Related to models of biogeography are models depicting species demographic or genetic responses to spatial configurations of vegetation and environments (e.g., Chambers 1995). Buechner (1987) and Stamps et al. (1987) modeled vertebrate dispersal across boundaries of vegetation patches. Fahrig and Merriam (1985) explored the effects of patch connectivity on population survival. Hassell (1987) explored how patchiness of an animal's distribution would contribute to overall population regulation. In recent years, there has been less emphasis on applying classic island biogeography theory to continental habitat conservation problems, with the advent and growth of geographically referenced population and biodiversity models and the use of remote sensing and geobased information system models.

Models of Ecosystems and Ecosystem Management

Pronouncements for what has been termed ecosystem management (EM) of federal public (and other) lands in the United States have prompted the revival of interest in nutrient cycling, energy flow, and other aspects of ecosystem dynamics in a variety of vegetation and environmental conditions (Yaffee et al. 1996), although some question the rigor of this new approach (Zeide 1996). To date, most EM models pertain to planning and assessment procedures (e.g., papers in Thompson 1995) or ecological conceptual frameworks (e.g., Haufler 1994), and they proceed on assumption rather than by use of validated, quantitative tools for evaluating ecosystem function and process.

One type of EM planning model used more and more extensively is that of the historic range of natural conditions, erroneously called the range of variability or variation (Morgan et al. 1994). The central assumption of this planning approach is that current land management is acceptable if it results in landscape and vegetation conditions that fall into historic ranges of conditions. In actuality, estimates of historic conditions and their ranges (e.g., Swanson et al. 1994) constitute a form of hindcasting. The appropriate metric for depicting past variation in such conditions as percentage of forest cover is a tolerance interval rather than an overall range or confidence interval of values. The use of hindcast tolerance intervals as a predictive or prescriptive model should entail additional validation steps which have not yet been taken in most of the literature on the subject. The modeling of historic range of conditions for prescribing land use planning is akin to the mean-variance approach to resource planning touted two decades ago by Regier (1978), which sets management goals as a mean condition within an empirically calculated or modeled variance limit.

There are several major problems with the use of historic range of conditions to set planning objectives, and these should be solved or at least tested before the approach is used. (1) The approach, as described in current literature, does not account for autocorrelation of data, that is, for time series correlation of ecological conditions. Site history can exert a strong ecological inertia; current conditions can greatly mediate and constrain near-future conditions. Therefore, temporal series of conditions should be worked out before asserting planning guidelines. Some of the vegetation succession and disturbance models reviewed above may aid in this. (2) The approach does not account for long-term time series trends underlying shorter-term "blips." For example, recent trends in successful dispersal of large-ranging carnivores may belie a longer-term population decline; or short-term weather trends might run counter to longer-term climatic change. (3) Use of overall range of conditions overstates the desirable conditions that can be self-sustaining or in which species can persist. What should be used instead are parametric estimates of tolerance intervals, or nonparametric measures of quartiles, or other nonparametric measures of central tendency that are insensitive to outlier conditions. (4) The range of historic conditions is purely a function of the time duration considered. Many environmental conditions may have considerable, overlapping variability among years, decades, centuries, and millennia. Each duration produces a new range of conditions. One could justify any range desired by extending the time period of the assessment. (5) The oft-used term *range of natural variation* is a misnomer; operationally, the measures involved in this term have not been of a range of "variation" (i.e., variance) but simply a range of conditions. Estimating a range of variation (variance) is a far more complicated and probably more meaningful approach, such as entailing analysis of covariance models. (6) It is often assumed that a range of historic conditions displays the set of environmental conditions under which wildlife species evolved (e.g., Covington and Moore 1994). This is fallacious; the range of historic conditions nearly always depicts only the conditions under which populations have recently persisted. And the set of conditions under which populations have persisted is not necessarily the same as the set that is optimal for providing for their long-term viability (e.g., see the nested climatic charts of the Quaternary by Tausch et al. 1993). (7) The historic range of natural conditions was affected by species- and land-disturbing activities of aboriginal, or indigenous, peoples. In the United States, these included prairie burning by plains Indians and harvesting of large ungulate game. Even the Polynesians devastated some of the native Hawaiian bird fauna prior to the arrival of European settlers. Thus, given these complicating factors, a more appropriate approach might be to depict the central tendency and spread (as suggested above) of both historic and prehistoric conditions over several specified time periods (on the order of years, decades, centuries, and if possible millennia) and on different spatial scales (on the order of subwatersheds to entire bioregions).

The following kinds of models may also be of interest to ecosystem managers.

Models of Indicator Species

The use of management indicator species in national forest planning is mandated by U.S. federal law (36 CFR 219 regulations pursuant to the National Forest Management Act of 1977). Management indicator species constitute categories of wildlife species of management significance, including species that are threatened, endangered, valuable economically or socially, and ecologically representative of particular environments or of other wildlife species also associated with those environments.

In this last category, ecological indicator species, there has been recent controversy and confusion. The controversy has focused on whether the population attributes of one wildlife species can be validly used to represent those of other wildlife species (Landres et al. 1988; Patton 1987). The confusion has arisen from equating the general concepts of management indicator species with the specific category of ecological indicators used in the precise context of one species indicating other species. The scientific arguments are sound regarding cautions that few true ecological species indicators can be demonstrated. Ecological indicators are seldom rigorously tested for their indicator function, and their utility varies by spatial scale (Weaver 1995). Rather, ecological indicator species are likely useful for monitoring general ecosystem conditions (Kremen 1992), not for inferring population viability of other species per se. Ecological indicators other than species can provide a useful template for modeling changes in overall biodiversity conditions (Noss 1990).

Models for Monitoring Species and Habitats

Models of species-habitat relationships may prove useful for the expensive task of monitoring species and habitats over time. Single-species models are probably best for monitoring management indicator species and other species of singular concern. Models depicting environmental associations, such as HSI, HC, HE, PATREC, and HEP models, are useful for species that are too expensive to monitor directly. However, it is vital to demonstrate the degree of reliability and validity of the model first. Also, population trends of species of high concern, especially those that are state- and federal-listed as threatened or endangered, can be better assessed if they are monitored directly in the field rather than inferred through habitat relationship models.

In monitoring, a useful model is the adaptive management paradigm (Walters 1986; Walters and Hilborn 1978; Holling 1984). Adaptive management entails: viewing a habitat management plan and its expected effects on wildlife species as a hypothesis; monitoring environments and species to ascertain how they respond to the management; and revising the direction of management if results so warrant. Monitoring is an essential step in the adaptive management feedback process, and it helps managers deal directly with limitations of biological uncertainty (Lee and Lawrence 1986; Ringold et al. 1996). Under some land management plans, specific land allocations—called adaptive management areas in at least one plan (Franklin 1994)—can be established to test management hypotheses.

Knowledge-based and Decision-aiding Models

Models that aid habitat evaluation and decision-making are another class of tools for assessing species-habitat relationships. Recent years have seen an explosion in computer-aided decision making, including in many fields of resource management. An example is to be found in entomology (Coulson and Saunders 1987). Such models help organize and document factors associated with habitat evaluation and planning, and assess and convey uncertainties in habitat management planning

(Cleaves 1995). These models include decision support models and expert system models for helping to evaluate and monitor species and environments.

Validating Wildlife-Habitat Models

Validating wildlife-habitat models should be part of every step in building and using such tools (Marcot et al. 1983). Model validation is best thought of as a general approach to developing as well as testing models, and should be conducted in a variety of ways throughout the model development and application process.

Aspects of validating a model (table 10.2) include:

1. model verification: verifying that mathematical equations are correct or that the computer program code has been written without bugs;
2. testing the audience: ensuring that the audience for whom the model is intended will accept and use the tool;
3. running the model: confirming that the model can be run with available or obtainable data;
4. assessing purpose and context: ensuring that the purpose of the model and the conditions in which it is to be used have been clearly stated and adhered to in its use;
5. testing the output: assessing whether the output of the model matches real-world biological conditions.

Model validation is typically associated with just the first and final items on this list. Each item, however, contributes to the successful development and application of a wildlife-habitat model.

Model Verification

Ensuring that the formulae and computer code are correctly written is a simple but important aspect of model validation. A similar task is documentation. Documentation refers to the following: explicitly explaining the development procedure used to create the model; writing down major assumptions and uncertainties inherent in the model; disclosing sources of information and analyses used to develop variables and their relationships in the model; and annotating any computer code. The more a model is verified, the more open it is to understanding—and critique.

Verification is an important aspect of modeling where meeting legal mandates is a concern, as in developing models for use in National Environmental Policy Act documents such as environmental impact statements. In this case, keeping careful records in process documents is paramount.

Testing the Audience

The best model in the world may fail to be used if it is too complex or esoteric. It will also be ignored if existing administrative organizations or policies do not provide for its use, or if for some reason it is not credible. In an operational sense, a model is valid, in part, if it is accepted (has face validity) and is usable in the intended work setting. Thus, developing models with teams combining managers and researchers helps enhance utility of such tools (Bunnell 1989). A team approach would help ensure that the model addresses the correct question, is based on data available from existing databases, is credible, and can be used in the everyday course of work.

Running the Model

In a typical management situation, a wildlife-habitat model is used to predict the response of wildlife species to potential environmental conditions created from alternative management activities. Such a model must run from information available from existing or easily obtainable inventories of vegetation and environments. Accuracy of model predictions may be limited, however, if inventories are dated, incomplete, or fail to include pertinent variables. In such cases, the models may be more useful for suggesting changes to inventory procedures.

At best, proxy variables might be used to represent the missing variables; for example, the size of a forest opening can represent a more complex habitat patch juxtaposition. If proxy variables are used, at least their degree of correlation with the intended variable should be evaluated.

Purpose and Context

The purpose of a model, often incompletely stated, should guide how the model is used. If a model is intended to be used to predict real-world environments and populations, it should be evaluated against a set of criteria different from that which should be used if it is intended to formalize our knowledge and understanding.

Also, the context of a model should be specified by the model builder. Context includes the range of environmental conditions (e.g., weather) and the types of environments and seasons in which the model was built and tested. It is the onus of the model user to adhere to that context. When a model is used outside that context, its accuracy and reliability are essentially unknown, unless these have been formally evaluated.

Testing the Output

Too often, models created for prediction are untested against real-world situations for a variety of reasons. Models are often developed without regard to validation until after they are built, and postconstruction validation requires too much time and money. Or, models are built mostly from theory and are difficult or impossible to test, even if they are used for prediction. Or, it is unclear or unspecified what the model output represents, as with habitat capability index (HSI, HEP) models.

Not all models require rigorous field testing. However, the validity of models used to help make decisions about irreversible or expensive losses of environments and populations should be known. In this case, the accuracy, bias, precision, and reliability of the model should be evaluated (e.g., see guidelines by Golbeck 1986).

A Type I error in prediction occurs when a model predicts species presence (or some other measure) and the species is actually absent. This error could occur because of inadequate or incorrect sampling for the species, because the field study was conducted during the wrong season, because the species is inherently rare and does not maximally occupy all suitable environments, or because the model was wrong and overstated the value of environmental parameters or failed to account for an environmental condition that deducts the presence of the species. The degree to which a model avoids Type I errors is given by the confidence coefficient P (where $P = 1 - \alpha$, and α is the significance level).

Conversely, a Type II error in prediction occurs at rate β when the model predicts absence and the species is actually present. This error could occur because the animals detected were wandering or their presence is not indicative of actual environmental quality because of sampling design, or because the model is wrong and does not include a vital parameter that affects presence of the species. The degree to which a model avoids Type II errors is given by the power of the model, $1 - \beta$ (Toft and Shea 1984). Power provides a means for selecting models for trend analysis (Gerrodette 1987), detecting environmental impacts (Osenberg et al. 1994), and determining population declines (Reed and Blaustein 1995; Taylor and Gerrodette 1993).

The ramifications of each type of error in model prediction depend on how the model is to be used. If the objective is to identify needs for mitigation, such as the purchase or trade of habitats with high opportunity costs or the restoration or enhancement of environmental conditions, then the model must accurately predict species presence. That is, frequencies of Type I errors should be minimized because costs of actions based on model predictions are high. On the other hand, if the model is to be used for predicting impacts, especially on rare or vulnerable species, then errors in predicting species presence or positive responses may be tolerable, but false predictions of species absence or negative responses might be of greater concern than in the case of mitigation. Thus, the power of a model and its ability to avoid Type II errors are critical.

Few wildlife-habitat models used to inform managers and decision makers have undergone formal and thorough statistical testing, because such studies are costly, tedious, unglamorous, and often viewed as unnecessary once a working model has been built. Notable exceptions include the work by Cook and Irwin (1985), who tested and revised an HSI model for the pronghorn. The model included five variables deemed important components for pronghorn winter range: shrub canopy cover, shrub height, shrub diversity, availability of winter wheat, and topographic cover. The HSI model converted field values of each of these variables to a 0 to 1 scale, and then produced a composite value of all five rescaled values. Cook and Irwin collected environmental data from 28 winter ranges in Montana, Idaho, Colorado, and Wyoming, calculated corresponding HSI values, and then correlated them with known pronghorn densities. The HSI model explained 39 percent of the variation in pronghorn density. By using simple linear regression, they identified that shrub cover was the most controlling variable. They then restructured the HSI model to maximize the correlation, which then increased to 50 percent.

In general, most HSI or habitat models can be expected to account for roughly half the variation in species density or abundance. Onsite environmental conditions generally account for even less variation in population density when migratory species are considered, especially Neotropical bird migrants. At first it might seem that the low correlations from Cook and Irwin's tests suggest that the model is not very useful, but if we consider the large array of other factors that affect population density, the model—especially their modified version—performed well. Cook and Irwin also recognized that density can be misleading as an index to suitability of an environment or to fitness of the individuals in the population.

This is also a lesson for the manager who will use the model for maintaining vegetation conditions. The manager must understand that most models that predict species presence, population density, or species richness from vegetation characteristics are likely to capture only a portion—typically half or less—of the variation in those species' parameters. This does not mean that habitat is unimportant; it is usually critical. It means that one cannot manage for vegetation conditions alone and be highly confident that the population will show a direct response. Another way of interpreting this is: By managing for (readily measurable) environmental conditions, we control only a portion of the factors that affect the occurrence and abundance of species.

Given these validation results, the appropriate use of habitat models then appears to be to help us recognize the degree (correlation) to which we can provide for species presence and abundance, and thus which environmental parameters under consideration are the more critical. Such models can also be used to assess potential (hypothetical) effects on species from alternative management scenarios. However, if such models are used specifically to predict population size, the predictions should be treated as hypotheses. Such predictions would assume that all factors not considered by the model—the 50 percent or more of unexplained variation in occurrence or abundance—are unimportant or are at optimal values. This assumption is invariably false.

Lancia et al. (1982) tested an HEP model for predicting the distribution of bobcats (*Lynx rufus*) in southeastern evergreen forests of North Carolina. They tested how well the model predicts frequency of use of vegetation patches by six radio telemetered bobcats. To do this, they constructed a map showing HEP values for each cell in the study area. Their results suggested a significant correlation between their habitat quality index values and bobcat frequency. However, although these trends suggested that the model is useful for prognosticating the occurrence of bobcats in locations within their study area, only 21 percent of predicted use levels exactly matched observed frequencies of use. By combining some use levels, the model correctly predicted 56 percent of bobcat frequencies. Some 12 percent of all cells showed a high frequency of bobcat occurrence but low predicted habitat quality values. This error suggests the inability of the model to capture all components of good habitat. On the other hand, some 32 percent of the cells had a low frequency of bobcat occurrence but a high predicted habitat quality. According to the authors, this error may have occurred because such sites usually were located adjacent to activity centers, and lower use of the sites may have occurred because of behavioral, population, or geometric effects, or occupancy of such sites by uninstrumented cats.

The study by Lancia and colleagues was essentially a test of how well their HEP model predicted habitat selection by individual bobcats within a study area, whereas the Cook and Irwin validation tested how well the pronghorn HSI model predicted population density in study areas.

Other validation studies of habitat relationships models include those of Laymon and Barrett (1986) and Laymon and Reid (1986), who tested a model (Laymon et al. 1985) predicting occurrence of spotted owls (*Strix occidentalis*) on the basis of the occurrence and distribution of old forests. Raphael and Marcot (1986) validated wildlife-habitat relationship models for errors of omission and commission for amphibians, reptiles, birds, and mammals in a Douglas-fir sere in California. Conroy et al. (1995) evaluated the reliability of spatially explicit population models. In each of these tests, different criteria were used to test various aspects of model prediction, including the robustness of the models when used in various ways, sensitivity of predictions to precision of input variables, and accuracy of predictions of species abundances when compared among different seral stages.

Overall, validating models is a many-faceted problem and should be done routinely as models are built and used. Validation should address the appropriateness of objectives and structure of the model, the utility, reliability, accuracy, and completeness of the model, and its credibility.

Literature Cited

Acevedo, M. F., D. L. Urban and M. Ablan. 1995. Transition and gap models of forest dynamics. *Ecological Applications* 5(4):1040–55.

Akçakaya, H. R., M. A. McCarthy, and J. L. Pearce. 1995. Linking landscape data with population viability analysis: Management options for the helmeted honeyeater. *Biological Conservation* 73:169–76.

Allen, T. F. H., and T. W. Hoekstra. 1984. Nested and non-nested hierarchies: A significant distinction for ecological systems. In *Proceedings of the Conference of the Society for General Systems Research* Vol. 1: *Systems methodologies and isomorphies,* ed. A. W. Smith, 175–80 N.p.: Intersystems Publications.

Allen, T. F. H., R. V. O'Neill, and T. W. Hoekstra. 1984. *Interlevel relations in ecological research and management: Some working principles from hierarchy theory.* USDA Forest Service General Technical Report RM-110. Fort Collins, Colo. 11 pp.

Arney, J. D. 1985. *User's guide for the Stand Projection System (SPS).* Report No. 1. Spokane, Wash.: Applied Biometrics. 9 pp.

Askins, R. A., M. J. Philbrick, and D. S. Sugeno. 1987. Relationship between the regional abundance of forest and the composition of forest bird communities. *Biological Conservation* 39:129–52.

Avery, T. E., and H. E. Burkhart. 1983. *Forest measurements.* 3d ed. New York: McGraw-Hill.

Barclay, H. J., and P. T. Gregory. 1981. An experimental test of models predicting life-history characteristics. *American Naturalist* 117:944–61.

Belcher, D. M., M. R. Holdaway, and G. J. Brand. 1982. *A description of STEMS—the Stand and Tree Evaluation Modeling System.* USDA Forest Service General Technical Report NC-79. North Central Forest Experiment Station, St. Paul, Minn. 18 pp.

Belovsky, G. E. 1987. Foraging and optimal body size: An overview, new data and a test of alternative models. *Journal of Theoretical Biology* 129(3):275–87.

Bender, L. C., G. J. Roloff, and J. B. Haufler. 1996. Evaluating confidence intervals for habitat suitability models. *Wildlife Society Bulletin* 24(2):347–52.

Biesterfeldt, R. C. 1984. System dynamics. *Forest Farmer* 43:10–11.

Blake, J. G., and J. R. Karr. 1987. Breeding birds of isolated woodlots: Area and habitat relationships. *Ecology* 68:1724–34.

Bledsoe, J. L., and D. A. Jamieson. 1969. Model structure of a grassland ecosystem. In: Dix and Biedlman, ed. *Grassland ecosystems: A preliminary synthesis.* Fort Collins: Colorado State University Press.

Block, W. M., L. A. Brennan, R. J. Gutierrez. 1987. Evaluation of guild-indicator species for use in resource management. *Environmental Management* 11:265–69.

Bojorquez-Tapia, L. A., I. Azuara, E. Ezcurra, and O. Flores-Villela. 1995. Identifying conservation priorities in Mexico through geographic information systems and modeling. *Ecological Applications* 5(1):215–31.

Boonstra, R., C. J. Krebs, and T. D. Beacham. 1980. Impact of botfly parasitism on *Microtus townsendii* populations. *Canadian Journal of Zoology* 58:1683–92.

Bourgeron, P. S., H. C. Humphries, and M. E. Jensen. 1994. Landscape characterization: A framework for ecological assessment at regional and local scales. *Journal of Sustainable Forestry* 2(3-4):267–81.

Boyce, S. G. 1980. *Management of forests for optimal benefits (DYNAST-OB).* USDA Forest Service Research Paper SE-204. Southeastern Forest Experiment Station, Asheville, N.C. 92 pp.

Boyce, S. G., and W. H. McNab. 1994. Management of forested landscapes: Simulations of three alternatives. *Journal of Forestry* 92(1):27–32.

Buechner, M. 1987. A geometric model of vertebrate dispersal: Tests and implications. *Ecology* 68:310–18.

Bunnell, F. L. 1989. *Alchemy and uncertainty: What good are models?* USDA Forest Service General Technical Report PNW-GTR-232. Portland, Oreg. 27 pp.

Burgman, M. A., H. R. Akcakaya, and S. S. Loew. 1988. The use of extinction models for species conservation. *Biological Conservation* 43(1):9–25.

Caicco, S. L., J. M. Scott, B. Butterfield, and B. Csuti. 1995. A gap analysis of the management status of the vegetation of Idaho (U.S.A.). *Conservation Biology* 9(3):498–511.

Cantwell, M., and R. Forman. 1993. Landscape graphs: Ecological modeling with graph theory to detect configurations common to diverse landscapes. *Landscape Ecology* 8:239–55.

Capen, D. E. 1981. *The use of multivariate statistics in studies of wildlife habitat.* USDA Forest Service General Technical Report RM-87. Fort Collins, Colo. 249 pp.

Caraco, T. 1979. Ecological response of animal group size frequencies. In *Statistical distributions in ecological work,* ed. J. K. Ord, G. P. Patil, and C. Tailie, 371–86. Statistical Ecology Series Vol. 4. Burtonsville, N.Y.: International Co-operative Publications.

Carey, A. B. 1984. A critical look at the issue of species-habitat dependency. In *New forests for a changing world,* 356–61. Proceedings of the 1983 National Convention of the Society of American Foresters, Bethesda, Md. N.p.

Chambers, S. M. 1995. Spatial structure, genetic variation, and the neighborhood adjustment to effective population size. *Conservation Biology* 9(5):1312–15.

Church, R. L., and A. T. Murray. 1994. Designing a hierarchical planning model for USDA Forest Service planning. In *Sixth Symposium on Systems Analysis and Management Decisions in Forestry.* Pacific Grove, Calif. (Papers are numbered separately.)

Cleaves, D. A. 1995. Assessing and communicating uncertainty in decision support systems: Lessons from an ecosystem policy analysis. *AI Applications* 9(3):87–102.

Clutter, J. L., J. C. Fortson, L. V. Pienaar, G. H. Brister, and R. L. Bailey. 1983. *Timber management: A quantitative approach.* New York: John Wiley and Sons.

Conner, R. N. 1988. Wildlife populations: Minimally viable or ecologically functional? *Wildlife Society Bulletin* 16:80–84.

Conroy, M. J., and B. R. Noon. 1996. Mapping of species richness for conservation of biological diversity: Conceptual and methodological issues. *Ecological Applications* 6(3):763–73.

Conroy, M. J., Y. Cohen, F. C. James, Y. G. Matsinos, and B. A. Maurer. 1995. Parameter estimation, reliability, and model improvement for spatially explicit models of animal populations. *Ecological Applications* 5(1):17–19.

Cook, J. G., and L. L. Irwin. 1985. Validation and modification of a habitat suitability model for pronghorns. *Wildlife Society Bulletin* 13:440–48.

Coulson, R. N., and M. C. Saunders. 1987. Computer-assisted decision-making as applied to entomology. *Annual Review of Entomology* 32:415–37.

Covington, W. W., and M. M. Moore. 1994. Postsettlement changes in natural fire regimes and forest structure: Ecological restoration of old-growth ponderosa pine forests. In *Assessing forest ecosystem health in the inland West,* ed. R. N. Sampson, D. L. Adams, and M. J. Enzer, 153–81. New York: Haworth Press.

Curtis, R. O., G. W. Clendenen, and D. J. DeMars. 1981. *A new stand simulator for coast Douglas-fir: DFSIM user's guide.* USDA Forest Service General Technical Report PNW-128, Washington, D.C. 79 pp.

Dale, V. H., and M. Hemstrom. 1984. *CLIMACS: A computer model of forest stand development for western Oregon and Washington.* USDA Forest Service Research Paper PNW-327, Pacific Northwest Forest and Range Experiment Station, Portland, Oreg. 60 pp.

DeGraaf, R. M., N. G. Tilghman, and S. H. Anderson. 1985. Foraging guilds of North American birds. *Environmental Management* 9:493–536.

Dennis, B., B. E. Brown, A. R. Stage, H. E. Burkhart, and S. Clark. 1985. Problems of modeling growth and yield of renewable resources. *American Statistician, Proceedings 8th Symposium on Statistics, Law, and the Environment* 39:374–83.

Dobson, A. P. 1988. Restoring island ecosystems: The potential of parasites to control introduced mammals. *Conservation Biology* 2:31–39.

Dueser, R. D., and J. H. Porter. 1986. Habitat use by insular small mammals: Relative effects of competition and habitat structure. *Ecology* 67:195–201.

Dunning, J. B., Jr., D. J. Stewart, B. J. Danielson, B. R. Noon, T. L. Root, R. H. Lamberson, and E. E. Stevens. 1995. Spatially explicit population models: Current forms and future uses. *Ecological Applications* 5(1):3–11.

Eberhardt, L. L. 1977. Applied systems ecology: Models, data, and statistical methods. In *New directions in the analysis of ecological systems,* ed. G. S. Innis, Part 1, 132. Simulations Councils Proceedings.

Ek, A. R., and R. A. Monserud. 1974. *FOREST: A computer model for simulating the growth and reproduction of mixed species forest stands.* University of Wisconsin, School of Natural Resources, Research Report A2635.

Ellison, A. M., and B. L. Bedford. 1995. Response of a wetland vascular plant community to disturbance: A simulation study. *Ecological Applications* 5(1):109–23.

Fahrig, L., and G. Merriam. 1985. Habitat patch connectivity and population survival. *Ecology* 66:1762–68.

Flood, B. S., M. E. Sangster, R. S. Sparrow, and T. S. Baskett. 1977. *A handbook for habitat evaluation procedure.* USDI Fish and Wildlife Service, Research Publication 132, Washington, D.C.

Franklin, J. F. 1994. Adaptive management areas. *Journal of Forestry* 92(4):50.

Franklin, J. F., and R. T. T. Forman. 1987. Creating landscape patterns by forest cutting: Ecological consequences and principles. *Landscape Ecology* 1:5–18.

Fröberg, C. 1969. *Introduction to numerical analysis.* 2d ed. Reading Mass.: Addison-Wesley Publishing Co.

Fujita, T. 1990. Habitat preference of birds for different vegetation structures in the Bikin River basin in the Primorskii Krai, Far East USSR. (In Japanese with English abstract.) *Strix* 9:159–66.

Garsd, A. 1984. Spurious correlation in ecological modelling. *Ecological Modelling* 23:191–201.

Gaston, K. J., and T. M. Blackburn. 1996a. Conservation implications of geographic range size-body size relationships. *Conservation Biology* 10(2):638–46.

Gaston, K. J., and T. M. Blackburn. 1996b. Rarity and body size: Importance of generality. *Conservation Biology* 10(4):1295–98.

Gazey, W. J., and M. J. Staley. 1986. Population estimation from mark-recapture experiments using a sequential Bayes algorithm. *Ecology* 67:941–51.

Gerrodette, T. 1987. A power analysis for detecting trends. *Ecology* 68:1364–72.

Golbeck, A. L. 1986. *Evaluating statistical validity of research reports: A guide for managers, planners and researchers.* USDA Forest Service General Technical Report PSW-87. Berkeley, Calif. 22 pp.

Grant, G. E., and F. Swanson. 1991. Cumulative effects of forest practices. *Forest Perspectives* 1(4):9–11.

Haas, T. C. 1992. A Bayes network model of district ranger decision making. *AI Applications* 6(3):72–88.

Haas, T. C., H. T. Mowrer, and W. D. Shepperd. 1994. Modeling aspen stand growth with a temporal Bayes network. *AI Applications* 8(1):15–28.

Hall, C. A. S., and J. W. Day. 1977. *Ecosystem modeling in theory and practice.* New York: Wiley Interscience.

Hall, C. A. S., J. W. Day, and H. T. Odum. 1977. A circuit language for energy and matter. In *Ecosystem modeling in theory and practice,* ed. C. A. S. Hall and J. W. Day, 38–48. New York: Wiley Interscience.

Halle, F., R. A. A. Oldeman, and P. B. Tomlinson. 1978. *Tropical trees and forests: An architectural analysis.* Berlin: Springer-Verlag.

Hariston, N. G. 1981. An experimental test of a guild: Salamander competition. *Ecology* 62:65–72.

Hassell, M. P. 1987. Detecting regulation in patchily distributed animal populations. *Journal of Animal Ecology* 56:705–13.

Hassell, M. P., H. N. Comins, and R. M. May. 1991. Spatial structure and chaos in insect population dynamics. *Nature* 353:255–58.

Haufler, J. B. 1994. An ecological framework for planning for forest health. In *Assessing forest ecosystem health in the inland West,* ed. R. N. Sampson, D. L. Adams, and M. J. Enzer, 307–16. New York: Haworth Press.

Haufler, J. B., C. A. Mehl, and G. J. Roloff. 1996. Using a coarse-filter approach with species assessment for ecosystem management. *Wildlife Society Bulletin* 24(2):200–208.

Holl, S. A. 1982. Evaluation of bighorn sheep habitat. *Desert Bighorn Council Transactions,* 47–49.

Holling, C. S. 1984. *Adaptive environmental assessment and management.* New York: Wiley and Sons.

Holling, C. S. 1992. Cross-scale morphology, geometry, and dynamics of ecosystems. *Ecological Monographs* 62(4):447–502.

Holm, E. 1988. Environmental restraints and life strategies: A habitat templet matrix. *Oecologia* 75(1):141–45.

Holmes, R. T., R. E. Bonney, and S. W. Pacala. 1979. Guild structure of the Hubbard Brook bird community: A multivariate approach. *Ecology* 60:512–20.

Holt, R. D., S. W. Pacala, T. W. Smith, and J. Liu. 1995. Linking contemporary vegetation models with spatially explicit animal population models. *Ecological Applications* 5(1):20–27.

Holthausen, R. S., M. J. Wisdom, J. Pierce, D. K. Edwards, and M. M. Rowland. 1994. *Using expert opinion to evaluate a habitat effectiveness model for elk in western Oregon and Washington.* USDA Forest Service Research Paper PNW-RP-479. Pacific Northwest Research Station, Portland, Oreg. 16 pp.

Horn, H. S. 1975. Markovian properties of forest succession. In *Ecology and evolution of communities,* ed. M. L. Cody and J. M. Diamond, 196–211. Cambridge, Mass.: Harvard University Press.

Hunter, M. L. 1991. Coping with ignorance: The coarse-filter strategy for maintaining biodiversity. In *Balancing on the brink of extinction,* ed. K. A. Kohm, 266–81. Covelo, Calif.: Island Press.

Johnson, D. H. 1985. Improved estimates from sample surveys with empirical Bayes methods. *Proceedings of the American Statistical Association,* 395–400.

Johnson, D. H. 1989. An empirical Bayes approach to analyzing recurring animal surveys. *Ecology* 70:945–52.

Juanes, F. 1986. Population density and body size in birds. *American Naturalist* 128:921–29.

Karieva, P., and U. Wennergren. 1995. Connecting landscape patterns to ecosystem and population processes. *Nature* 373:299–302.

Karr, J. R., and K. E. Freemark. 1985. Disturbance and vertebrates: An integrative perspective. In *The ecology of natural disturbance and patch dynamics,* ed. S. T. A. Pickett and P. S. White, 153–68. Orlando, Fla.: Academic Press.

Keane, R. E., S. F. Arno, and J. K. Brown. 1990. Simulating cumulative fire effects in Ponderosa pine/Douglas-fir forests. *Ecology* 71:189–203.

Kimmins, J. P. 1987. *Forest ecology.* New York: Macmillan.

Kingston, T. 1995. RAMAS/GIS: Linking landscape data with population viability analysis (software review). *Conservation Biology* 9(4):966–68.

Kirkman, R. L., J. A. Eberly, W. R. Porath, and R. R. Titus. 1984. A process for integrating wildlife needs into forest management planning. In *Wildlife 2000: Modeling habitat relationships of terrestrial vertebrates,* ed. J. Verner, M. L. Morrison, and C. J. Ralph, 347–50. Madison: University of Wisconsin Press.

Knopf, F. L., J. A. Sedgwick, and R. W. Cannon. 1988. Guild structure of a riparian avifauna relative to seasonal cattle grazing. *Journal of Wildlife Management* 52:280–90.

Kolasa, J. 1989. Ecological systems in hierarchical perspective: Breaks in community structure and other consequences. *Ecology* 70:36–47.

Kotz, S., N. L. Johnson, and C. B. Read, eds. 1982. *Encyclopedia of statistical sciences.* Vol. 2: *Classification—eye estimate.* New York: John Wiley and Sons.

Kremen, C. 1992. Assessing the indicator properties of species assemblages for natural areas monitoring. *Ecological Applications* 2(2):203–17.

Lancia, R. A., S. D. Miller, D. A. Adams, and D. W. Hazel. 1982. Validating habitat quality assessment: An example. *Transactions of the North American Wildlife and Natural Resources Conference* 47:96–110.

Landres, P. B., J. Verner, and J. W. Thomas. 1988. Ecological uses of vertebrate indicator species: a critique. *Conservation Biology* 2:316–28.

Law, A. M., and W. D. Kelton. 1982. *Simulation modeling and analysis.* New York: McGraw-Hill.

Laymon, S. A., and R. H. Barrett. 1986. Developing and testing habitat-capability models: Pitfalls and recommendations. In *Wildlife 2000: Modeling habitat relationships of terrestrial vertebrates,* ed. J. Verner, M. L. Morrison, and C. J. Ralph, 87–91. Madison: University of Wisconsin Press.

Laymon, S. A., and J. F. Reid. 1986. Effects of grid-cell size on tests of a spotted owl HSI model. In *Wildlife 2000: Modeling habitat relationships of terrestrial vertebrates,* ed. J. Verner, M. L. Morrison, and C. J. Ralph, 93–96. Madison: University of Wisconsin Press.

Laymon, S. A., H. Salwasser, and R. H. Barrett. 1985. *Habitat suitability index models: Spotted owl.* USDI Fish and Wildlife Service Biological Report 82(10.113). Washington, D.C.

Leary, R. A. 1979. *A generalized forest growth projection system applied to the lakes states region.* USDA Forest Service General Technical Report NC-49. St. Paul, Minn.

Lee, K. N., and J. Lawrence. 1986. Adaptive management: Learning from the Columbia River Basin Fish and Wildlife Program. *Environmental Law* 16:431–60.

McGarigal, K., and B. J. Marks. 1995. *FRAGSTATS: Spatial pattern analysis program for quantifying landscape structure.* USDA Forest Service General Technical Report PNW-GTR-351. Pacific Northwest Research Station, Portland, Oreg. 122 pp.

Mannan, R. W., R. N. Conner, B. G. Marcot, and J. M. Peek. 1994. Managing forestlands for wildlife. In *Research and management techniques for wildlife and habitats,* ed. T. A. Bookhout, 689–721. The Wildlife Society, Washington, D.C.

Marcot, B. G. 1985. Habitat relationships of birds and young-growth Douglas-fir in northwestern California. Ph.D. dissertation, Oregon State University, Corvallis.

Marcot, B. G., and P. Z. Chinn. 1982. Use of graph theory measures for assessing diversity of wildlife habitat. In *Mathematical models of renewable resources,* ed. R. Lamberson, 69–70. Proceedings of the 1st Pacific Coast Conference on Mathematical Models of Renewable Resources. Arcata, Calif.: Humboldt State University.

Marcot, B. G., M. G. Raphael, and K. H. Berry. 1983. Monitoring wildlife habitat and validation of wildlife-habitat relationships models. *Transactions of the North American Wildlife and Natural Resources Conference* 48:315–29.

Marks, H. M., and S. M. Woodruff. 1979. Allocations using empirical Bayes estimators. In *American Statistical Association proceedings of section on survey research methods,* 309–13. Alexandria, Va.: American Statistical Association.

Martin, T. E. 1988. Habitat and area effects on forest bird assemblages: Is nest predation an influence? *Ecology* 69:74–84.

Maurer, B. A. 1990. Extensions of optimal foraging theory for insectivorous birds: Implications for community structure. *Studies in Avian Biology* 13:455–61.

Maxwell, J. R., C. J. Edwards, M. E. Jensen, S. J. Paustian, H. Parrott, and D. M. Hill. 1995. *A hierarchical framework of aquatic ecological units in North America (Nearctic zone).* USDA Forest Service General Technical Report NC-176. North Central Forest Experiment Station, St. Paul, Minn. 72 pp.

Meldahl, R. 1986. Alternative modeling methodologies for growth and yield projection systems. In *Data management issues in forestry,* 27–31. Florence, Ala.: Forest Resources System Institute. (Proceedings of a computer conference held 7–9 April 1986 in Atlanta, Ga.)

Miller, R. I., ed. 1994. *Mapping the diversity of nature.* London: Chapman and Hall.

Milliken, G. A., and D. E. Johnson. 1992. *Analysis of messy data.* Volume 1: *Designed experiments.* New York: Chapman and Hall.

Mills, L. S. 1995. Edge effects and isolation: Red-backed voles on forest remnants. *Conservation Biology* 9(2):395–403.

Mills, L. S., S. G. Hayes, C. Baldwin, M. J. Wisdom, J. Citta, D. J. Mattson, and K. Murphy. 1996. Factors leading to different viability predictions for a grizzly bear data set. *Conservation Biology* 10(3):863–73.

Moeur, M. 1985. *COVER: A user's guide to the CANOPY and SHRUBS extension of the stand prognosis model.* USDA Forest Service General Technical Report INT-190. Ogden, Utah. 49 pp.

Morgan, P., G. H. Aplet, J. B. Haufler, H. C. Humphries, M. M. Moore, and W. D. Wilson. 1994. Historical range of variability: A useful tool for evaluating ecosystem change. In *Assessing forest ecosystem health in the inland West,* ed. R. N. Sampson, D. L. Adams, and M. J. Enzer, 87–111. New York: Haworth Press.

Morris, C. N. 1983. Parametric empirical Bayes inference: Theory and applications. *Journal of the American Statistical Association* 78:47–55.

Morris, W. F. 1990. Problems in detecting chaotic behavior in natural populations by fitting simple discrete models. *Ecology* 71:1849–62.

Neitro, W. A., V. W. Binkley, S. P. Cline, R. W. Mannan, B. G. Marcot, D. Taylor, and F. F. Wagner. 1985. Snags (wildlife trees). In *Management of wildlife and fish habitats in forests of western Oregon and Washington,* Part I—Chapter narratives, ed. E. R. Brown, 129–69. USDA Forest Service, Pacific Northwest Region, Portland, Oreg.

Noss, R. F. 1990. Indicators for monitoring biodiversity: A hierarchical approach. *Conservation Biology* 4:355–64.

Osenberg, C. W., R. J. Schmitt, S. J. Holbrook, K. E. Abu-Saba, and A. R. Flegal. 1994. Detection of environmental impacts: Natural variability, effect size, and power analysis. *Ecological Applications* 4(1):16–30.

Overton, W. S. 1971. *Sensitivity analysis as "propagation of error" and model evaluation: Modeling notebook.* Internal Report No. 1, Coniferous Forest Biome, International Biological Program. Department of Forestry, Oregon State University, Corvallis.

Parsons, W. F. J., D. H. Knight, and S. L. Miller. 1994. Root gap dynamics in lodgepole pine forest: Nitrogen transformations in gaps of different size. *Ecological Applications* 4(2):354–62.

Patton, D. R. 1987. Is the use of "management indicator species" feasible? *Western Journal of Forestry* 2(1):33–34.

Pickett, S. T. A., and P. S. White. 1985. *The ecology of natural disturbance and patch dynamics.* Orlando, Fla.: Academic Press.

Poiani, K. A., and W. C. Johnson. 1993. A spatial simulation model of hydrology and vegetation dynamics in semi-permanent prairie wetlands. *Ecological Applications* 3(2):279–93.

Porter, W. F., and K. E. Church. 1987. Effects of environmental pattern on habitat preference analysis. *Journal of Wildlife Management* 51:681–85.

Powell, G. V. N., R. D. Bjork, M. Rodriguez S., and J. Barborak. 1995–1996. Life zones at risk: Gap analysis in Costa Rica. *Wild Earth* 5(4):46–51.

Pulliam, H. R., and J. B. Dunning. 1995. Spatially explicit population models. *Ecological Applications* 5(1):2.

Pyke, G. H. 1984. Optimal foraging theory: A critical review. *Annual Review of Ecology and Systematics* 15:523–75.

Quinn, J. F., and A. Hastings. 1987. Extinction in subdivided habitats. *Conservation Biology* 1:198–209.

Ramm, C. W., and C. L. Miner. 1986. Growth and yield programs used on microcomputers in the North Central Region. *Northern Journal of Applied Forestry* 3:44–45.

Raphael, M. G., and B. G. Marcot. 1986. Validation of a wildlife-habitat-relationships model: Vertebrates in a Douglas-fir sere. *Wildlife 2000: Modeling habitat relationships of terrestrial vertebrates,* ed. J. Verner, M. L. Morrison, and C. J. Ralph, 129–38. Madison: University of Wisconsin Press.

Rastetter, E. B., A. W. King, B. J. Cosby, G. M. Hornberger, R. V. O'Neill, and J. E. Hobbie. 1992. Aggregating fine-scale ecological knowledge to model coarser-scale attributes of ecosystems. *Ecological Applications* 2:55–70.

Reed, J. M., and A. R. Blaustein. 1995. Assessment of "nondeclining" amphibian populations using power analysis. *Conservation Biology* 9(5):1299–1300.

Regier, H. A. 1978. *A balanced science of renewable resources.* Washington Sea Grant Publication, University of Washington, Seattle. 108 pp.

Rewa, C. A., and E. D. Michael. 1984. Use of habitat evaluation procedures (HEP) in assessing guild habitat value. *Transactions of the Northeast Section of the Wildlife Society* 41:122–29.

Richards, C., and P. L. Leberg. 1996. Temporal changes in allele frequencies and a population's history of severe bottlenecks. *Conservation Biology* 10(3):832–39.

Ringold, P. L., J. Alegria, R. L. Czaplewski, B. S. Mulder, T. Tolle, and K. Burnett. 1996. Adaptive monitoring design for ecosystem management. *Ecological Applications* 6(3):745–47.

Roberts, R. H., and L. J. O'Neil. 1985. Species selection for habitat assessments. *Transactions of the North American Wildlife and Natural Resources Conference* 50:352–62.

Roozen, H. 1987. Equilibrium and extinction in stochastic population dynamics. *Bulletin Mathematical Biology* 49(6):671–96.

Rosenzweig, M. L. 1987. Habitat selection as a source of biological diversity. *Evolutionary Biology* 1(4):315–30.

Rosenzweig, M. L., and C. W. Clark. 1994. Island extinction rates from regular censuses. *Conservation Biology* 8(2):491–94.

Rubinstein, M. F. 1975. *Patterns of problem solving.* Englewood Cliffs, N.J.: Prentice-Hall.

Ruggiero, L. F., R. S. Holthausen, B. G. Marcot, K. B. Aubry, J. W. Thomas, and E. C. Meslow. 1988. Ecological dependency: The concept and its implications for research and management. *North American Wildlife and Natural Resources Conference* 53:115–26.

Running, S. W. 1994. Testing forest-BGC ecosystem process simulations across a climatic gradient in Oregon. *Ecological Applications* 4(2):238–47.

Salwasser, H., and F. B. Samson. 1985. Cumulative effects analysis: An advance in wildlife planning and management. *Transactions of the North American Wildlife Natural Resources Conference* 50:313–21.

Schamberger, M., A. H. Farmer, and J. W. Terrell. 1982. *Habitat suitability index models: Introduction.* USDI Fish and Wildlife Service FWS/OBS-82/10. Washington, D.C. 2 pp.

Schlosser, I. J. 1990. Environmental variation, life history attributes, and community structure in stream fishes: Implications for environmental management and assessment. *Environmental Management* 14(5):621–28.

Schroeder, R. L. 1983. *Habitat suitability index models: Yellow warbler.* USDI Fish and Wildlife Service FWS/OBS-82/10.27. Washington, D.C. 8 pp.

Scott, J. M., B. Csuti, J. D. Jacobi, and J. E. Estes. 1987. Species richness: A geographic approach to protecting future biological diversity. *BioScience* 37(11):782–88.

Scott, J. M., F. Davis, B. Csuti, R. Noss, B. Butterfield, C. Groves, H. Anderson, S. Caicco, F. D'erchia, T. C. Edwards, Jr, J. Ulliman, and R. G. Wright. 1993. Gap analysis: A geographic approach to protection of biological diversity. *Wildlife Monographs* 123:1–41.

Short, H. L. 1983. *Wildlife guilds in Arizona desert habitats.* USDI Bureau of Land Management Technical Note 362, Fort Collins, Colo. 258 pp.

Shugart, H. H., Jr. 1984. *A theory of forest dynamics.* New York: Springer-Verlag.

Shugart, H. H., and S. W. Seagle. 1985. Modeling forest landscapes and the role of disturbance in ecosystems and communities. In *The ecology of natural disturbance and patch dynamics,* ed. S. T. A. Pickett and P. S. White, 353–68. Orlando, Fla.: Academic Press.

Smith, T. M., H. H. Shugart, and D. C. West. 1981. Use of forest simulation models to integrate timber harvest and nongame bird management. *Transactions of the North American Wildlife and Natural Resources Conference* 46:501–10.

Stamps, J. A., M. Buechner, and V. V. Krishnan. 1987. The effects of edge permeability and habitat geometry on emigration from patches of habitat. *American Naturalist* 129:533–52.

Strebel, D. E. 1985. Environmental fluctuations and extinction—single species. *Theoretical Population Biology* 27:1–26.

Swanson, F. J., J. A. Jones, D. O. Wallin, and J. H. Cissel. 1994. Natural variability—implications for ecosystem management. In *Eastside forest ecosystem health assessment,* Vol. 2: *Ecosystem management: Principles and applications,* ed. M. E. Jensen and P. S. Bourgeron, 80–94. USDA Forest Service General Technical Report PNW-GTR-318. Portland, Oreg. 376 pp.

Tausch, R. J., P. E. Wigand, and J. W. Burkhardt. 1993. Viewpoint: Plant community threshold, multiple steady states, and multiple successional pathways: Legacy of the Quaternary? *Journal of Range Management* 46(5):439–47.

Taylor, A. D. 1990. Metapopulations, dispersal, and predator-prey dynamics: An overview. *Ecology* 71:429–33.

Taylor, B. L., and T. Gerrodette. 1993. The uses of statistical power in conservation biology: The vaquita and northern spotted owl. *Conservation Biology* 7(3):489–500.

Thompson, J. E., comp. 1995. *Analysis in support of ecosystem management.* Analysis Workshop III, 10–13 April 1995, Fort Collins, Colo. USDA Forest Service, Washington, D.C. 360 pp.

Toft, C. A., and P. J. Shea. 1984. Detecting community-wide patterns: Estimating power strengthens statistical inference. In *Ecology and evolutionary biology,* ed. G. W. Salt, 38–45. Chicago: University of Chicago Press.

Turner, M. B., G. J. Arthaud, R. T. Engstrom, S. J. Hejl, J. Liu, S. Loeb, and K. McKelvey. 1995. The usefulness of spatially explicit population models in land management. *Ecological Applications* 5(1):12–16.

U.S. Fish and Wildlife Service. 1980. *Habitat evaluation procedures (HEP).* Washington, D.C.: Division of Ecological Services ESM 102.

Verner, J., and A. S. Boss. 1980. *California wildlife and their habitats: Western Sierra Nevada.* USDA Forest Service General Technical Report PSW-37. Berkeley, Calif. 439 pp.

Wakeley, J. S., and L. J. O'Neil. 1988. *Techniques to increase efficiency and reduce effort in applications of the habitat evaluation procedures (HEP).* Environmental Impact Research Program Technical Report EL-88-13, Department of the Army, U.S. Army Corps of Engineers, Vicksburg, Miss. 52 pp. + 4 appendices.

Wahlberg, N., A. Moilanen, and I. Hanski. 1996. Predicting the occurrence of endangered species in fragmented landscapes. *Science* 273(5281):1536–38.

Walters, C. 1986. *Adaptive management of renewable resources.* New York: Macmillan.

Walters, C., and R. Hilborn. 1978. Ecological optimization and adaptive management. *Annual Review of Ecology and Systematics* 9:157–88.

Watt, K. E. F. 1966. *Systems analysis in ecology.* New York: Academic Press.

Weaver, J. C. 1995. Indicator species and scale of observation. *Conservation Biology* 9(4):939–42.

Weaver, J., R. Escano, D. Mattson, T. Puchlerz, and D. Despain. 1985. A cumulative effects model for grizzly bear management in the Yellowstone ecosystem. In *Proceedings—grizzly bear habitat symposium,* comp. G. P. Contreras and K. E. Evans, 234–46. 30 April–2 May 1985, Missoula, Mont. USDA Forest Service General Technical Report INT-207, Ogden, Utah.

Williams. G. L., D. R. Russell, and W. K. Seitz. 1977. Pattern recognition as a tool in the ecological analysis of habitat. In *Classification, inventory, and analysis of fish and wildlife habitat,* 521–31. USDI Fish and Wildlife Service, FWS/OBS-78/76. Washington, D.C. 604 pp.

Williams, P., D. Gibbons, C. Margules, A. Rebelo, C. Humphries, and R. Pressey. 1996. A comparison of richness hotspots, rarity hotspots, and complementary areas for conserving diversity of British birds. *Conservation Biology* 10(1):155–74.

Williamson, J. F. 1983. Woodplan: Microcomputer programs for forest management. In *Proceedings of a national workshop on computer uses in fish and wildlife programs,* 128–30. 5–7 December 1983. Blacksburg, Pa.: Virginia Polytechnic Institute and State University.

Wilson, M. F., D. A. Graff, and C. J. Whelan. 1990. Color preferences of frugivorous birds in relation to the colors of fleshy fruits. *Condor* 92:545–55.

Wisdom, M. J., L. R. Bright, C. G. Carey, W. W. Hines, R. J. Pedersen, D. A. Smithey, J. W. Thomas, and G. W. Witmer. 1986. *A model to evaluate elk habitat in western Oregon.* USDA Forest Service Publication No. R6-F&WL-216-1986. Pacific Northwest Region, Portland, Oreg. 33 pp.

Woods, G. R., D. C. Guynn, W. E. Hammitt, and M. E. Patterson. 1996. Determinants of participant satisfaction with quality deer management. *Wildlife Society Bulletin* 24(2):318–24.

Wykoff, W. R., N. L. Crookston, and A. R. Stage. 1982. *User's guide to the stand prognosis model.* USDA Forest Service General Technical Report INT-133. Ogden, Utah.

Yaffee, S. L., A. F. Phillips, I. C. Frentz, P. W. Hardy, S. M. Maleki, and B. E. Thorpe. 1996. *Ecosystem management in the United States.* Covelo, Calif.: Island Press.

Zeide, B. 1996. Is "the scientific basis" of ecosystem management indeed scientific? *Bulletin of the Ecological Society of America* 77(2):123–24.

PART 3

The Management of Wildlife Habitat

11 Wildlife Management in a New Era: Managing Habitat for Animals in an Evolutionary and Ecosystem Context

Introduction

Concepts and theories, statistical analysis tools, and modeling technologies for habitat assessment have come a long way since the early days. The focus of habitat management has continued to evolve in response to new environmental problems and new scientific concepts and findings. In some cases, what seemed to be utterly sufficient and immutable axioms of habitat management, such as the value of forest openings and edges to wildlife (e.g., Lay 1938), has given way to different perspectives and new knowledge gleaned from changing landscapes, such as the more recent concern for excessive fragmentation of old- or native-forest cover (e.g., Hagan et al. 1996).

Learning from traditional approaches to wildlife management, including their successes and failures, helps us prepare for management challenges to come. The basic tenets of habitat management provided by early wildlife biologists such as Leopold (1933) still prove useful today: wild animals need the essentials of cover, food, and water for survival of individuals and for the chance of the population's continuation. But to this foundation we provide some additional tenets to aid future assessments and management.

We first discuss managing wildlife *in situ* in an ecosystem context. Borrowing from the old

German concept of *Umwelt,* we view wildlife organisms, populations, species, and communities as a function of more than just cover, food, and water ("habitat" in the traditional sense). They also respond to numerous other biotic and abiotic factors. In this regard, wildlife is a function of (1) the full set of environmental factors—including but extending beyond those essential three traditional habitat factors of cover, food, and water—which together influence realized fitness of organisms, (2) the ecological roles of other species, and (3) abiotic conditions and events, including what we may term systematic or chronic changes, and acute environmental disturbances or perturbations. In this fuller context, wildlife managers should attend not just to conserving taxonomic species and maintaining viability of populations (particularly threatened ones), but also to providing the rich complex of organisms' ecological functions, the full set of ecosystem processes, and all *Umwelt* conditions collectively required for their persistence and development.

We raise the question, What is wildlife? and propose that the term *wildlife* and the approaches currently taken for their management be extended to consider the full array of all biota present in an ecosystem. We explore the question from a variety of dimensions, all affecting conservation policy and management success on the ground. We propose an enhanced approach to depicting, modeling, and predicting the status and condition of wildlife organisms in an ecosystem, by extending beyond the traditional wildlife-habitat relationships approach to a fuller species-environment relations framework.

Ultimately, we hope that successful wildlife management will provide global opportunities to maintain conditions to support long-term evolutionary processes of species lineages. In this context, we propose some checkpoints and tenets for managing wildlife in context of ecological domains and evolutionary time frames.

Since much of land management in the twenty-first century will likely entail unprecedented changes in environmental conditions, we advocate an aggressive and rigorous use of adaptive habitat management. In this approach—really, at heart a fundamental philosophy of the habits of land use and personal resource use—we should learn from experiences, including planned and unplanned natural experiments, and then modify our management actions and resource-use behaviors accordingly to meet wildlife conservation goals better. Much has been written on ideal procedures to adaptive management, but more needs to be said about real-world problems in implementing these procedures. In the spirit of learning from history and experience, we state some basic tenets of the adaptive management approach and cite cases of both success and failure. Instead of prognosticating future declines of scarce habitat and endangered species, which can be found in a variety of other sources, we focus on exploring a more optimistic future of the science and sociology of habitat management.

Managing Wildlife in an Ecosystem Context

What Is Wildlife?

More than just a means of focusing the work of agencies and the subject of this book, the question of what is wildlife has recently been evolving along scientific and management, social, cultural, legal, and even ethical and aesthetic dimensions (box 11.1). Traditional views have focused on wildlife solely as terrestrial vertebrates—initially, game birds and mam-

Box 11.1. What is wildlife?

Different dimensions of the question that wildlife biologists, managers, lawyers, judges, economists, social scientists, political scientists, politicians, indigenous peoples, hunters, conservationists, educators, and artists need to ponder. The answers lie not in specific responses to each question (that's the easy way out) but rather in how the needs expressed in the multiple dimensions of all the questions can be met at the same time.

What is wildlife?

- the scientific dimension
  - Which taxonomic classes pertain?
  - Which conditions of evolutionary significance pertain?
  - What ecological functional groups of organisms pertain?
- the management dimension
  - What is a population?
  - What deserves specific management attention, and what is assumed to derive secondary benefit?
  - What taxa have agency or landholder status for management focus (sensitive, rare, game, etc.)?
  - Is the focus on populations, species, habitats, or ecosystems?
- the legal dimension
  - What is threatened or endangered (or rare *sensu* Red Data Book listings used outside the United States)?
  - What is a candidate for listing?
  - What is mandated for ensuring viability, biodiversity, and persistence (nonextinction)?
  - What is legally hunted or gathered, and is there a mandate for sustained harvest above and beyond minimal viability?
  - What are obligations to tribes and indigenous peoples?
  - What are the paralegal obligations for representing the rights of nonhuman living entities in judicial and land use arenas?
- the economic dimension
  - How should use of wildlife resources help support local private economies, such as through fur trapping and game ranching?
  - What role should questions of economic impact play in determining wildlife habitat management mandates on private and on public lands?
- the cultural dimension
  - What is necessary to provide for the rights and persistence of indigenous peoples?
  - How should patterns of other traditional use of wildlife resources be managed?
  - What are the obligations for providing for future generations of all peoples?
- the ethical dimension
  - What takes precedence—individual organisms, populations, species, or systems?
  - What should fall first in a triage approach?
  - How should human habits of land use and resource consumption be met or modified?
- the aesthetic dimension
  - What should science, management, politics, and publics learn from artistic perspectives?
  - How should the needs and interests of nature artists and educators be met?

In pluralistic societies designed to resolve conflicting interests through confrontational means, such as courts of law, the process of finding the difficult answers that integrate across these dimensions is often tedious and expensive and sometimes imperfect. Rarely do litigative compromises satisfy everyone fully. Newer approaches to conflict resolution (e.g., Maguire 1991a) are meant to circumvent purely judicial and litigative solutions. In a world of increasing scarcity of wildlife resources, it is probably wise to begin to explore and refine such new approaches to provide better for future generations, if only the next one.

mals, and later also organisms of conservation concern, principally threatened and endangered species. Although much of wildlife management today still focuses on these elements, because indeed they are still important concerns for conservation action, the term *wildlife* is being broadened along several fronts.

For example, thanks to increasing interest in cumulative effects analysis and integrating evolutionary perspectives into wildlife science, the question of what is wildlife in turn prompts asking which conditions of conservation interest, evolutionary significance, and ecological function should be included. Should relatively scarce subspecies, particularly increasingly uncommon local endemics such as the Hawaiian coot, or 'alae ke 'oke'o (*Fulica americana alai*), the Hawaiian gallinule, or 'alae 'ula (*Gallinula chloropus sandvicensis*), and the black-crowned night-heron, or 'auku'u (*Nycticorax nycticorax hoactli*), of the Hawaiian Islands, be considered on equal footing with full species, for scientific, management, and legal priorities? Similar questions pertain to considering ecological functional groups of organisms as subjects of wildlife investigations and regulatory mandates.

The question of what is wildlife raises important management concerns. Wildlife management—whether on private lands for game ranching, on state or federal public lands, on lands of indigenous peoples, or on resource industry lands including farms and commercial forests—has been defined mostly in terms of economic impacts and legal mandates. Empirically, much of wildlife management focuses on meeting regulatory edicts and litigative exigencies. The focus is often on the lists of threatened, endangered, and rare or sensitive species. While other species are not excluded from the realm of wildlife management, they often receive diminished or no formal management attention. Arguments have been made that it is too complicated to address the full ecological community and that selected signposts or indicators, largely species of legal concern or consumptive interest, must be chosen.

As such, some federal land management agencies and other landholders bank on "coarse-filter" approaches. The basic tenet of this approach is that managing generally defined or broad-scale habitat conditions will provide for the needs of all associated native species (Hunter 1991). This tenet has seldom been formally tested. Our experience (Marcot 1997) suggests that more often than not it is likely to be proved dangerously wrong. In the coarse-filter approach, wildlife is operationally defined as the often unspecified wildlife community that is associated with some general macrohabitat condition for one or a few species (the indicator species approach—e.g., Sidle and Suring 1986), or it is defined in some other general way, such as by reconstructing historic conditions (the range of natural variations approach—e.g., Morgan et al. 1994; Swanson et al. 1994). However, at least in forest management in the western United States, the environmental requirements of many species, particularly nonvertebrate taxa, are not necessarily met by this approach (Landres et al. 1988). At best, it is wise to check the validity of the coarse-filter approach to biodiversity conservation, lest some elements or species be excluded, as was discovered in such an approach to management of old-growth forests for spotted owls (Marcot 1997).

The legal dimension raises other questions regarding what wildlife is. Legal listings can include wildlife listed as rare, threatened, or endangered, such as in international Red Data Books on rare species (e.g., King and Warren 1981) or by government agencies such as the USDI Fish and Wildlife Service (FWS) following the Endangered Species Act (ESA). Additional listings with potential litigative implications may include those suggested by the International Union for Conservation of Nature and Natural Resources (IUCN) and associated organizations (McNeely et al. 1990). Typically included in such compendia are species, geographically defined subspecies, and species in portions of their range ("populations" in the loose sense). Most of these lists, and thus what qualifies as wildlife in the legal or quasi-legal sense, change often. For instance, recently the FWS drastically reduced its formal list of candidate species by virtually doing away with the candidate 2 and 3 categories in the listing status and trimming species from the remaining candidate 1 category. Quite literally, then, many of the taxa now excluded from these candidate lists have no federal legal status as potentially threatened or endangered wildlife under current ESA operations. However, at least in the United States, individual states can and often do offer some protection.

Other federal legislation for management of wildlife in the United States has put further legal boundaries on the definition of wildlife, such as by specifying organisms for conservation in sundry acts (e.g., the Marine Mammal Protection Act, the Migratory Bird Treaty Act). The federal regulations implementing the National Forest Management Act of 1976 (36 CFR 219.19) mandated that the USDA Forest Service provide for viability of all native and desired nonnative vertebrates on Forest Service lands. The regulations provide other mandates for maintaining biodiversity and therein refer to plant and animal species and communities, but they do not further specify such organisms by taxonomic group (e.g., vertebrates). Thus, it has been a point of ongoing legal challenge and internal debate as to which wildlife qualify under these regulations: just vertebrates and federally listed plants or, in the spirit of broadly based ecosystem management, all individual species or species groups of fungi, lichens, bryophytes, vascular plants, invertebrates, and vertebrates?

Additional questions of legal definitions of wildlife pertain to organisms hunted and fished by the general public, as well as to obligations for providing native tribes and indigenous peoples with specific plants and animals for hunting and gathering purposes. In some judicial systems, organisms have been represented in court cases by interest groups, which has raised the question of the legal standing of nonhuman living entities.

A number of other questions about what qualifies as wildlife pertain to economic, cultural, ethical, and even aesthetic dimensions (box 11.1). Economic issues focus on questions of crop depredation, disease and pathogen organisms, forest insect pests, and similar direct assaults on natural or food resources and human health. They also focus on indirect effects of providing habitat for threatened or endangered organisms that are then seen to have adverse economic effects on human communities and resource-use interests. Cultural issues focus on rights of indigenous peoples to use particular lands and habitats and to hunt or gather plants and animals. Other cultural issues deal with traditional, although not necessarily indigenous, use of lands and wildlife populations for hunting purposes. Ethical issues can pertain to general use of resources or to specific animal groups. Animal rights interest groups tend to focus on species that seem closer to humans in their degree of sentience (and in their expression of pain) and tend to exclude most fish, herps, invertebrates, and plants. Artistic questions may deal with a sense of place or with organisms of specific aesthetic interest, and thereby exclude habitats and organisms not meeting such standards (e.g., some dangerous or venomous organisms such as snakes, spiders, and microorganisms).

It is clear that such a simple question as What is wildlife? has deceptively vast ramifications for guiding social desires, legal decisions, directions for management, public education, and ultimately the future of organisms and habitats across the land. It is equally clear that no one definition of wildlife satisfies all such interests. We propose that the science of wildlife ecology and management of wildlife habitat expand to encompass fully all organisms collectively addressed by these dimensions in an ecosystem context. Effectively, wildlife should ultimately be defined as no less than the full array of living organisms of all taxonomic groups in terrestrial, riparian, and aquatic environments, including marine, estuarine, lacustrine, and below-ground ecosystems.

What Is the Controversy?

Much of the controversy surrounding ecosystems as a unit of study and management is really about an ecosystem *as a definable unit.* Here we will briefly review this, because it is important that students understand the ongoing controversy prior to using the term *ecosystem,* and especially before designing an "ecosystem" study.

Defining ecosystems and choosing indicators for them would be possible if the science of ecology were able to provide us with simple, rigorous models for describing and predicting the status of ecosystems. However, at this point our knowledge is largely inadequate to identify the most important variables (Keddy et al. 1993). Likewise, Tracy and Brussard (1994) argued that we do not yet know which few ecosystem processes are the most important to study, to index, and to preserve. They noted that ecosystems are arbitrarily defined study units that range from water droplets to the entire biosphere. And as summarized by Peters (1991:91), *ecosystem* as currently used indicates a "multidimensional, unlimited, relativistic" entity representing the environment.

Ecologists have not yet adopted a single, general classification system for (nonsystematic) ecological units above the species (the Linnaean classification system), except perhaps for the biome, which is too large and crude to be a useful scale of study and management (Orians 1993). Any classification system of ecological units is certain to be highly contentious both within the biological community and among policy makers. The controversies surrounding the identification of old-growth forest in the Pacific Northwest and the delineation of wetlands across the United States are recent examples. Because of the lack of the necessary, rigorous ecological models, we need to identify an ecosystem in an operational way.

Does this mean that the ecosystem concept has no value in our analyses of wildlife habitat? We think not, because this concept embodies all the interactions that influence what an animal does and the context in which lineages evolve. The concept thus reminds us that the animal lives in a complex situation and that changes in discrete factors can have a cascading effect on the individual as well as on the population and the community. Any advancement in our understanding of these interactions thus improves our ability to manage the species. The ecosystem concept forces us to view animals in a broader context of space and time than previously witnessed in wildlife management.

Now, how do we move beyond the standard ecological rhetoric—espoused by us in the preceding paragraph—and actually implement the ecosystem concept in studies of wildlife habitat?

Much of this book has emphasized the conducting of studies within clearly elucidated spatial scales, although they have not been explicitly identified. We did not previously adduce the ecosystem concept because we feared it would distract us from the sampling and analytical issues at hand, but below we make this attempt. Further, much of our wildlife research will, and should, remain at or near the microscale end of the spectrum (e.g., Dueser and Shugart 1978), although ultimately both micro- and macroscale research should be integrated (e.g., Balcom and Yahner 1996; Paszkowski 1984).

Regardless of the scale of study, organisms are influenced by factors that operate across broader geographic areas but cannot be effectively measured at the relatively smaller geographic area of study. For example, studies of microhabitat seldom measure the abundance of predators, although predators can influence the presence or absence of other animal species at a specific site and time, and the distribution or range of predators may extend well beyond the geographic area covered by the study. Likewise, microhabitat studies are seldom able to measure habitat quality (e.g., some measure of fecundity; see chapter 6), although quality is probably the most appropriate measure of habitat.

Thus, viewing a habitat study in light of the larger ecosystem should help determine what factors to measure and on what scale to measure them. A rather traditional principle that was first advanced in the general systems theory literature of the 1960s (e.g., Lance and Williams 1967) and that applies to studies of wildlife in an ecosystem context is to look at least at two levels of spatial scale or organization smaller than that of the entity of central interest, to determine the specific mechanisms responsible for observed patterns, and two levels larger, to describe the environmental context and emergent behaviors or patterns.

If the goal is to determine why an animal uses a specific site, the ecosystem concept tells us that it will usually be necessary to measure things on many scales. The ecosystem concept thus places a habitat study within a hierarchy (Allen and Hoekstra 1984; Kolasa 1989), not unlike the overall hierarchic concept of habitat selection, developed in chapter 7. Yes, it would be helpful to draw a line around an ecosystem on a map, but we do not think it is likely that this could be done with full consensus. However, it is possible to draw lines between many of the factors that we know influence an animal, such as specific vegetation conditions, presence of cliffs and talus, and areas of moderate winter climate. Food web theory, predator-prey relationships, competitive interactions, and so on, have well-developed bodies of theory and empirical studies that lend insight into how animals perceive their surroundings. All these interactions are part of what we perceive as "ecosystem functions." Our discussion of the debate surrounding keystone species in chapter 2 exemplifies how autecology supports understanding of ecosystem functions.

What this tells us is that the days of measuring some vegetation plots and calling the measurements a habitat study are rapidly falling behind us. There is a still need for such work, especially for little-known species and habitats. This work has formed a solid foundation upon which we can now build a better understanding of ecological relationships. Next we describe how this understanding might be achieved.

What Is an Ecosystem Context?

For effective conservation, we advocate studying and managing wildlife in an ecosystem context. What is meant by an ecosystem context? We alluded to this above by adducing the concept of *Umwelt* and in our discussions of ecosystem functions, but operationally we propose the following approaches to describing species in an ecosystem context.

An ecosystem consists of organisms of various taxonomic designations, along with their interactions with each other and with abiotic conditions and processes. This is more than a mere collection of populations (organisms of the same species in a given area), species assemblages (groups of species of particular taxa), or communities (species with their interactions). Understanding wildlife in an ecosystem context entails understanding (1) population dynamics, including demographic and genetic variations; (2) the evolutionary context of organisms, populations, and species, including the contribution of genetic variation to persistence of species lineages, mechanisms of speciation and hybridization, and selection for adaptive traits; (3) interactions between species that affect their persistence and that influence community structure, including obligate mutualisms such as pollination or dispersal vectors, predation, and competition; and (4) the influence of the abiotic environment on the vitality of organisms (organism health and realized fitness) and populations (viability), including how disturbance mechanisms operate and how organisms respond. Habitat ecology plays a key role in many of these facets of an ecosystem context but itself needs to be subsumed in a broader ecological tapestry.

An ecosystem context also necessitates understanding the role of humans in modifying environments, habitats, and wildlife populations. The question of whether humans are a part of the ecosystem is not a useful focus for debate. Rather, evidence clearly shows that humans can greatly affect ecosystems. A more useful and challenging question is, To what extent do we want our actions to modify environments, habitats, and wildlife populations? To what extent do we need to change our own resource and land use habits to meet our goals for conservation? And, perhaps most difficult, How many humans should occupy a specific area? An ecosystem context for studying and managing wildlife should prompt such questions.

Depicting and Managing Key Environmental Correlates

Understanding and managing wildlife in an ecosystem context also raises the question of what characteristics of ecosystems to study and identify for management and planning. The traditional approach to wildlife habitat management focused on habitat elements of water, food, and cover. These are essential elements for maintaining the health of organisms and persistence of populations. But they do not adequately describe all facets of ecosystems vital to ensuring realized fitness of individuals and viability of populations and species. In previous chapters we have explored many approaches to empirical studies and modeling of a wide variety of other aspects of species' *Umwelts.*

Recently, a broader approach has been taken in an ecological assessment of wildlife in the western United States (Marcot et al., 1997). In this assessment, species (and selected subspecies and plant varieties) were described by use of *key environmental correlates* (KECs). KECs are biotic or abiotic conditions of a species' environment that proximately influence the realized fitness of individuals and viability of populations. KECs can include the biophysical attributes traditionally considered as habitat elements. They can also include other biotic or abiotic factors not traditionally considered as habitat elements, such as use of roads, air quality, hunting or collection pressure, and interspecific interactions.

The purpose of extolling the use of KECs is to extend a focus on environmental factors beyond simple descriptions of vegetation types and their structural or successional stages, so commonly used for wildlife habitat assessment and management. The use of KECs can help shed light on effects (positive or negative) of human activities and other dynamic aspects of ecosystems beyond those affecting just vegetation conditions.

In one assessment in the interior Columbia River basin in the western United States, Marcot et al. (1997) depicted KECs for a wide variety of selected taxa and species groups of macrofungi, lichens, bryophytes, rare vascular plants, selected soil microorganisms, arthropods, and mollusks, and for all vertebrates. A single, hierarchic classification of KECs (table 11.1) was developed for use with all these organisms. A database of KECs of each wildlife taxon was used to determine which species shared common correlates, how management of some correlates (e.g., large down wood in forests) affect suites of species, and what collective set of correlates should be recognized for managing the full set of species in an area. The KEC classification was developed hierarchically, so that groups of organisms could be identified sharing various levels of specificity of correlates.

Table 11.1. A hierarchic classification of key environmental correlates for wildlife species

Code	Element
1	Vegetation elements
1.1	cover types
1.2	structural stages
1.3	forest or woodland vegetation substrates
1.3.1	down wood
1.3.2	snags (entire tree dead)
1.3.2.1	bark piles at base of snag
1.3.3	mistletoe brooms
1.3.4	litter
1.3.5	duff
1.3.6	shrubs
1.3.7	fruits, seeds, mast
1.3.8	dead parts of live trees
1.3.9	moss
1.3.10	live trees
1.3.10.1	exfoliating bark
1.3.11	flowers
1.3.12	lichens
1.3.13	bark
1.3.14	forbs (including grass)
1.3.15	cactus
1.3.16	fungi
1.3.17	roots, tubers, underground plant parts
1.3.18	peatlands
1.4	herbaceous vegetation elements or substrates
1.4.1	herbaceous vegetation cover
1.4.1.1	aquatic submergent vegetation
1.4.2	fruits, seeds
1.4.3	moss
1.4.4	cactus
1.4.5	flowers
1.4.6	shrubs
1.4.7	fungi
1.4.8	forbs
1.4.9	bulbs, tubers
1.4.10	cryptogamic crusts
1.5	diversity of vegetation cover types
1.6	edges
1.6.1	openings
1.6.2	meadows
1.7	mycorrhizal associations
2	Biological (nonvegetation) elements
2.1	presence of prey species
2.1.1	carrion
2.2	presence of predators
2.2.1	absence of predator
2.3	presence of exotic species
2.3.1	exotic plants
2.3.2	exotic animals

(table continued on following page)

Table 11.1 Continued

- 2.4 insect irruption areas
  - 2.4.1 mountain pine beetle
  - 2.4.2 spruce budworm
  - 2.4.3 gypsy moth
- 2.5 presence of burrows or presence of burrowing mammals
- 2.6 grazing
  - 2.6.1 direct effects (trampling, consumption)
  - 2.6.2 indirect effects (habitat degradation)
  - 2.6.3 seasonality of grazing
- 2.7 presence of beaver or muskrat ponds or lodges
- 2.8 presence of nesting structures
  - 2.8.1 cavities
  - 2.8.2 platforms
- 2.9 presence of other species (specify)
- 2.10 forest pathogens
- 2.11 colonial nester

3 Nonvegetation terrestrial substrates

- 3.1 rocks
  - 3.1.1 gravel
- 3.2 soils
  - 3.2.1 soil class
  - 3.2.2 soil depth
  - 3.2.3 soil texture
    - 3.2.3.1 sand, dunes
    - 3.2.3.2 soil suitable for burrowing vertebrates
    - 3.2.3.3 soil suitable for burrowing invertebrates
  - 3.2.4 soil pH
  - 3.2.5 soil temperature
  - 3.2.6 soil moisture
  - 3.2.7 soil chemistry
  - 3.2.8 soil organic matter
- 3.3 lithic (rock) substrates
  - 3.3.1 lithic series or types (including lithic formations)
  - 3.3.2 avalanche chutes
  - 3.3.3 cliffs
  - 3.3.4 talus
  - 3.3.5 boulders, large rocks
  - 3.3.6 caves
  - 3.3.7 rock outcrops, crevices
  - 3.3.8 lava flows
  - 3.3.9 lava tubes
  - 3.3.10 canyons
  - 3.3.11 barren ground
  - 3.3.12 rugged terrain
  - 3.3.13 rocky ridges
  - 3.3.14 ravines
  - 3.3.15 cirques or basins (also see entry 5.7 below)
- 3.4 snow
  - 3.4.1 snow depth (winter)
  - 3.4.2 glaciers, snow fields
- 3.5 water characteristics
  - 3.5.1 dissolved oxygen
  - 3.5.2 water depth
  - 3.5.3 dissolved solids
  - 3.5.4 water pH
  - 3.5.5 water temperature
  - 3.5.6 water velocity
  - 3.5.7 water turbidity
- 3.6 environment space (typically for foraging) above tree canopy

4 Riparian and aquatic bodies

- 4.1 rivers
  - 4.1.1 riverine wetlands
  - 4.1.2 oxbows
- 4.2 streams (permanent or seasonal)
  - 4.2.1 intermittent
  - 4.2.2 rocks in streams
- 4.3 seeps or springs (including warm seeps or springs)
- 4.4 exposed mudflats, sandbars
- 4.5 sandbars, unconsolidated shore
- 4.6 gravel bars
- 4.7 shallow water
- 4.8 lakes or reservoirs (lacustrine)
  - 4.8.1 lakes with submergent vegetation
  - 4.8.2 lakes with floating mats
  - 4.8.3 lakes with silt or mud bottom
  - 4.8.4 lakes with emergent vegetation
  - 4.8.5 alkaline lake beds
- 4.9 ponds (permanent or seasonal)
  - 4.9.1 ponds with submergent vegetation
  - 4.9.2 ponds with floating mats
  - 4.9.3 ponds with silt or mud bottoms
  - 4.9.4 ponds with emergent vegetation
- 4.10 wetlands, marshes, or wet meadows (palustrine)
  - 4.10.1 bulbs or tubers in wetlands, marshes, or wet meadows
  - 4.10.2 *Phragmites*
- 4.11 bogs or fens
- 4.12 swamps
- 4.13 islands

(*table continued on following page*)

Table 11.1 Continued

- 4.14 waterfalls
- 4.15 hyporheic zone
- 4.16 irrigation ditches
- 4.17 ephemeral pools
- 4.18 deciduous riparian areas, including willow and cottonwood
- 4.19 vernal or seasonal flooding or flood plains
- 4.20 bottomlands
- 4.21 water table

5 Topographic or physiographic elements
- 5.1 elevation
- 5.2 slopes
- 5.3 aspect
- 5.4 slope position
- 5.5 ridgetops
- 5.6 plateaus
- 5.7 convex or concave basins (also see entry 3.3.15 above)
- 5.8 flats
- 5.9 mima mounds

6 Climate
- 6.1 precipitation (amount, pattern, seasonality)
- 6.2 Mediterranean influence (dry summers)
- 6.3 maritime influence (higher humidity and more moisture)
- 6.4 temperature
- 6.5 humidity
- 6.6 wind

7 Fire
- 7.1 fire recency
  - 7.1.1 recent fire
  - 7.1.2 old fire
- 7.2 effects of fire-suppression activities
- 7.3 fire frequency
- 7.4 fire intensity
  - 7.4.1 overstory lethal
  - 7.4.2 overstory nonlethal
- 7.5 prescribed fire
  - 7.5.1 spring prescribed fire
  - 7.5.2 late summer or fall prescribed fire
- 7.6 historic fire suppression

8 Human disturbance elements (positive or negative effects)
- 8.1 recreation areas and activities (including dispersed camping areas)
- 8.2 roads or trails
- 8.3 residential developments
- 8.4 buildings
- 8.5 bridges
- 8.6 tunnels
- 8.7 agriculture and croplands
- 8.8 livestock (disease)
- 8.9 mines and mining activities
- 8.10 harvest (including legal hunting, legal trapping, and illegal poaching of animals)
- 8.11 fences
- 8.12 bird feeders
- 8.13 winter recreation
- 8.14 garbage
- 8.15 logging
- 8.16 nest boxes
- 8.17 perch structures
- 8.18 platforms
- 8.19 guzzlers
- 8.20 pesticide use
- 8.21 exotic plant effects
  - 8.21.1 direct displacement
  - 8.21.2 indirect competition
  - 8.21.3 inhibited recruitment
  - 8.21.4 habitat structure change
- 8.22 livestock grazing strategies
  - 8.22.1 season-long
  - 8.22.2 spring grazing
  - 8.22.3 summer grazing
  - 8.22.4 fall grazing

9 Barriers to movement
- 9.1 forest management (clearcuts)
- 9.2 canopy closure
- 9.3 agriculture

10 Other natural disturbances—floods, scouring, openings in forests

Source: modified from Marcot et al., 1997

Note: This classification was developed from taxon-specific information on fungi, lichens, bryophytes, vascular plants, invertebrates, and vertebrates in the inland West of the United States. The numbered codes in this classification are strictly hierarchic (e.g., item 1.1.2 is one element of 1.1, which is one element of the broadest-level category 1) and can be used in species-environment relations databases.

To our knowledge, this is the first time that such a classification system has been developed for so wide an array of organisms. It may prove useful for other studies of biodiversity and development of other wildlife management plans.

Depicting Species' Key Ecological Functions

Yet another facet of managing wildlife in an ecosystem context pertains to understanding the ecological roles played by species. The traditional approach to habitat management has assumed simply that wildlife are a function of habitat and that managing wildlife entails providing the right kinds of habitat. Instead, the array of key ecological functions (KEFs) of individual species can be explicitly depicted. In this context, *key ecological functions* refers to species' or a taxon's main ecological roles that influence diversity, productivity, or sustainability of ecosystems. A given KEF can be shared by many species, and a given species can have several KEFs. These can be depicted, along with KECs, in databases and models of species-environment relations.

Table 11.2 presents a classification of KEFs that was developed for the recent wildlife assessment in the U.S. inland West (Marcot et al., 1997). Main categories of KEFs include trophic relations; herbivory; nutrient cycling; interspecies relations; disease, pathogen, and parasite relations; soil relations; wood relations; water relations; weather, climate, insolation relations; and vegetation structure and composition relations. Each category was divided into a number of hierarchic subcategories. As with the KEC classification, this may be the first such classification system developed for the variety of plant and animal taxa addressed in this study.

The KEF classification was used in a database to determine species sharing specific ecological functions and the array of functions performed by specified species. Species with the same KEFs were called ecological functional groups of species. Information on KEFs of species was cross-linked to that on KECs and range distributions of species, as discussed below. The classification of KEFs we present here may be useful for guiding other studies of wildlife biodiversity by helping to focus on ecological roles of species as a complement to the more traditional assessments of biodiversity as species richness.

A species ecological function may often appear in more than one subsystem. As an example, consider the ecological function of a parasite carrier or transmitter (KEF 5.3 in table 11.2). Species with this ecological function in the inland West of the United States include the least bittern (*Ixobrychus exilis*), which is a host to ecto- and endoparasites; the western sage grouse (*Centrocercus urophasianus phaios*), which is a host for protozoan, helminth, and bacterial parasites; and the snowshoe hare (*Lepus americanus*), which supports a variety of ecto- and endoparasites and is a reservoir of several viruses and bacterial pathogens. Even this simple ecological functional group of three species occupies a collective set of key environmental correlates—including vegetation cover types, structural stages, and other environmental factors—that spans terrestrial forest, shrubland, grassland, and riparian (wetland) ecological subsystems. In this way, ecological processes for each subsystem can be identified.

Ecological processes, as used here, are those groups of key ecological functions of species that pertain to each specific subsystem. For example, ecological processes associated with soil subsystems include organic matter decomposition, nutrient pooling and cycling, and provision of conditions for mesoinvertebrates

Table 11.2. A hierarchic classification of key ecological functions of wildlife species

1 Trophic relations
- 1.1 primary producer (chlorophyllous vascular plant)
  - 1.1.1 autotrophe (fully independent chlorophyllous plant)
  - 1.1.2 hemiparasite (chlorophyllous plant that also partly derives nutrients through attachment to other chlorophyllous plants)
- 1.2 heterotrophic consumer
  - 1.2.1 primary consumer (herbivore) (also see below under Herbivory)
    - 1.2.1.1 foliovore (leaf eater)
    - 1.2.1.2 spermivore (seed eater)
    - 1.2.1.3 browser
    - 1.2.1.4 grazer
    - 1.2.1.5 frugivore (fruit eater)
    - 1.2.1.6 sap feeder (sucking insect)
    - 1.2.1.7 root feeder (invertebrate)
    - 1.2.1.8 sequestration of plant metabolites
  - 1.2.2 secondary consumer (primary predator or carnivore)
    - 1.2.2.1 consumer or predator of invertebrates, potentially including insects (insectivore)
    - 1.2.2.2 consumer or predator of vertebrates (species other than itself)
  - 1.2.3 tertiary consumer (secondary predator or carnivore)
    - 1.2.3.1 consumer of soil microorganisms
  - 1.2.4 largely omnivore (consumer of plants and animals)
  - 1.2.5 carrion feeder
  - 1.2.6 cannibal
  - 1.2.7 coprophagist (consumer of fecal material)
  - 1.2.8 aquatic herbivore (invertebrate)
  - 1.2.9 consumer of algae, ooze, and plankton in water (invertebrate)
- 1.3 achlorophyllous vascular plant (see 1.9 below for nonvasculars)
  - 1.3.1 mycotrophe (indirectly parasitic, nongreen plant that derives nutrients from mycorrhizal fungi that is also associated with a chlorophyllous species that serves as the indirect host)
  - 1.3.2 saprophyte (derives nutrients from decaying organic matter through mycorrhizal fungi)
  - 1.3.3 parasite (derives nutrients through direct attachment to chlorophyllous plants)
    - 1.3.3.1 root parasite
    - 1.3.3.2 stem parasite
- 1.4 detritovore (direct consumer of dead organic material)
- 1.5 decomposer (consumer of byproducts of decaying organic material)
- 1.6 comminutor (chewing insect, typically feeding on wood or vegetation)
- 1.7 forage or prey relations
  - 1.7.1 forage for animals
  - 1.7.2 prey for secondary or tertiary consumer (primary or secondary predator or carnivore)
  - 1.7.3 carrion source
  - 1.7.4 forage for invertebrates
- 1.8 major biomass
- 1.9 achlorophyllous nonvascular plants (see 1.3 above for vasculars)
  - 1.9.1 mycorrhizal fungus
  - 1.9.2 saprophyte
  - 1.9.3 parasite
  - 1.9.4 decomposer
- 1.10 moss feeder (invertebrate)

2 Herbivory
- 2.1 ungulate herbivore (may influence rate or trajectory of vegetation succession and presence of plant species)
  - 2.1.1 herbivore of tree or shrub species (browser)
  - 2.1.2 herbivore of grasses or forbs (grazer)
- 2.2 insect herbivore (may influence rate or trajectory of vegetation succession or presence of plant species)
  - 2.2.1 defoliator
  - 2.2.2 bark beetle
  - 2.2.3 tree bole feeder

3 Nutrient cycling relationships (see number 6 below for nutrient cycling relationships in soil)
- 3.1 aids in physical transfer of substances for nutrient cycling (C, N, P, other)
- 3.2 nitrogen relationships
  - 3.2.1 N-fixer
  - 3.2.2 N-immobilizer
  - 3.2.3 source for N mineralization
- 3.3 carbon relationships
  - 3.3.1 sequestration of atmospheric carbon

(table continued on following page)

Table 11.2 Continued

- 4 Interspecies relationships
  - 4.1 insect control
  - 4.2 ungulate or other vertebrate population control
  - 4.3 pollination vector
  - 4.4 transportation of seed, spores, plant or animal disseminules
    - 4.4.1 disperses fungi
    - 4.4.2 disperses lichens
    - 4.4.3 disperses bryophytes, including mosses
    - 4.4.4 disperses insects
    - 4.4.5 disperses seeds and fruits
    - 4.4.6 disperses plants
  - 4.5 commensal or mutualist with other species
  - 4.6 provides substrates or cover for animals
    - 4.6.1 nesting or breeding substrate (e.g., nesting material)
    - 4.6.2 thermal, hiding cover, loafing or den site
    - 4.6.3 provides microhabitat (as for invertebrates)
      - 4.6.3.1 aquatic or riparian environments
      - 4.6.3.2 terrestrial environments
      - 4.6.3.3 canopy environments
      - 4.6.3.4 tree bole environments
    - 4.6.4 creates "sap wells" in trees
  - 4.7 nest parasite
    - 4.7.1 nest parasite species (viz., cowbird)
    - 4.7.2 host for nest parasitism
  - 4.8 primary cavity excavator in snags or live trees
  - 4.9 primary burrow excavator (fossorial)
    - 4.9.1 creates large burrows (rabbit, badger size)
    - 4.9.2 creates small burrows (smaller than rabbit size)
  - 4.10 competitor
  - 4.11 uses burrows dug by other species (secondary burrow user)
  - 4.12 uses cavities excavated by other species (secondary cavity user)
  - 4.13 endo- or ectoparasite (invertebrate) (also see number 5, Disease, pathogen, and parasite relationships)
- 5 Disease, pathogen, and parasite relationships
  - 5.1 carrier, transmitter, or reservoir of vertebrate diseases (including rabies)
  - 5.2 acts as pathogen or disease
  - 5.3 parasite carrier or transmitter
  - 5.4 carrier, transmitter, or reservoir of plant diseases (invertebrate)
  - 5.5 activity increases host susceptibility to plant diseases (invertebrate)
- 6 Soil relationships
  - 6.1 physically affects (improves) soil structure, aeration (typically by digging)
  - 6.2 aids general turnover of soil nutrients and layers
  - 6.3 aids N retention or uptake in soil
  - 6.4 aids soil stabilization
  - 6.5 aids rock weathering
  - 6.6 detoxifies xenobiotics (invertebrate)
  - 6.7 metal accumulator (sequesters heavy metals)
  - 6.8 soil (invertebrate) organisms which influence rate or trajectory of vegetation succession and presence of plant species
- 7 Wood relationships
  - 7.1 physically breaks down wood
    - 7.1.1 large logs
    - 7.1.2 smaller wood pieces
  - 7.2 chemically breaks down wood
- 8 Water relationships
  - 8.1 impounds water (e.g., beaver)
  - 8.2 bioindicator of water quality
  - 8.3 hydrological buffer
  - 8.4 improves water quality
  - 8.5 contributes to short-term increase in stream flow (invertebrate)
- 9 Weather, climate, insolation relationships
  - 9.1 affects albedo (as of soil, rock, or soil)
- 10 Vegetation structure and composition relationships
  - 10.1 creates canopy gap openings (tree death) (invertebrate)
  - 10.2 creates standing dead trees (snags) (invertebrate)

Source: modified from Marcot et al., 1997

Note: This classification was developed from taxon-specific information on fungi, lichens, bryophytes, vascular plants, invertebrates, and vertebrates in the inland West of the United States. The numbered codes in this classification are strictly hierarchic (e.g., item 1.1.2 is one element of 1.1, which is one element of the broadest-level category 1) and can be used in species-environment relations databases.

and fungi critical to vascular plant productivity. Species ecological functions associated with such processes in soil subsystems include soil aeration, turnover of soil nutrients and layers, nitrogen retention and uptake, and soil stabilization.

Species' key ecological functions and ecological processes all contribute to diversity, sustainability, and productivity over time. For example, the ecological processes and species' key ecological functions associated with the soil subsystem of forest ecosystems all contribute to the following: the biodiversity of fungi, lichens, plants, mesoinvertebrates (e.g., soil mites), macroinvertebrates (e.g., earthworms), and fossorial vertebrates (e.g., pocket gophers); and the productivity of plant and animal populations, including tree growth and the sustainability of resource growth and use (e.g., sustained timber production and harvest) over the long term.

Collectively, KECs and KEFs describe what has been traditionally referred to as habitat and niche dimensions—respectively, an organisms' "address" and "occupation." However, we advocate expanding the address to include facets of the environment not traditionally considered in habitat studies and expanding the occupation to include the full array of ecological functions that an organism or taxonomic group performs. KECs and KEFs can be described from a conceptual basis of the expected range of addresses and occupations, corresponding to carrying capacity and fundamental niche dimensions; or they can be described from an empirical basis of observed addresses and occupations, corresponding to actual distribution and abundance, and realized niche dimensions.

Beyond Habitat: Species-Environment Relations

A relational database can be developed that lists wildlife species along with their KECs, KEFs, and range distributions. Such a database extends beyond the traditional wildlife-habitat relationship approach by considering nonhabitat environmental elements and species' ecological functions. With queries devised to address the span of species and their correlates, functions, and distributions, such a database can be termed a *species-environment relations model.*

A species-environment relations model can be used to pose new questions about the roles of species in ecosystems and how managers might provide for species' ecological functions through habitat management. New questions to ask can include: Which species provide specific ecological functions, such as nutrient cycling, soil turnover, or insect population control? What are the collective habitat and environmental requirements of such species? Where do they occur? How do such ecological functions affect productivity and diversity of ecosystems?

In the context of a species-environment relations model, habitats can be viewed as part of ecological subsystems, and species along with their ecological functions can be identified as part of each subsystem. Subsystems include below-ground, surface, and arboreal components of terrestrial, riparian, and aquatic environments. Each subsystem has associated species and ecological processes which contribute to the overall functioning of the ecosystem. A list of terrestrial subsystems is presented in table 11.3. The general approach may be (1) to identify a specific habitat, substrate, or KEC of interest, and identify the ecological subsystem in which it pertains; (2) to use a species database to list all species associated with that habitat, substrate, and KEC, by subsystem; and (3) to list the collective set of key ecological functions of all species by subsystem. An example of such an approach is presented in box 11.2.

Table 11.3. A simple classification of ecological subsystems and some of their components

1. Terrestrial above-ground subsystem
   - (a) Upland terrestrial conditions of forests
     - tree canopy, subcanopy components
     - shrub-layer components
     - grass/forb-layer components
   - (b) Upland terrestrial conditions of shrublands and grasslands
     - tree components
     - shrub components
     - grass/forb-layer components
     - cryptogamic crust (soil surface) components
2. Terrestrial below-ground subsystem
   - (a) at- and below-surface zones: coarse wood, litter, duff components
   - (b) soil (O, A horizons) components
   - (c) root sphere influence zone components
3. Riparian influence zone subsystem
   - (a) riparian shade components
   - (b) leaf litter throughput components
   - (c) large wood components
   - (d) hyporheic zone
4. Aquatic subsystems
   - (a) Lotic (moving water) subsystem
     - Bank and channel stability components (e.g., side slope angle and erosion potential)
     - Reach components (e.g., reach type, habitat characteristics, processes such as edge effects, storage of organisms, hydrology, and particle size sorting)
   - (b) Lentic (still water) subsystem
     - Water body size components
     - Water permanence components
       - lake or pond
       - astatic lake or pond
       - intermittent pond
       - wet meadow, seep, bog, fen

Note: Each subsystem consists of unique sets of species along with their ecological functions and other abiotic ecological processes. Species-habitat relations models can help identify associated species and how their attributes contribute to functioning, diversity, and productivity of the overall ecosystem.

Alternately, one could specify an ecological function and identify all species having this function along with their collective set of habitat requirements and ecological correlates by subsystem. This would help identify the set of conditions that habitat managers might provide to help maintain habitat conditions for species with specific ecological functions. Other approaches are also possible, such as overlay-mapping all species associated with a particular ecological function as a way to represent the full geographic distribution of (and potential management risks to) specific functions. In a sense, this would extend the current approaches of spatially-explicit modeling of wildlife populations to include species ecological functions.

Classification of land units, including ecoregions or bioregions, existing or potential vegetation communities, and habitat types, is useful and often essential, but land units alone do not tell us much about the status and trend of associated wildlife species or communities. Instead of striving to map wildlife communities or ecosystems per se, we suggest that abiotic and biotic components of ecosystems that most affect wildlife species can be more readily mapped and analyzed.

Modeling Species-Environment Relations with Species-Influence Diagrams

One approach to depicting and modeling species-environment relations is by use of what may be called species-influence (SI) diagrams. The generic form of an SI diagram (fig. 11.1) depicts the major relations discussed above: the

Box 11.2. Wildlife, down wood, and ecological functions: beyond simple species-habitat associations

Presented here is an example of how species, their ecological functions, their use of key environmental correlates and habitats, and their occurrence in ecological subsystems can be combined to provide a greater understanding of ecological processes and a broader basis for ecosystem management.

Species and Their Ecological Functions Associated with Down Wood

Habitat managers are often concerned with providing down wood for wildlife during management activities such as timber harvesting or habitat rehabilitation. If the management question is, What wildlife uses down wood? the typical answer comes in the form of a list of species associated with down wood in specific habitats. And the premise is "If you provide it, they will come," so the management focus needs to be mostly on the down wood itself and the beneficiaries are the associated wildlife species.

In addition, however, we ought to be asking: (1) What is the collective set of environmental and habitat requirements of species using down wood, beyond the simple presence of down wood? and (2) What fuller set of ecological functions and services are provided by species associated with down wood? The answers are surprising.

For purposes of this example, we will assume that the questions pertain to managing Ponderosa pine (*Pinus ponderosa*) forest communities of the U.S. inland West. We draw on a species-environment relations (SER) database model developed for ecosystem management planning in the interior Columbia River basin (Marcot et al., 1997).

Step One: Identify the Habitat Components

In this example, we focus on down wood ("coarse woody debris") in Ponderosa pine forest communities of the U.S. inland West. The ecological subsystem of interest as listed in table 11.3 is thus the terrestrial below-ground subsystem and the at-surface zones of coarse wood components within this particular vegetation community.

Step Two: List All Associated Species and Their Collective Habitat and Environmental Correlates

The SER database provides a list of species meeting the criteria of occurring in interior Ponderosa pine forest communities and using down wood as at least one of their key environmental correlates. In the U.S. inland West, these species number at least 47 and include 5 amphibians, 8 reptiles, 10 birds, and 24 mammals. For the sake of simplicity, we will focus on the mammals. These include 8 sciurids, 4 mustelids, 3 other carnivores, a porcupine, 4 rodents, and 4 shrews, as listed below. Doubtless, other species may use down wood, but for these species it is a key habitat requirement. The list includes:

Erethizon dorsatum	common porcupine	*Sorex hoyi*	pygmy shrew
Glaucomys sabrinus	northern flying squirrel	*Sorex trowbridgii*	Trowbridge's shrew
Lutra canadensis	river otter	*Sorex vagrans*	vagrant shrew
Lynx rufus	bobcat	*Spermophilus saturatus*	golden-mantled ground squirrel
Mustela erminea	ermine	*Tamias amoenus*	yellow-pine chipmunk
Mustela frenata	long-tailed weasel	*Tamias minimus*	least chipmunk
Mustela vison	mink	*Tamias ruficaudus*	red-tailed chipmunk
Neotoma cinerea	bushy-tailed woodrat	*Tamias umbrinus*	Uinta chipmunk
Peromyscus keenii	Columbian mouse	*Tamiasciurus douglasii*	Douglas' squirrel
Peromyscus maniculatus	deer mouse	*Tamiasciurus hudsonicus*	red squirrel
Reithrodontomys megalotis	western harvest mouse	*Ursus americanus*	black bear
Sorex cinereus	masked shrew	*Ursus arctos horribilus*	grizzly bear

(*box continued on following page*)

Box 11.2 Continued

Step Three: List the Collective Key Ecological Functions

The next step entails asking the question, What is the full set of ecological functions of these species? In other words, in what way does providing down wood for mammals in this forest type support a range of species ecological functions? This question extends beyond the general set of ecological services provided by the down wood itself, which include soil stabilization on slopes, sources of organic matter, sources of in-stream structure for fish habitat, sources of soil macronutrients, habitat for invertebrates and microorganisms, and other services.

The traditional approach would have us focus just on such functions associated with down wood, but we extend this approach. As it turns out, the set of 24 mammal species collectively performs some 22 categories of key ecological functions, including trophic relations, nutrient cycling, interspecies relations, and disease, soil, and wood relations. Most of these functions extend far beyond the physical down wood substrate itself (see box table 11.2A).

Among trophic relations functions, the mammal species include primary consumers (herbivores) and secondary consumers (primary predators or carnivores), and the wide-ranging carnivores are also tertiary consumers (secondary predators or carnivores). Food habits run the range of herbivory (plant eating), carnivory (animal eating), spermivory (seed eating), frugivory (fruit eating), and insectivory (insect eating). At least one species contributes to soil nutrient cycling, another may control insect populations, and a number may contribute to dispersal of fungi, lichens, seeds, and fruits. Two species physically dismember down wood. Six species excavate burrows used by at least three other of these mammal species using down wood (a number of other burrow excavators and burrow users occur in this community, but these nine species are mammals specifically associated with down wood). Other ecological functions of mammals using down wood in this forest community include carriers of vertebrate diseases, pathogens, or parasites.

Box Table 11.2A. The collective set of key ecological functions performed by 24 mammals associated with down wood in Ponderosa pine forests of the interior Columbia River basin, U.S. inland West.

Key ecological function code[a]	Key ecological function description	Number of species[b]
1.2.1	Trophic relations: primary consumer (herbivore)	7
1.2.1.2	Spermivore (seed eater)	4
1.2.1.5	Frugivore (fruit eater)	1
1.2.2	Secondary consumer (primary predator or carnivore)	3
1.2.2.1	Consumer or predator of invertebrates, potentially including insects (insectivore)	7
1.2.2.2	Consumer or predator of vertebrates (species other than itself)	5
1.2.4	Largely omnivore (consumer of plants and animals)	3
1.7.2	Prey for secondary or tertiary consumer (primary or secondary predator or carnivore)	17
3.1	Nutrient cycling relationships: aids in physical transfer of substances for nutrient cycling (C, N, P, other)	1
4.1	Interspecies relationships: insect control	1
4.2	Ungulate or other vertebrate population control	2
4.4	Transportation of seed, spores, plant or animal disseminules	2
4.4.1	Dispersal of fungi	4
4.4.2	Dispersal of lichens	1
4.4.5	Dispersal of seeds and fruits	11
4.9.2	Primary burrow excavator (fossorial): creates small burrows (smaller than rabbit size)	6

(box continued on following page)

Box 11.2 Continued

Key ecological function code[a]	Key ecological function description	Number of species[b]
4.10	Key competitor	1
4.11	Uses burrows dug by other species (secondary burrow user)	3
5.1	Disease, pathogen, and parasite relationships: carrier, transmitter, or reservoir of vertebrate diseases (including rabies)	1
5.3	Disease, pathogen, and parasite relationships: parasite carrier or transmitter	1
6.1	Soil relationships: physically affects (improves) soil structure, aeration (typically by digging)	1
7.1	Wood relationships: physically breaks down wood	2

[a]The key ecological function code refers to the classification system presented in table 11.2
[b]The number of species is the number of mammal species (maximum = 24) having each function

Overall, the range of ecological functions provided by species associated with just this one habitat component is probably far broader than most managers would suspect. The links between down wood, associated species, and functions such as dispersal of plant disseminules, nutrient cycling, and burrow excavation and use typically have not been previously acknowledged in ecological assessments and management plans for down wood.

Similar assessments can be conducted for other taxonomic groups of organisms including amphibians, reptiles, and birds, as well as invertebrates and even plants themselves; for other vegetation communities; for other ecological subsystems and habitat components; and for other sets of key ecological functions.

Identifying Other Data Needs

Of course, in this example most of the mammal species require habitat components beyond down wood. These are identified in box table 11.2A, which lists the collective set of key environmental correlates of all species. This full set of conditions includes other vegetation, lithic (rock), and aquatic substrates and components, as well as biological elements such as the presence of burrows, beaver or muskrat ponds or lodges, snag cavities, nesting platforms, and even roads and buildings. The full set of conditions could constitute the habitat and environmental basis for managing habitats for mammal species associated with down wood and their collective ecological functions in interior Ponderosa pine communities.

The approach exemplified here helps to frame new questions about the contribution of wildlife associated with specific substrates or habitat components, to overall ecological processes within ecosystems. Most of such ecological relations need far greater study to determine actual rates and frequencies of occurrence.

In addition, in our example, data are usually lacking on the specific amount (volume or mass) and arrangement of down wood that is used and selected by wildlife species and on the frequency with which down wood should be provided, given wood decay rates in various environments. Only careful experimental studies can provide such information. The overall approach illustrated here can help to identify the specific ecological links and functions for which studies on rates would be most beneficial.

The example presented here could be expanded also to deal with wildlife habitat relations with all wood and organic matter in terrestrial below-ground and forest floor subsystems. Such assessments could address not just large down wood but also branches and litterfall; living and dead wood; standing and down wood; and the movement of wood between subsystems, particularly from terrestrial above-ground environments into below-ground, riparian, and aquatic environments. Such comprehensive assessments should also address maintaining suites of organisms—not just vertebrates—and their ecological functions of organic matter decay and recycling.

(box continued on following page)

Box 11.2 Continued

Box Table 11.2B. The collective set of key environmental correlates (other than vegetation cover types) used by all 24 mammal species associated with down wood in Ponderosa pine forests of the interior Columbia River basin, U.S. inland West

Key environmental correlate code[a]	Key environmental correlate description	Number of species[b]
1.3.1	Forest or woodland vegetation substrates: down wood	24
1.3.2	Snags (entire tree dead)	3
1.3.5	Duff	3
1.3.6	Shrubs	1
1.3.7	Fruits/seeds/mast	3
1.3.10	Live trees	3
1.3.11	Flowers	1
1.3.12	Lichens	1
1.3.13	Bark	1
1.3.14	Forbs and grasses	3
1.3.16	Fungi	3
1.3.17	Roots, tubers, underground plant parts	1
2.1	Biological (nonvegetation) elements: presence of prey species	12
2.1.1	Carrion	2
2.5	Presence of burrows or presence of burrowing mammals	8
2.7	Presence of beaver or muskrat ponds or lodges	1
2.8.1	Presence of cavities	4
2.8.2	Presence of nesting platforms	1
3.1	Nonvegetation terrestrial substrates: rocks	1
3.3.3	Cliffs	4
3.3.4	Talus	7
3.3.5	Boulders and large rocks	1
3.3.6	Caves	5
3.3.7	Rock outcrops/crevices	4
3.4.1	Snow depth (winter)	1
4	Riparian and aquatic bodies	5
4.1	Rivers	1
4.2	Streams (permanent or seasonal)	2
4.8	Lakes or reservoirs (lacustrine)	2
4.9	Ponds (permanent or seasonal)	1
4.9.4	Ponds with emergent vegetation	1
4.10	Wetlands, marshes, or wet meadows (palustrine)	5
4.11	Bogs or fens	4
4.18	Deciduous riparian areas, including willow and cottonwood	1
5.1	Topographic or physiographic elements: elevation	3
5.3	Aspect	1
8.2	Human disturbance elements and activities: negative effects of roads or trails	1
8.4	Positive effects of buildings and structures	1

[a]The key environmental correlate code refers to the classification system presented in table 11.1

[b]The number of species is the number of mammal species (maximum = 24) having each correlate

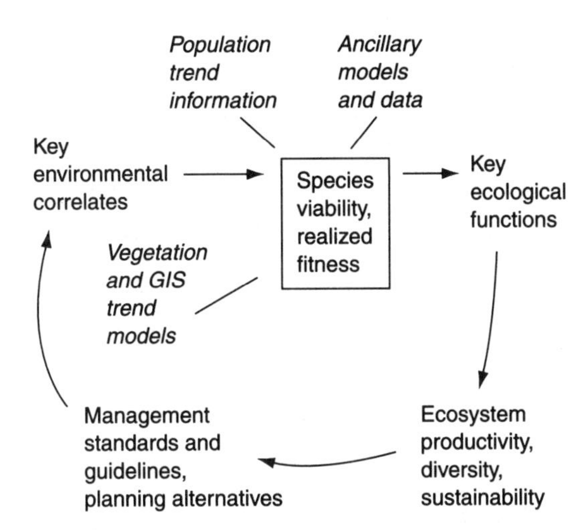

Figure 11.1. The general form of a species-influence diagram. This is the basic arrangement of information used in species-environment relations (SER) databases and models. The central box represents the realized fitness of selected individual species or species groups (and, by extension, viability of populations). Fitness is influenced by a series of key environmental correlates (KECs), which include vegetation cover types and structural stages, as well as other substrate-specific aspects of the species habitat and other environmental features not traditionally included in habitat evaluations. KECs can be represented, in part, by vegetation change simulation models and GIS models. Along with additional information on population trends, and with ancillary models and data sources, the KECs provide a means by which an indirect inference can be made on fitness and viability by tracking patterns and trends in potentially suitable environments for the species. KECs are influenced by management activities. Each species in turn performs a set of key ecological functions (KEFs) such as aiding nutrient cycling, providing a source of carrion, or excavating burrows. These KEFs ultimately influence the sustainability, productivity, and biodiversity of the ecosystem. Under a paradigm of ecosystem management, understanding these relations should help develop management standards, guidelines, and planning alternatives designed to restore or maintain ecosystem productivity, sustainability, and biodiversity.

effects of key environmental correlates on species viability and realized fitness of organisms; the key ecological functions of species; and the influence of ecological functions on ecosystem productivity, diversity, and sustainability. An SI diagram can also depict management guidelines to maintain key environmental correlates and the design of such guidelines based on goals for ecosystem productivity, diversity, and sustainability. In this way, the circle is complete and the SI diagram can form the basis for articulating testable management hypotheses (similar to "research hypotheses" discussed in chapter 4) regarding species-habitat relations, the effects of providing environments on maintaining species, species functions, and the effects of functions on productivity and diversity of ecosystems and on sustainability of resource production. Overall, this provides an ecological basis and research approach to ecosystem management. An example of an SI diagram for one species is presented in figure 11.2.

The advantage of the species-influence diagram approach is that it can easily be turned into a quantitative model of expected influences by use of Bayesian statistics. Analogous to

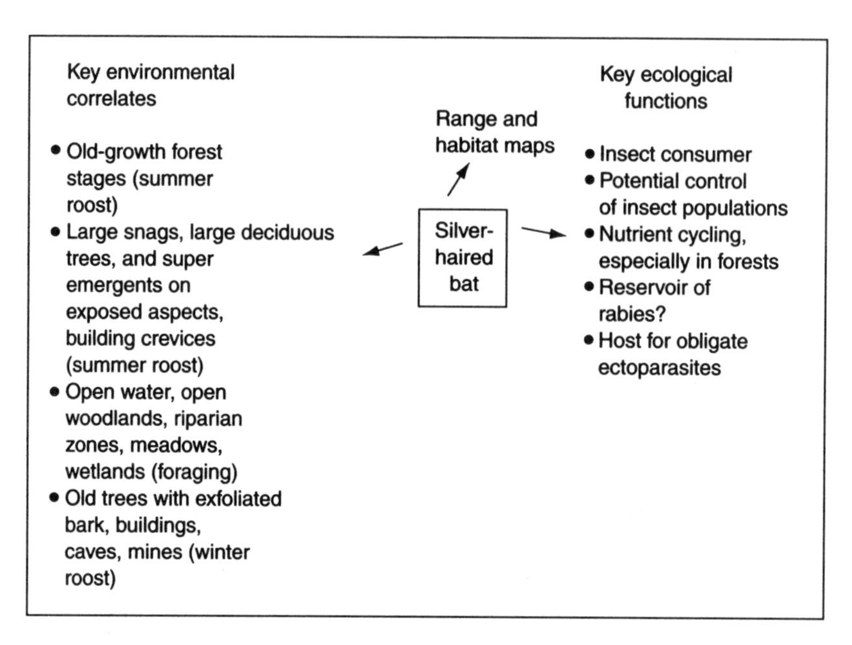

Figure 11.2. An example of a species-influence diagram for the silver-haired bat (*Lasionycteris noctivagans*), listing its key environmental correlates (KECs) and key ecological functions (KEFs).

the path regression approach discussed in chapter 10, the Bayesian approach assigns prior (often best-guess) probabilities to each relation in the model as conditional likelihoods of outcomes. For example, the sample model for the silver-haired bat shown in figure 11.2 can be developed into a Bayesian model by linking key environmental correlates and assigning likelihoods for each link (fig. 11.3). With silver-haired bats, the overall likelihood of a species outcome (perhaps measured in categories of viability level or realized fitness levels) can be calculated as a joint probability of each key environmental correlate being present in an area, such as appropriate vegetation cover types, vegetation structural stages, and substrates for roosting or maternity colonies. Effects of individual key environmental correlates given the influence of all others, as well as relative effects of unknown influences, can be partialed out. Similar models can depict the influence of species' key ecological functions on ecosystem diversity, productivity, and sustainability, and the influence of management activities on levels of key environmental correlates. In this way, explicit models of likely or expected outcomes and effects on ecosystem management goals can be articulated and tested in an adaptive management approach.

Beyond Viability: Managing Populations in a Community and Ecosystem Context

Much has been written recently on whether wildlife management should focus on individual species or on ecosystems (e.g., Franklin 1993). We do not view this argument as par-

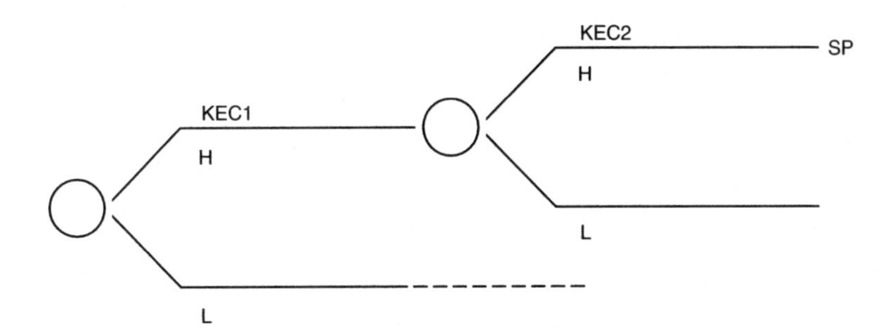

Figure 11.3. Schematic for a Bayesian belief model of species-environment relations. KEC = key environmental correlate; H = high value; L = low value; SP = species outcome; circles = chance events.

ticularly fruitful, because both are necessary for ensuring the long-term viability and evolutionary potential of species, provision of habitats, and integrity of ecological communities and systems. An approach that incorporates habitat components and other environmental correlates, as well as explicitly acknowledges species' functions and ecosystem processes, should strive to encompass several levels of biological organization. Simple coarse-filter approaches to wildlife habitat management may be one useful starting point, but they cannot account for requirements of all ecological entities (see table 3.6).

In a more comprehensive approach as outlined above, quandaries of defining minimum viable population levels and prioritizing individual species for recovery efforts as in a triage approach would be reduced. Ultimately, we are all part of the ecological systems we try to manage; species-influence diagrams and Bayesian-type models that can be made for depicting wildlife species can just as well be built for the human species. In this way, species, ecosystems, ecological functions, environmental conditions, management goals, and management activities can all be merged into one coherent approach for assessment and management. The utility of this comprehensive approach should become more evident for managing wildlife in an evolutionary context and for use in adaptive management.

Managing Wildlife in an Evolutionary Context

Much of wildlife habitat management focuses on more immediate concerns of habitat mitigation, conservation, and restoration. But environments will continue to change, climates will vary stochastically, and disturbance events such as major fires will recur; nature is not static. Thus, one aspect of management should view wildlife in an evolutionary context. The aim is not to predict future evolution; genetic engineering aside, this is not possible given our current scientific understanding. Instead, the purpose would be to provide for conditions that permit natural evolutionary processes, including species interactions and the role of abiotic factors in selection pressures. The ebb and flow of taxonomic lineages—including emergence of subspecies, species, and higher taxonomic groupings—are afforded only by providing a diverse and stochastic array of environmental conditions and gene pools of organisms.

Managing for Taxonomic Lineages

A few general guidelines can be offered for managing for future taxonomic lineages. First is to represent native ecosystem types across their natural range of conditions and variation. *Range of conditions* pertains to the array of various kinds of habitats, vegetation, and abiotic cover types that may naturally occur in a region. *Range of variation* pertains to the natural fluctuations in environmental conditions that may occur in a region over a specified time period, such as the number and intensity of wildfires in a prairie ecosystem over the span of a century, as a result of which many vegetation and wildlife species are coadapted. Understanding and providing for ranges of conditions is simpler than doing so for ranges of variation, although the idea is hardly new (Regier 1978).

A second guideline is to recognize and provide for novel genetic entities in wildlife populations. Several of these were presented as examples in chapter 3. The mere presence of a species is insufficient to ensure continuation of taxonomic lineages.

A third guideline is to maintain evolutionary and ecological processes by which new taxonomic lineages emerge and develop. Some of these processes were discussed in chapters 2 and 3, and they can be explicitly worked into planning and management guidelines. Some examples include ensuring provision of species' peripheral ranges, isolated and disjunct populations, novel morphs and ecotypes, and environmental conditions for organisms at the limits of their ecological range of physiological tolerance. In such conditions new lineages are apt to arise.

Managing for Ecological Domains and Evolutionary Time Frames

Certainly, managing for evolutionary time frames can seem a daunting task for managers concerned with relatively short-term resource management objectives. We are not suggesting that evolutionary changes be predicted; this is not yet possible given current ecological knowledge. Rather, we are suggesting that the base conditions from which evolution proceeds be provided, per the guidelines discussed above. In this way, the array of ecological domains—the spectra of conditions and variations thereof—can be provided over time to form the foundation for a long-term evolutionary change in organisms.

Adaptive Management: Learning from Experience

The Adaptive Management Approach

We have advocated the use of adaptive management as a means of conducting better habitat management over time. The approach entails identifying areas of scientific uncertainty, devising field management activities as real-world experiments to test that uncertainty, learning from the outcome of such experiments, and recrafting management guidelines on the basis of the knowledge gained (Holling 1978; Irwin and Wigley 1993; Walters 1986). Modeling can play a key role in formalizing our current knowledge and identifying important areas of uncertainty (Barrett and Salwasser 1982). In an ideal situation, management guidelines equate to the creation of testable hypotheses; monitoring and adaptive management studies equate to conducting the experiment; and revision of the management guidelines equates to reevaluation and interpretation of the study results in terms of testing the validity of the initial hypothesis.

Adaptive management, however, has seldom been applied successfully or fully in managing wildlife habitat and ecosystems (Lee 1993; Hilborn 1992). Problems are largely technical and administrative (Lee and Lawrence 1986). Following are some basic tenets of a successful approach to adaptive management. They can be used as a checklist for specific programs to ensure successful application.

Tenets of a Real-World Approach to Adaptive Management

1. The administration has the political will, and explicit procedures, to accept change. This first tenet summarizes a great deal of experience in the administrative behavior and political reality of resource management organizations and institutions. Despite good intentions and explicit promises in management plans, if an organization lacks the political motivation or the actual process for accepting new knowledge, then management activities will remain unchanged. What is needed is a clear political mandate to accept change in a timely manner and a formal protocol for weighing and incorporating new information for potential use in updating or reaffirming management decisions.

A corollary to this tenet is that *an explicit risk management framework exists to accept and weigh new information in a timely fashion.* To use new information, resource management organizations must have a basis for acknowledging and incorporating uncertainty into their decision-making processes (e.g., Starr 1985; Stout and Streeter 1992). Often, however, new information is treated as antithetical to carrying out the management or planning guidelines already chosen, and it is used only under duress of litigation or during intermittent updates to planning in predefined planning cycles (Clark et al. 1996). For federal resource agencies in the United States, such planning cycles run years or decades long.

Another corollary to this tenet is that *performance standards for managers or decision makers should explicitly address environmental conservation.* Here is the heart of the problem. If individual managers or decision makers are not held individually responsible for meeting the objectives of adaptive management (and wildlife habitat management), then new information will not be used and management will not change. In this context, environmental conservation refers to wise sustainable use; one facet of this may, but will not necessarily, include strict preservation of habitats or environments.

An example appeared with a scientific panel report on the management of habitat for the northern spotted owl (*Strix occidentalis caurina*) in the Pacific Northwest of the United States. Dawson et al. (1987) posited that the lack of performance standards for federal agency decision makers clearly defining their responsibilities in habitat management for the subspecies would ultimately be the cause of failure of any spotted owl habitat management plan. Indeed, since this monition, federal agencies have conducted several additional scientific assessments of the species and have issued at least three environmental impact reports, using rather massive amounts of expertise and agency resources in the absence of such performance standards.

A third corollary is that *what is "at risk" is the condition of the land, not the status of one's career or the political decision space of the management directives.* Researchers of wildlife risk analysis and risk management typically assume that the very concept of what is at risk is shared by biologists, managers, decision makers, and politicians alike (e.g., Graham et al. 1991; Maguire 1986, 1991a, b). In reality, this is far from the truth. Biologists explicitly speak of risk in the sense of the likelihood of species extirpation. Managers and decision makers may implicitly use the concept of risk but more in reference to how a decision may impact their career status or the meeting of overall organizational directives, which may extend beyond species or habitat objectives. Upper-management decision-makers and politicians may view risk in terms of the political decision space—what is politically acceptable to their peers, to special interest groups, and to the funding or voting constituency. Each of these aspects of risk is authentic but quite different. Only through clearly specified performance standards for meeting adaptive management goals can specialists, managers, and decision makers be assured of viewing the concept of risk in the same way. In the absence of such standards, the next best approach is to articulate clearly the bases for risk in a given assessment or decision so that it is clear to all how the term and concepts were defined and weighed in any decision affecting public lands and wildlife.

Use of decision modeling techniques can aid managers in choosing an optimal course of action and in articulating their decision criteria. For example, the decision model of McNay et al. (1987) provided a means of prioritizing management for deer populations in coastal British Columbia, Canada, and making explicit the environmental and management conditions that form the basis for the priorities. However, such models are of limited utility if the decision maker does not wish to follow such a rigorous procedure or to expose his or her decision-making criteria because of political risk.

2. Options for change exist. This may seem to be an obvious point, but it is often overlooked or deliberately not addressed. In some cases, extensive funding has been provided for research or monitoring while options vanish for changing conditions on the land, such as through protection or restoration of dwindling habitats. One example may be that of the Mt. Graham Red squirrel (*Tamiasciuris hudsonicus grahamensis*), a potentially threatened subspecies occurring on a mountaintop in the southwestern United States that is also coveted as a development site for an astronomical observatory (Warshall 1995). In this example, ongoing population studies are being conducted during initial development of the observatory site. The development is designed to retain options for more- or less-restrictive habitat conservation depending on the findings, although the degree of risk to the population is still subject to some debate. In other cases, however, monitoring may proceed while scarce habitat continues to be changed, such as monitoring populations of the endangered Lanyu scops owl (*Otus elegans botolensis*) (Severinghaus 1992) while tropical forests there are felled or converted to other uses.

An adaptive management approach needs to determine the rate of change in habitats, species, or populations of dire concern, as compared with the pace of the information-gathering and decision-making or decision-changing process, to ensure that critical conservation options are not lost during the information-gathering process. More fundamentally, monitoring and research studies should not be used in place of making difficult decisions about allocation of scarce resources, or as a smokescreen to permit the continuation of management activities that eliminate options for conservation of a scarce and dwindling resource. Monitoring, research, and funding support *but do not substitute for* sound resource management decisions and actions.

A corollary to this tenet is that *irreversible losses of resources or environmental conditions should not be incurred during the "testing period."* For example, monitoring population dynamics or demography of an endangered species while its habitat continues to be adversely altered violates this corollary and ensures serious problems for meeting conservation goals. In some cases, adaptive management experiments can deliberately sacrifice or seriously alter some environmental conditions for a habitat or species of interest, in the name of quantifying effects of management activities. But the pace of such experiments and the degree of reversibility of losses must still permit options for changing management activities, such as for *in situ* protection or recovery of threatened species.

Another corollary to this tenet is that *changes in environmental conditions from human activities together with those from natural disturbances should not outstrip the pace of monitoring and learning and the potential to change management activities.* In other words, adaptive management approaches should attend to the additive effects of both human activities (ongoing management or new experiments) and natural disturbances, and therefrom gauge likelihoods of being able to change activities in time to ensure meeting conservation objectives. For example, if management experiments are testing the effect of various degrees of draining of some rare wetland type over a specified time period, the likelihood of environmental catastrophes such as prolonged drought over the same time period should be factored into the experiment to help ensure that options for change still exist at the end of the experiment.

3. Indicator variables—environmental parameters—can indeed be identified and realistically monitored in a cost-effective way. Many adaptive management studies can be thought up but cannot be realistically carried out. One reason for this is that indicator variables cannot be readily identified or measured; another is that the experiment is so complex or costly it cannot be completed with adequate sample sizes or intensity of study. This is often the case, for example, in studies of the response of carnivore populations to management activities or studies of ecological processes across entire landscapes that cannot be replicated. If this proves to be the case, then the directives for management and the focus for associated adaptive management studies need to be more tightly specified and new statistical approaches considered (e.g., Reckhow 1990).

Another corollary to this tenet is that *a statistical sampling frame should be established by which to distinguish effects of human activities from background changes.* One premise of adaptive management is that we can distinguish changes caused by human activities from those caused by natural variation or background noise. We can do so by using carefully designed manipulation experiments—less often by using "natural experiments" or passive observation studies. Chapter 4 discusses appropriate study designs. This is vital for determining when we can and cannot effect change through management activities. Adaptive management studies need to pay close attention to sampling design, to the use of controls and treatments, and to ensuring specific confidence and power levels in statistical tests.

Often, conservation questions pertain to conditions that occur across very large landscape areas and thus in contexts that cannot be replicated with adequate controls and in sufficient number to meet the assumptions of traditional statistical techniques. In such cases, we must look to other statistical designs, including use of comparative time series and spectral analysis, empirical Bayesian statistics, and optimization approaches (Walters and Hilborn 1978; Walters and Holling 1990; Williams et al. 1996; Reckhow 1990), although some of these approaches are controversial. In some cases, entire landscapes can be devoted to demonstration experiments (Franklin 1994). We advocate that a plurality of approaches be taken to provide the widest possible means of learning when unique conditions violate assumptions of traditional statistics.

A third corollary to this tenet is that *objectives and expected effects should be clearly articulated and quantified.* This may seem an obvious necessity, but it is often overlooked, particularly in observation (nonexperimental) studies that result in post hoc "fishing expeditions" for patterns and management effects and in demonstration studies designed mostly to justify and conduct *a priori* desired activities.

Overall, we make the following suggestions to help ensure a successful approach to adaptive management of habitats for wildlife:

First, review the above tenets and their corollaries as a checklist.

Second, clearly separate risk analysis from risk management. Risk analysis should entail estimating (and partitioning) likelihoods of outcomes resulting from potential management actions and from natural conditions and changes ("chance events"). Risk management should entail articulating criteria used for reaching a management decision, including explicating risk attitudes in light of uncertain projections and incomplete information.

Third, in the future we will need to learn from all kinds of information-gathering approaches, including observational or correlative studies, controlled field experiments, laboratory experiments, uncontrolled field trials, and anecdotal experience, as well as theoretical models. All these approaches can be useful and should be complementary in an adaptive management framework. The challenge is in appropriately combining information gathered from disparate studies and approaches for an overall understanding of the real-world system.

And fourth, technical staffs can help managers interpret and understand ramifications of uncertainties and probabilities of outcomes. Many resource decision makers may not be particularly adept in dealing with scientific uncertainty (Policansky 1993). For example, they may view scientific uncertainty as lack of "proof" of any particular effect, so that it can be discounted and ignored in, or even touted as supporting evidence for, management decisions. This is not an appropriate interpretation of scientific uncertainty. And it is not a correct understanding of the scientific process; that is, to "verify" a hypothesis, we do not seek proof but rather corroboration or, more accurately, lack of statistical falsification.

Also—and this is greatly misunderstood by many policy makers—uncertainty is not the same as complete lack of knowledge. An uncertain outcome can be expressed as a low likelihood or a high variation in potential outcomes, whereas lack of knowledge simply means we do not know and cannot estimate outcome likelihoods. That is, lack of knowledge is not the same as high variance. Ecosystems and their component communities, populations, processes, and disturbance regimes can be highly variable through space and time, and the variance can be precisely measured. In a sense, we can be certain of high variance in some systems. What may be uncertain is a specific future population level or resource productivity level (e.g., the population "standing crop" of some big game species in a specific future decade). Even this might be expressed as an expected level with some degree of associated variance (such as the mean herd size of harvestable bull elk plus or minus a standard error of prediction). We can do better to express inventories on the basis of samples and predictions on the basis of projections, as likelihoods and variances, and to help managers and decision makers better understand and interpret uncertainty and, where appropriate, lack of knowledge.

Real-World Circumstances and Use of Complementary Studies

Ideally, adaptive management would proceed as rigorously defined experiments adhering to all the assumptions and tenets of well-designed scientific studies (chapter 4). In reality, adaptive management studies have to contend with a number of problems, including:

- landscape-scale studies with few or, more often, no replicates
- "experiments" with no controls or with controls in vastly different situations (such as higher-elevation wilderness areas)
- little or no time to collect baseline data
- loss of selected samples and declining sample sizes over the course of the study because of changes in administrative or management direction
- unannounced and undirected treatment of controls
- overall short duration of the study with few truly long-term studies to determine lag and secondary effects
- changes in management objectives, treatments, and sometimes even land ownership over the course of the study

What can be done in the face of such ruinous circumstances? The answers may be found in using multiple studies of various kinds and in taking advantage of prior knowledge to establish study objectives, management activities to test, and analysis of results. In some cases, the real-world circumstances listed above degrade an otherwise rigorously defined experiment to the status of an observational or demonstration study. Such studies still have value in incrementally adding evidence to help corroborate or refute management hypotheses. But outcomes of observational or demonstration studies should not be taken as hard evidence of the correctness of assumptions underlying a management approach in the absence of supporting rigorous investigation. It is just too easy to find situations and to craft (inadvertently or otherwise) observational or demonstration studies to provide specific answers regardless of actual effects. An example is locating species thought to be closely associated with specific environments in situations other than what would truly be the modal condition of the population, such as observing a few errant spotted owls in very young-growth landscapes. This provides no "proof" (or corroboration) of habitat requirement.

Another kind of investigation that can greatly complement the adaptive management approach is that of retrospective studies. The use of retrospective studies is often of great value for understanding the developmental history and past conditions of habitats and wildlife. Retrospective studies are not experiments; rather, they are hindcast reconstructions of prehistoric or historic events and conditions to help us better understand long-term or recent change. Retrospective studies can borrow from a variety of tools and techniques, including dendrochronology, palynology, paleoecology, archaeology, analysis of historic documents, use of historic photopoints, and other information sources or methods.

Retrospective studies can help inform us on long-term changes in environments, climates, and biota and complex effects of land management actions. For example, Crumley (1993:377) provided examples of retrospective studies to unravel "complex chains of mutual causation in human-environment relations" by tracing past human-environment interactions on global, regional, and local scales (compare this with the discussion on scales of ecological disciplines and management issues presented in chapter 8). Covington et al. (1994) analyzed historic changes in forest ecosystems in the U.S. inland West to help project future changes. Retrospective studies also have been useful in documenting patterns and potential reasons for declines in songbirds (e.g., Briggs and Criswell 1978) and in assessing forest health and insect pest irruptions (Harvey 1994) and conditions of biodiversity (McNeely 1994).

Closing Remark

How vertebrate-centric is the traditional study of wildlife, including many of the examples in this book! Had humans evolved as poikilotherms, or as invertebrates, or in an aquatic environment, or as clonal organisms, or as canopy-dwelling or flying creatures, our definitions of wildlife for scientific, management, political, legal, ethical, and aesthetic interests would doubtless be far different from what they are. We urge our community of wildlife scientists and managers to reexamine our species and life-form biases and open the doors to all organisms and environments in our work. In this way, we envision a future era of wildlife management that weighs equally, in studies as well as in management, the host of mostly unknown creatures, both for their own sake and for their influence on vertebrates. These largely unknown creatures include soil, aquatic, and canopy microbes and mesoinvertebrates, and cryptogams and vascular plants. Collectively, they are critical to the *Umwelt* and health of vertebrates and to the productivity of crops, forests, and grasslands, which we hope to sustain.

Literature Cited

Allen, T. F. H., and T. W. Hoekstra. 1984. Nested and non-nested hierarchies: A significant distinction for ecological systems. In *Proceedings of the Conference of the Society for General Systems Research, Vol. 1: Systems methodologies and isomorphies,* ed. A. W. Smith, 175–80. N.P.: Intersystems Publications.

Balcom, B. J., and R. H. Yahner. 1996. Microhabitat and landscape characteristics associated with the threatened Allegheny woodrat. *Conservation Biology* 10(2):515–25.

Barrett, R. H., and H. Salwasser. 1982. Adaptive management of timber and wildlife habitat using DYNAST and wildlife-habitat relationships models. Presented at the Joint Annual Conference of Western Association of Fish and Wildlife Agencies and the Western Division of the American Fish Society at Las Vegas, Nev., 21 July 1982.

Briggs, S. A., and J. H. Criswell. 1978. Gradual silencing of spring in Washington: Selective reduction of species of birds found in three woodland areas over the past 30 years. *Atlantic Naturalist* 32:19–26.

Clark, T. W., R. P. Reading, and A. L. Clarke. 1996. *Endangered species recovery: Finding the lessons, improving the process.* Covelo, Calif.: Island Press.

Covington, W. W., R. L. Everett, R. Steele, L. L. Irwin, T. A. Daer, and A. N. D. Auclair. 1994. Historical and anticipated changes in forest ecosystems of the inland west of the United States. In *Assessing forest ecosystem health in the inland West,* ed. R. N. Sampson, D. L. Adams, and M. J. Enzer, 13–63. New York: Haworth Press.

Crumley, C. L. 1993. Analyzing historic ecotonal shifts. *Ecological Applications* 3(3):377–84.

Dawson, W. R., J. D. Ligon, J. R. Murphy, J. P. Myers, D. Simberloff, and J. Verner. 1987. Report of the scientific advisory panel on the spotted owl. *Condor* 89:205–29.

Dueser, R. D., and H. H. Shugart. 1978. Microhabitats in a forest-floor small mammal fauna. *Ecology* 59:89–98.

Franklin, J. F. 1993. Preserving biodiversity: Species, ecosystems, or landscapes? *Ecological Applications* 3(2):202–5.

Franklin, J. F. 1994. Adaptive management areas. *Journal of Forestry* 92(4):50.

Graham, R. L., C. T. Hunsaker, R. V. O'Neill, and B. L. Jackson. 1991. Ecological risk assessment at the regional scale. *Ecological Applications* 1(2):196–206.

Hagan, J. M., W. M. Vander Haegen, and P. S. McKinley. 1996. The early development of forest fragmentation effects on birds. *Conservation Biology* 10(1):188–202.

Harvey, A. E. 1994. Integrated roles for insects, diseases and decomposers in fire dominated forests of the inland western United States: Past, present and future forest health. In *Assessing forest ecosystem health in the inland west*, ed. R. N. Sampson, D. L. Adams, and M. J. Enzer, 211–20. New York: Haworth Press.

Hilborn, R. 1992. Can fisheries learn from experience? *Fisheries* 17(4):6–14.

Holling, C. S. 1978. *Adaptive environmental assessment and management.* New York: John Wiley and Sons.

Hunter, M. L. 1991. Coping with ignorance: The coarse-filter strategy for maintaining biodiversity. In *Balancing on the brink of extinction*, ed. K. A. Kohm 266–81. Covelo, Calif.: Island Press.

Irwin, L. L., and T. B. Wigley. 1993. Toward an experimental basis for protecting forest wildlife. *Ecological Applications* 3(2):213–17.

Keddy, P. A., H. T. Lee, and I. C. Wisheu. 1993. Choosing indicators of ecosystem integrity: Wetlands as a model system. In *Ecological integrity and the management of ecosystems*, ed. S. Woodley, J. Kay, and G. Francis, 61–79. Delray Beach, Fla.: St. Lucie Press.

King, B., and B. Warren. 1981. *Endangered birds of the world—the ICBP Bird Red Data Book.* Washington, D.C.: Smithsonian Institution Press.

Kolasa, J. 1989. Ecological systems in hierarchical perspective: Breaks in community structure and other consequences. *Ecology* 70:36–47.

Lance, G. N., and W. T. Williams. 1967. A general theory of classificatory sorting strategies. 1. Hierarchial systems. *Computing Journal* 9:373–80.

Landres. P. B., J. Verner, and J. W. Thomas. 1988. Ecological uses of vertebrate indicator species: A critique. *Conservation Biology* 2:316–28.

Lay, D. W. 1938. How valuable are woodland clearings to wildlife? *Wilson Bulletin* 50:254–56.

Lee, K. N. 1993. *Compass and gyroscope: Integrating science and politics for the environment.* Washington, D.C.: Island Press.

Lee, K. N., and J. Lawrence. 1986. Adaptive management: Learning from the Columbia River Basin Fish and Wildlife Program. *Environmental Law* 16:431–60.

Leopold, A. 1933. *Game management.* New York: Charles Scribner's Sons.

McNay, R. S., R. E. Page, and A. Campbell. 1987. Application of expert-based decision models to promote integrated management of forests. *Transactions of the North American Wildlife Natural Resources Conference* 52:82–91.

McNeely, J. A. 1994. Lessons from the past: Forests and biodiversity. *Biodiversity and Conservation* 3:3–20.

McNeely, J. A., K. R. Miller, W. V. Reid, R. A. Mittermeier, and T. B. Werner. 1990. *Conserving the world's biological diversity.* IUCN, the World Bank, World Resources Institute, Conservation International, and World Wildlife Fund, Gland, Switzerland, and Washington, D.C. 193 pp.

Maguire, L. A. 1986. Using decision analysis to manage endangered species populations. *Journal of Environmental Management* 22:345–60.

Maguire, L. A. 1991a. Decision analysis and environmental dispute resolution: Partners in resolving resource management conflicts. Presented at the Society for Risk Analysis Annual Meeting, 8–11 December 1991, Baltimore, Md. Abstracts distributed by the U.S. Department of Agriculture, Beltsville, Md.

Maguire, L. A. 1991b. Risk analysis for conservation biologists. *Conservation Biology* 5(1):123–25.

Marcot, B. G. 1997. Biodiversity of old forests of the west: A lesson from our elders. In *Creating a forestry for the 21st century: The science of ecosystem management,* ed. K. A. Kohm and J. F. Franklin, 88–105. Washington, D.C.: Island Press.

Marcot, B. G., M. A. Castellano, J. A. Christy, L. K. Croft, J. F. Lehmkuhl, R. H. Naney, R. E. Rosentreter, R. E. Sandquist, and E. Zieroth. 1997. Terrestrial ecology assessment. In *An assessment of ecosystem components in the interior Columbia Basin and portions of the Klamath and Great Basins,* ed. T. M. Quigley, S. J. Arbelbide, and S. F. McCool, 1497–1713. USDA Forest Service General Technical Report PNW-GTR–405. Pacific Northwest Research Station, Portland, Oreg. 1713 pp.

Morgan, P., G. H. Aplet, J. B. Haufler, H. C. Humphries, M. M. Moore, and W. D. Wilson. 1994. Historical range of variability: A useful tool for evaluating ecosystem change. In *Assessing forest ecosystem health in the inland West,* ed. R. N. Sampson, D. L. Adams, and M. J. Enzer, 87–111. New York: Haworth Press.

Orians, G. H. 1993. Endangered at what level? *Ecological Applications* 3:206–8.

Paszkowski, C. A. 1984. Macrohabitat use, microhabitat use, and foraging behavior of the hermit thrush and veery in a northern Wisconsin forest. *Wilson Bulletin* 96:286–92.

Peters, R. H. 1991. *A critique for ecology.* Cambridge: Cambridge University Press.

Policansky, D. 1993. Uncertainty, knowledge, and resource management. *Ecological Applications* 3(4):583–84.

Reckhow, K. H. 1990. Bayesian inference in non-replicated ecological studies. *Ecology* 71:2053–59.

Regier, H. A. 1978. *A balanced science of renewable resources.* Washington Sea Grant Publication. University of Washington, Seattle.

Severinghaus, L. L. 1992. Monitoring the population of the endangered Lanyu scops owl (*Otus elegans botolensis*). In *Wildlife 2001: Populations,* ed. D. McCullough and R. H. Barrett, 790–802. London and New York: Elevier Applied Science.

Sidle, W. B., and L. H. Suring. 1986. *Wildlife and fisheries habitat management notes: Management indicator species for the national forest lands in Alaska.* USDA Forest Service Alaska Region Technical Publication R10-TP-2. 62 pp.

Starr, C. 1985. Risk management, assessment, and acceptability. *Risk Analysis* 5:97–102.

Stout, D. J., and R. A. Streeter. 1992. Ecological risk assessment: Its role in risk management. *Environmental Professional* 14(3):197–203.

Swanson, F. J., J. A. Jones, D. O. Wallin, and J. H. Cissel. 1994. Natural variability—implications for ecosystem management. In *Eastside forest ecosystem health assessment,* Vol. 2: *Ecosystem management: Principles and applications,* ed. M. E. Jensen and P. S. Bourgeron, 80–94. USDA Forest Service General Technical Report PNW-GTR-318. Portland, Oreg. 376 pp.

Tracy, C. R., and P. F. Brussard. 1994. Preserving biodiversity: Species in landscapes. *Ecological Applications* 4:205–7.

Walters, C. 1986. *Adaptive management of renewable resources.* New York: Macmillan Publishing.

Walters, C., and R. Hilborn. 1978. Ecological optimization and adaptive management. *Annual Review of Ecology and Systematics* 9:157–88.

Walters, C. J., and C. S. Holling. 1990. Large-scale management experiments and learning by doing. *Ecology* 71:2060–68.

Warshall, P. 1995. The biopolitics of the Mt. Graham red squirrel (*Tamiasciuris hudsonicus grahamensis*). *Conservation Biology* 8(4):977–88.

Williams, B. K., F. A. Johnson, and K. Wilkins. 1996. Uncertainty and the adaptive management of waterfowl harvests. *Journal of Wildlife Management* 60(2):223–32.

12 The Future: New Initiatives and Advancing Education

Introduction

In our book we have reviewed the "state of the art" of wildlife-habitat relationships, emphasizing the need to conduct all studies within a clear conceptual framework and with the necessary sampling rigor. But we live in a time of rapid technological advancements, including miniaturization of monitoring devices, growth in our ability to study genetics and physiology, computers that allow us to construct detailed population models, and so forth. These advances are occurring within the context of a rapidly growing human population that is exerting both direct and indirect pressures on the land and its inhabitants. Concomitant with these increasing human pressures is an increase in extinctions and worries about human-induced changes in the environment.

These changes present us with several closely related challenges. Wildlife scientists are being asked to conduct studies that more directly address large-scale, land-use problems. This means that we must study and understand ecological relationships both within and among a variety of scales. Further, we must be able to keep abreast of the latest technological advances and understand the latest in computer software and modeling capabilities. These requirements necessitate a higher level of education in ecology and mathematics-statistics than previously seen in the wildlife profession.

In this chapter we want to discuss several problems that will be confronting wildlife scientists in the coming years and the types of information we will need to address them adequately. We will not attempt to solve these problems here; that would take another book and a lot of luck. Rather, we simply wish to highlight areas that need study, mention some of the pitfalls inherent in their study, and suggest some directions for further work. Because this is a book on habitat relationships, we will focus on this area of research, although much of our discussion has broader applicability.

Scale of Conservation

A substantial change in the spatial scale at which field scientists conduct their studies occurred during the 1980s. The historical concentration on single-season, site-specific studies conducted on the microhabitat scale began to give way to multiseasonal, multisite studies that examined ecological relationships on broader spatial scales. These shifts were driven by both ecological and practical considerations. Ecologically, scientists began to understand better the general factors that drive habitat selection in terrestrial vertebrates, that is, the hierarchic nature of habitat selection as developed by Johnson (1980) and Hutto (1985) (see chapter 7). In addition, people began to see the mistakes that had been made in previous studies of habitat, including gross mismatching of spatial scales within a single analysis (Wiens 1989:227–33; see also chapter 7). At the same time and from a practical standpoint, land managers were becoming increasingly frustrated with the lack of guidelines available from field scientists for application to management of properties under their jurisdiction. In particular, land managers were asking for studies that considered multiple species on larger spatial scales than had been historically conducted. Pressures from an ever-expanding human population were increasing the need to view land management on larger, multispecies scales because of limited land areas available for conservation.

Changes in the scales on which studies are being conducted are also a reflection of the natural progression of our knowledge. That is, the study of the ecology of animals usually begins with species-specific, detailed studies of life history parameters. Truly, nothing replaces basic field zoology. Remember that studies of animals in North America began in earnest only over the past 100 years or so. As our knowledge of animals grows, so does our ability to study interactions between them along with the host of factors influencing their survival and behavior. As such, eventually it becomes necessary to begin increasing the spatial scale on which we view an animal.

Thus, what we see is a merging of both ecological interest and knowledge with a practical need for solving environmental problems en masse. Relative to our recent past (since the 1970s), a plethora of papers has appeared on "landscape level" analyses (figure 12.1). In fact, the need for analyses of large geographic areas was in part responsible for the formation of the Society for Conservation Biology and initiation of the journals *Conservation Biology* and *Landscape Ecology*.

Regardless of our need for such analyses, we must ask, Is our knowledge of spatial relationships adequate to meet the conservation challenge? Below we review the current thoughts on this question and outline steps we can take to advance our knowledge in this arena of study.

As noted above, researchers have been turning increasingly to studies at the landscape and ecosystem levels. "Landscape management," which is a reflection of our new interest in expanding our spatial scale of analysis, was discussed in chapter 9. Of particular interest among both scientists and land managers is the controversial area of "ecosystem management." In chapter 11, we presented an approach for studying wildlife in an ecosystem context. Certainly, additional focus on disturbance dynamics and habitat configuration (chapter 9) is needed to augment autecological studies.

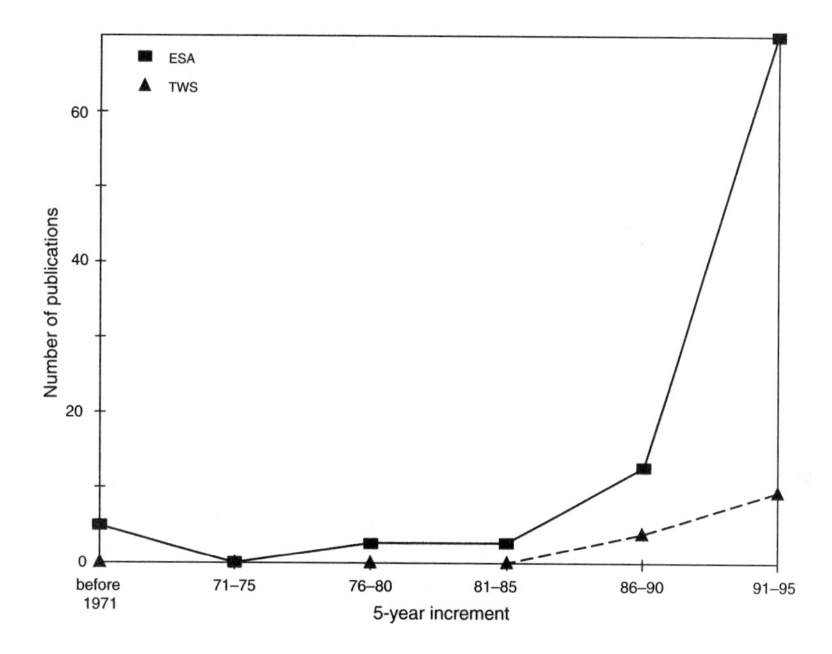

Figure 12.1. The growth of landscape studies in American ecological and wildlife literature. Plotted are the number of publications of the Ecological Society of America and the Wildlife Society that have the word *landscape* in their title or listed as a key word.

Focused Initiatives in Habitat Research

In this section we present some thoughts on how we might want to pursue studies of habitat relationships. There are many approaches to such efforts; we certainly do not present our ideas as a prescription for curing all the ills of wildlife research. Our ideas are primarily a synthesis of the ideas of other thinkers; we trust we have captured their thoughts correctly. We hope that our ideas stimulate further discussion on how we can advance the study of habitat relationships.

Habitat Selection: An Extensive Concept

Slobodkin (1992) made the critical distinction between extensive and intensive variables and noted the differing perceptions of them by individuals and the resulting impact on theory. Intensive variables are those that can be directly evaluated by an individual organism at any point in a system, whereas extensive variables can be evaluated only by considering an overview of an entire system. Systems here could be a study area, a community, the ecosystem, and the like. For example, Slobodkin argued that the extensive variable population size means less to an individual in the population than do the intensive variables that are directly associated with food. In fact, Johnson (1980) implied this relationship when he noted that habitat usage studies and studies of feeding are of different orders. Food is of a lower order than (or is a subset of) habitat and, as such, is a relatively more intensive variable. These comments relate directly to our call above for hierarchic studies of habitat use.

Failure to realize the distinction between extensive and intensive variables results in models that may have little direct connection with how animals perceive their surroundings and, thus, may have little predictive value. That is, by focusing on extensive variables, we are asking individuals "to do impossible things" (Slobodkin 1992). Given that vegetation and other geographic features are not randomly distributed, some heterogeneity will always exist within the range of all animals. If we then simply draw various boundaries around their range, we will proclaim that animal distribution is related to scale; this is an extensive approach, and it is trivial in many regards. These comments are not criticism of Johnson's (1980) framework but, rather, are directed at the ways in which animals are studied. Earlier (in chapter 7) we discussed the major conceptual and practical differences between studies on relatively microhabitat versus macrohabitat scales and the need to avoid mismatching these and other scales.

Active, hierarchic habitat selection implies that fitness is affected at each level. Must an organism, then, be able to perceive the effect that each level has on fitness for an area to be occupied, or is there some summation process going on? "Macrohabitat" or "landscape" patterns may be only the accumulated picture of many individuals selecting resources on the smallest scale. It does not mean that individual animals recognize distributions of resources or constraints on population scales or, if they do, that these are of primary importance to the individual. The analogy is the operation of natural selection itself: natural selection operates on individuals, or more specifically on variation among individuals, and the accumulated picture of this action is expressed in resultant population parameters.

Developing theories based on assumptions, constraints, and variables that cannot in principle be sensed by individuals perpetuates theoretical incompleteness, and thus stymies advancement. Witness the approach taken in applied studies of habitat ecology: Rather than studies being framed in terms of testable theory and then designed to understand the process of selection, it has become standard practice to beg the issue and study habitat "use." Here, *use* does not imply any active process, any perception by the animal, or any consequences for survival and fitness (e.g., Morse 1980:89–90; Hutto 1985; Hall et al. 1997). Asking what habitat type a species uses is neither a "how" nor a "why" question (Gavin 1991). Hobbs and Hanley (1990) showed that measures of habitat use–availability reveal little about the value of the habitat unless the underlying resource distributions are understood mechanistically. In fact, such a mechanistic understanding negates the need for indices in any case. Searching for ultimate causation should help end the current reliance on proximate supposition.

Studies of density-dependent habitat selection are framed in the context of searching for the role of competition in structuring communities. "Selection" is seen when animals are rare and thus are not likely to be competing for resources. Animals will, therefore, actually appear *not* to be selecting habitat from time to time (Owen-Smith 1989; Rosenzweig 1989; Morris 1992). Similarly, applied studies of habitat usually equate selection with nonrandom use of some item, often vegetation (e.g., see reviews by Johnson 1980; Thomas and Taylor 1990; Alldredge and Ratti 1992). Again, animals can be nonselective under such designs.

We argue, however, that a more consistent and reasonable approach is to view habitat selection as *always* occurring (it is a process; Hall et al., 1997), as a fundamental expression on the scale of the individual, and as something to be viewed in terms of intensive variables. This, at least, establishes a more consistent framework because of its more direct relation to fundamental, mechanistic currencies. We need to separate clearly the application of extensive studies of population phenomena (e.g., density-dependency, animal-vegetation relationships) from intensive studies of resource acquisition.

Thus, we are not arguing against trying to understand the interactions going on within some larger spatial area, or ecosystem. Rather, our point is that different questions are best answered on different scales. In addition, many studies would likely need to incorporate both intensive and extensive variables. For example, although determining a species' food requirements (intensive) is often necessary, the availability of food might be determined by intra- or interspecific population densities (extensive). Our argument is that extreme care must be taken when choosing variables, care which includes the knowledge of how animals perceive their environment; experimentation will likely be warranted in many cases.

Habitat and the Niche

Interrelated and central to our discussion of habitat is the concept of the niche. Like habitat, the concept of the niche has been the subject of much debate, and certainly falls under the same criticisms noted above by Peters (1991). Much of the problem with the study of the niche is the lack of an accepted operational definition. Arthur (1987) overviewed the niche concept and noted the gamut of definitions ranging from Grinnell's (1917) description of niche as the ultimate distributional unit, to Elton's (1927) primarily behavioral concept, to Hutchinson's (1978) *n*-dimensional construct (for behavior and distribution). Leibold (1995) expanded on these definitions and the historical development of the niche, concluding that Elton's view and that of MacArthur and Levins (1967) were oriented toward describing how an organism affects the environment by consuming resources and by serving as a resource for higher trophic levels, in contrast with Grinnell's view of the niche, in which he emphasized the environmental requirements of the organism.

Arthur (1987) concluded that, to be useful, the accepted concept must be simple and quantifiable. As such, he chose to use MacArthur's (1968) description of the niche, which plots utilization against some quantifiable resource variable—the resource utilization function (RUF). The hypervolume is an "infinitely large set of properties" that cannot be operationalized (Peters 1991:91). Arthur argued that it is better to build complexity as needed, as with RUFs, than to dissect it as necessary when using a hypervolume. This rationale appears to meet Peters', Romesburg's (1981), and Slobodkin's (1992) requirements for a testable theory of fundamental causation. In concept, RUFs describe the choice of resources by animals. Choices can be *constrained* by predators, competitors, and various other factors, all of which can be tested.

From this rationale, then, we should focus our studies on the level of what we classically call a resource axis of the niche. Habitat is largely attendant to the underlying reasons why an animal is present at a certain time and place, but is certainly a population phenomenon under most of our study designs. Animals thus *select habitat* on only the broadest geographic scales, and they *select resources* on the finest scales. We need to recognize that niche and habitat reside, respectively, at opposite ends of the intensive-extensive scale and to cast future studies with this in mind.

Thus, we suggest a focus on the basic, fundamental currency that allows individuals to survive and reproduce. This currency is popularly termed *resources*. Below we will develop some fundamental underpinnings of this theory and proceed with more specific examples. Perhaps this approach will help us to focus our efforts and lead to more rapid advancement of our understanding. Leibold (1995) distinguished between environmental requirements and environmental impacts of species, and related these to the "habitat" and "functional" aspects of the niche as developed by Grinnell and Elton, respectively. Leibold went on to develop a framework for merging these two aspects of the niche into a single concept that could be analyzed through various mechanistic models. Here we are focusing on the "habitat," or "environmental requirements," aspect of the niche, while acknowledging that Liebold presents a comprehensive framework for more advanced study.

Resources and Their Identification

Peters (1991:91) dismissed the niche as theory because there are likely many resources, which are usually defined relatively and thus are difficult to test. Arthur's (1987) approach, as outlined above, addresses much of this criticism by concentrating on RUFs. Thus, to be useful in advancing knowledge, the resource currency must have relevance to the fitness of the animal *and* be within their perceptive abilities.

Abrams (1988) recognized the critical importance of the proper identification of resources and noted the confusion that exists over their definition and enumeration. He identified two central issues: (1) the separation of entities that serve as resources versus those that do not; and (2) the determination of how a set of resource entities can be divided into distinct resources. Further, he developed the theory that consumption rates exist where per capita population growth is an increasing function of the rate of resource consumption. Various constraints act to reduce the rate of consumption and, thus, the rate of per capita population growth.

Abrams separated "entities" into resources according to these tenets: (1) the densities of the entities cannot be related by a function that is independent of the consumption rate of any type; and (2) the resource types must be identifiable by a consumer. Thus, resources cannot be counted separately from how consumers use them. For example, researchers should realize that a single biological *species* usually cannot serve as the "resource," because a species is composed of individuals with different behaviors (based on sex, age, life histories) and/or phenotypes, and thus a consumer may be selecting a specific subset of the species. Bird diets and foraging behavior vary widely, for example, depending upon the availability of various arthropod developmental stages (e.g., Wolda 1990). Likewise, there is no *a priori* basis for dividing chemical and physical properties into categories (based on size, shape, or content) without regard to how the consumer uses them. "Resources" can be documented only when consumption rates and subsequent effects on fitness are known.

What Should We Do?

Most of our advances in the study of wildlife-habitat relationships have involved refinements of analytical techniques (e.g., see reviews by Alldredge and Ratti 1986, 1992; Thomas and Taylor 1990), despite clear warnings that our fundamental approaches are flawed (e.g., Van Horne 1983; Hobbs and Hanley 1990). The analytic tool we select is of little importance if the question we ask is trivial or if the approach taken to address a meaningful question is flawed.

As noted by Arthur (1987) building complexity is conceptually easier than deciphering it. Thus, we suggest that workers concentrate on identifying and analyzing the separate roles of critical resources and the factor(s) constraining their use. We would rather learn about the animal's behavior along a single resource axis (intensive approach) than read yet another "community-level" study (extensive approach) that degenerates into a string of ad hoc suppositions about observed phenomena. This "bottom-up" approach will eventually lead, we think, to studies encompassing more complexity and, especially, food webs (see review by Hunter and Price 1992). This also supports our contention concerning the role that ecosystem processes might play in designing studies of habitat. Separation and quantification of resources and constraints have clear operational utility. Although complete operationalism is probably impossible, standardized operational definitions are essential if different people are to make similar measurements of similar entities (Peters 1991:77).

Perhaps we can divide our future approaches into several levels of increasing intensity and relatedness to fitness. There are several different levels to the study of biology. We must recognize the limitations of purely functional studies of wildlife and avoid unwarranted reliance on their results in basic and applied situations. But, at the same time, it is unreasonable to suppose that a researcher, especially a graduate student, will be able to complete the entire picture of an animal's relation to the environment, even over the short term. Instead, we think wildlife science would be well served if we were to narrow the scope of our questions, concentrate on even a single resource or constraint, and avoid top-down, tautological approaches to studying habitat, community, ecosystem, and other nebulous concepts per se. Extensive studies are worthwhile and necessary, but mostly in a screening context.

The resources that an animal *encounters* are the parameters in which we should be interested. Models of foraging, for example, view resource selection as a "search-encounter-decide" sequence. This means that we must know about the forager's sensory abilities if we are to identify properly when an encounter occurs, which is a necessary step in identifying what a resource is to an animal. It is certainly not enough simply to measure "abundance" of some resource (Stephens and Krebs 1986:13; Hobbs and Hanley 1990). In fact, it is actually the rate at which an animal encounters resources, not the density (or abundance) of the resources, that is of importance. Density is determined by the *researcher,* who counts items and expresses them by some unit of area. Density is thus not necessarily biologically relevant. As such, we should not assume that density of a resource is an appropriate surrogate to the encounter rate of a resource. Studies of diet, food selection, and foraging behavior have a well-developed methodological base (e.g., see reviews in Morrison et al. 1990) and will be useful in the design of studies seeking to determine the fundamental mechanisms of survival and fitness.

Stephens and Krebs (1986:182) recommended that we must first clearly elucidate and separate resources and constraints. Their approach: If the interest is in determining the resource, then study systems with well-defined constraints; but if the interest is in constraints, then study systems in which the resources are well-known. Obviously, experimental designs will be necessary in many cases (see chapter 9).

Laboratory studies within the broad framework of physiological ecology will allow us to put bounds on the possible behaviors and physiological responses of animals to variations in resources and constraints. Below we list a few laboratory and field considerations that might help in placing those bounds. The study design used will depend upon the specific consideration in point and the cleverness of the researcher.

1. To see what currency gives the best accounting of an animal's behavior, make a comparison of behaviors with a range of resource choices presented to the animal. Foraging models dealing with risk-sensitivity are especially appropriate here. Different resource levels have been shown to lead to variation in behavior along a risk-averse to risk-prone continuum (Stephens and Krebs 1986). A general experimental design would determine prey selected at varying ratios of prey types and at varying renewal rates. Such a design could be used to determine if foraging is based on qualities of individual food types or the relative frequency of distribution (Real 1990).

2. To determine which constraints affect an animal most, make a comparison of behaviors within a range of constraints introduced into the animal's environment. Here, the different solutions the animal employs are reflected in different behaviors, resource uses, physiological correlations, and the like. Constraints need to be identified through experimentation: for example, switching behavior is a useful determinant of response (Real 1990). Further study would elucidate (1) changes in an animal's foraging place, (2) its ability to distinguish among different places, (3) an alteration of the search path, (4) changes in resource handling time, (5) its aversion to certain prey types, and (6) its use of specialized foraging techniques. Individual-based, spatially explicit models governed by "local rules" (i.e., based on factors sensed by the individual in some finite area around its current location) can also be developed (e.g., Johnson et al. 1992).

Research has focused on six broad categories of constraints: time, learning and memory, perception ability, genetic and developmental, and nutritional (Real 1990). Herbivores, for example, have notable constraints including mineral requirements, rumen capacity, and available foraging time (Belovsky 1981). A threat of predation can result in partial preferences when choosing among different prey types (e.g., Godin 1990), but this is rarely considered in our studies. Morphological (e.g., Winkler and Leisler 1985) and physiological (e.g., Walsberg 1985) limitations and innate behaviors (e.g., Klopfer and Ganzhorn 1985) are other examples of constraints known to set bounds on resource use, but are seldomly explicitly studied.

3. Develop bioenergetics models. The major physiological adaptations of an animal are related to water balance, thermal regulation, and nutritional intake. Each activity of an animal carries with it some net physiological consequence. Determining the quantity and quality of food eaten under varying abiotic (e.g., weather) and biotic (e.g., competition, cover) constraints will allow models of energy balance to be developed. Such models will lead directly to predictions concerning survival and fitness under varying conditions. Such studies will link the mechanisms regulating resource selection with the observed distribution of animals (e.g., Hobbs 1989).

For example, habitat use by moose (*Alces alces*) has been predicted with the use of linear programing that considers the caloric value of aquatic and terrestrial plants, the maintenance requirements of moose for survival, moose digestive limitations, and sodium requirements

(Belovsky 1978, 1981, 1984; see also Stephens and Krebs 1986:118–22 for review). Belovsky's work showed *why* moose use aquatic vegetation, as well as what is the correct amount needed in their diet for survival. This type of approach should be emphasized in future studies of wildlife-habitat relationships.

4. Conduct studies of habitat quality. Although the concept of density as a misleading indicator of habitat quality (*sensu* Van Horne 1983) is well known, few studies directly measure quality. Studies of the relationship between habitat characteristics, especially resources, and the survival of adults and their offspring are badly needed. This is what we were getting at in adducing "key environmental correlates" in chapter 11. Recent studies such as those by Henson and Cooper (1993), Loegering and Fraser (1995), and Paradis and Croset (1995) are good examples of the direction that future habitat studies of this type should take.

5. We need to determine the influence of study area size (the geographic-extent dimension of scale) on estimates of population abundance. As developed in chapter 7, tests of ecological theory, as well as applications in resource management and conservation, depend upon reliable estimates of numeric abundance (Smallwood and Schonewald 1996). However, virtually no researcher examines these potential bias when designing a study. Interesting tests would include sampling abundance over areas that are much larger than those usually used in ecological studies and then subsampling in various-sized grids from within that area. An example would be determining small mammal density in a 25-ha grid or spot-mapping birds in a 100-ha grid. Such efforts would involve many personnel and much equipment but could result in extremely useful data, depending of course on the study objectives.

6. In chapter 11 we discussed the need to place some studies of wildlife in an ecosystem context.

Summary

Wiens (1992) concluded that our study of ecology has been largely phenomenological: A pattern is observed and matched with the prediction of a theory that postulates a certain linkage between pattern and process. Here, while the pattern has been empirically determined, the underlying process is still largely inferential. Our theories of habitat selection, including questions of scale, certainly fall into this category. Such ad hoc hypotheses have little predictive power and tell us nothing about *why* the process has occurred. Thus, Wiens calls for a turn to mechanistic ecology, under which observed phenomena are considered in terms of their underlying causes or mechanisms. As Wiens noted, resources are the foundation of ecological processes ("you are what you eat") and, as such, are the basic mechanistic element of biology.

Our advocacy of a search for ecological mechanisms is not equivalent with reductionistic biology. We argue that we should assume simplicity, search for the underlying, dominant factors (resources) driving fitness, and determine the constraints on their use. As concluded by Schultz (1992), "Reductionist studies are needed to prevent obese generalizations. . . . Creative conceptual development is needed to provide appropriate context for mechanistic studies." Habitat as a concept is ingrained in both scientific and popular usage. But this fact is not justification for the study of habitat per se. There are different levels of inquiry, but answers to the fundamental "why" questions hold the most promise for scientific understanding and will, in the long run, provide the most generality.

Table 12.1. Desiderata for a more rigorous community ecology

1. Be more explicit about defining the "community" studied and justifying that definition.
2. Deemphasize community macroparameters and focus on individuals, especially aspects relating to energetics, density effects, and habitat selection.
3. Use resource-defined guilds as a framework for intensive comparative studies.
4. Consider both ecological and evolutionary constraints on community patterns.
5. Consider all life stages in community analyses, and evaluate the effects of community openness versus closure.
6. Conduct studies, interpret the results, and generalize from them within the appropriate domains of scales in space and time for the phenomena or biotas investigated.
7. Avoid thinking of communities as either in equilibrium or in nonequilibrium, but examine the dynamics and variability of community measures as features of interest in their own right.
8. Conduct long-term observational and experimental studies.
9. View communities in a landscape context, considering the effects of habitat-mosaic patterns and abandoning notions based on assumptions of spatial homogeneity.
10. Focus on the factors influencing community assembly as a conceptual framework for community studies.
11. Deal with the effects of multiple causes on community patterns.
12. Emphasize the importance of defining and measuring resources and testing the assumption of resource limitation.
13. Develop specific, mechanistically based theory.
14. Frame hypotheses in precise, testable terms whenever possible.
15. Take into account the effects of feedback relationships, indirect interactions, time lags, and nonlinear responses.
16. Avoid extrapolating from particular taxa or habitats to other taxa or habitats, and avoid especially a "north-temperate bias" in thinking about communities.
17. Recognize the importance of replication in both observational and experimental studies.
18. Do not shun or avoid criticism and controversy.

Source: From J. A. Wiens, *The Ecology of Bird Communities,* Vol. 2: *Processes and Variations* (Cambridge: Cambridge University Press © 1989), 258, table 6.1; reprinted with the permission of Cambridge University Press

Similarly, Wiens argued (point 2, 1989:257–59) that macroparameters such as species richness, diversity, and niche overlap should be deemphasized in favor of studies concentrating on expressing patterns in terms of physiology, behavior, and life-history traits. We might focus on attributes of species that influence their use of energy rather than on the effects of interactions on their population dynamics. He also argued (point 13, 1989:262) for development of specific, mechanistically based theories as an approach to advancing understanding of ecological relationships.

Wiens developed a list of 18 specific steps that avian ecologists might take to develop a more rigorous understanding of community ecology (table 12.1); his list can certainly be generalized to ecology as a whole. We find it satisfying that all of Wien's points have been discussed to some extent in our book—in particular, his call for more explicit definitions of terms (point 1) and resources (points 3, 12), the consideration of evolutionary as well as ecological constraints (point 4), the framing of studies in the appropriate spatiotemporal context (points 6, 9), and the use of rigorous designs and experimentation (points 8, 14, 15, 17).

Management Implications: Restoration Ecology

As human populations continue to increase in size and distribution over the landscape, there will be fewer opportunities to preserve existing areas in a relatively natural condition. Wildlife conservation will thus depend more and more on the modification of existing reserves, the management of lands between reserves, and the restoration of degraded environments. Although countless papers have been published on various species of wildlife and their habitats, neither wildlife ecologists nor restorationists have made much effort to apply this information systematically to the work of restoration.

Traditionally, wildlife biologists have not dealt with ecological restoration per se. Rather, they have concentrated on modifying vegetation and other environmental features for the benefit of specific species. Conservation biologists have been especially interested in community-level analyses and the development of landscape-level reserve designs, but they too have paid little attention to the details of restoration. Here, restorationists have been called on for assistance. As discussed by Jordan (1989), however, it has generally been assumed that ecologists are the source of information, ideas, and insight, whereas restorationists are the practitioners, the doers. Restoration projects are usually carried out for highly practical reasons in a nonacademic setting, so there has been a tendency to overlook the scientific value of the work. Thus, despite attempts to draw attention to its heuristic value (Jordan et al. 1987), restoration has not been recognized as a basic science. Restorationists have only recently started to consider their profession explicitly as a science, as indicated by the initiation of the journal *Restoration Ecology* in 1993. And the link between restoration and basic ecology remains weak.

What we want to discuss here is that this weak link is unfortunate and that we could do better with wildlife conservation if we could find a way to integrate more effectively the principles, experiences, and viewpoints of wildlife biology, conservation biology, and restoration. Certainly there are both empirical and academic grounds for such integration. Developing a restoration plan is, or ought to be, the same as developing a plan for an ecological study in wildlife biology or conservation biology (Morrison 1994). Both wildlife and conservation biology offer basic ideas and research techniques, while restoration expands the scope to include vegetation and whole systems, along with practical methods. In fact, the goal of restoration transcends the disciplines and might offer a way of integrating them. After all, one primary stated or implied goal of wildlife and conservation biologists is the restoration of something.

Wildlife biologists are, however, currently engaged in debate over what constitutes reliable knowledge and "good science" and how the profession can continue to grow and advance biological understanding (Romesberg 1981, 1991; Matter and Mannan 1989; Nudds and Morrison 1991; this topic is developed below in "Education for the Future"). Overall, we are getting better at designing and implementing studies of wildlife and wildlife-resource requirements, demographics, genetics, and so on. Our work usually ends, however, with a list of general "management implications"; we are good at telling resource managers what the goals should be, but this is seldom accompanied by instructions on how these goals can be reached or how to determine whether they *have* been reached. Unfortunately, the more academically oriented wildlife-conservation biologists often know little about such matters. Further, even if the plans are implemented, often very little postrestoration monitoring is attempted. Frankly, we get the distinct impression that many of our colleagues in wildlife biology feel that implementation of our recommendations is not our responsibility and that any failure to do so is the fault of "the managers." It is the managers who, in effect, really test ideas by putting them into practice and who are often in the best position to develop new ideas, to expose the weaknesses of existing ideas, and to redefine research priorities. However, wildlife conservation biologists can play more central roles by advising managers on risk analysis, aiding them with decisions based on risk management, and helping to craft integrated applications and monitoring studies in an adaptive management framework.

Clearly, restorationists have a crucial role to play here. This becomes especially clear when we consider specific questions that arise in the course of restoration efforts, whether these are primarily concerned with wildlife or whole systems. Morrison (1995) listed some of these and briefly discussed weaknesses in understanding that could be strengthened by closer collaboration between wildlife conservationists and restorationists.

Directly or indirectly, all restoration projects modify wildlife habitat to some degree. Restoration projects often fail for lack of a clear understanding of how wildlife-habitat relationships are developed. In fact, most reports of restoration projects either do not mention wildlife habitat at all, or if they do mention it, they do so only in the collective sense. For example, most of the articles in *Biological Habitat Reconstruction* (Buckley 1989), while describing the reestablishment of wildlife habitat as a primary goal of restoration, do not incorporate species-specific (or other specific) considerations of habitat. Clearly, the habit of loosely using the term *wildlife habitat,* a common practice in wildlife and conservation circles (see Hall et al. 1997), has been transferred to the restoration field.

The point here is that restoration provides unique opportunities to test and refine our ideas about these matters. The endangered least Bell's vireo (*Vireo bellii pusillus*), for example, has been induced to breed in several areas of southern California following implementation of a management plan that included the restoration of nesting and feeding sites and the reduction of populations of brown-headed cowbirds (*Molothrus ater*), a nest parasite (e.g., Franzreb 1989, 1990). Whether successful or not (sometimes, especially when *not* successful), well-designed restoration projects such as this one can teach us a lot about the critical parameters that define a particular community or ecosystem. The vireo example incorporates a lesson about both the resource needs of the species and how a nest parasite can negate the best-designed plans for restoration of the resources. It is also an example of how the study of intensive variables (nest and food requirements) can be set in the larger-scale context of the influence of extensive variables (cowbird density) on the species; moreover, it is an example of how manipulation of extensive variables alone cannot effect the recovery of the vireo population.

Wildlife habitats, especially those in regions undergoing development, often occur in a landscape largely dominated by human use.

Habitat loss in such a landscape is a complex process that typically separates populations into fragments with different kinds and levels of linkage between them. Since the characteristics of these fragments and linkages have a profound effect on the persistence of wildlife, landscape considerations are crucial to any kind of wildlife habitat restoration in such a setting (see chapters 8 and 9). For example, the influence of vegetation patch size and edge-to-interior area is poorly understood, even after decades of research by wildlife biologists (see review in Paton 1994). Here again, restoration provides opportunities to test ideas and deepen our understanding of these complex interactions, and this is another incentive for wildlife-conservation biologists and restorationists to work together more closely.

One of the most controversial topics in environmental law is the effect of specific human activities on wildlife species. The way that individual animals adjust their behavior to human influence or manipulation and how these adjustments influence population processes are poorly understood. Wildlife biologists have not worked extensively in this area, and conservation biologists, for their part, have tended to view human influences as a form of disruption that disorganizes systems to the point where studies are unproductive. Neither of these perspectives is appropriate now that virtually all systems are subject to some form of human influence. Though the topic of human influence on wildlife is beginning to receive more attention (see, e.g., Holthuijzen et al. 1990; Griffiths and Van Schaik 1993; Truett et al. 1994), this work still lacks a unified approach or theoretical basis.

Here again, restorationists clearly have an important contribution to make. They work extensively in human-dominated landscapes and have a wealth of knowledge concerning the development of plan communities under stress. Wildlife biologists, on the other hand, can contribute detailed knowledge of wildlife resource use based on work in relatively natural areas, an essential for the proper design of habitat restoration projects in impacted areas. Collaboration between these two groups can be extremely fruitful. In southern California, for example, wildlife biologists and restorationists assisted land-use planners in developing a plan to enhance the habitat of many animal species in several urban parks (Morrison, Scott, and Tennant 1994; Morrison, Tennant, and Scott 1994). These projects, once implemented, will serve as tests of the ideas about wildlife-habitat relationships on which the plans were based.

Jordan et al. (1987) saw the heuristic value of restoration efforts as the basis for an intimate, two-way relationship between ecological theory and ecological practice. We urge restorationists to increase communication with their sibling societies. There is no reason why restorationists must be tagged onto a project for the sole purpose of carrying out someone else's recommendations. The results of restoration efforts will be much enhanced by involvement of restorationists in all aspects of a project. They can play a crucial role in developing adaptive management experiments to help managers learn by trial. To participate effectively in such work, however, restorationists will need to expand their knowledge of study design, statistics, and sampling methodologies.

Translating Wildlife-Habitat Research into Management

Risser (1993) wrote a plea for theoretical ecologists to accept responsibility for clarifying the domain in which results can be used by those who have the responsibility for managing. He argued that it is unfair to publish papers in the conventional scientific format and then criticize managers for not finding the information and/or not using it correctly. He called for journals such as *Ecological Applications* to include in each article a closing section that describes results in ways that could be used by resource managers. Such a section is, of course, a standard feature in many applied ecological journals (most notably the *Journal of Wildlife Management* and *Wildlife Society Bulletin*). Even within the resource management community, critics have often called for a closer tie between wildlife researchers and managers (e.g., Baskett 1985).

However, as we discuss in the following section, we think that it is also the responsibility of the wildlife community to become educated in the theoretical underpinnings of the ecological systems within which we work. Although we are not calling for all wildlife biologists and managers to become theoreticians per se, it is nonetheless our responsibility to review and possess at least a basic understanding of and appreciation for the theories upon which our management decisions are or should be made. Although it may be, as Risser (1993) encouraged, the responsibility of theoreticians to place their work in an applied context, it is not their responsibility to educate wildlifers in the basics of theoretical ecology and the scientific method.

During the initial phases of study design, when the researcher is reviewing the literature and deciding on the variables to measure, thought should be given to how study results might be specifically used in management contexts and who would make such use. As developed in chapter 7, it is standard practice for researchers to divide vegetation measurements into extremely fine categories. For example, often, vegetation cover is divided into 1- or 2-m vertical intervals; dbh classes are made for every 10-cm increment and are further divided by numerous tree vigor classes; and ground cover is classified into many down wood decay classes, substrate particle size classes, and so forth. But the question seldom asked is, How will my results be translated into management practices? That is, how does a person effectively manage for vegetation cover in the 4–6-m height interval (or more generally, in the understory) or for small stones? Wildlife researchers are usually consumed by generating the highest possible R^2 or best P-value rather than by determining the best result in relation to possible implementation in the field. Further, land managers rightfully become frustrated when asked to understand and implement recommendations based on log(canopy cover) and $(\text{shrub cover})^3$, even though the researcher can be complimented for transforming nonlinear data (see chapter 7). Researchers should also translate their results into practical terms.

Risser (1993), using a paper by Painter and Belsky (1993), further exemplifies our point. The Painter and Belsky work discussed compensation-overcompensation in grassland ecosystems, but artificially eliminated from consideration the issues of changes in species composition and in nutritive values of the plants because there is ostensibly little debate on these topics among ecologists. Species composition and nutritive value are, however, central components in range managers' determination of grassland status. Risser noted that by arbitrarily dismissing these considerations, ecologists are reinforcing the perception that they are trying to set the conditions of discussion rather than trying to identify the practical considerations of the end users.

Although beyond our scope, greater emphasis on adaptive resource management (ARM)

(Walters 1986) should help researchers, land managers, and administrators to achieve their goals. ARM is done whenever the goals of achieving management objectives and gaining reliable knowledge are accomplished simultaneously. ARM provides the opportunity to test hypotheses on large geographic scales while allowing some level of management to proceed (Lancia et al. 1993). For example, models of wildlife-habitat relationships developed from descriptive studies can be simultaneously implemented and tested under ARM. In addition to C. J. Walters' (1986) text, students can review the articles published in a special section of the *Transactions of the North American Wildlife and Natural Resources Conference* (1993; see Lancia et al. 1993). We discussed some critical and practical tenets of adaptive management in chapter 11.

Education for the Future

University Education

"Our current understanding amounts to bungalows. To continue on the present course will only lead to better bungalows. . . . it is within our capacity to take a revolutionary course and develop understanding to the state of Taj Mahals" (Romesburg 1991). Romesburg used this analogy to introduce his argument that we are failing to advance our understanding of ecological relationships and thus are failing to advance wildlife management as rapidly as we could. Further, he argued strongly that we are failing to educate our students in a manner that will lead to such advancements in the future. Although not everyone agrees with his conclusions (Knight 1993; but see response by Romesburg 1993), we think there are certainly many truths in Romesberg's writings that we as professionals would be wise to follow. Romesberg is not alone, nor was he the first to raise serious questions regarding the rigor with which we educate our wildlife students (e.g., Gavin 1989; Hunter 1989).

Classically, we fill our students with known facts and figures. As more and more literature is published and as our technology improves, we are forced to fill their minds with even more facts (Keppie 1990; Romesburg 1991). But little attention is given to training students formally to analyze and synthesize, to be creative. We are not training students to advance the science per se; rather, we are training them to know what we know. Along the way, bright students are able to reorganize and synthesize these facts, challenge established dogmas, and in some manner advance our understanding of what has gone on before us. But the vast majority remain ignorant of how to think.

Romesburg (1991) argued that separate undergraduate and graduate courses in research philosophy are vital to teaching students how to think; he provided specific topics that such courses should contain. Such courses do indeed teach students how science is structured, how the hypothetico-deductive method works, and so on (see chapter 4). Beyond the teaching of the processes of science, however, he argued persuasively that we should teach students how to invent new theories. These courses would introduce students to the diverse structure of theories from many disciplines and how such theories can be applied to natural resources.

In addition, we should educate students on how to identify significant problems (Keppie 1990; Romesberg 1991). Wildlifers have been good at studying the same phenomenon in different locations, at different times, and with different species, without ever understanding why the phenomenon is occurring in the first place. The role that a published study might play in advancing science can probably be judged by the firmness with which the author draws conclusions, even if (*particularly if*) conclusions derive from negative results. Typically, our conclusions are nothing more than a list of ad hoc hypotheses about what our study results might mean. This does not advance science, nor does it advance management.

We need to teach our students to push beyond such standard procedures, identify ways in which we can advance understanding, and then pursue those ways. Such an effort might require the learning of a new discipline, be it physiology, genetics, or mathematical modeling. Although the realities of time and money usually prevent a student from becoming an accomplished physiologist *and* finishing a thesis in a few years, we can at least start the process that will lead him or her to higher knowledge and more creative research projects in the future. For example, advances in technology have made it possible for a master's-level student to incorporate analyses of reproductive hormones into a field study of small mammals for a few hundred dollars, given that his or her university has the necessary analytic equipment (which is likely). As developed in our chapter 7, for example, the study of time-energy budgets (physiological ecology) should greatly enhance the field of habitat relationships. We also need to encourage more debate on how we approach habitat studies. The argument we made above regarding intensive-extensive variables is an example of the type of debates needed.

Although it is easy to talk about advances in education, effecting these advances requires a faculty who are determined to advance their own knowledge. Unfortunately, changes are slow to come in the university setting. Part of this slowness is understandable: It is difficult for faculty who have not been trained in research philosophy, statistics, or modeling to teach these subjects. In addition, some simply do not agree that substantial changes are needed (Knight 1993). We largely agree with Romesburg (1991, 1993) and choose not to be picky about our few disagreements. Rather, we think that wildlife science must do a much better job of educating our students and, further, that current wildlife professionals must do a much better job of continuing their education.

In addition to more training in specific courses (see below in "Professional Training"), faculty should increase the rigor with which they teach their courses. For example, in our graduate courses we have found that students resent being challenged to think on a higher order to solve a problem and expect, instead, to be told answers, as if wildlife ecology and management consist merely of fixed facts. This response is certainly a holdover from their fact-filled undergraduate educations. For example, one of us makes it a practice to start an advanced course by asking students to review an issue—say, ecosystem management or indicator species—and find ways to solve its inherent problems. This requires the students to explore the literature, synthesize the arguments, understand the theory, and so forth, in other words, to be their own fact finders, synthesizers, and interpreters. The resentment comes from the frustration they feel over the lack of training they have received in independent, critical, and creative thinking.

Specifically, wildlife programs should require that all their students receive training in the following areas prior to leaving the master's program:

- the history of science and the links between disciplines
- research philosophy in the sciences

- how theories and hypotheses are developed
- the rigorous development of study designs, including impact assessment
- statistics through nonparametric and multivariate analyses
- mathematical and computer modeling

In addition, undergraduate wildlife students should be well schooled in genetics, physiology, and anatomy.

Professional Training

Federal land managers in the United States are required to develop inventory and management plans for all land, soil, timber, forage, water, air, fish, wildlife aesthetics, recreation, wilderness, and energy and mineral resources in all areas under their jurisdiction (e.g., Forest Service 1990). This is an enormous task which most field personnel, for a variety of reasons, are unable to accomplish adequately. Advanced academic training and extensive research experience are usually necessary to design inventories, analyze inventory data, and establish an effective monitoring program (Morrison and Marcot 1995; see also Garcia 1989; Schreuder et al. 1993).

Unfortunately, few field personnel in any federal or state agency have received this training in their university studies, nor are they adequately trained by their agency. There appears to be a genuine reluctance to advance their education on their own time, including something as simple as keeping current on the literature. For example, during an address one of us made to Forest Service regional biologists a few years ago, a show of hands indicated that only about 15 percent of the 100 or so biologists in attendance belonged to the Wildlife Society. Additionally, many people expressed frustration over not being allowed to read professional journals during working hours but being required to do so on their own time. Clearly, such attitudes show a lack of professionalism that is reflected in poor job performance.

We think that, in addition to the changes in the university educational systems discussed above, a substantial change in workplace education is required. Additional training is ultimately the responsibility of each individual. After all, simply graduating from an accredited institution does not a professional make. However, supervisors can create an atmosphere where staying current, creatively thinking about new ways to approach old problems, auditing advanced courses, and so forth, lead to new responsibilities, job advancement, and, most important, sound resource management as a legacy for future generations. Managers and other nonresearchers must keep current on the basics of theory, study design, and statistics if they are to make rational decisions. They must be able to evaluate independently the research that is published.

Conclusions

Will these proposals be implemented? Romesburg (1993) noted that members of a profession will instinctively bristle at suggestions that they do not adequately educate and, especially, that their intellectual capabilities are inadequate. Further, administrators view creativity in terms of grant dollars and new buildings, not in terms of restructuring education to emphasize thinking rather than memorization. But as Romesburg explained, none of these ideas threatens or degrades anyone; no one will lose his or her job or be dismissed from school. Professors who challenge themselves and try new ideas, and then encourage their students to do the same, will be the builders of the foundation of resource management. Such changes should take place in a structured manner, with developed feedback loops and evaluations so that modifications can be made as experience is gathered. What we are trying to do is lay the foundation for future advancements. The highest possible legacy any of us can leave is this foundation.

Postlude

In this book we have tried to review the foundations of wildlife-habitat relationships, describe methods of conducting rigorous studies, and suggest ways in which we can advance our profession. The study of wildlife-habitat relationships encompasses far more than the classic view of habitat as vegetation. It includes consideration of physiology, behavior, ecological history, genetics, and evolution. In essence, it forms the basis of wildlife research. As such, it is imperative that we approach it correctly. Although we hope that our book will help advance the field, we also hope that it will serve as a basis for more critical debate on how we should proceed.

Much has changed in the realms of wildlife habitat assessment and management since our first edition of this text. If we go out on a limb, we can extrapolate from these trends and portrend the near future for the science and management of wildlife-habitat relationships.

We foresee several general trends in habitat assessment, including continued growth in habitat analysis technologies. Expanding technologies will include more sophisticated computer simulation modeling, improvements to geographically referenced ("spatially explicit") population modeling, and broad-scale analysis of disturbance dynamics aided by new remote sensing technologies (e.g., SIR-C radar imaging from space shuttle flights). Doubtless, the growth of networked computer systems and living databases, such as those shared over the Internet and World Wide Web, will bring a new democracy to the availability of data and findings. The public will have increasingly easy access to complex information through such tools and should be able to participate more effectively in resource management decisions. However, it will likely be only the more advantaged portion of the public that can do so, unless local libraries, public agency offices, and other educational and governmental facilities help provide training and access.

We also anticipate increasing use of scientific visualization techniques, including the use of remote sensing data for detecting and displaying change, as well as new analysis methods for discovering and presenting trends in data on habitat conditions.

Human population growth, particularly in developing nations, will lead to escalating conflicts in land use allocations and resource conservation. There is much to learn about wildlife conservation plights and successes among nations of the world, particularly from nations long struggling with high human densities and severe resource scarcities that others of us too often assume will never happen to us. We hope that the trend toward globalization of information will be matched by an earnestness to learn from each other. It is up to us, as professionals, students, and responsible citizens, to talk with one another.

The future of wildlife-habitat relationship assessments and management might broaden to include nonvertebrate species and other taxonomic groups and ecological entities of conservation concern and to consider the ecosystem context of ecological functions and processes. We hope that broad-scale assessments (e.g., WRI et al. 1992) and policy decisions (e.g., Caldwell et al. 1994) can provide a framework for considering habitat management in an evolutionary context. In the future of ecosystems lies the future of ourselves. Management policies that rely less on short-term profit (e.g., maximizing net present value of resources) and at least equally on the needs of future generations would do well to recognize the economic and social benefits—the economic and sociological resiliency and stability—of long-term conservation planning.

Sustainable resource use of the more distant future can be attained by beginning now to articulate an ideal, long-term, resource-use scenario. Under such a scenario, humans would learn to adjust their resource use patterns and habits to ensure resource productivity and sustained use in perpetuity. This is an idealized future where resource production, such as timber growth and availability of fishery resources and grazing potentials, is in harmony with rates of use and extraction. Much as been written on "sustainable futures" over the past several decades (e.g., O'Neill et al. 1996; papers in Bissonette and Krausman 1995; Sandlund et al. 1992).

We can help outline the primary elements of such a sustainable future, as listed in table 12.2. This outline involves a long-term ideal that would entail difficult changes in habits of personal resource use, population centers of growth, resource extraction industries and infrastructures, and even administrative, economic, political, and educational systems. It may entail use of conservation reserves (Franklin 1993) as well as actively managed landscapes (e.g., Everett et al. 1994). However, even if the ideal can never be reached, we feel it is valuable to articulate its parameters to help determine which components are feasible or desirable ecologically, socially, politically, and economically. It can also help determine future generations' costs and benefits if we do *not* reach for an environmental ideal.

Finally, we might propose two overall principles for habitat managers. The first borrows from a popular and idealistic modern myth, and may be called the *prime habitat management directive:* Insofar as possible, do not interfere with the normal course of development of native ecosystems and populations. We have learned this lesson from many trials: by suppressing wildfires without considering subsequent adverse changes in forest community structures and increasing susceptibility to catastrophic fire events (Graham 1994); by removing biological diversity from forests, prairies, and grasslands without concern for sources of largely unknown biological elements critical to long-term soil health, resource productivity, and resistance of ecosystems to rampant disease and pathogens (Koopowitz et al. 1994; Robinson 1993); by damming rivers without considering the dispersal blockades they impose to migrating salmon and indigenous peoples (Anderson 1993); by introducing exotic species without first testing or considering their potential for dispersal and ecological havoc on native ecosystems (Coblentz 1990); and by numerous similar lessons.

The second habitat management principle would pertain where the first proves impossible or infeasible. We can borrow from the medical profession and propose a *habitat management Hippocratic Oath:* Do no ecological harm. That is, where "invasive" management is necessary to provide for human resource needs, activities should be designed to do no harm to native ecosystems and communities. As an example, manipulation of soils for agriculture or forestry objectives should also strive to avoid wind and water erosion and to ensure, at least on a

Table 12.2. Primary elements and assumptions of a resource-planning scenario designed for long-term sustainability of habitats for wildlife and humans

Planning Elements
1. Humans will amend their habits of resource use to ensure future, sustained production and availability of resources from forest, grassland, wetland, riparian, and aquatic ecosystems.
2. Design of human occupation of the land is one of the most important facets of this scenario. This includes consideration of urban-forest interfaces, grazing, mining, timber management, the locating of urban and rural communities and transportation infrastructures, resource recycling, solid waste management, toxic waste management, air quality control, and changes in habits of resource use and lifestyles.
3. Land allocations and patterns will be coordinated among private, state, federal, and other kinds of ownerships. Each has its own specific, albeit different, contribution to maintaining habitats, wildlife, and biodiversity.
4. Land use activities will focus on the amounts, flows, consumption rates, and renewal rates of material resources. However, additional goals such as recovery of threatened species and recovery of ecological communities at risk will also be addressed.
5. Humans will amend their habits of resource use and patterns of land occupation according to what will provide for a more aboriginal or natural set of conditions, rather than invest energy and effort into changing natural disturbance regimes and ecosystems to meet predefined desired levels of resource use. The amount and flows of resources used (consumptive or nonconsumptive) will result from the capability of the land to maintain itself.
6. Indigenous peoples will engage in their traditional hunting, gathering, and cultural activities, using desired species, resources, and environmental conditions in perpetuity.
7. Social and economic systems, as well as nonindigenous cultural conditions, habits and patterns of resource use, and individual lifestyles of resource use will be subject to change to ensure long-term, sustained resource use conditions. Administrative boundaries affecting human populations and patterns of resource use will be amended along ecological lines.
8. Interregional and international conditions and effects will be explicitly considered in making local resource management decisions.
Assumptions
1. Given present conditions of forests, grasslands, wetlands, and riparian and aquatic systems, it is possible to recover at-risk elements and adjust levels of resource use and land allocations to eventually achieve a steady state of resource use and land occupation. There are no elements that cannot be recovered; or if some elements are moribund, their loss will not jeopardize overall mission success.
2. Social, cultural, lifestyle, and political institutions will change where necessary to ensure successfully meeting the goals of this scenario.
3. Human populations and their patterns of land use, including placement and kinds of habitations and infrastructures, will change to ensure successfully meeting the goals of this scenario.
4. All legal and aboriginal citizens will participate in and support the attainment of a future that ensures resource availability in perpetuity. Local participation by citizens will aid in transition to a sustainable future.
5. This scenario will provide a conservation benchmark and leadership for considering similar planning for all land administrations, ownerships, regions, and even other nations.

portion of the land base, the full complement of beneficial soil mesoinvertebrates, organic matter, microfungi, invertebrate pollinators, and other components essential to long-term productivity.

In the end, the future of wildlife is the future of us all.

Literature Cited

Abrams, P. A. 1988. How should resources be counted? *Theoretical Population Biology* 33:226–42.

Alldredge, J. R., and J. T. Ratti. 1986. Comparison of some statistical techniques for analysis of resource selection. *Journal of Wildlife Management* 50:157–65.

Alldredge, J. R., and J. T. Ratti. 1992. Further comparison of some statistical techniques for analysis of resource selection. *Journal of Wildlife Management* 56:1–9.

Anderson, M. 1993. *The living landscape,* Vol. 2: *Pacific salmon and federal lands.* Washington, D.C.: The Wilderness Society.

Arthur, W. 1987. *The niche in competition and evolution.* New York: John Wiley and Sons.

Baskett, T. S. 1985. Quality control in wildlife science. *Wildlife Society Bulletin* 13:189–96.

Belovsky, G. E. 1978. Diet optimization in a generalists herbivore: The moose. *Theoretical Population Biology* 14:105–34.

Belovsky, G. E. 1981. Food selection by a generalist herbivore: The moose. *Ecology* 62:1020–30.

Belovsky, G. E. 1984. Herbivore optimal foraging: A comparative test of three models. *American Naturalist* 124:97–115.

Bissonette, J. A., and P. R. Krausman, eds. 1995. *Integrating people and wildlife for a sustainable future.* Bethesda, Md.: The Wildlife Society.

Buckley, G. P., ed. 1989. *Biological habitat reconstruction.* New York: Belhaven Press.

Caldwell, L. K., C. F. Wilkinson, and M. A. Shannon. 1994. Making ecosystem policy: Three decades of change. *Journal of Forestry* 92(4):7–10.

Coblentz, B. E. 1990, Exotic organisms: A dilemma for conservation biology. *Conservation Biology* 4:261–65.

Elton, C. S. 1927. *Animal ecology.* London: Sidgwick and Jackson.

Everett, R. L., P. F. Hessburg, and T. R. Lillybridge. 1994. Emphasis areas as an alternative to buffer zones and reserved areas in the conservation of biodiversity and ecosystem processes. *Journal of Sustainable Forestry* 2(3–4):283–92.

Forest Service. 1990. *Resource inventory handbook.* USDA Forest Service, FSH 1909.14, Amendment No. 1, 3/29/90, Washington, D.C.

Franklin, J. F. 1993. Preserving biodiversity: Species, ecosystems, or landscapes? *Ecological Applications* 3(2):202–5.

Franzreb, K. E. 1989. Ecology and conservation of the endangered least Bell's vireo. *U.S. Fish and Wildlife Service Biological Report* 89:1–17.

Franzreb, K. E. 1990. An analysis of options for reintroducing a migratory native passerine, the endangered least Bell's vireo *Vireo bellii pusillus* in the Central Valley, California. *Biological Conservation* 53:105–23.

Garcia, M. W. 1989. Forest Service experience with interdisciplinary teams developing integrated resource management plans. *Environmental Management* 13:583–92.

Gavin, T. A. 1989. What's wrong with the questions we ask in wildlife research? *Wildlife Society Bulletin* 17:345–50.

Gavin, T. A. 1991. Why ask "why": The importance of evolutionary biology in wildlife science. *Journal of Wildlife Management* 55:760–66.

Godin, J.-G. J. 1990. Diet selection under the risk of predation. In *Behavioural mechanisms of food selection,* ed. R. N. Hughes, 739–69. Berlin: Springer-Verlag.

Graham, R. T. 1994. Silviculture, fire and ecosystem management. In *Assessing forest ecosystem health in the inland West,* ed. R. N. Sampson and D. L. Adams, 339–51. Binghamton, N.Y.: Haworth Press.

Griffiths, M., and C. P. Van Schaik. 1993. The impact of human traffic on the abundance and activity periods of Sumatran rain forest wildlife. *Conservation Biology* 7:623–26.

Grinnell, J. 1917. The niche-relationships of the California thrasher. *Auk* 34:427–33.
Hall, L. S., P. R. Krausman, and M. L. Morrison. 1997. The habitat concept and a plea for standard terminology. *Wildlife Society Bulletin* 25:173–182.
Henson, P., and J. A. Cooper. 1993. Trumpeter swan incubation in areas of differing food quality. *Journal of Wildlife Management* 57:709–16.
Hobbs, N. T. 1989. *Linking energy balance to survival in mule deer: Development and test of a simulation model.* Wildlife Monograph No. 101. Bethesda, Md.: The Wildlife Society. 39 pp.
Hobbs, N. T., and T. A. Hanley. 1990. Habitat evaluation: Do use/availability data reflect carrying capacity? *Journal of Wildlife Management* 54:515–22.
Holthuijzen, A. M. A., W. G. Eastland, A. R. Ansell, M. N. Kochrt, R. D. Williams, and L. S. Young. 1990. Effects of blasting on behavior and productivity of nesting prairie falcons. *Wildlife Society Bulletin* 18:270–81.
Hunter, M. L. 1989. Aardvarks and Arcadia: Two principles of wildlife research. *Wildlife Society Bulletin* 17:350–51.
Hunter, M. D., and P. W. Price. 1992. Playing chutes and ladders: Heterogeneity and the relative roles of bottom-up and top-down forces in natural communities. *Ecology* 73:724–32.
Hutchinson, G. E. 1978. *An introduction to population ecology.* New Haven, Conn.: Yale University Press.
Hutto, R. L. 1985. Habitat selection by nonbreeding, migratory land birds. In *Habitat selection in birds,* ed. M. L. Cody, 455–76. San Diego, Calif.: Academic Press.
Johnson, A. R., J. A. Wiens, B. T. Milne, and T. O. Crist. 1992. Animal movements and population dynamics in heterogeneous landscapes. *Landscape Ecology* 7:63–75.
Johnson, D. H. 1980. The comparison of usage and availability measurements for evaluating resource preference. *Ecology* 61:65–71.
Jordan, W. R., III (with S. Packard). 1989. Just a few oddball species: Restoration practice and ecological theory. In *Biological habitat reconstruction,* ed. G. P. Buckley, 18–26. New York: Belhaven Press.
Jordan, W. R., III, M. E. Gilpin, and J. D. Aber, eds. 1987. *Restoration ecology: A synthetic approach to ecological research.* New York: Cambridge University Press.
Keddy, P. A., H. T. Lee, and I. C. Wisheu. 1993. Choosing indicators of ecosystem integrity: Wetlands as a model system. In *Ecological integrity and the management of ecosystems,* ed. S. Woodley, J. Kay, and G. Francis, 61–79. Delray Beach, Fl.: St. Lucie Press.
Keppie, D. M. 1990. To improve graduate student research in wildlife education. *Wildlife Society Bulletin* 18:453–58.
Klopfer, P. H., and J. U. Ganzhorn. 1985. Habitat selection: Behavioral aspects. In *Habitat selection in birds,* ed. M. L. Cody, 436–53. San Diego, Calif.: Academic Press.
Knight, R. L. 1993. On improving the natural resources and environmental sciences: A comment. *Journal of Wildlife Management* 57:182–83.
Koopowitz, H., A. D. Thornhill, and M. Andersen. 1994. A general stochastic model for the prediction of biodiversity losses based on habitat conversion. *Conservation Biology* 8(2):425–38.
Lancia, R. A., T. D. Nudds, and M. L. Morrison. 1993. Opening comments: Slaying slippery shibboleths. *Transactions of the North American Wildlife and Natural Resources Conference* 58:505–8.
Leibold, M. A. 1995. The niche concept revisited: Mechanistic models and community context. *Ecology* 76:1371–82.
Loegering, J. P., and J. D. Fraser. 1995. Factors affecting piping plover chick survival in different brood-rearing habitats. *Journal of Wildlife Management* 59:646–55.
MacArthur, R. H. 1968. The theory of the niche. In *Population biology and evolution,* ed. R. C. Lewontin, 159–76. Syracuse, N.Y.: Syracuse University Press.
MacArthur, R. H., and R. Levins. 1967. The limiting similarity, convergence, and divergence of coexisting species. *American Naturalist* 101:377–95.

Matter, W. J., and R. W. Mannan. 1989. More on gaining reliable knowledge: A comment. *Journal of Wildlife Management* 53:1172–76.

Morris, D. W. 1992. Scales and costs of habitat selection in heterogeneous landscapes. *Evolutionary Ecology* 6:412–32.

Morrison, M. L. 1994. Resource inventory and monitoring: Concepts and applications for ecological restoration. *Restoration and Management Notes* 12:179–83.

Morrison, M. L. 1995. Wildlife conservation and restoration ecology. *Restoration and Management Notes* 13:203–208.

Morrison, M. L., and B. G. Marcot. 1995. An evaluation of resource inventory and monitoring program used in national forest planning. *Environmental Management* 19:147–56.

Morrison, M. L., T. A. Scott, and T. Tennant. 1994. Wildlife-habitat restoration in an urban park in southern California. *Restoration Ecology* 2:17–30.

Morrison, M. L., T. Tennant, and T. A. Scott. 1994. Laying the foundation for a comprehensive program of restoration for wildlife habitat in a riparian floodplain. *Environmental Management* 18:939–55.

Morrison, M. L., C. J. Ralph, J. Verner, and J. R. Jehl, Jr., eds. 1990. *Avian foraging: Theory, methodology, and applications.* Studies in Avian Biology no. 13. Los Angeles: Cooper Ornithological Society.

Morse, D. H. 1980. *Behavior mechanisms in ecology.* Cambridge, Mass.: Harvard University Press.

Nudds, T. D., and M. L. Morrison. 1991. Ten years after "reliable knowledge": Are we gaining? *Journal of Wildlife Management* 55:757–60.

O'Neill, R. V., J. R. Kahn, J. R. Duncan, S. Elliott, R. Efroymson, H. Cardwell, and D. W. Jones. 1996. Economic growth and sustainability: A new challenge. *Ecological Applications* 6(1):23–24.

Orians, G. H. 1993. Endangered at what level? *Ecological Applications* 3:206–8.

Owen-Smith, N. 1989. Morphological factors and their consequences for resource partitioning among African savanna ungulates: A simulation modelling approach, In *Patterns in the structure of mammalian communities,* ed. D. W. Morris, Z. Abramsky, B. J. Fox, and M. R. Willig, 155–65. Lubbock: Texas Technological University Press.

Painter, E. L., and A. J. Belsky. 1993. Application of herbivore optimization theory to rangelands of the western United States. *Ecological Applications* 3:2–9.

Paradis, E., and H. Croset. 1995. Assessment of habitat quality in the Mediterranean pine vole (*Microtus duodecimcostatus*) by the study of survival rates. *Canadian Journal of Zoology* 73:1511–18.

Paton, P. W. C. 1994. The effect of edge on avian nest success: How strong is the evidence? *Conservation Biology* 8:17–26.

Peters, R. H. 1991. *A critique for ecology.* Cambridge: Cambridge University Press.

Real, L. A. 1990. Predator switching and the interpretation of animal choice behavior: The case for constrained optimization. In *Behavioural mechanisms of food selection,* ed. R. N. Hughes, 1–20. Berlin: Springer-Verlag.

Risser, P. G. 1993. Making ecological information practical for resource managers. *Ecological Applications* 3:37–38.

Robinson, J. G. 1993. The limits to caring: Sustainable living and the loss of biodiversity. *Conservation Biology* 7(1):20–28.

Romesburg, H. C. 1981. Wildlife science: Gaining reliable knowledge. *Journal of Wildlife Management* 45:293–313.

Romesburg, H. C. 1991. On improving the natural resources and environmental sciences. *Journal of Wildlife Management* 55:744–56.

Romesburg, H. C. 1993. On improving the natural resources and environmental sciences: A reply. *Journal of Wildlife Management* 57:184–89.

Rosenzweig, M. L. 1989. Habitat selection, community organization, and small mammal studies. In *Patterns in the structure of mammalian communities,* ed. D. W. Morris, Z Abramsky, B. J. Fox, and M. R. Willig, 5–21. Lubbock: Texas Technological University.

Sandlund, O. T., K. Hindar, and A. H. D. Brown, eds. 1992. *Conservation of biodiversity for sustainable development.* New York: Oxford University Press.

Schreuder, H. T., T. G. Gregoire, and G. B. Wood. 1993. *Sampling methods for multiresource forest inventory.* New York: John Wiley and Sons.

Schultz, J. C. 1992. Factoring natural enemies into plant tissue availability to herbivores. In *Effects of resource distribution on animal-plant interactions,* ed. M. D. Hunter, T. Ohgushi, and P. W. Price, 175–97. San Diego, Calif.: Academic Press.

Slobodkin, L. B. 1992. A summary of the special feature and comments on its theoretical context and importance. *Ecology* 73:1564–66.

Smallwood, K. S., and C. Schonewald. 1996. Scaling population density and spatial pattern for terrestrial, mammalian carnivores. *Oecologia* 105:329–35.

Stephens, D. W., and J. R. Krebs. 1986. *Foraging theory.* Princeton, N.J.: Princeton University Press.

Thomas, D. L., and E. J. Taylor. 1990. Study designs and tests for comparing resource use and availability. *Journal of Wildlife Management* 54:322–30.

Tracy, C. R., and P. F. Brussard. 1994. Preserving biodiversity: Species in landscapes. *Ecological Applications* 4:205–7.

Truett, J. C., R. G. B. Senner, K. Kertell, R. Rodrigues, and R. H. Pollard. 1994. Wildlife responses to small-scale disturbances in Arctic tundra. *Wildlife Society Bulletin* 22:317–24.

Van Horne, B. 1983. Density as a misleading indicator of habit quality. *Journal of Wildlife Management* 47:893–901.

Walsberg, G. E. 1985. Physiological consequences of microhabitat selection. In *Habitat selection in birds,* ed. M. L. Cody, 389–413. San Diego, Calif.: Academic Press.

Walters, C. J. 1986. *Adaptive management of renewable resources.* New York: Macmillian.

Wiens, J. A. 1989. *The ecology of bird communities,* Vol. 2: *Processes and variations.* Cambridge: Cambridge University Press.

Wiens, J. A. 1992. Ecology 2000: An essay on future directions in ecology. *Bulletin of the Ecological Society of America* 73:165–70.

Winkler, H., and B. Leisler. 1985. Morphological aspects of habitat selection in birds. In *Habitat selection in birds,* ed. M. L. Cody, 415–34. San Diego, Calif.: Academic Press.

Wolda, H. 1990. Food availability for an insectivore and how to measure it. *Studies in Avian Biology* 13:38–43.

WRI (World Resources Institute), IUCN (International Union for Conservation of Nature and Natural Resources), and UNEP (United Nations Environmental Program). 1992. *Global biodiversity strategy: Guidelines for action to save, study, and use Earth's biotic wealth sustainably and equitably.* Washington, D.C.: World Resources Institute.

Author Index

Subject Index

Author Index

Author Index

Author Index

Author Index

Author Index

Author Index

Subject Index